# POLYURETHANES

# WILEY SERIES ON POLYMER ENGINEERING AND TECHNOLOGY
## Richard F. Grossman and Domasius Nwabunma, Series Editors

**Polyolefin Blends /** Edited by Domasius Nwabunma and Thein Kyu

**Polyolefin Composites /** Edited by Domasius Nwabunma and Thein Kyu

**Handbook of Vinyl Formulating, Second Edition /** Edited by Richard F. Grossman

**Total Quality Process Control for Injection Molding, Second Edition /** M. Joseph Gordon, Jr.

**Microcellular Injection Molding /** Jingyi Xu

**Poly(lactic acid): Synthesis, Structures, Properties, Processing, and Applications /** Edited by Rafael Auras, Loong-Tak Lim, Susan E.M. Selke, and Hideto Tsuji

**Hyperbranched Polymers: Synthesis, Properties, and Applications /** Edited by Deyue Yan, Chao Gao, and Holger Frey

**Advanced Thermoforming: Methods, Machines and Materials, Applications and Automation /** Sven Engelmann

**Biopolymer Nanocomposites: Processing, Properties and Applications /** Alain Dufresne, Sabu Thomas, Laly A. Pothan

**Polymers for PEM Fuel Cells /** Hongting Pu

**Polyurethanes: Science, Technology, Markets, and Trends /** Mark F. Sonnenschein

**Functional Polymer Coatings: Principles, Methods, and Applications /** Edited by Limin Wu and Jamil Baghdachi

**Polyurethanes: Science, Technology, Markets, and Trends, Second Edition /** Mark F. Sonnenschein

# POLYURETHANES

## Science, Technology, Markets, and Trends

SECOND EDITION

MARK F. SONNENSCHEIN, Ph.D.
The Dow Chemical Company
Midland, MI, USA

WILEY

This edition first published 2021


*Edition History*
John Wiley & Sons (1e, 2014)



*Registered Office*
John Wiley & Sons, Inc., 111 River Street, Hoboken, NJ 07030, USA

*Editorial Office*
111 River Street, Hoboken, NJ 07030, USA

For details of our global editorial offices, customer services, and more information about Wiley products visit us at www.wiley.com.

Wiley also publishes its books in a variety of electronic formats and by print-on-demand. Some content that appears in standard print versions of this book may not be available in other formats.



*Library of Congress Cataloging-in-Publication Data*

Names: Sonnenschein, Mark F., author.
Title: Polyurethanes : science, technology, markets, and trends / Mark F. Sonnenschein, Ph.D., The Dow Chemical Company, Midland, Michigan, US.
Description: Second edition. | Hoboken : Wiley, 2021. | Series: Wiley series on polymer engineering and technology | Includes bibliographical references and index.
Identifiers: LCCN 2020024254 (print) | LCCN 2020024255 (ebook) | ISBN 9781119669418 (hardback) | ISBN 9781119669463 (adobe pdf) | ISBN 9781119669470 (epub)
Subjects: LCSH: Polyurethanes.
Classification: LCC TP1180.P8 S56 2021 (print) | LCC TP1180.P8 (ebook) | DDC 668.4/239–dc23
LC record available at https://lccn.loc.gov/2020024254
LC ebook record available at https://lccn.loc.gov/2020024255

Cover Design: Wiley
Cover Image: Courtesy of The Dow Chemical Company

Set in 10/12pt Times by SPi Global, Pondicherry, India

Printed in the United States of America

SKY10025954_032921

*Dedicated to my wife, Geraldine Franklin Sonnenschein, for her beauty, kindness, and endless support and to my children, Matthew, Anne, and Susan, for the inspiration and the laughs. My canines, Arlo, Lark, Bugle, and Sprite, are acknowledged for patiently and cheerfully keeping me company throughout the writing process.*

# CONTENTS

# PREFACE

I want to begin by thanking the people and organizations that bought the first edition of *Polyurethanes: Science, Technology, Markets, and Trends*. I have received many encouraging and constructive comments since it appeared at the end of 2014. At the same time I have also had a certain amount of angst about the number of errors in the book, most of them typographical, but not all of them! In addition, I have been acutely aware as time has passed, even over just 6 years, that the polyurethane industry is continuously changing and evolving. The "Science" and "Technology" parts of the first edition title have changed somewhat, but the "Markets" and "Trends" components of the first edition were badly in need of updating. What was also striking was how third-party industry forecasts made in 2014 about subsequent years were in most cases wrong, demonstrating how poorly crystal ball analyses of economic activity perform. In this second edition there is much less attention paid to these prognostications.

An interesting macrotrend at the time of this writing is the rapid expansion of global capacity with only moderate growth occurring now and in the future. This assures that producer prices will remain under pressure and help maintain low price inflation for the foreseeable future. This may also result in additional consolidation in the polyurethanes industry, or the spinning out of polyurethane businesses from chemical conglomerates as was see when Bayer AG spun off its polyurethane assets into a separate company called Covestro AG. An additional market complication is the introduction of large chemical concerns that represent and depend on national subsidies, such as those appearing in China and Saudi Arabia. As a statement of a societal commitment apart from market forces, they are capable of distorting market supply and pricing. Whether by consolidation or spinning out of assets, the ability of the surviving companies to efficiently and effectively allocate and spend capital in the prevailing economic environment could be a continuing element in the polyurethanes industry story.

This edition adds two new special chapters at the end: Chapter 13 on polyurethane hybrid polymers and Chapter 14 on polyurethane recycle. As the reader will see, both topics cover nascent technologies in terms of the scientific progress and industrial intent to commercialize in these technology fields. Polymer hybrids are often conceived in the belief that if two different polymers are good, then their combination (or hybridization) will be even better. While this has not historically been found true for a number of reasons, there are a number of instances for polyurethanes where polymer backbone hybridizations are beneficial, making this technology area potentially ripe for growth in the future.

Polyurethane recycle – the subject of Chapter 14 – is a subset of the much larger topic of plastics recycle. However, unlike soda straws, water bottles, and disposable food packaging, polyurethanes are usually not found in articles that are commonly thrown away. There is no developed municipal collection and distribution system for acquiring polyurethanes, separating them from the objects they are a part of, and readying them for chemical cleanup and reversion back to useful building blocks. However, regulations, especially those emanating from the European Union, requiring end-of-life stewardship of things such as mattresses and automotive seats – containing significant volumes of polyurethanes – are forcing producers to develop technology and commercial partnerships. This industrial evolution will permit polyurethane recycle either through low-value physical incorporation or by chemical transformation of the polyurethane polymer into useable feedstocks for making new materials. If governmental regulations create price inelastic demand for these materials, then the methods of their production and the value they bring will become a topic of increasing importance into the future.

Mark F. Sonnenschein
Midland, MI

# ACKNOWLEDGMENTS

I would like to express my deep gratitude to the people who have helped me through the years and provided fertile ground for growth. Particularly I would like to mention my colleagues whom I have worked with in the field of polyurethanes over the past 20 years. First, I would like to mention my constant collaborator Benjamin Wendt, who has worked with me closely in the lab for many years and excelled at making hard things work easily. Many people have provided guidance, encouragement, and excellent collaboration over the years. Especially I would like to mention Dr. Alan Schrock, Dr. Justin Virgili, Dr. Mark Cox, Dr. Jack Kruper, Dr. Chris Christenson, Dr. Valeriy Ginzburg, Dr. Jozef Bicerano, Mr. Will Koonce, Dr. Juan-Carlos Medina, Prof. Tony Ryan, Dr. David Babb, Dr. Robbyn Prange, Dr. Nelson Rondan, Dr. Maria Pollard, Dr. Jai Venkatesan, Dr. David Bem, Dr. Florian Schattenmann, Dr. Andre Argenton, Dr. Jorge Jimenez, Dr. Kaoru Aou, Dr. Kshitish Patankar, Dr. Steve Guillaudeu, Dr. Cecile Boyer, Dr. Steve Montgomery, Dr. Brian Landes, Dr. Steve Webb, Dr. Phillipe Knaub, Dr. Hamdy Khalil, Dr. Tirtha Chatterjee, Dr. Lotus Huang, Dr. Cathy Tway, Dr. Shawn Feist, Prof. John Klier, Prof. Craig Hawker, and Dr. John Kramer.

I would also like to recognize the great support I received from The Dow Chemical Company in writing this book by giving me the encouragement, time, resources, and freedom to realize this vision.

I would be remiss to not acknowledge the contributions of my copy-editor, Lindsey Williams. Her patient and meticulous contributions have made this a more readable and useful text.

Lastly, I would like to acknowledge the people who gave me my scientific foundations and inspired in me a love of experiment and a respect for theory. Particularly I would like to mention Prof. Richard G. Weiss (Georgetown University), Dr. C. Michael Roland (The United States Naval Research Lab), and Prof. Gordon Johnson (Kenyon College) for putting up with me in my early years.

# 1

# INTRODUCTION

In the early 1900s there were very few of the synthetic polymers we have grown accustomed to now. During succeeding years polymer science experienced explosive growth with the invention of polyvinyl chloride (PVC, 1913), polyethylene (1933), polyvinylidene chloride (Saran, 1933), polyamides (nylon, 1934), and polytetrafluoroethylene (Teflon, 1938). In addition, during the 1930s the polymer family known as polyurethanes was invented. Now, of course, polyurethanes, and all the polymers developed during this period, have become an integral part of modern life. As you read this you may not be aware of how many ways polyurethanes surround you. They are present in the shoes you stand in, the seat cushion you sit upon, the carpet backing and foam pad underlay you walk upon, in the fibers of your clothing, insulation of your walls and roof, in your refrigerator, dishwasher, water heater, automotive seating, automotive structural foam, automotive paints and coatings, furniture coatings, your bed mattress, the adhesive holding this book together – the list just goes on. This book's purpose is to explain polyurethane science, technology, applications, trends, and markets in virtually all of its forms and relate those structures to the properties that make them so suited for so many uses. It is not an overstatement to say that if polyurethanes are not *the* most versatile class of materials, then they are certainly one of the most versatile polymer categories in existence.

Discovery of polyurethane chemistry is attributed to the efforts of Otto Bayer and the research team he led at the now defunct I.G. Farben AG chemical company. The first patent associated with polyurethanes was filed in 1937 and numerous other

*Polyurethanes: Science, Technology, Markets, and Trends*, Second Edition. Mark F. Sonnenschein.

patents, most notably the production of flexible foams resulting from isocyanate–water reactions, were filed thereafter. I.G. Farben was broken up following World War II for complicity in war crimes and the company's top leaders were convicted of crimes against humanity (exploitation of slave labor and production of nerve gas). The largest surviving components of I.G. Farben – Bayer AG and BASF SE – remain very large and respected global industrial concerns. While BASF continues to engage with and maintain a significant presence within the polyurethane industry, Bayer spun off its polyurethane business and the rest of its industrial chemicals concerns into a new company called Covestro.

After the initial discovery and expositions of basic chemistry, mostly based on short-chain diols and polyester polyols, industrial polyurethanes saw immense growth following the development of polyether polyols by E.I. du Pont de Nemours and Company (now known as DuPont) and The Dow Chemical Company. While Dow Chemical remains one of the world's largest manufacturers of polyurethane chemicals, DuPont has exited its polyurethanes businesses, which were primarily textile and coatings related. While polyesters remain prominent components of polyurethane chemistry, it was the superior processing, low-temperature flexibility, and hydrolytic stability of polyether polyols that expanded polyurethane polymers into their current acceptance in every aspect of modern life.

As ubiquitous as polyurethanes are, it is perhaps surprising that they represent a relatively minor (but still significant) fraction of the overall global consumption of plastics by volume (Figure 1.1).

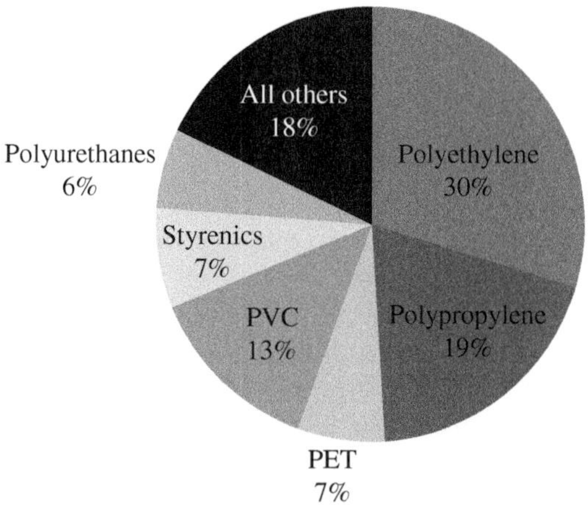


**FIGURE 1.1** Percentage global consumption of plastics in 2018. Polyethylene encompasses all densities; styrenics includes all copolymers along with atactic polystyrene. These relative values are similar to those in the first edition using 2012 data. The consumption of many plastics grows at a rate relative to economic activity plus a small accelerator or decelerator for each given plastic's role in the market. PET = polyethyene terephthate.

The structures of the listed commodity polymers are relatively simple repeating units (Figure 1.2). Their simplicity is in part responsible for their high level of utility and low-cost positions. The plastics industry has generated variants of the structures shown in Figure 1.2 by, for instance, introducing branches, but these complexities do not fundamentally alter the basic polymer structure.

Polyurethane is the largest volume commodity polymer that cannot be characterized by a simple structure such as that shown in Figure 1.2. Instead, polyurethane represents a class of polymers, and *any* polymer with a urethane repeat unit is classified as a polyurethane regardless of the other functional or polymer structures incorporated (Figure 1.3).

Specific polyurethane structures used for making mattress foam, insulation foam, or shoe foam can be significantly different from one another and cannot be neatly represented like the structures in Figure 1.2. In fact, even structures of different insulation foams can vary so widely that they also cannot be easily represented by a

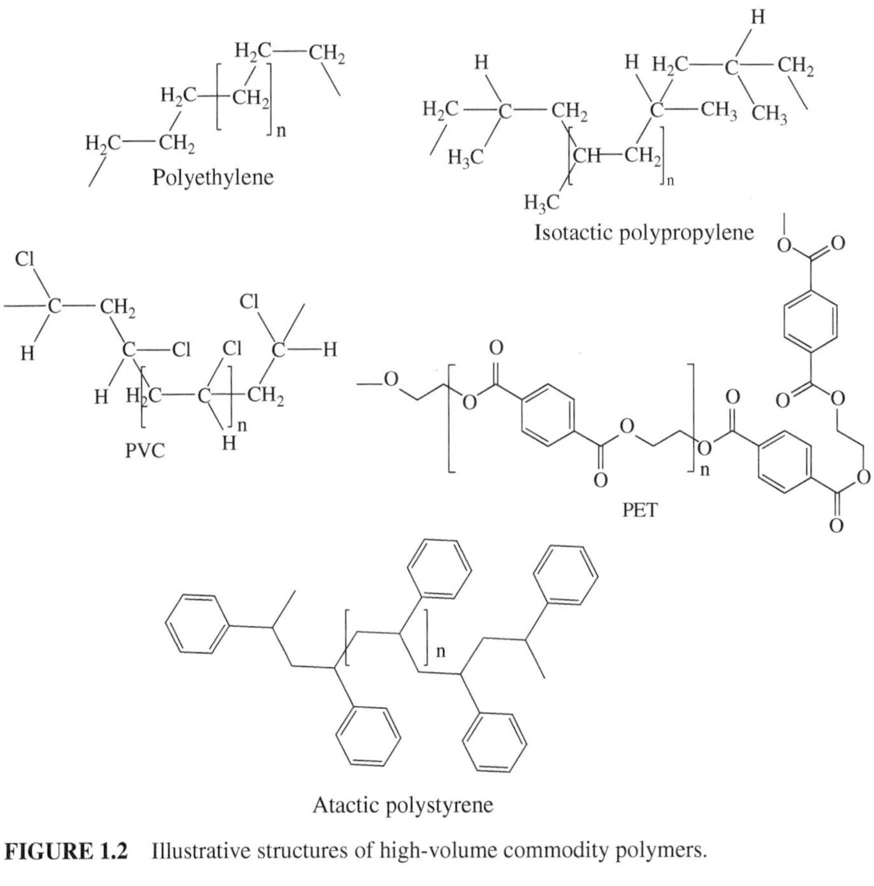


**FIGURE 1.2** Illustrative structures of high-volume commodity polymers.

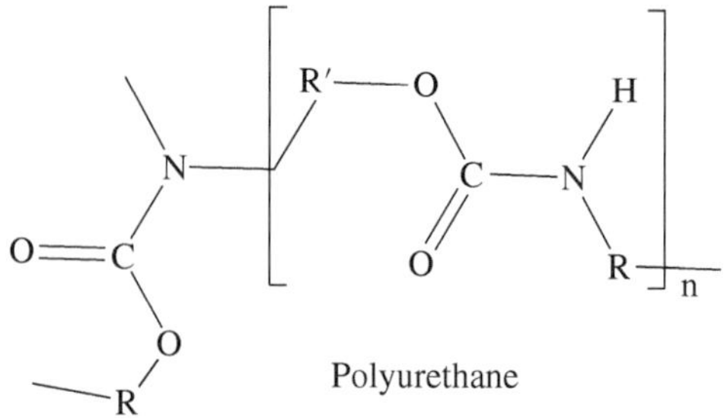


**FIGURE 1.3** The urethane unit within a polyurethane polymer chain.

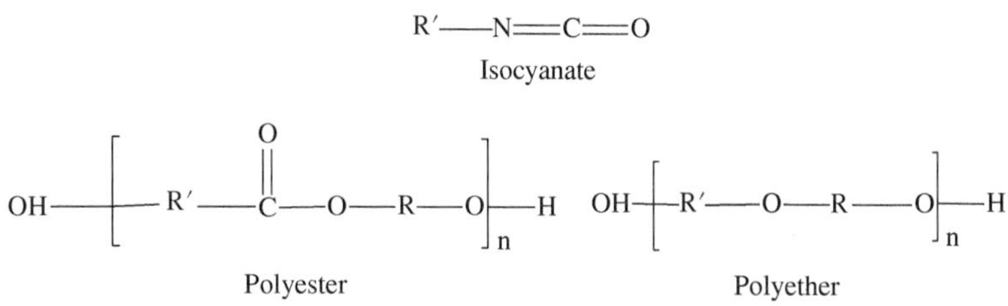


**FIGURE 1.4** Chemical structures of isocyanate, polyester, and polyether. To make a polyurethane the R′ of the isocyanate structure must also have an isocyanate function [1].

single structure. Another difference with other commodity polymers is that large-volume polyurethane applications require the mixing of two reactive liquid components rather than the processing of a pellet into a molded or extruded object. Given these complexities it is remarkable that polyurethanes have developed into a commodity plastic category, and it is testament to the versatility and performance of polyurethanes that they are so difficult to replace in their favored applications.

Polyurethane polymers as a class are made from commodity building block reagents and short-chain polymers (or oligomers). These building blocks include, for example, the following categories: polyisocyanates, polyethers, polyesters, water, and amines (Figure 1.4). As building block categories they also cannot be represented by unique structures and are denoted by "R" to allow designers to insert any conceivable chemically allowable unit.

The polyurethane unit is easily mistaken for the related polyester, polyurea, or polyamide (nylon) structures (Figure 1.5). In fact, polyureas, polyesters, and polyurethanes are often joined into polyurethane materials and still broadly classified as polyurethane. (Polyamides were not previously a part of polyurethane chemistry because of their vastly different processing characteristics. However, recent literature indicates nascent explorations of urethane–amide hybrids; see Chapter 13.)

As commodity products, polyurethanes have achieved a certain establishment status in academic science. However, activity in polyurethane science shows no sign of abating owing to its high potential for design and innovation [1–18]. Figure 1.6

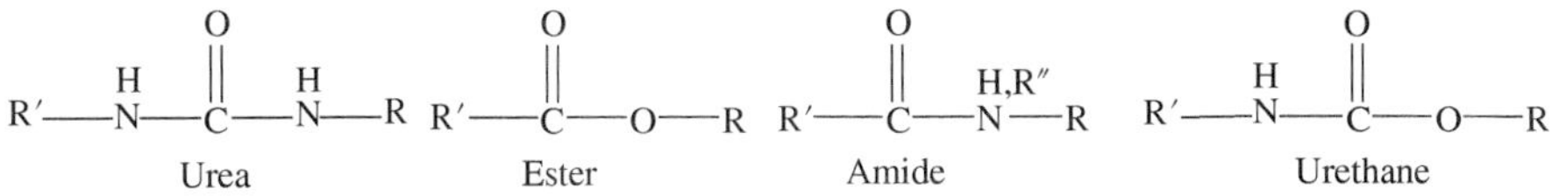


**FIGURE 1.5** Structures of urea, ester, amide, and urethane functionalities.

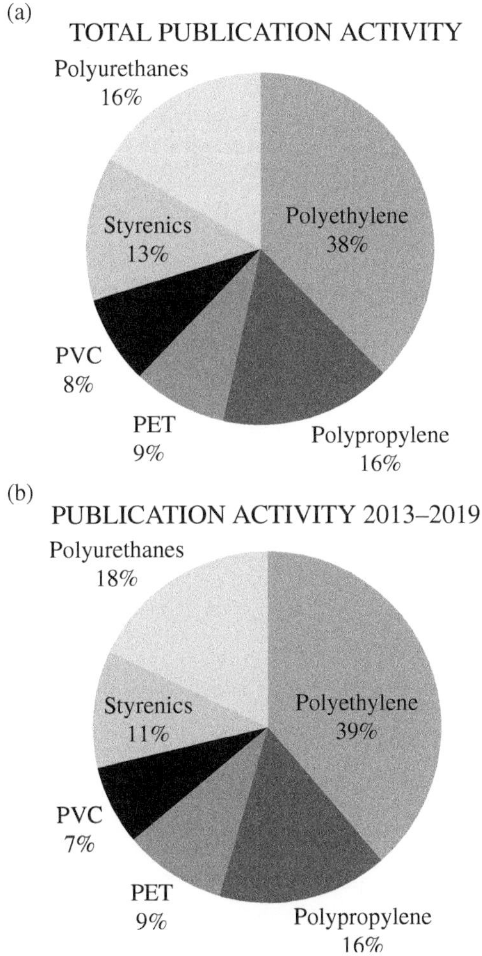


**FIGURE 1.6** Publication activity focused on commodity plastics for (a) 1954–2019 and (b) 2013–2019. The total includes all public literature, patent filings, conference proceedings, and books where the subject focus is the plastic. The total number of publications for all listed plastics is 2.5 million.

shows total global publication activity, including patents, journal articles, reviews, meeting abstracts, governmental documents, etc., for the years 1954–2019 and for the period 2013–2019 for all of the commodity plastics named in Figure 1.1. While many plastics exhibit publication activity approximately in proportion to their production, polyurethane publication activity is more than double its production. Figure 1.7 shows polyurethane publishing activity by language for the same periods, demonstrating the explosive growth in materials research in China. The steady growth of activity appears independent of general global economic activity. Figure 1.8 quantifies the kinds of publications over this time period, showing that patent publications predominate but open literature activity is nearly as prevalent. In the first edition of this book it was found that open literature and patent literature were produced at very similar levels.

Further exploration of these trends shows (Figure 1.9) that patent activity is broad across technologies and focused on specific applications in the areas of coatings and

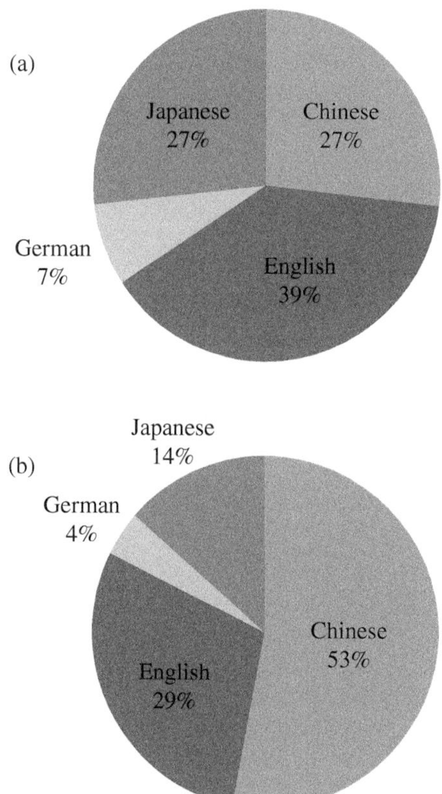


**FIGURE 1.7** Publication activity in polyurethane science by language from (a) 1954–2013 and (b) 2014–2019. Data demonstrate the recent explosive growth in materials research in China.

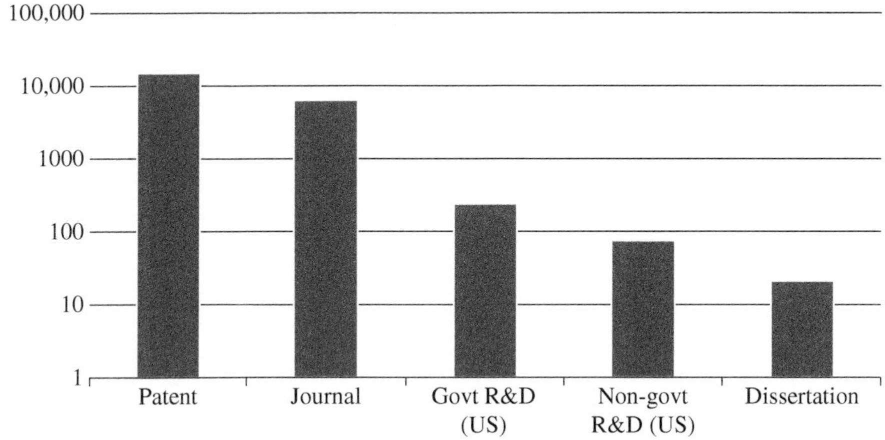


**FIGURE 1.8** Types of publication (all languages) where the focus of the work is polyurethane polymer properties for 2013–2019. The high level of open literature and patent activity demonstrates the continuing intellectual and commercial interest in these materials. The logarithmic scale may exaggerate the importance of items at the low end of the distribution.

fibers, and is related to performance in construction applications and fire resistance. Exploration of the coatings applications shows broad use of polyurethanes for fiber sizings, hair colorants, wound dressings, traffic paints, golf ball coatings, and numerous other unrelated fields. In contrast, categorization of topics studied in the open literature related to polyurethane properties tends more toward fundamental physical materials science issues.

This book also covers markets and commercial aspects of the polyurethane industry. The market and commercial activities overlap but they are not synonymous. The overlap in how the words "marketing" and "commercial" are used reflects their conflation in meaning. Polyurethane market concepts are broader, more strategic, and more theoretical than commercial concepts. Marketing encompasses the equilibrium and nonequilibrium driving forces that make one material attractive to a consumer and unacceptable to another. They take into consideration regional preferences and regional access to feedstocks, and the underlying cultural and societal influences that make a product useful, or desirable, or possess value. They also include the advantages a particular competitor in a market may possess from all facets, including intellectual (i.e. patents).

The commercial aspects of an industrial product include those that are important to specific customers or groups of customers, such as the advantages or value a product may have versus a competitive material or a competitive company in specific cases. Probably the most prominent commercial aspect, especially for commoditized products, is price and price movement. Without doubt, thorough and confident knowledge of pricing in commerce is essential and can distinguish a profitable

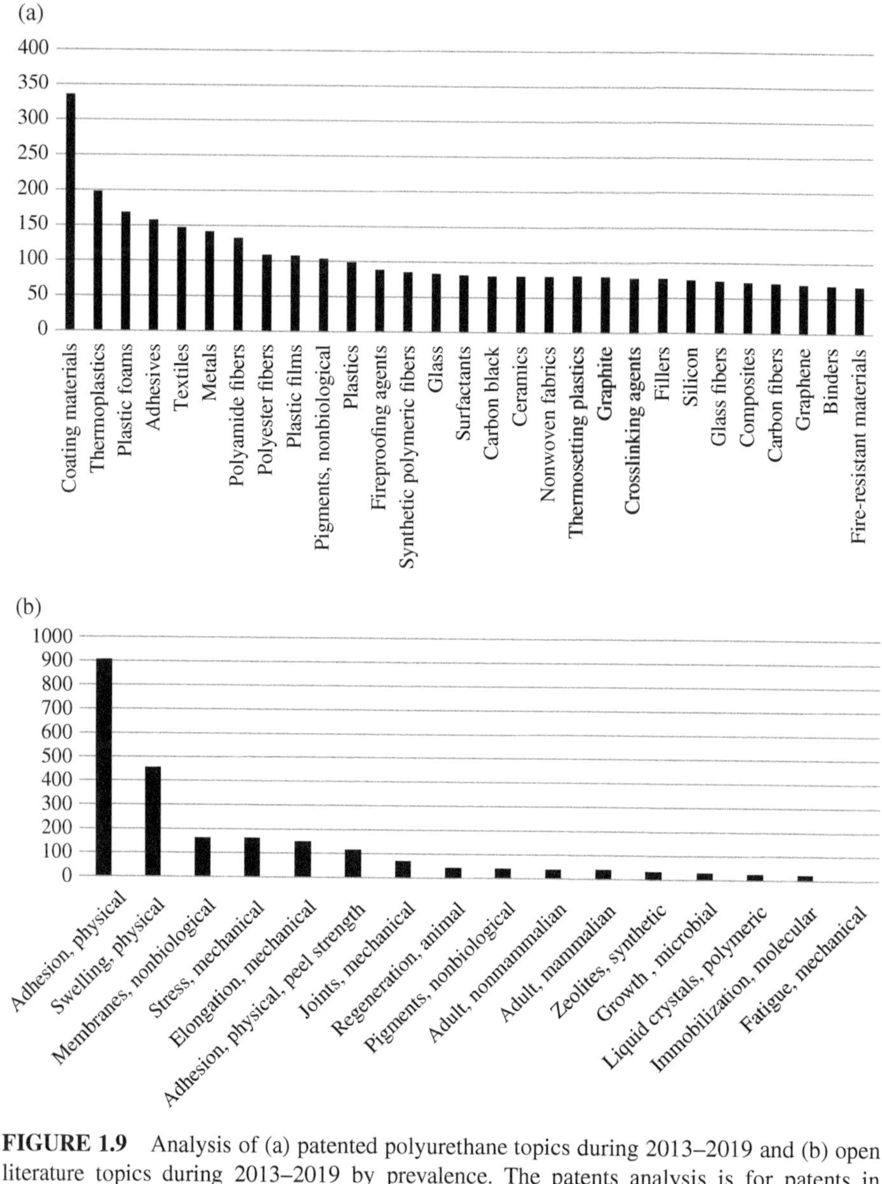


**FIGURE 1.9** Analysis of (a) patented polyurethane topics during 2013–2019 and (b) open literature topics during 2013–2019 by prevalence. The patents analysis is for patents in English, whereas the open literature analysis reflects activity in all languages.

enterprise from one that fails. A commercial move to raise prices when there is excess capacity or failure to raise prices when there are shortages or feedstock prices are rising are common commercial failures, and the ability to shrewdly navigate price movements is the hallmark of well-run companies.

In recent years, the polyurethane industry has been subject to significant macroeconomic forces. The overriding force has been the expansion of polyurethane feedstock manufacturing capacity globally and especially in China. In particular, this capacity growth has injected chemical production during a period of global economic growth but uncertain economics in the future. Figure 1.10 shows the extent of isocyanate overproduction. The demand/capacity ratio can have a very material impact on price expectations and influence decisions on additional capacity expansions. It is not always the case that manufacturers flee a market in response to temporary price declines resulting from overexpansion. In the past manufacturers with a strong financial base have decided to wait out the failure of financially weaker producers. The closure of these weak assets will reduce production volumes, called "tightening the market." It is also the case that in the past manufacturers with a particularly strong financial position and a strong commitment to the market will increase production in the face of overcapacity to further drive down prices, and drive weak manufacturers into untenable production economics. The anticipated response is for weak producers to shutter poorly performing manufacturing assets or to sell their business to one of their competitors. Following these closings, production can regain a capacity/demand balance and prices can rise. This kind of "game" is not seen very often now in the chemical industry. In part this is because of the relatively small number of global manufacturers and their similar financial strengths, maturity, and experience. Additionally, regulation of monopolistic behavior has become more stringent, and the potential gain by this kind of predatory practice may not be worth the potential reputational risk. In that same spirit, many manufacturers see the benefit

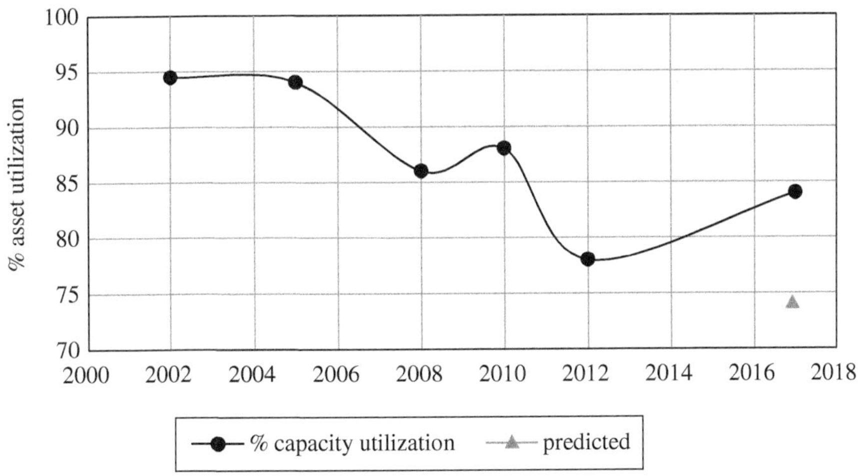


**FIGURE 1.10** Percentage isocyanate plant capacity utilization. The triangle denotes a published expectation for 2017 from the first edition while the circles show the actual. Predictions of future economic activity should always be viewed skeptically. MDI is methylene bisphenyl diisocyanates and its oligomers.

of strong and rational competition in the marketplace. Rational and mature competition can provide ballast to minimize market fluctuations and provide a stimulus to improve business performance. Lastly, it has recently been dramatically demonstrated that temporary occurrences, such as plant disruptions, can have significant effects on supply. From the middle of 2016 until the middle of 2018 commodity isocyanates nearly doubled in price as a confluence of planned and unplanned plant shutdowns during a time of vigorous economic growth put significant constraint on supply, allowing unaffected producers to greatly increase market share and profitability during this period. Upon resolution of these temporary issues, prices fell to historical trend levels.

Trends in polyurethane manufacturing reflect global competitive pressures and global opportunities. This has resulted in expansion of manufacturing assets close to raw material feedstocks and also close to geographies with increasing economic growth. For example, during 2017–2019 there were polyurethane feedstock expansions in Saudi Arabia and the US Gulf Coast (areas of high petrochemical resources) as well as in China and Europe (areas with lower petrochemical resources). It is not immediately clear whether it is cheaper to ship commodity feedstocks to centers of economic activity or ship finished polyurethane chemicals from low-cost manufacturing sites. However, low feedstock cost manufacturing is probably less prone to political factors and will always maintain a low-cost position. On the other hand, market and commercial flexibility is enhanced by proximity to customers.

There is continuing movement towards manufacturing innovation using processes that reduce usage of solvents and reagents and involve less purification and environmental impact. There is probably little incentive for production of new families of polyurethane building blocks, particularly for new polyisocyanates. It would appear that the regulatory burden of new isocyanate production inhibits innovation, and currently available products perform adequately and at acceptable cost. Innovation has rather focused on the development of gas-phase phosgenation processes that reduce solvent and energy consumption. While the large majority of polyols are produced by conventional KOH catalysis there has been moderately increasing production of polyols derived from new double-metal cyanide (DMC) catalysts. While DMC catalysis offers significantly improved production economics, it has been limited to primary utility making slab foam polyols and has been excluded from molded foaming operations because of performance issues (see Chapter 2). On the other hand, improvements in established products, such as production of copolymer polyols with ever higher solids content, lower viscosity, smaller particle size, and improved production operations, will undoubtedly continue and find success in the market. There has been increasing attention paid toward circular economy issues related to polyurethanes, particularly as components of large visible items such as mattresses. There has been progress in processing these materials back to useable feedstocks (see Chapter 14); however, the economics of polyurethane recycle collection and conversion of finished product back to useable polyols is still questionable. Meaningful progress in the circular economy of polyurethanes may await organized municipal collection and rational recovery processes to handle the waste. Lastly, there is a growing initiative in the development of polyurethane structures hybridized with backbones that would normally be thermodynamically immiscible. The goal in this development is to obtain desirable properties of polymer backbones while minimizing any negative attributes that may evolve from thermodynamic incompatibility (see Chapter 13).

The trend for polyurethane applications is being driven by overriding trends in the industries in which polyurethanes find purpose. Thus, automotive trends toward lighter weight dictate a trend toward higher performance at lower foam density. Higher performance includes achieving required comfort factors with lower vibration and noise transmission. In construction markets the trend is toward improved thermal insulation with new blowing agents that exhibit lower ozone depletion potential, and now lower global warming potential as well. Restrictions on acceptable flame-retardant packages for both flexible foams and rigid foams are also a driver of polyurethane industrial innovation. Thus, blowing agents and flame retardants score highly in the intensity of industrial activity associated with polyurethanes. Industrially, reactive catalyst innovation has been consistently pursued (to reduce fugitive catalyst emissions). This trend may intensify in the future as a result of governmental and consumer pressures, particularly in Europe. The trend toward the use of renewable feedstocks has been slow and, based on patent activity, will probably remain so for the near future.

The science of polyurethanes is ongoing and will continue a high level of activity in the future. While a great deal is known about the fundamentals of polyurethane structure–property relationships, the control of these relationships is still being actively pursued. Most understanding of polyurethanes is based on equilibrium properties; however, because of kinetic limitations of reaction-induced phase separation, theory and reality are often in conflict. The exponential increase in computing power allows for finer grained simulations of larger volumes that can be harnessed by modern molecular dynamics, self-consistent field, and coarse-grained theoretical techniques. Additionally, advances in predictive intelligence from massive dataset analysis now allows researchers to better predict or simulate experimental results. Such advances have resulted in commercial software such as Materials Studio® and GeoDict® achieving wide use for prediction of polyurethane properties industrially.

## REFERENCES

[1] M Sonnenschein and W Koonce, "Polyurethanes," in *Encyclopedia of Polymer Science and Technology*, H Mark, Ed., 4th Ed., Wiley, Hoboken, NJ, 2012, 1–63.

[2] R Herrington and K Hock, Eds., *Flexible Polyurethane Foams*, 2nd Ed., The Dow Chemical Co., Midland, MI, 1997.

[3] D Randall and S Lee, *The Polyurethanes Book*, Huntsman International/Wiley, Everberg/Hoboken, NJ, 2002.

[4] M Szycher, *Szycher's Handbook of Polyurethanes*, 2nd Ed., CRC Press, Boca Raton, FL, 2013.

[5] C Prisacariu, *Polyurethane Elastomers: From Morphology to Mechanical Aspects*, Springer Verlag, New York, NY, 2011.
[6] N Lamba, K Woodhouse, and S Cooper, *Polyurethanes in Biomedical Applications*, CRC Press, Boca Raton, FL, 1998.
[7] G Oertel, *Polyurethane Handbook*, Hanser, Munich, Germany, 1985.
[8] P Krol, *Linear Polyurethanes: Synthesis Methods, Chemical Structures, Properties and Applications*, CRC Press, Boca Raton, FL, 2008.
[9] I Clemitson, *Castable Polyurethane Elastomers*, CRC Press, Boca Raton, FL, 2008.
[10] J Brandrup, E Immergut, and E Grulke, Eds., *Polymer Handbook*, 4th Ed., Wiley, Hoboken, NJ, 2003.
[11] W Gum, W Riese, and H Ulrich, Eds., *Reaction Polymers: Chemistry, Technology, Applications, and Markets*, Hanser, Munich, Germany, 1992.
[12] L Gibson and M Ashby, *Cellular Solids – Structure and Properties*, 2nd Ed., Cambridge University Press, Cambridge, UK, 1997.
[13] D van Krevelen and K te Niejenhuis, *Properties of Polymers*, 4th Ed., Elsevier, Amsterdam, the Netherlands, 2009.
[14] J Bicerano, *Prediction of Polymer Properties*, 3rd Ed., Marcel Dekker, New York, NY, 2002.
[15] MF Sonnenschein, *Polyurethanes: Science, Technology, Markets, and Trends*, Wiley, Hoboken, NJ, 2014.
[16] S Thomas, A Rane, K Kanny, G Thomas, Eds., *Recycling of Polyurethane Foams*, William Andrew, Oxford, UK, 2018.
[17] S Cooper, J Guan, Eds., *Advances in Polyurethane Biomaterials*, Elsevier, Amsterdam, the Netherlands, 2016.
[18] B Obi, *Polymeric Foams: Structure–Property–Performance. A Design Guide*, William Andrew, Oxford, UK, 2017.

# 2

# POLYURETHANE BUILDING BLOCKS

The versatility of polyurethanes is derived in large part from the wide selection of building blocks available to materials designers. The growth in the use of polyurethanes has been highly dependent on the cheap and available feedstock polyisocyanates, polyols, and chain extender coreactants such as water, alcohols, and amines. However, this paradigm has recently been challenged as substitutions are made for the purposes of avoiding health and safety issues associated with isocyanates. A detailed discussion of substitute chemistries to urethane structures will be handled in Chapter 12 [1–3]. This chapter provides an overview of building blocks for conventional polyurethane polymerization.

In principle, there is as much potential for the design of isocyanate structures as there is for alcohol and amine coreactants. In reality, while there are numerous polyisocyanates to choose from, most of the innovation in polyurethane performance comes from the broad range of choices available in the coreactant alcohols and amines. To a great extent this reflects complications (both industrial and regulatory) associated with making isocyanates, and the comparative ease of making polyol and polyamine structures. It also reflects the fact that, for most polyurethanes, nonisocyanate reactants constitute more than 50% of the polyurethane volume. Thus, for the purpose of obtaining any particular outcome, varying the polyol and the chain extender components is the most straightforward way to begin.

*Polyurethanes: Science, Technology, Markets, and Trends*, Second Edition. Mark F. Sonnenschein.

## 2.1 POLYOLS

The term polyol refers simply to polymer backbones containing nominally two or more hydroxyl groups. Polyols are the largest volume raw material used in polyurethane applications with weight fractions in applications ranging from 90 wt% in low-modulus flexible sealants, 70 wt% in flexible foams, and as low as 30 wt% in rigid insulation foams. As implied by these values, the polyols in urethane formulations tend to provide softness and flexibility, while isocyanates and low-molecular-weight chain extenders provide hardness and stiffness to the resulting polymer structures.

Polyols are produced with a range of backbones and hydroxyl functionalities that can be tailored to best meet application processing and property requirements. The most industrially significant polyol backbones are ether and ester based, while a number of specialty backbones including carbonate, acrylic, and ethers derived from tetrahydrofuran (THF) are used in high-performance coating, adhesive, and elastomer applications. Figure 2.1 shows their relative global volumes as of 2017 [4, 5]. For comparison the relative global volume for polyether polyols in 1992 was 2654 thousand metric tons and for polytetramethylene ether glycol (PTMEG) it was 100 thousand metric tons. In the case of polyether polyols this translates to an approximately 4% annual growth rate. The volume growth of PTMEG has also been a long-term trend of approximately 4% [6]. Despite continuity of the long-term trend, the growth in these categories has not been linear over time and not uniform over all geographies. For instance, it is estimated that while the current growth rate for polyether polyols in China is nearly 8% it is only about 2.5% in North America. While the

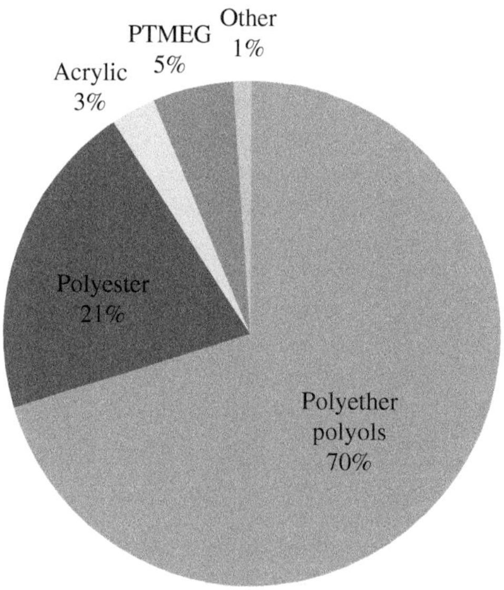


**FIGURE 2.1** Relative volumes of polyols produced in 2017. Within the error of the sources used for this figure, the data are only marginally different from 2012.

polyol industry as a whole has grown, the relative size of each category reflects the integrated underlying dependence on global economic growth in various regions.

Polyols vary markedly in structure, manufacture, function, and price. While each occupies a particular price/performance niche, for many of them there is overlap and the choice of which polyol to use in any particular application depends on the history and experience of the chemist and the end-user. In general, they all begin with low-cost commoditized building blocks. Final cost depends on factors related to volume of production, cost of polymerization process, and perceived value in an application. The structures of several commonly utilized polyols are provided in Figure 2.2. The structure to a great extent defines the properties of the resulting polyurethane, as well the compromises each structure demands of the final product performance. For instance, while polyethers generally provide good low-temperature performance, an easily processed backbone, and a good cost position, the low ceiling temperature (the temperature at which the rate of depolymerization is equal to the rate of polymerization) is relatively low for these structures [7]. While many factors may influence the exact degradation temperature, polyether polyols should not be considered stable above 220 °C. Polyether polyol flammability results from the high volatility of the monomers, the combination of oxygen and hydrocarbon fuel that the monomer represents, and is a fundamental limitation to these structures [8, 9]. This problem has in turn created a demand for flame retardants in many polyether-containing systems, especially foam systems having high surface to volume ratios [10]. Similarly, the well-known hydrolytic instability of polyesters is a compromise that must be accommodated if taking advantage of their low cost (in some instances) and high thermal stability (Table 2.1). Details in these regards will be supplied for each of the structures in Figure 2.2.

### 2.1.1 Polyethers

***2.1.1.1 Building Blocks*** Polyether polyols are the class of polymers formed from the exothermic reaction of ethylene oxide (oxirane; EO), 1,2-propylene oxide (methoxirane), or 1,2-butylene oxide (ethyloxirane). Other materials could be considered (i.e. epichlorohydrin) but these three monomers by far constitute the largest volume of this class and are of the most use in polyurethane chemistry. Additionally, the related cyclic ether THF is another important building block for polyether synthesis (Figure 2.3). Although the molecules are structurally related, each one is prepared by significantly different procedures.

EO is prepared directly from the catalyzed oxidation of ethylene [11, 12]. 1,2-Propylene oxide and 1,2-butylene oxide are prepared from a two-step reaction consisting of (i) reaction of the 1,2-alkene with chlorine and water to form the chlorohydrin followed by (ii) ring closure to form the cyclic ether by reaction with base [13, 14]. Figure 2.4 is an illustration of the synthesis of 1,2 butylene oxide.

A more recent innovation produces propylene oxide (PO) by direct oxidation of propylene using hydrogen peroxide (HPPO process) [15, 16] (Figure 2.5). The HPPO process is simpler and uses less energy, but requires the coproduction of hydrogen peroxide on a massive scale to make PO economically.

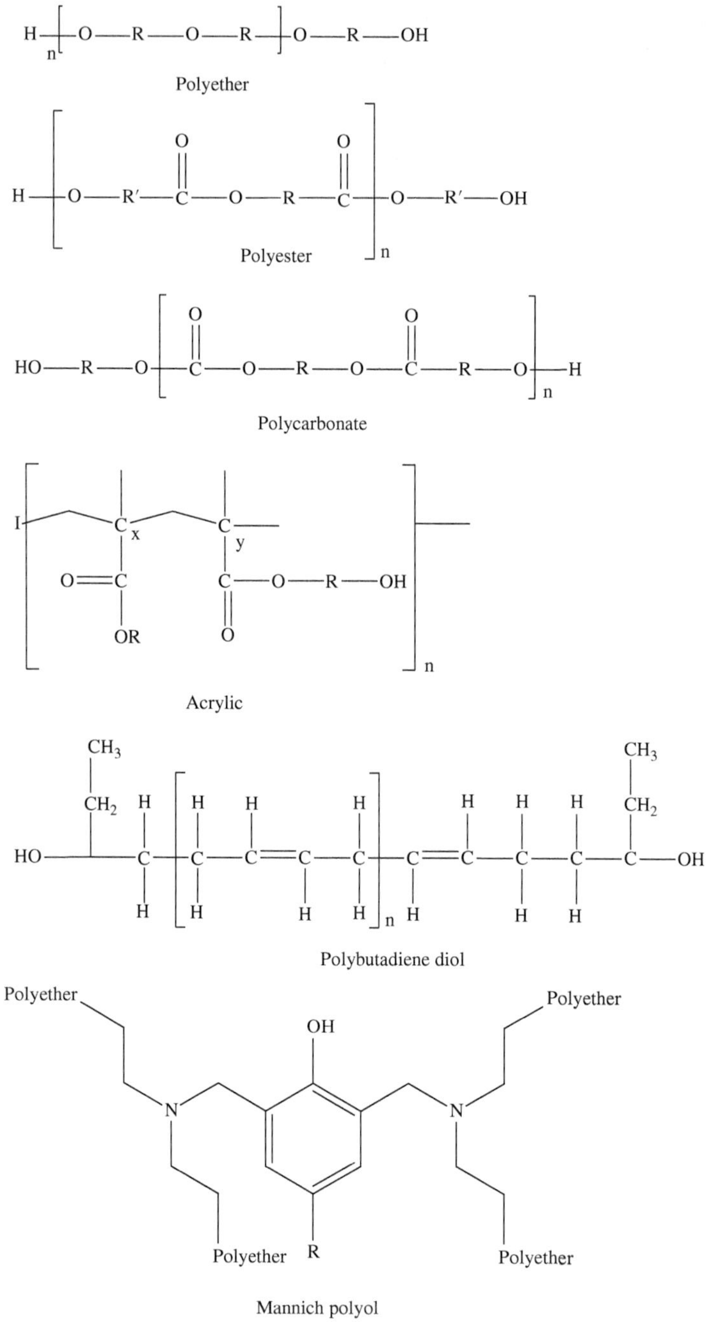


**FIGURE 2.2** Comparison of basic polyol structures.

**TABLE 2.1 Comparison of polyol general polyol properties**

| Polyol backbone | Advantages | Disadvantages |
|---|---|---|
| Polyether polyols based on propylene oxide and ethylene oxide | Hydrolytic stability, cost, viscosity, flexibility | Oxidative stability, modulus/strength, thermal instability, flammability |
| Aliphatic polyester polyol | Oxidative stability, modulus/strength | Viscosity, hydrolytic stability |
| Aromatic polyester polyol | Flame retardance, modulus/stiffness | Viscosity, low flexibility |
| Polyether polyols based on tetrahydrofuran | Hydrolytic stability, modulus/strength | Oxidative stability, viscosity, cost |
| Polycarbonate polyols | Hydrolytic stability, oxidative stability, modulus/strength | Viscosity, cost |
| Acrylic polyols | Hydrolytic/oxidative stability, hardness | Viscosity, cost, low flexibility |
| Polybutadiene polyol | Low-temperature flexibility, solvent resistance | Viscosity, thermally oxidizable (unless hydrogenated), cost |

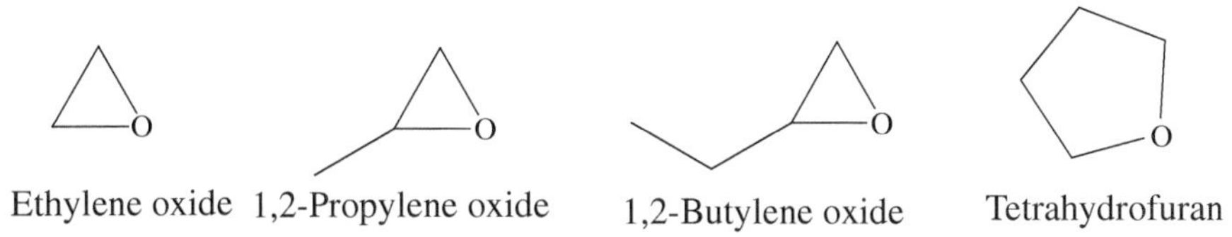


**FIGURE 2.3** Building blocks for polyether polyol synthesis.

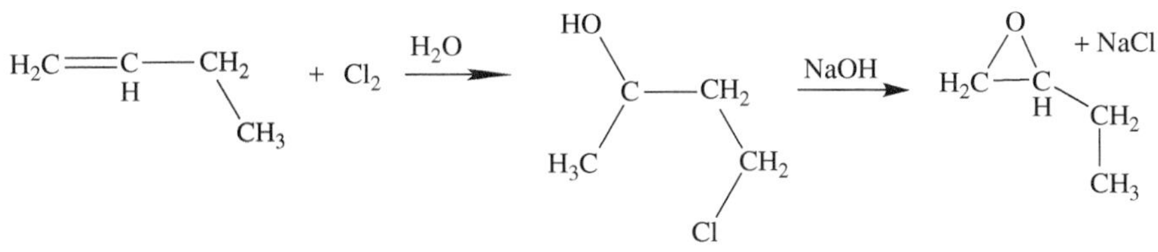


**FIGURE 2.4** Synthesis of butylene oxide via the chlorohydrin route.

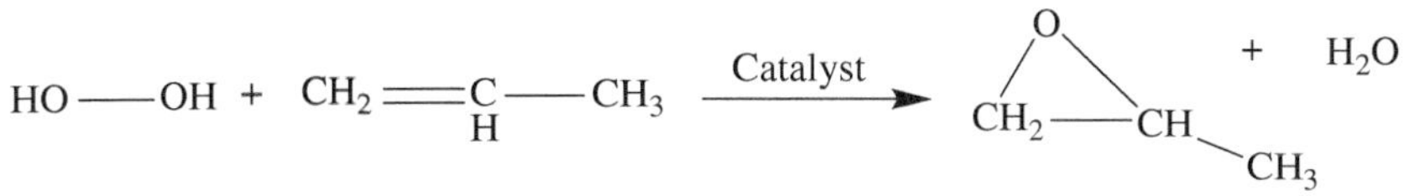


**FIGURE 2.5** Synthesis of propylene oxide via the hydrogen peroxide (HPPO) process.

A third industrial route to PO involves the peroxidation of ethyl benzene to form ethyl benzene hydroperoxide followed by reaction with propylene to produce the desired PO and the side product methyl phenyl carbitol, which can be dehydrated to styrene [17]. Other methods are also practiced commercially [18]. Figure 2.6 details the share of each process to PO as of 2018.

The last feedstock for polyether production is THF. The great majority of THF is produced by dehydration of 1,4-butanediol [19] (Figure 2.7).

Two alternative processes are practiced industrially. One route oxidizes butane to maleic anhydride followed by hydrogenation of the anhydride to THF [20]. A second route reacts butadiene with acetic acid and oxygen to form the 1,4-diacetoxy butane, which can be hydrolyzed to 1,4-butanediol or THF by deacetoxycyclization [21, 22]. While this route has the benefit of flexibly producing either the diol or the furan product, it creates an increased demand for the valuable butadiene monomer with implications for those products requiring synthetic rubber. A new process produces THF from biosourced sugars by separation of succinic acid from the bacterial broth followed by reduction to butanediol and THF from the dehydration of butanediol [23].

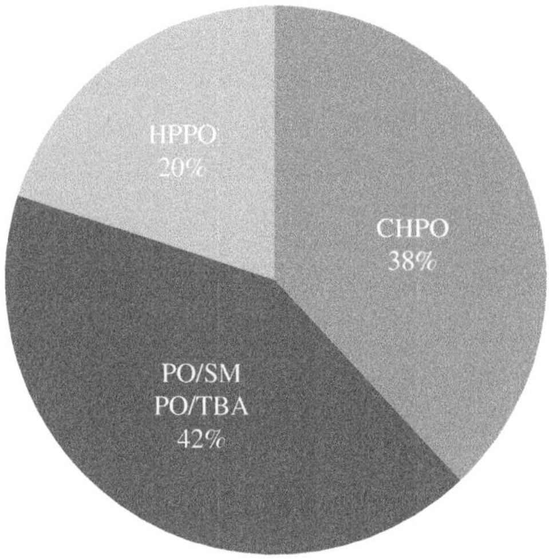


**FIGURE 2.6** Volume of PO produced by the numerous routes available in 2018. CHPO, chlorohydrin process; HPPO, PO by peroxidation of propylene; PO/SM, PO co-styrene monomer; PO/TBA, PO co-tert-butyl alcohol.

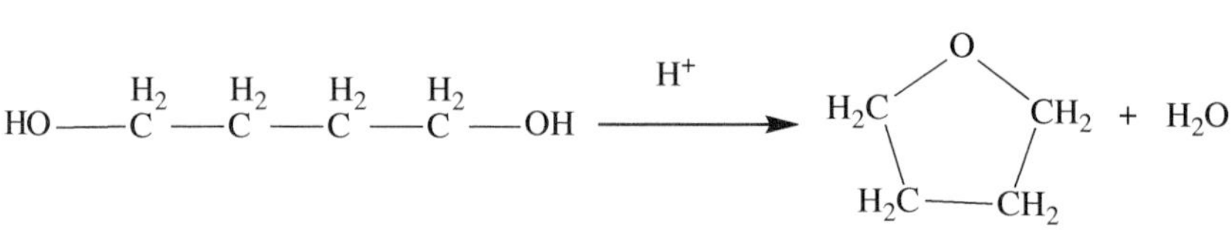

**FIGURE 2.7** Production of THF via dehydration of butanediol.

***2.1.1.2 Polymerization of Alkoxides to Polyethers*** While EO is an important building block for synthesis of polyurethane polyether soft segments, there are only a few exceptions where polyethylene glycol (PEG) is utilized as the sole soft segment. This is because of the crystallinity of PEG and because of its affinity for water. The affinity for water limits PEG use since the properties of the polyurethane can be strongly dependent on the humidity of the environment. Instead, for polyurethanes, EO is usually used as a coreactant with other alkylene oxides, usually PO, to create block or random copolymers.

PO is polymerized to polypropylene oxide by two industrially dominant processes. Both processes are defined by the catalyst that is used. The dominant method employs an initiator or starter molecule with active hydrogens such as water, ethylene glycol, glycerine, ethylene diamine, or sorbitol, for instance, and a base catalyst such as potassium hydroxide [24]. This approach is usually referred to as the base-catalyzed process. The other process to polypropylene oxide uses a class of heterogeneous metal-based catalysts termed "double-metal cyanide" (DMC) catalysts [25]. Each process has unique attributes, including differences in cost, process flexibility, and even structure of the final product.

The base-catalyzed process follows the simplified path given in Figure 2.8.

The functionality of the final product is determined by the functionality of the initiating alcohol (Table 2.2), and the final molecular weight is influenced by the ratio of initiating alcohol to oxirane monomer. The base resides in a molecularly dispersed state with the oxirane and is termed a homogeneous catalyst. In actuality the process is more complicated than expressed in Figure 2.8. The reaction mechanism of Figure 2.8 suggests that the result of the reaction is all secondary hydroxyl end groups. However, this is not in fact observed. In the synthesis of polypropylene oxide there are also observed a population of primary hydroxyl end groups and a population with vinyl termination. The vinyl termination is a particular concern since it is not reactive with the other components of polyurethane chemistry and so degrades the polymer network structure. The actual chemistry that occurs within the reactor is explicitly given in Figure 2.9. The reaction resulting in secondary hydroxyls predominates but the other structures do form, and the process and specific product influence the product distribution.

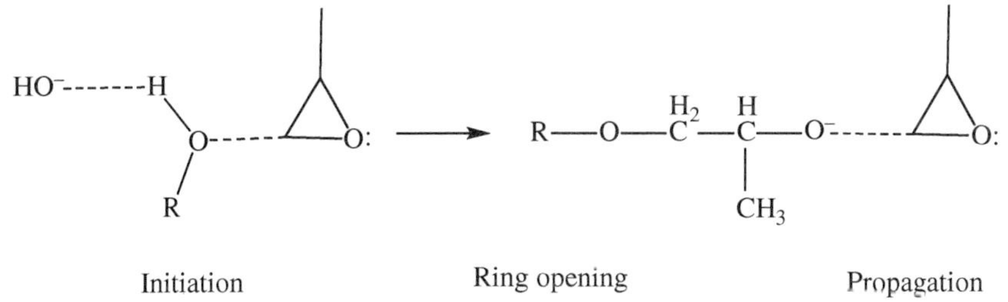


**FIGURE 2.8** Production of polypropylene oxide via base-catalyzed ring-opening polymerization.

**TABLE 2.2 Common polymerization starters or initiators for polyether polymerization**

| Initiator | Initiator functionality | Structure |
|---|---|---|
| 2 | Water | 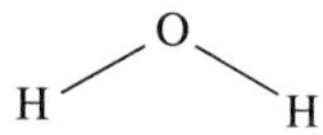 |
| 2 | Ethylene glycol |  |
| 2 | 1,2-Propylene glycol | 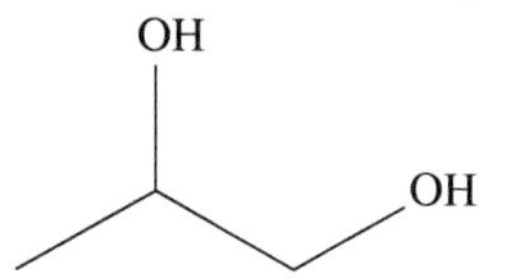 |
| 3 | Glycerine | 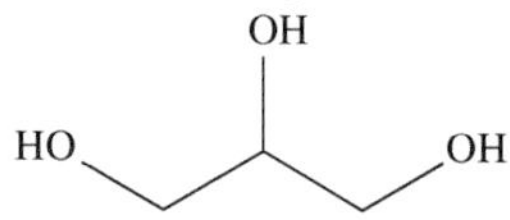 |
| 3 | Ethoxylated glycerine | 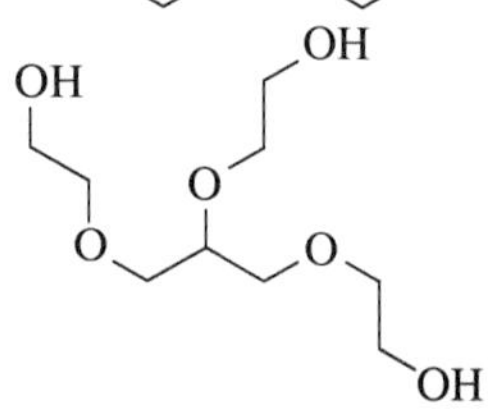 |
| 4 | Ethylene diamine | 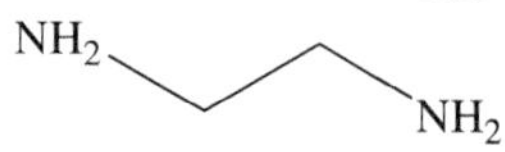 |
| 4 | *N*-methyl-amino-dipropylamine | 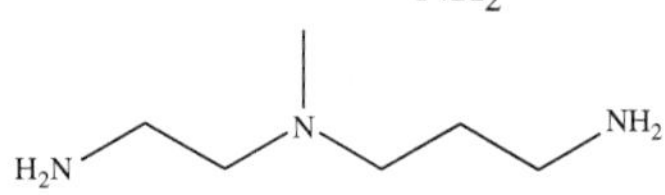 |
| 4 | Pentaerythritol | 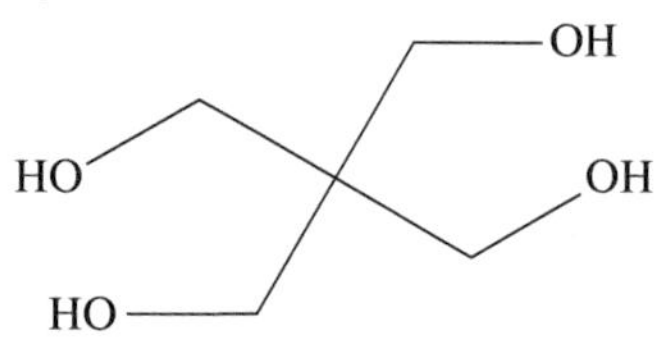 |
| 6 | Sorbitol | 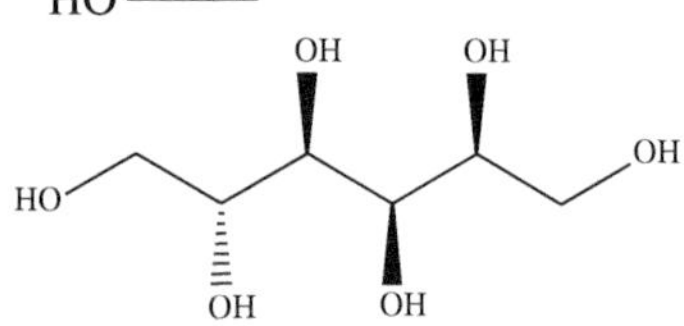 |
| 8 | Sucrose | 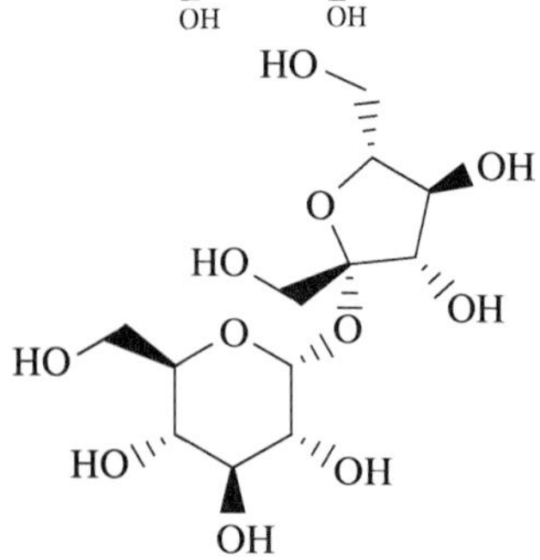 |

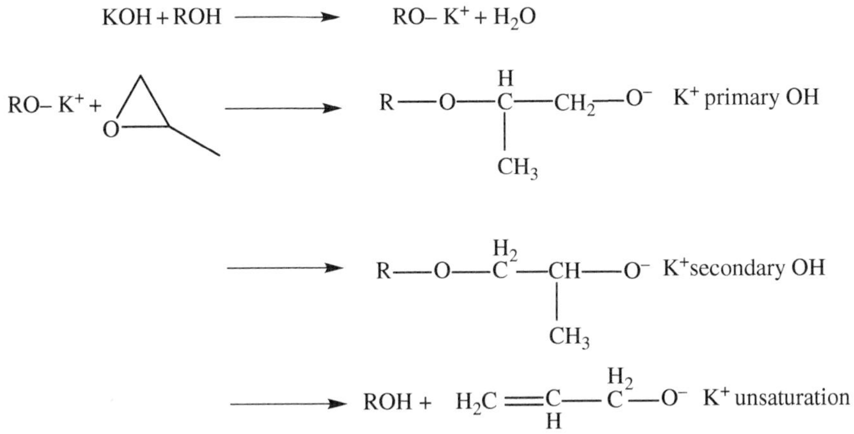

**FIGURE 2.9** Possible pathways by which PO polymerization can traverse to products. The route to secondary hydroxyls is the most prevalent.

The primary product in base-catalyzed (or explicit in these reactions *base initiated*) polymerization is the product possessing secondary hydroxyls because of the relative ease of nucleophilic attack on the less hindered carbon of the oxirane. Alkoxide attack on any of the methyl hydrogens results in the unsaturation given mechanistically in Figure 2.10 for the formation of allyl alcohol; this can then serve as the initiating alcohol for monohydroxy polyether formation, which is referred to as "monol" [26].

The amount of monol formed in the polymerization is a function of several process details increasing primarily with temperature.

An empirical equation for estimation of polyol functionality taking into account the effect of monol is given by Equation 2.1 [27]:

$$\text{Functionality} = \frac{\%\text{OH}/1.7}{\text{unsat} + \dfrac{\left(\dfrac{\%\text{OH}}{1.7}\right) - \text{unsat}}{\text{nominal functionality} - \text{mol\% diol}}} \tag{2.1}$$

Alternatively, the functionality can be calculated from the measured OH number (ASTM D4274) [28] and the ASTM measured allyl concentration using Equation 2.2:

$$\text{Functionality} = \frac{\text{OH}/56.1}{\big((\text{OH}/56.1) - \text{unsat}\big) \times \big(1/\text{nominal functionality}\big) + \text{unsat}} \tag{2.2}$$

The amount of unsaturation is measured by a standard method such as ASTM D4671 [29]. Unsaturation is quantified in units of "milliequivalents unsaturation per gram of polyol," and the diol content is obtained either from chemical analysis or by calculation. Figure 2.11 shows the potential loss of network connectivity resulting from

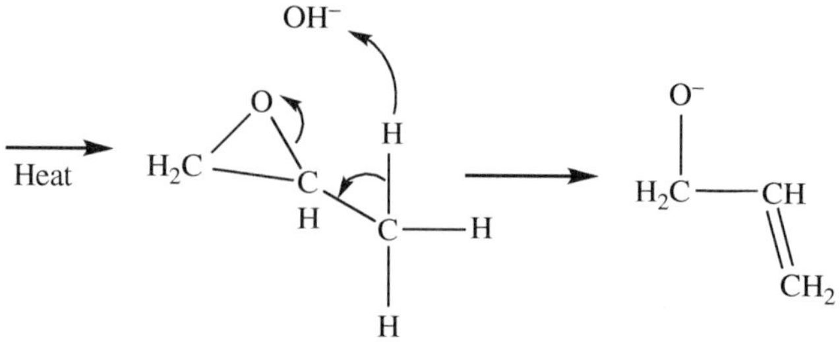


**FIGURE 2.10** Mechanism for formation of unsaturation in polyether polyols. PEGs cannot do this owing to the absence of pendant carbons for attack.

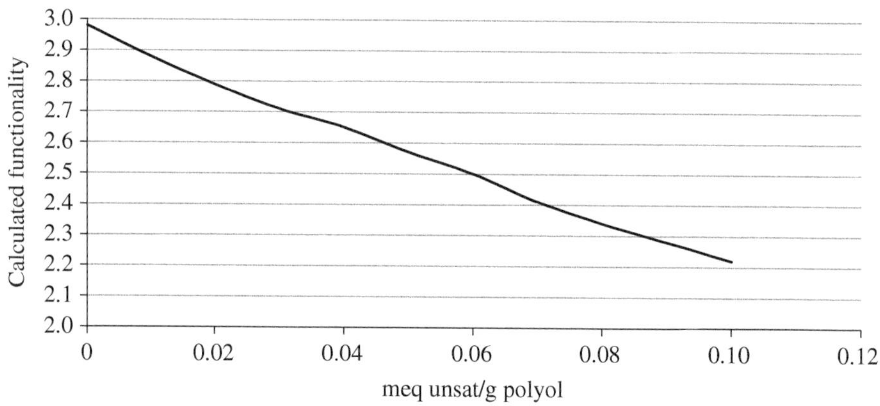


**FIGURE 2.11** The effect of monol on polyol functionality. The calculated actual functionality for a 5000 molecular weight (1666 equivalent weight) triol reflecting the effects of end-group unsaturation developed during polymerization.

formation of unsaturation of a nominally 5000 molecular weight triol. Similarly a 4000 molecular weight diol can have a true functionality of 1.7 because of monol formation.

EO cannot form unsaturation in the manner of PO since there is no beta carbon from which to abstract a hydrogen. Butylene oxide is capable of forming an unsaturated monol initiator in a manner analogous to PO.

The problems associated with monol formation have stimulated research for alternative polymerization catalysts less prone to formation of unsaturation. One approach has been to vary the complexing counterion. It has been shown that the tendency for the counterion to promote formation of allyl alcohol decreases in order down the periodic table from $Li^+ > Na^+ > K^+ > Rb^+ > Cs^+$. It has been reported that the use of CsOH instead of KOH can reduce the formation of unsaturated end groups by 50% [30, 31]. Alkaline earth salts bearing $Ba^{2+}$ and $Sr^{2+}$ have also been reported to result in significantly reduced formation of unsaturation in polyether synthesis [32].

As with all homogeneous catalysts, the base must be either extracted or neutralized after the polymerization step. If the polyol is sufficiently immiscible with water the catalyst can be extracted with water. The polyol will coalesce as a separate phase and be easily removed and, if necessary, the catalyst recovered. In general, the rate of base-catalyzed reaction is considered to be first order in catalyst and in oxirane:

$$\text{Rate} = K_2 \left[\text{oxirane}\right]\left[\text{catalyst}\right] \tag{2.3}$$

where $K_2$ is the second-order rate constant. However, since the catalyst is not consumed the rate can be further simplified to Equation 2.4:

$$\text{Rate} = K_1 \left[\text{oxirane}\right] \tag{2.4}$$

where $K_1$ is a pseudo-first-order rate constant equaling the product of [catalyst] × $K_2$.

Separate from the homogeneous base initiation of oxirane polymerization, a heterogeneous catalyst system has been developed. The most successful approach has been the development of heterogeneous catalysts described collectively as DMC catalysts [33]. DMC catalyst describes any of the class of catalysts of general structure:

$$M^1{}_a\left[M^2\left(CN\right)_b A_c\right]_d$$

where M refers to a metal, CN refers to the CN ligand, and A is an additive.

The effectiveness of DMC catalysis in polymerization of epoxies was first noticed in the early 1960s and rediscovered for the purpose of polypropylene oxide production by Shell, Arco, Asahi Glass, Union Carbide, and others in the 1980s. Considering the amount of time and effort that has been put into this very enticing class of catalysts, it remains safe to say that there are many open questions about how they operate. While DMC catalysts show a high level of reproducibility and reliability in their function, this technology has largely advanced through empirical observation and what can only be termed as "art."

The most efficient DMC catalysts for polymerization of polypropylene oxide contain $Zn_3[Co(CN)_6]_2$ scaffolding in their structure. However, it is well established that this structure alone is inactive for polymerization of PO and that additional additives must be incorporated for desired results. DMC additives typically include metal salts such as $Zn(Cl)_2$ and low-molecular-weight solvents such as tert-butanol [34]. Preparation of the optimized active catalyst is an involved and precise procedure. The mechanism of the catalyst is highly uncertain. The catalyst is sometimes referred to as a "precatalyst" since there is usually an induction time of the order of 30 minutes to several hours between when reaction conditions are reached and when molecular weight build due to polymerization commences. Several theories exist for this behavior; however, given the fundamental limitations of precisely determining the surface conditions during reaction, full understanding remains an area open for research.

One mechanism suggested for the action of DMC catalysts involves the initial substitution of the Cl ligand of $ZnCl_2$ with the alcohol (i.e. tert-butanol) stated above to be necessary for catalyst activity [35]. Another substitution of Cl on the Zn metal center coordinates the reactive species to a single site in a geometry potentially effective for an addition reaction to occur. Proton transfer from the alcohol coordination site to the Zn-PO bond ring opening would be facilitated. This model is consistent with the observation that too much alcohol (R′OH in Figure 2.12) depresses the activity of the catalyst. In this scheme, too much R′OH would possibly compete with the alkoxide for coordination to the Zn [36]. It is not immediately clear how this model accounts for the induction period, though it is possible to rationalize that attaining the condition of cocoordinate alcohol *and* alkoxide may take time, and may depend on such experimental details as mixing, temperature, the population of active catalyst sites on the surface, etc. Similarly, it is not immediately clear why the ratio of Zn to Co in the catalyst package has such a significant effect on catalyst activity.

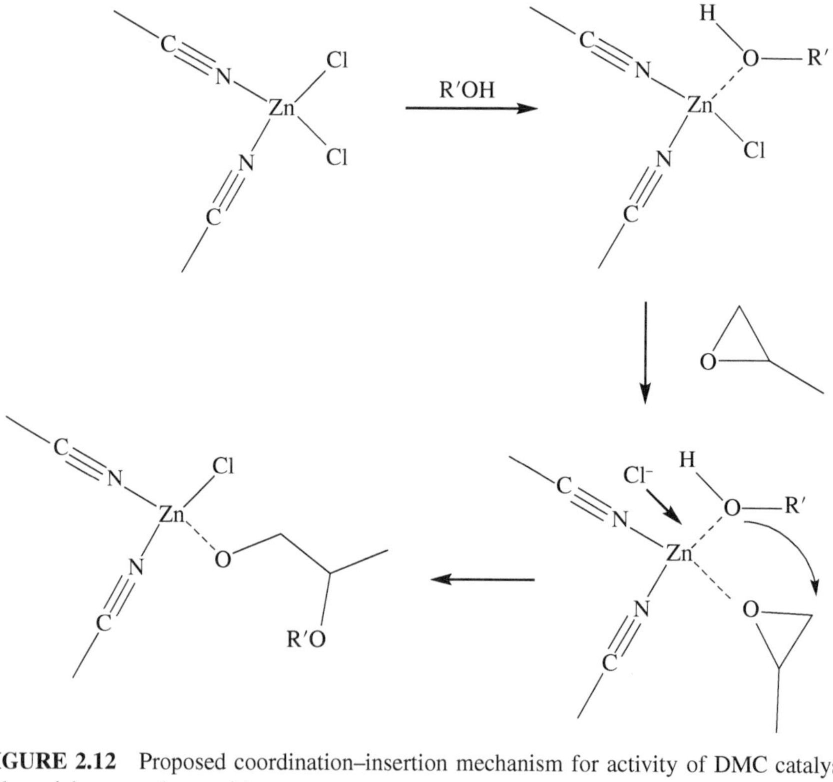


**FIGURE 2.12** Proposed coordination–insertion mechanism for activity of DMC catalysts polymerizing propylene oxide.

An alternative (but somewhat related) mechanism also relies on coordination of reactants to the surface but takes into account the totality of the catalyst morphology, including its layered structure and the coordination geometry. As mentioned previously, when the entire catalyst package including added metal salts such as $ZnCl_2$ is taken into account the total formula for the catalyst can be generalized from $Zn_3[Co(CN)_6]_2$ to

$$Zn_3\left[Co(CN)_6\right] \cdot xZnCl_2 \cdot yH_2O \cdot z\text{Ligand}$$

For a Zn/Co ratio of 1.5 each Zn is coordinated to four cyanide ions and lacks a coordination site (assuming 4-coordinate Zn). In fact it is established that DMC catalyst activity for this series of catalysts requires a Zn/Co ratio greater than 1.5. For a Zn/Co of 2.0, zinc obtains a population of coordination conditions with an average of three cyanide ions and one chloride ion for every two zinc ions. By this model, a surface fragment of a DMC catalyst may have two Zn sites: one acid, which is capable of binding to an alcohol function, and the other basic, which is capable of binding to an alkoxide. The binding of the alcohol to the acid can activate the proton as suggested in the model described by Figure 2.12. Figure 2.13 describes this model for the acid–base-regulated coordination chemistry.

Regardless of the details of the mechanism of DMC polymerization of polyols, there are several advantages associated with this process. These include higher

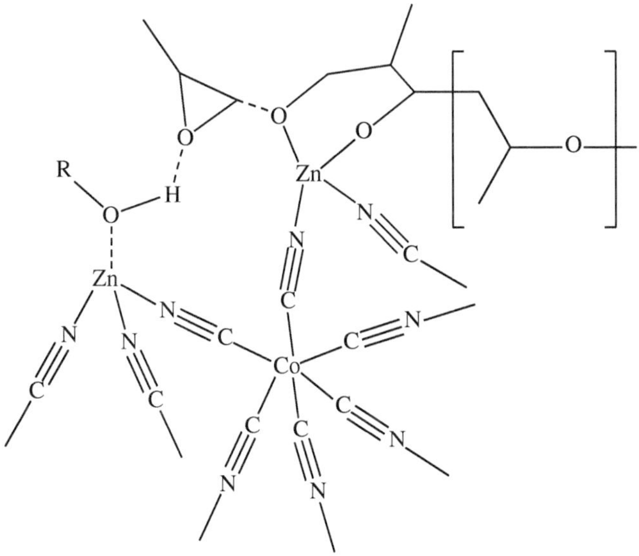


**FIGURE 2.13** Proposed acid–base coordination mechanism for activity of DMC catalysts polymerizing polypropylene oxide.

activity at any given temperature (allowing low catalyst loadings), narrow molecular weight distributions, a highly atactic product (low tendency to crystallize), and very low levels of unsaturation, well under 0.01 meq/g (see Figure 2.14 for comparison with KOH). The meaning of the lower unsaturation to the polymer structure can be seen in Figure 2.14. Since the number of end groups decreases rapidly with increasing molecular weight, and the formation of unsaturation is dependent on reaction conditions but is otherwise fixed, the effects of unsaturation on polyurethane network connectivity will intensify as the molecular weight increases. The differences between KOH- and DMC-catalyzed polymerizations become more noticeable at equivalent weights greater than 1000 g/eq [37].

The obvious advantages associated with DMC catalysis in the final product come with a price. A well-known process caution is associated with the induction period. Given the very high activity of DMC-catalyzed ring opening and the high exotherm of the reaction (~25 kcal/mol), the induction results in a very significant and potentially unpredictable exotherm produced during the polymerization. Process optimization has dealt with this on an industrial scale by standardization of materials and process [38].

A larger and more intractable problem is the inability to develop block copolymer structures of EO and PO. As will be discussed later, there is a significant reactivity difference between the secondary hydroxyls of polypropylene oxide and the primary hydroxyls of polyethylene oxide. One processing compromise to obtain primary hydroxyls without the drawbacks of all PEG structures is to polymerize PO to polypropylene oxide to a given molecular weight and subsequently add EO. When

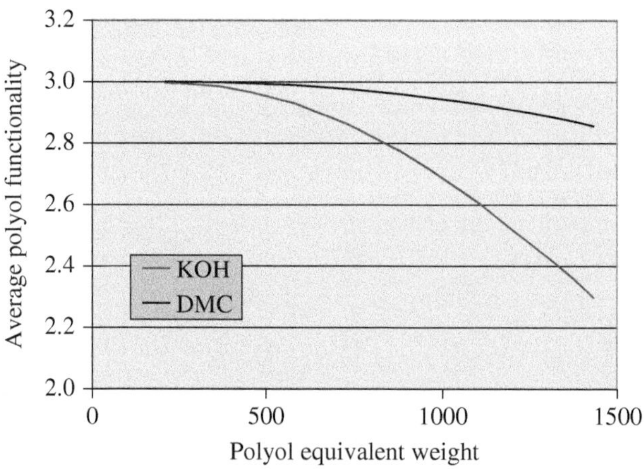


**FIGURE 2.14** Difference in measured polyol functionality attained by KOH and DMC catalysis for a triol of variable equivalent weights. The difference becomes appreciable at higher equivalent weights as the number of end groups decreases [37]. (*See insert for color representation of the figure.*)

handled appropriately, homopolymerization of the EO can be minimized and EO blocks can grow from the polypropylene oxide polymer ends. With KOH catalysis this is not a problem since PO and EO can be equally polymerized via the mechanism of Figure 2.9. A similar process has not been achieved with DMC catalysis. The reason for this has not been well established. One hypothesis has been that the growing PEG block has a much higher affinity for the DMC catalyst surface, causing the initially formed end blocks to grow inhomogeneously. The ever-growing PEG blocks become ever less miscible with the polypropylene oxide bulk, and so polymer chain exchange off the surface to the bulk becomes increasingly less favorable. Instead, obtaining EO-capped polypropylene oxide polyols prepared by DMC catalysis has required deactivation of the DMC catalyst followed by EO end capping using conventional KOH catalysis. A process in which EO mixed with the PO in the reacting mixture has been developed and proven effective at producing random PO/EO polymer chains [39].

A product-related problem with DMC catalysis is ironically caused by the high level of patent activity that has been pursued by a few competitors. The result has been that only a few industrial concerns have developed or licensed patent coverage for making DMC-derived polyols. At the same time there have been advances in traditional base catalysis using varied counterions (Figure 2.15). The motive for further innovation in heterogeneous coordination catalysis has also been dampened by the observation that polyurethane foam properties, accounting for a large volume of polyurethane usage, are not noticeably affected by the undeniable polyol structural improvements derived by DMC catalysis. This has relegated DMC polyols to non-foam applications, i.e. those applications where the low reactivity of secondary

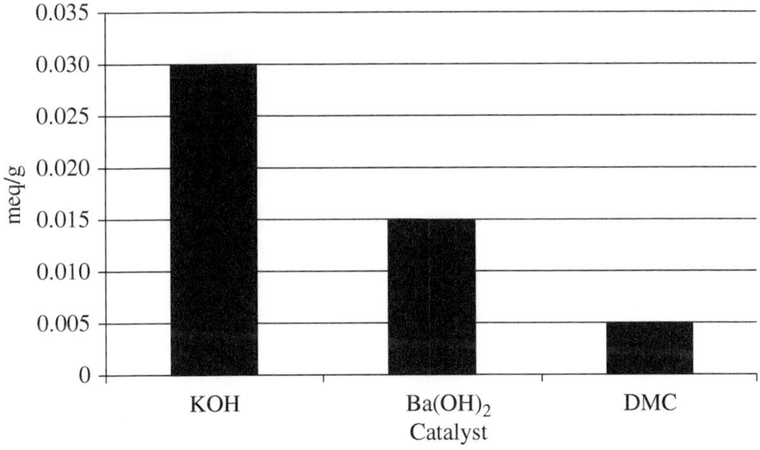


**FIGURE 2.15** Measured unsaturation levels by titration of three polyols for a 2000 molecular weight diol. The lower the value, the fewer vinyl end groups found in the system. *Source:* Modified from [40].

hydroxyls is irrelevant. DMC-catalyzed polyols are also preferred for higher equivalent weight polyols since, at lower equivalent weights, the structural differences are minimized as the fixed number of vinyl chain ends is less important owing to the large number of total chain ends. Since publication of the first edition of this book, the economic advantages associated with PO polymerization using DMC catalysis have resulted in expansion of their use into low-cost applications, such as making polyols for slab foam production [4]. Plants are now being dedicated to production by DMC catalysis and this trend may continue into the future.

More recently there has been industrial interest in the use of Lewis acids, particularly strong acids based on perfluoro-triaryl boranes for production of polypropylene oxides [41]. These polyols exhibit relatively high levels of primary hydroxyl end groups because of electronic and steric factors associated with the catalyst structure. Polymerization rates are reported to be intermediate between KOH and DMC catalysis, and could provide a route to increasing polypropylene oxide reactivity without having to resort to EO end capping.

As of the writing of this second edition, the majority of industrially produced polyols are still prepared by KOH-catalyzed polymerization of PO or PO/EO monomers. The process in concept is remarkably simple, but the details associated with producing a commoditized product can create significant amounts of redundancy, controls, and safety additions. To make polyols efficiently and competitively the production volumes must be very large and the accompanying capital costs very large as well. Plant capacities range from about 50 thousand metric tons to about 500 thousand metric tons. For the purpose of scale, a 50 thousand metric ton per year plant can produce about 12 thousand pounds per hour of polyol. Smaller quantities are sometimes produced by enterprises that use the entirety of their product captively. Figure 2.16 provides a very simplified view of a polyol plant process to nominally produce a three-functional (vis-à-vis glycerine) 12.5% EO-capped PO or a random PO/EO polyether polyol by a batch process. Also provided are the approximate unit ratios required to make that polyol 1000 g/OH equivalent. An important and unappreciated component of the process is the treatment stage at which unreacted monomer may be driven off and the KOH removed by filtering with magnesium silicate, or neutralized with acidified water, and the polyol coalesced as a separate phase from the water and decanted. These neutralization and filtering steps are important aspects of the process, and important components of process research when introducing a new product.

The polyether polyols from THF (referred to commonly as PTMEG) are treated separately here and in many other places for several reasons. One is that, although it is a relatively high-volume polymer (~2 billion pounds/year, 2018) [6, 42], PTMEG occupies a specialty position within polyurethane chemistry as a whole. As a specialty it is produced at smaller volumes, by a significantly different process, and at much higher costs to manufacture and prices to the consumer. Growth rates are in line with most other polyols (~4% per annum), reflecting growth in the overall global economy. Capacity expansion has recently accelerated and asset utilization has been reported at less than 70%. This apparent oversupply may put significant pressure on

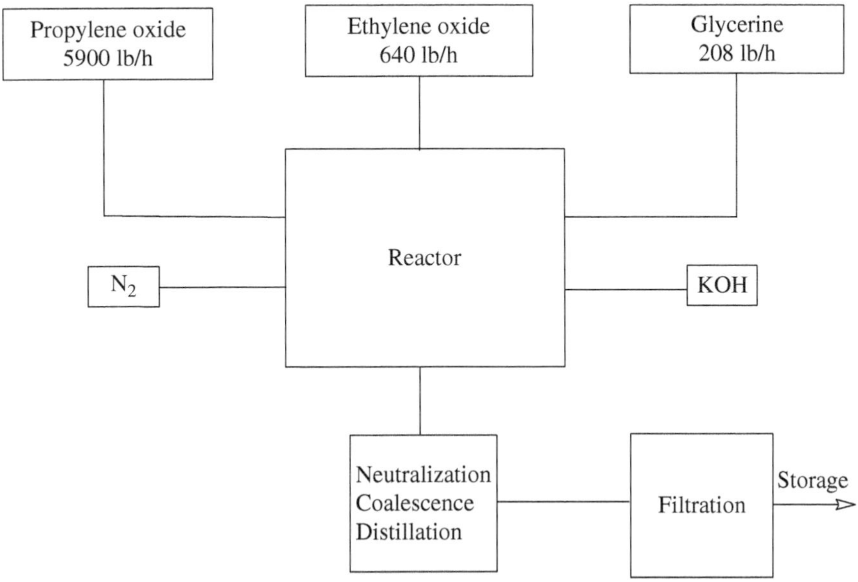


**FIGURE 2.16** Simplified block diagram for a small polyol production of a glycerine-initiated (triol) KOH-catalyzed PO/EO block copolymer (12.5% EO) with total molecular weight 3000 (equivalent weight 1000 g/eq).

merchant pricing until consumption catches up or weak producers exit the market. Major uses by volume for PTMEG are polyurethane elastomers, sealants, adhesives, polyesters, and spandex fibers.

The production of PTMEG is achieved by two routes [43]. The most common way is via acid-catalyzed ring-opening polymerization (Figure 2.17). This path has the advantages of simplicity. The most common acid is fluorosulfonic acid, although other variations are practiced as well. Disadvantages of this process include inability to recover the acid, the potential for the acid to degrade the product, and difficulties in molecular weight control because of the limited, but still finite, ability of terminal hydroxyl groups to be deprotonated and initiate polymerization. Molecular weights can be narrowed through a water wash step, which can remove relatively low-molecular-weight polyols and the acid catalyst.

An alternative process has been established to produce the final polyether product by preparation of an intermediate, but stable, diacetate ester, which can subsequently be converted to the diol by alcoholysis (Figure 2.18) [44]. The greater stability of the ester end group in this environment allows for better molecular weight control.

While many older plants utilize the acid-catalyzed polymerization route, there has been significant optimization of the acetic anhydride polymerization process now used by many newly constructed plants. Industrial production always entails separation of unwanted side products by either volatilization or filtration and recyclization

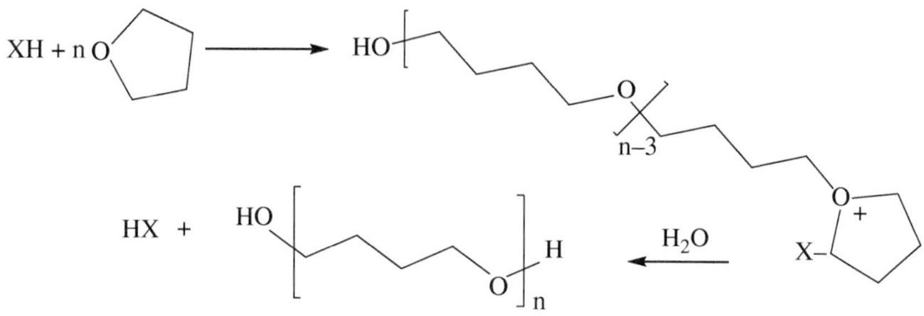


**FIGURE 2.17** Acid-catalyzed polymerization of THF to make polytetramethylene ether glycol (PTMEG).

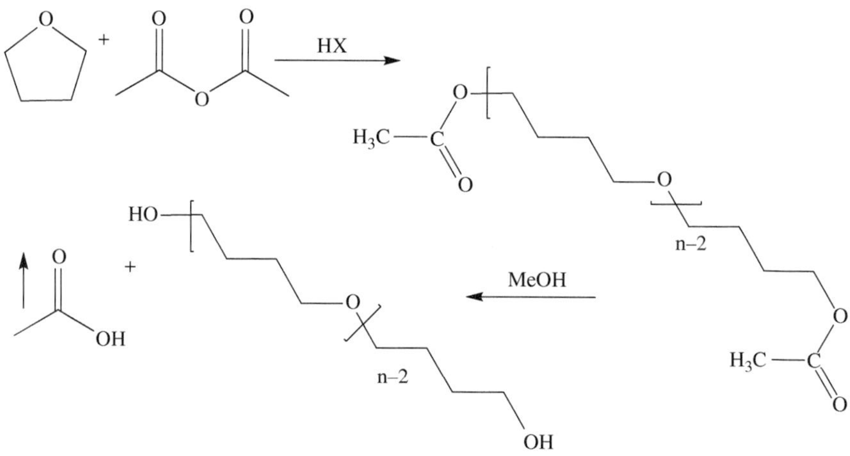


**FIGURE 2.18** Process for preparation of PTMEG by reaction of THF with acetic anhydride to form the stable ester.

of unused reagents back to earlier stages of the reaction. Figure 2.19 is a highly simplified version of what would be included in a real plant but does provide detail of the primary unit operations.

Because of the similarity of polymerization processes, monomers of polyether polyol synthesis are very amenable to copolymer design. The properties of virtually any imaginable design and combination of oxirane monomers have been examined to look for advantageous structures. Figure 2.20 illustrates the various combinations that are imaginable with these building blocks.

The most common type of block copolymer structure is the internal block polyol in which a polypropylene oxide block is reacted at either end with EO to balance the low glass transition temperature ($T_g$) and low tendency to crystallize of polypropylene oxide with the relatively high reactivity of the polyethylene oxide primary hydroxyl end group.

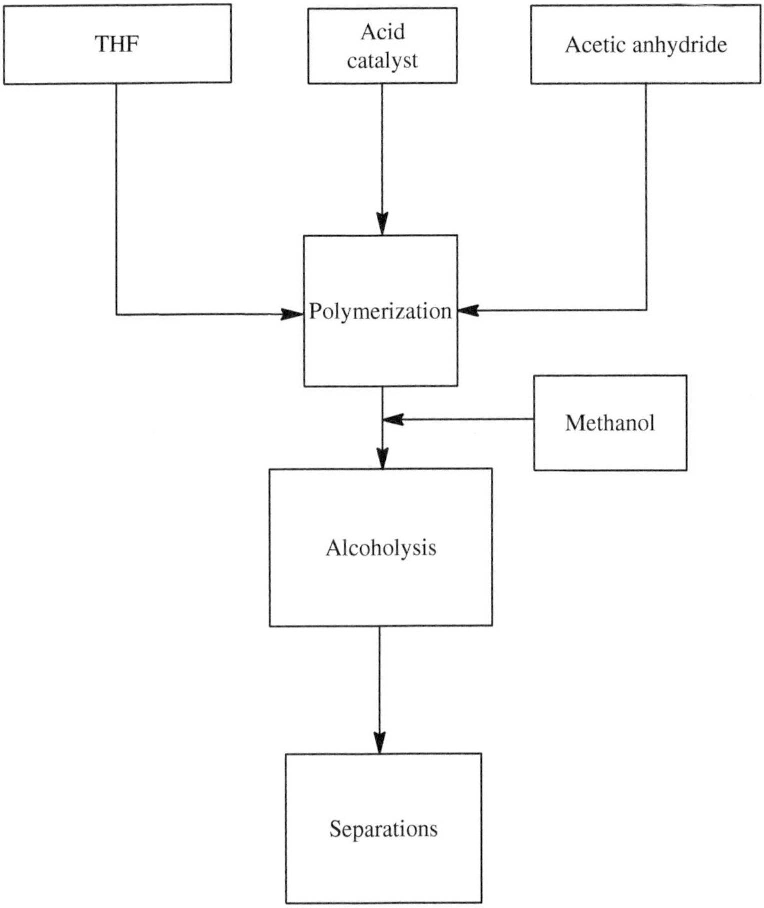


**FIGURE 2.19** Simplified block diagram for production of PTMEG using the acetic anhydride/alcoholysis process.

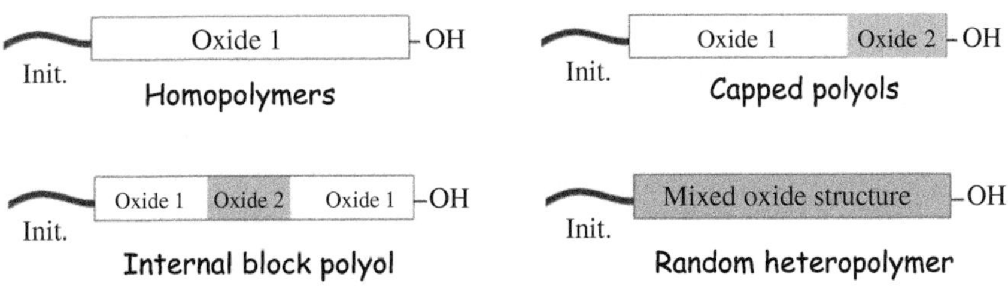


**FIGURE 2.20** Cartoon representations of the types of comonomer organization available to produce polyalkylene glycol soft-segment structures [37].

### 2.1.2 Polyester Polyols

As shown in Figure 2.1, polyester polyol production is only about 20% of total polyol production; however, given the scale of the industry, and the importance and ubiquity of polyester polyols, the impression from the pie chart could be misleading. Indeed, the world capacity for polyester polyols in 2017 was about 2500 thousand metric tons and actual consumption was about 1900 thousand metric tons (~5.0 billion pounds!). The annual growth rate of polyester polyol consumption is about 4%, similar to that for polyether polyols, and approximately mirrors the growth of the world and local economies. Within polyurethane production, polyester polyols are a key feedstock in coatings, adhesives, elastomers, and rigid foam formulations. Depending on the application, an aromatic or an aliphatic diacid will be employed to deliver the desired suite of properties. As will be discussed in detail, aromatic polyesters are usually deployed for the purpose of enhancing flame retardance [45], increasing $T_g$, and reducing gas permeability, so are widely used in polyurethane insulation foams. Aliphatic polyesters used in urethane formulations enhance tensile strength, abrasion resistance, ultraviolet (UV) resistance, and low-temperature flexibility (low $T_g$), and find primary use in coatings and elastomers. Although the synthetic procedures for the preparation of aromatic polyester and aliphatic polyester polyols are relatively similar, the properties and applications are diverse enough to warrant their separate treatment.

Figure 2.21 sets out the approximate differential consumptions of aromatic and aliphatic polyester polyols in the USA, Western Europe, and China in 2017. The usage of aromatic and aliphatic polyester polyols is clearly not uniform across geographies. Aromatic polyols are used to a much greater extent in the USA because of consumer expectations for the superior flame retardance, structural integrity, and thermal insulation properties found in polyurethane and polyisocyanurate insulation panels [5]. The usage of aliphatic polyester polyols is much higher in China because of the relatively high manufacturing of coatings, elastomers, artificial leather, etc. Western Europe is intermediate.

*2.1.2.1 Polyester Polyol Building Blocks* Polyesters are a class of polymers allowing for an enormous amount of structural and property design. Polyesters are the products of condensation polymerization between polyhydroxy and polyacid monomers. In some cases a polyester or anhydride monomer is substituted for the polyacid, usually dependent on processing parameters related to difficulties associated with insoluble acids, inconvenience associated with relative cost, methanol toxicity, or incompatibility of manufacturing materials with the condensation water product (Figure 2.22). Polyester chemistry is relatively straightforward, the monomers are usually not expensive, and the polymer properties are reasonably easy to anticipate [46]. There are a large number of building blocks and a large number of commercial polyesters to choose from when designing a specific polyurethane. Table 2.3 is an abbreviated but representative list of common building blocks for construction of polyester soft segments for polyurethane synthesis.

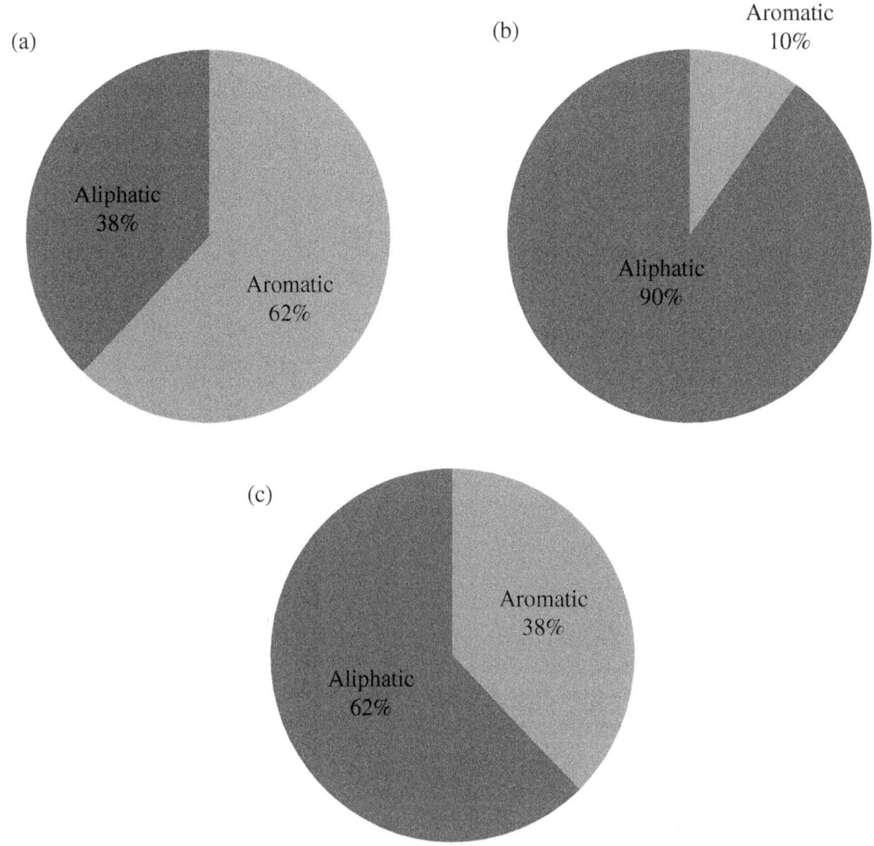


**FIGURE 2.21** Fractional consumption of polyester polyols in 2017. (a) USA; (b) China; (c) Western Europe.

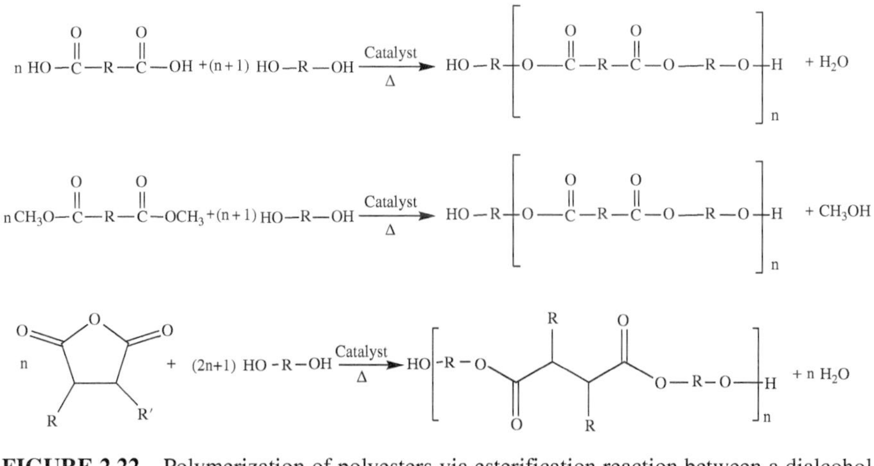


**FIGURE 2.22** Polymerization of polyesters via esterification reaction between a dialcohol and a diacid, transesterification between a diester and a dialcohol, and ring-opening polymerization between an acid anhydride and a dialcohol.

**TABLE 2.3 Structures of acids and alcohols common to polyester polyol synthesis for making polyurethanes**

| Examples of acids and anhydrides | Examples of glycols and diols |
|---|---|
| 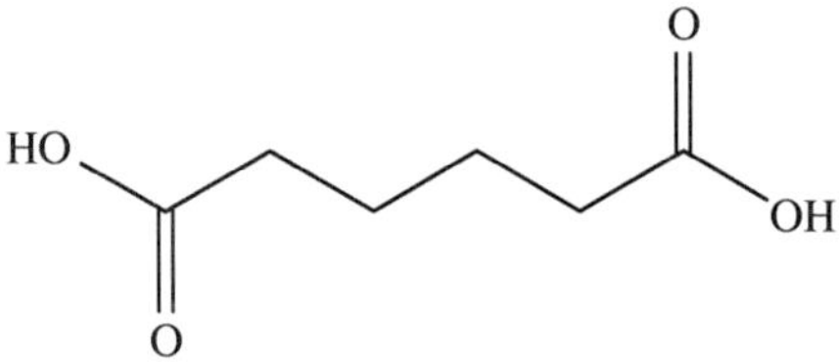 Adipic acid | 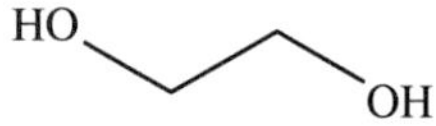 Ethylene glycol |
| 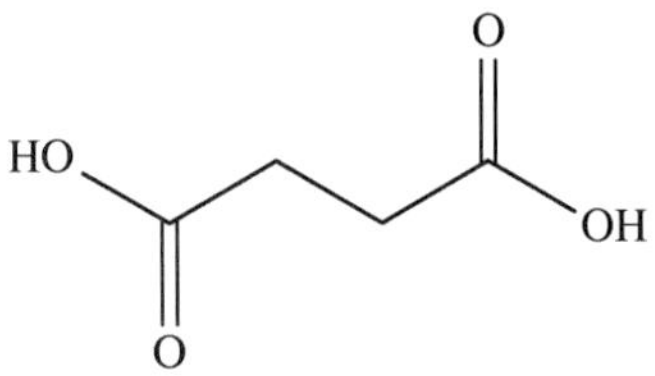 Succinic acid | 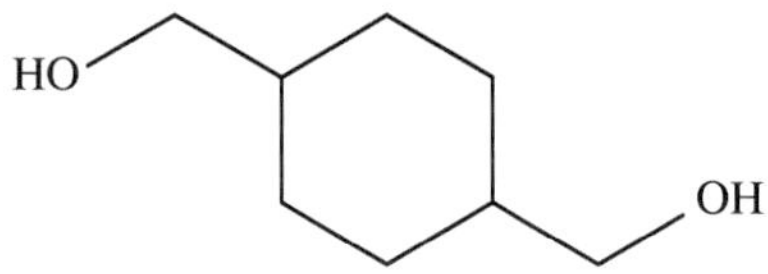 1,4-Cyclohexane dimethanol |
| 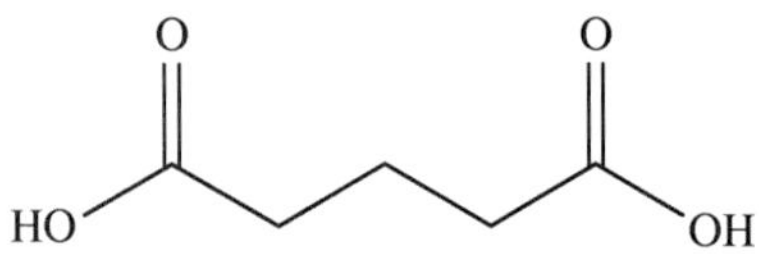 Glutaric acid | 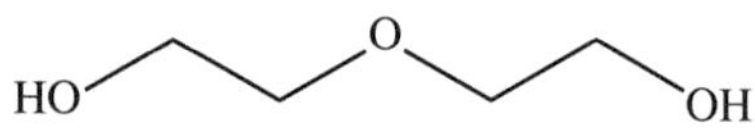 Diethylene glycol |
| 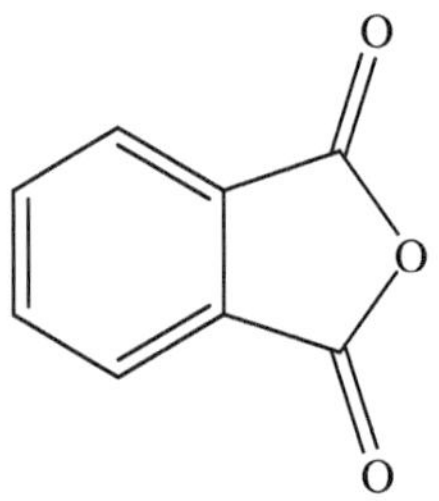 Phthalic anhydride |  1,4-Butanediol |
| 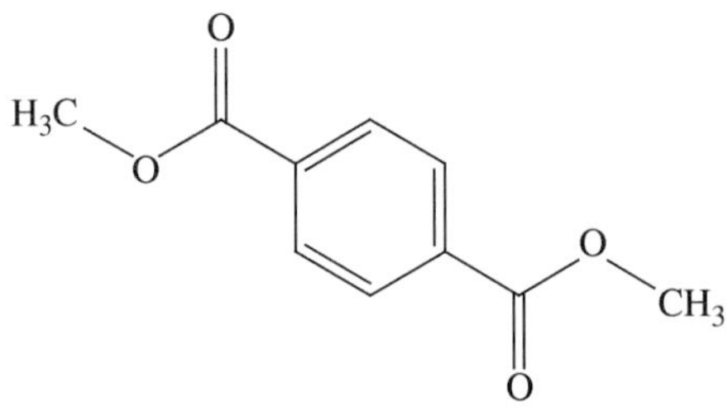 Dimethyl terephthalate (DMT) | 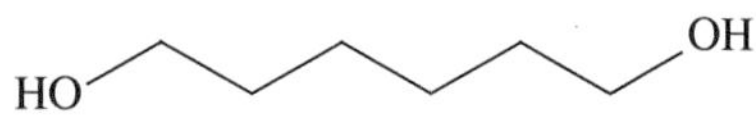 1,6-Hexanediol |

**TABLE 2.3 (Continued)**

| Examples of acids and anhydrides | Examples of glycols and diols |
|---|---|
| 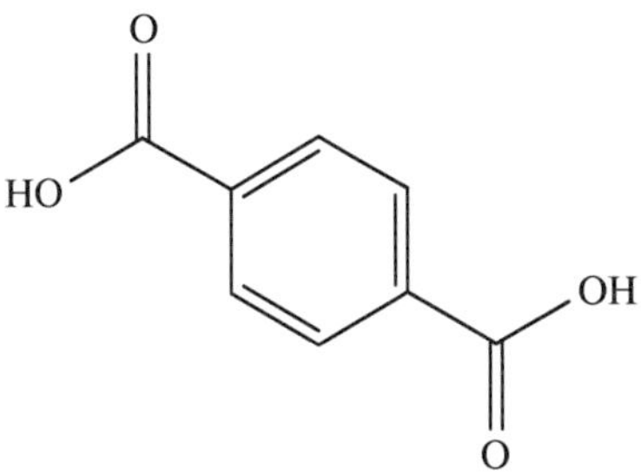 Terephthalic acid |  Neopentyl glycol |
| 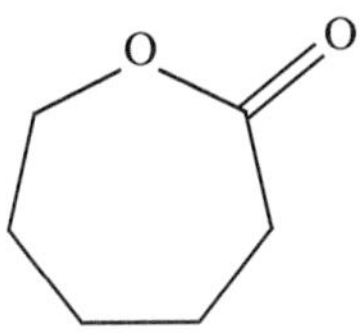 Caprolactone | |
| 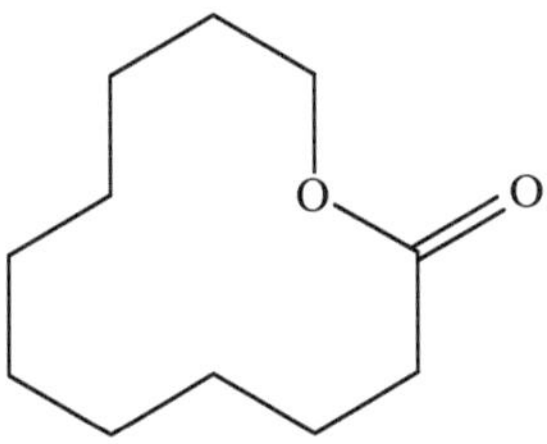 Undecalactone | |
| 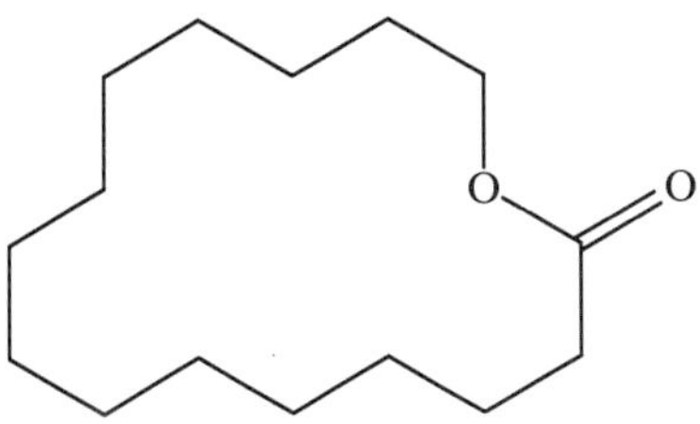 Pentadecalactone | |

Although the process of combining polyacids and polyalcohols to form polyesters is generically similar in every case, the individual monomers are obtained from vastly different building blocks and sources. In fact each building block can be obtained by multiple routes, each one having advantages, disadvantages, and economic sensitivities. For instance adipic acid can be obtained from oxidized derivatives of cyclohexane or from butadiene [47]. Butanediol can be obtained from acetylene and formaldehyde or from fermentation of sugar to succinic acid followed by reduction [48]. Terephthalic acid is produced by oxidation of xylene [47]. Thus each of these monomers will vary in price independently of all the others, their manufacturers will be independent of one another, and the macroeconomic pressures may vary independently. In contrast to polyether manufacturers, who generally are back integrated into the building blocks (EO, PO, THF), polyester polyol manufacturers purchase building blocks such as those in Table 2.3 from several sources and will qualify multiple sources as primary suppliers using contracts and futures strategies. This also results in polyester manufacturers being "middle man" manufacturers, primarily suppliers to other businesses. Thus, the three largest polyester manufacturers in the world as of 2018 are, in order, with the largest first, Huafeng Group, Stepan Corporation, and Coim Corporation, which are all certainly well-regarded companies but are not generally well known or in the popular consciousness. The lone exception in this regard would be BASF, which is a major chemical manufacturer with extensive back integration into polyester building blocks and forward integration into products that use polyester polyols. On the other hand, BASF produces less than half the volume of either of the two largest polyester polyol producers [5].

***2.1.2.2 Preparation of Polyester Polyols*** The desired product of polyester polyol synthesis is a polyester with high levels of terminal hydroxyl functionality. Achieving this goal is facilitated by applying the Carothers equation for determining the ratio of acidic and alcoholic monomers to get the desired molecular weight and hydroxyl functionality [49, 50]. In preparation of a simple linear polyester, this would take the form of

$$\frac{\text{Moles diol}}{\text{Moles acid}} = 1 + \frac{1}{n} \tag{2.5}$$

where $n$ is determined by considering one unit of diol as a chain initiator and the monomer unit to be the condensation unit of (diacid+diol – $2H_2O$). Thus the target number average molecular weight ($M_n$) is calculated as

$$M_{\text{n}} = M_{\text{Wdiol}} + n \cdot M_{\text{Wmonomer}} \tag{2.6}$$

As an example, if the desired result was a 2000 molecular weight polyester polyol of polybutylene succinate the $M_{\text{Wmonomer}}$ is

$$118\left(M_{\text{Wsuccinic acid}}\right) + 90\left(M_{\text{WBDO}}\right) - 36\left(2 \times M_{\text{WH}_2\text{O}}\right) = 172 \tag{2.7}$$

where $n$ = moles monomer/moles initiator = moles succinic acid/(moles 1,4-butane diol [BDO] – moles succinic acid) and a molar ratio of diol to diacid of 1.09. In practice this result should be used as a theoretical guide and not as strictly predictive. The final result will reflect details of the polymerization procedure and systematic errors resulting in loss or decomposition of monomer or polymer over time. Nuclear magnetic resonance and chromatography are commonly applied to determine molecular weight, and titration of acid end groups by well-known techniques is applied to determine the level of hydroxyl functionality.

Polyester polymerization can often be done without the need for a solvent since many of the monomers melt and phase mix prior to reaction. Polyester formation involves evolution of a condensation product (e.g. water when using a diacid or methanol when employing a dimethyl ester) and the polyester products are in equilibrium with the reactants [51]. For this reason water (for instance) must be removed as ester is formed, or high-molecular-weight polymers cannot be made. This is usually achieved by sweeping a nonreactive gas such as $N_2$ over the surface of the polymer surface. As the polymer forms, the molecular weight increases and the viscosity of the melt increases; linearly at first, and to the 3.4 power above the entanglement molecular weight. The increase in melt viscosity can interfere with removal of the condensation product and slow the reaction down markedly. There is a temptation to increase the melt temperature to reduce viscosity, which in principle will facilitate removal of the water or alcohol. But increasing temperature may also facilitate the differential evaporation of the monomers. This will upset the calculation of monomer ratios required to obtain a specific molecular weight with terminal hydroxyl functionality, and result in a stubborn lack of molecular weight advancement or sudden increases in molecular weight and the number of acid terminal groups far above that expected based on calculation. Such occurrences, as often happen in the laboratory in the absence of well-established procedures, will initiate the addition of diacid or diol in order to adjust molecular weight and/or reduce the prevalence of acid end groups.

The production of polyester polyols by ring-opening polymerization of lactones has also been explored extensively and many synthetic procedures and catalyst alternatives have been developed. One of the most common approaches employs a small polyol (diol or higher) as an initiator and propagates chain extension through a coordination–insertion mechanism (Figure 2.23) [52, 53]. The catalyst of choice for this approach is usually a tin or titanium alkoxide or other common catalyst used for polyester polymerization. The case where R is a triol results in a polymer with three polymer arms and this can be used to control the final molecular weight of the product.

*2.1.2.3* ***Aliphatic Polyester Polyols*** Many aliphatic polyester polyols have been prepared and at one time or another been offered on the market. Despite this, the largest volume aliphatic polyester polyol used in polyurethane applications by far is polybutylene adipate, followed by polycaprolactone. The ratio of polybutylene adipate to polycaprolactone production is approximately 100:1 [5]. The properties of a polyester polyol are of course a function of the diacid and diol building blocks used

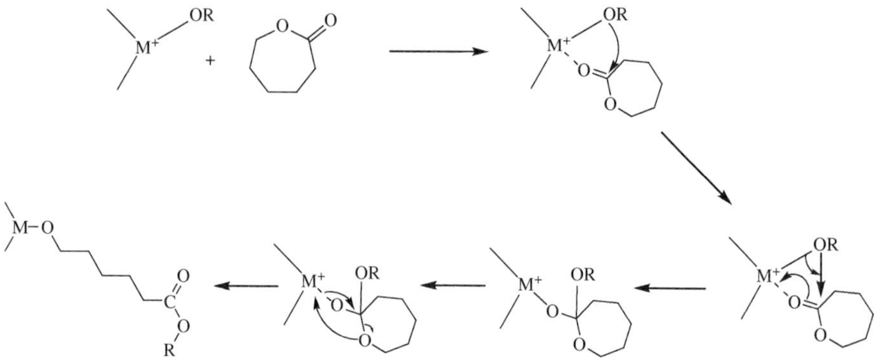


**FIGURE 2.23** Accepted mechanism for coordination–insertion ring-opening polymerization of ε-caprolactone to polycaprolactone.

to produce the polymer. The properties of a polyester polyol are most easily accessed by their thermal properties. Table 2.4 provides a limited view of the thermal properties based on the polyester components.

A simple analysis of trends in the polylactone sequence shows that the carbon length between ester linkages is strongly influential on the crystalline melting point and the crystalline melt enthalpy, asymptotically approaching that of polyethylene (Figure 2.24). For polyesters prepared from diacids and diols, a simple analysis of crystalline melting point and enthalpy shows virtually no correlation relative to the length of the diol, with a square of the correlation coefficient ($r^2$) 0.1 or less. With the exception of polybutylene succinate, there is a positive correlation of increasing diacid length with melting point and melt enthalpy ($r^2 = 0.8$ and 0.5, respectively).

***2.1.2.4 Aromatic Polyester Polyols*** Aromatic polyester polyols are primarily used in the production of polyurethane and polyisocyanurate foams for construction and appliance applications. There are additional applications in slabstock, spray foams, elastomers, and adhesives. Commercially, aromatic polyester polyols are based on diethylene glycol (DEG) or ethylene glycol, and low-molecular-weight PEGs, polymerized with either dimethyl terephthalate, terephthalic acid, phthalic anhydride, or crude recycle scrap of polyester terephthalate (PETr) [54]. In some cases, polyol functionality is increased by the addition of trimethylolpropane and pentaerythritol. Of course any combination of aromatic and aliphatic monomer could be used for making these materials, but price and performance constraints also limit the menu of employable building blocks. The nonnegotiable and established performance and process attributes coupled with high price competition has led to commoditization of this technology [55].

Although the functions of construction and appliance insulation foams overlap, the details differ in some important ways and result in divisions in polyol specifications unique to the two markets. For the appliance market, aromatic polyester polyols

**TABLE 2.4 Properties of selected polyester polyols useful in polyurethane synthesis and comparison with high-density polyethylene**

| Polyester | Acid | Diol | $T_m$ (°C) | Enthalpy of melting (J/g) |
|---|---|---|---|---|
| Polycaprolactone | na | na | 50 | 105 |
| Polypentadecalactone | na | na | 92 | 182 |
| Polydodecanlactone | na | | 83 | 70 |
| Polyundecanlactone | na | na | 73 | 150 |
| Polybutylene adipate | adipic acid | 1,4-Butanediol | 55 | 96 |
| Polyethyldodacanoate | 1,12-Dodecandicarboxylic | Ethylene glycol | 87 | 137 |
| Polyhexylnonanoate | Azelaic | 1,4-Butanediol | 65 | 141 |
| Polyhexanehexanoic | Sebacic | 1,6-Hexanediol | 64 | 123 |
| Polyhexandodecanoic | 1,2-Dodecanoic | 1,6-Hexanediol | 74 | 125 |
| Polybutylene succinate | Succinic | 1,4-Butanediol | 115 | 96 |
| Polyethylene high density | na | na | 133 | 285 |

na, not applicable; $T_m$, melting temperature.

are expected to contribute positively to maximizing thermal insulation, minimizing demold post expansion, and maintaining structural integrity [56]. These attributes are detailed in Table 2.5

The overlap is significant only in those specific applications where the requirements are the same. In spray construction foams it is imperative that the sprayed components foam and build strength very quickly, in a manner similar to that required for appliance foams, but for very different reasons. In the appliance sector there is criticality in being able to process more refrigerators, water heaters, or dishwashers in a shorter time. It is also important to appliance manufacturers that the foam blown into an appliance mold builds strength quickly to prevent deformation from the growing insulation internal gas pressure, and not deform from shrinkage upon cooling. For spray construction foams, designed rapid strength development is just a reflection of the desirability that the foam stick to vertical surfaces when necessary.

A critical requirement of construction foams is their flame retardance. It is not expected that the aromatic polyester polyol supply all of this property. The aromatic isocyanate and materials specifically employed to inhibit fire are also used [57, 58]. However, fire retardance is also considered a desirable feature of an aromatic polyester polyol in the construction foam market. The need to spray foam over distances can also limit the foam viscosity of a formulation that a specific application can tolerate. This can in turn dictate the viscosity (and so the structure) of the aromatic polyester polyol used in this application [59]. It also requires in many cases that spray foam applications use significant amounts of low-viscosity polyether polyols

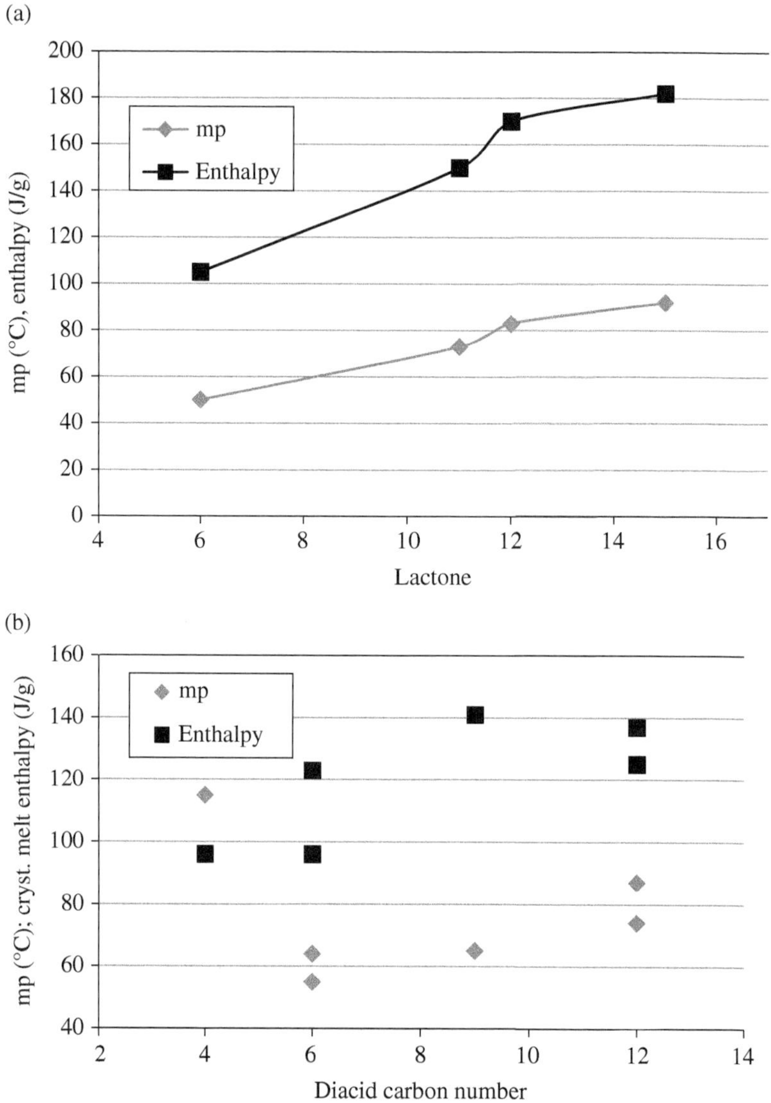


**FIGURE 2.24** (a) Melting point and enthalpy of melting of polyester polyols derived from lactone ring-opening polymerization as a function of ring size. Properties asymptotically approach those of polyethylene. (b) Melting point and enthalpy of melting for polyester polyols made from condensation polymerization as a function of carbons in the diacid. The trends are somewhat weaker than for the lactone series.

**TABLE 2.5 Comparison of polyols used for rigid appliance foam and rigid construction foam**

| Appliance foams | | Construction foams | |
|---|---|---|---|
| Property | Specification | Property | Specification |
| Functionality | 2–2.7 | Functionality | 2–2.5 |
| $M_n$ (g/mol) | 300–600 | $M_n$ (g/mol) | 300–600 |
| OH (mg KOH/g) | 250–315 | OH (mg KOH/g) | 230–350 |
| Viscosity at 25 °C (cP) | 2700–15,000 | Viscosity at 25 °C (cP) | 2000–5000 |

Along with differences associated with gel time optimization, construction foams must also meet stringent flammability requirements.
$M_n$, number average molecular weight.

**TABLE 2.6 Example of an aromatic polyester polyol for construction application**

| Component/property | Specification |
|---|---|
| Diacid | Phthalic anhydride |
| Diol | DEG |
| Functionality | 2 |
| $M_n$ (g/mol) | 348 |
| OH (g KOH/g diol) | 330 |
| Viscosity (cP) | 2700 |

DEG, diethylene glycol; $M_n$, number average molecular weight.

in combination with polyester polyols to meet viscosity requirements. Terephthalic acid-based polyester polyols have an increased tendency to crystallize, so using terephthalic acid-based polyesters may be limited by application temperature, or may require polyol design considerations that inhibit crystal formation. Table 2.6 is an example of a commercial aromatic polyester polyol available on the market.

Unlike aliphatic polyester polyols, the synthesis of an aromatic polyester polyol may not follow principles for obtaining a defined product of condensation polymerization, but may instead target the properties required for a result. The final product may constitute a designed amount of free DEG, for instance, to obtain the required viscosity or reactivity.

Table 2.7 shows the unit ratios for two commercial aromatic polyester polyols. They are, in principle, the proportions for a diacid reacted with two diols. However, as actually practiced, the desired structure cannot be the calculated structure since a polyester product represents an equilibrium, with structures from oligomers to free glycols and acids. The actual distribution of components can be calculated using a Flory–Schulz distribution or it can be measured chromatographically. These methods are usually in good agreement [60]. To meet required properties it is common to

**TABLE 2.7 Raw materials and unit ratios for two commercial aromatic polyester polyol formulations**

| Component | Polyol no. 1 | Polyol no. 2 |
|---|---|---|
| Phthalic anhydride | 0.42 | |
| Terephthalic acid | | 0.32 |
| DEG | 0.58 | 0.26 |
| PEG 200 ($M_n$ = 200 g/mol) | | 0.42 |
| Free glycol calculated (%) | 14.5 | 24.2 |
| Free glycol measured (%) | 15.7 | 26.2 |
| Viscosity (cP) | 2700 | 630 |
| OH | 330 | 317 |

DEG, diethylene glycol; $M_n$, number average molecular weight; PEG, polyethylene glycol.

adjust the final polyol with free glycol. In fact, between the two polyols described in Table 2.7 the substantial difference in viscosity is a function of (i) the lower unit ratio of diacid resulting in fewer ester linkages, but primarily it is a function of (ii) the higher volume of free glycol in the final product [61, 62].

Since aromatic polyester polyols are typically lower in molecular weight than aliphatic polyester polyols, there is a higher probability of finding free monomer in the final product. The free monomers can have a significant effect on the efficacy of the polyol in the final foam product. Product polydispersity in polyester synthesis is an expected phenomenon, but it is wise to measure, or at least to calculate, the final product oligomer distribution. This is most easily done using the gamma distribution, or, more commonly, the Flory–Schulz distribution. The distribution is easily calculated based on knowledge of inputs such as the molecular weight of the repeat unit. The weight average of the Flory–Schulz distribution is calculated from Equation 2.8:

$$W_x = x(1-p)p^{(x-1)} \tag{2.8}$$

where $W_x$ = (weight of all oligomers of weight $x$/weight of all oligomers) and $p$ is essentially the fraction of converted reactive groups. While the mass of free monomers present as a fraction of total product weight may be small, their *number* may be relatively large even at fairly high monomer conversion to product. The number of oligomers as a function of conversion is calculated from Equation 2.9:

$$N_x = N_o(1-p)p^{(x-1)} \tag{2.9}$$

The meaning of this distribution is shown in Figure 2.25, where it is seen that, except at very high conversion, the mole fraction of monomers in the product polyol can be significant, even when their weight fraction is quite small. Given the very low

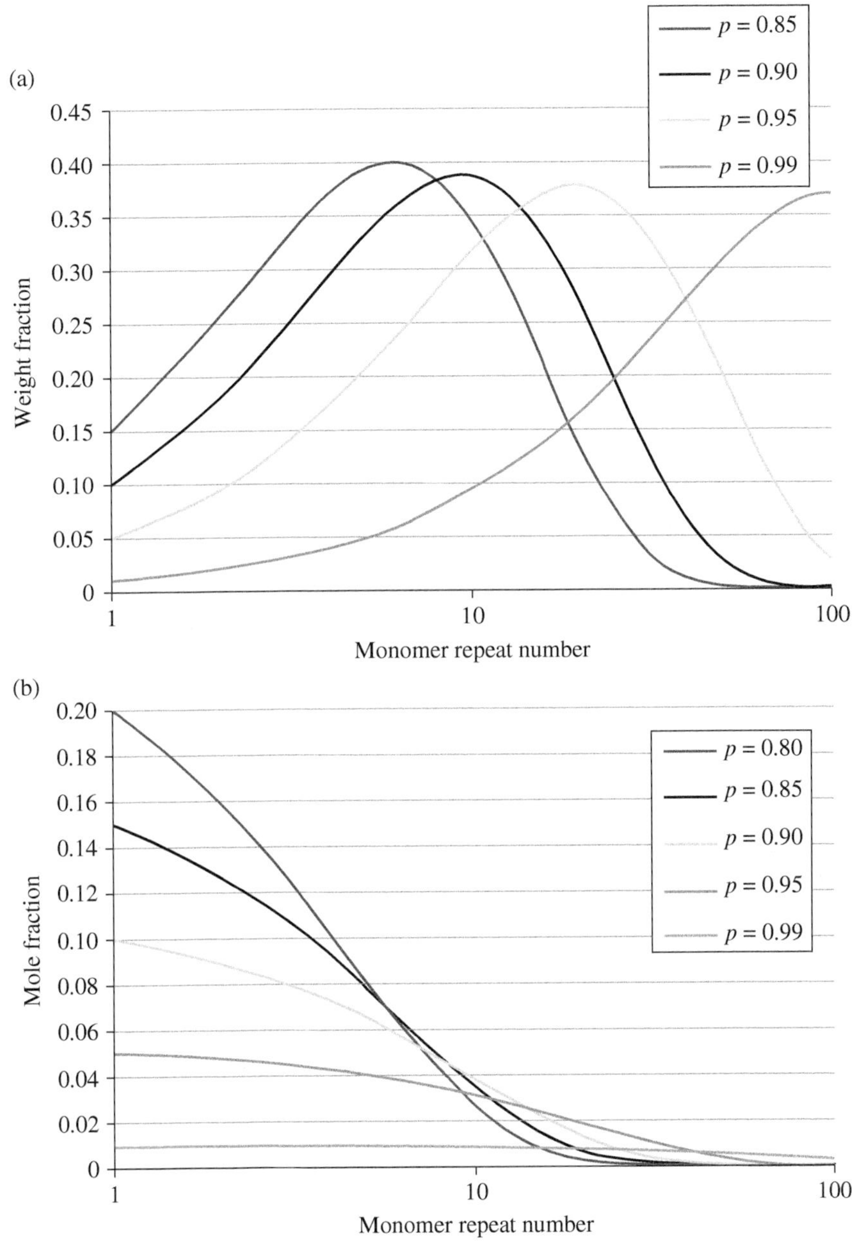


**FIGURE 2.25** (a) Weight and (b) mole fraction distributions of oligomer sizes present in a polyester synthesis based on the Flory–Schulz distribution.

equivalent weight, their role in the final polyurethane properties can be important. The figure also illustrates the importance of high conversion percentages in obtaining high-molecular-weight polymers.

The free monomer diol components have a clear and unambiguous role in these polyols. They produce reactive components for the aromatic diacid of the polyol and for the isocyanate component of the subsequent polyurethane foam formulation. Additionally they provide rheological tuning to the polyols. Beyond that they also provide a means of improving compatibility of the components of the formulation. Lastly, as the cost of the diols is usually lower than that of the diacid, the diols can provide a means of managing overall product cost. Artful combinations of diols can, in principle, provide a means of performance innovation by improving properties at reduced cost.

The diacid components, terephthalic acid (or dimethyl terephthalate) and phthalic anhydride, have significant effects on cost as well (isophthalic acid is rarely used because of cost considerations). In turn they also exert a subtle effect on properties owing to their different structures. Cost is affected by the so-called unit ratio of the building block – essentially a determination of mass equivalence for the different molecules within an equivalent final product. Although all final products are either identical (using terephthalic acid or dimethyl terephthalate) or isomeric (using phthalic anhydride), the differences relate to the molecular weight of the building block diacid and the mass of condensation product produced (Figure 2.26). By this measure phthalic anhydride is the most cost-effective aromatic unit; however, it is usually almost 20% more expensive than terephthalic acid, resulting in supply issues usually being the deciding factor for which is used.

In addition, there are differences between the terephthalate residues and orthophthalic residues from phthalic anhydride (Figure 2.27). The substitution pattern has a substantial effect on free volume requirements of the polyol component, which in turn has an effect on subsequent rheological properties. Figure 2.28 shows energy-minimized structures for the aromatic polyester polyol of DEG with terephthalic (Figure 2.28a) and orthophthalic (Figure 2.28b) acid with approximately equal energies accounted for by hydrogen-bonding interactions and overall conformational energies. It is clear that the aromatic polyester polyol from orthophthalic acid requires additional free volume because of the pendant benzylic residue, but, in turn, allows more intramolecular van der Waals interaction between ether moieties. While the aromatic positional isomer is known to affect foam properties, it has not previously been associated with fundamental chain physics and the causal relationship has been lacking [63].

Lastly, aromatic positional isomers are also associated with small but real differences in performance (Table 2.8). While the origins of these differences are not established, they emphasize the role of positional isomers in establishing structure–property relationships [64].

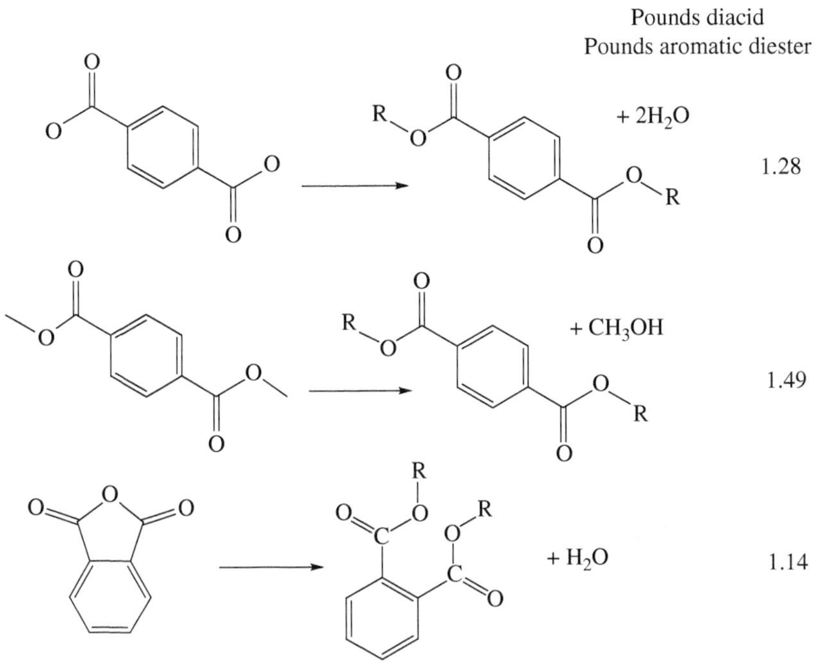


**FIGURE 2.26** Unit ratios for diacids or diesters used in preparation of aromatic polyester polyol synthesis.

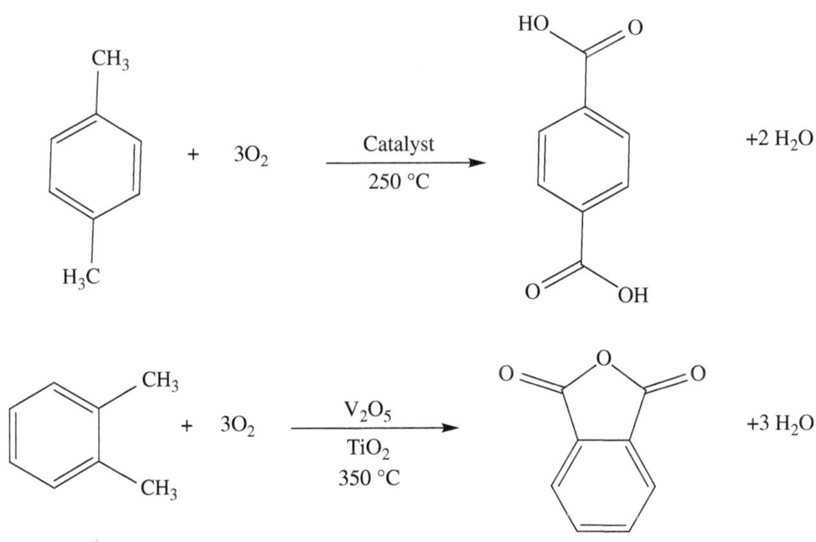


**FIGURE 2.27** Production of terephthalic acid and phthalic anhydride.

(a) (b)

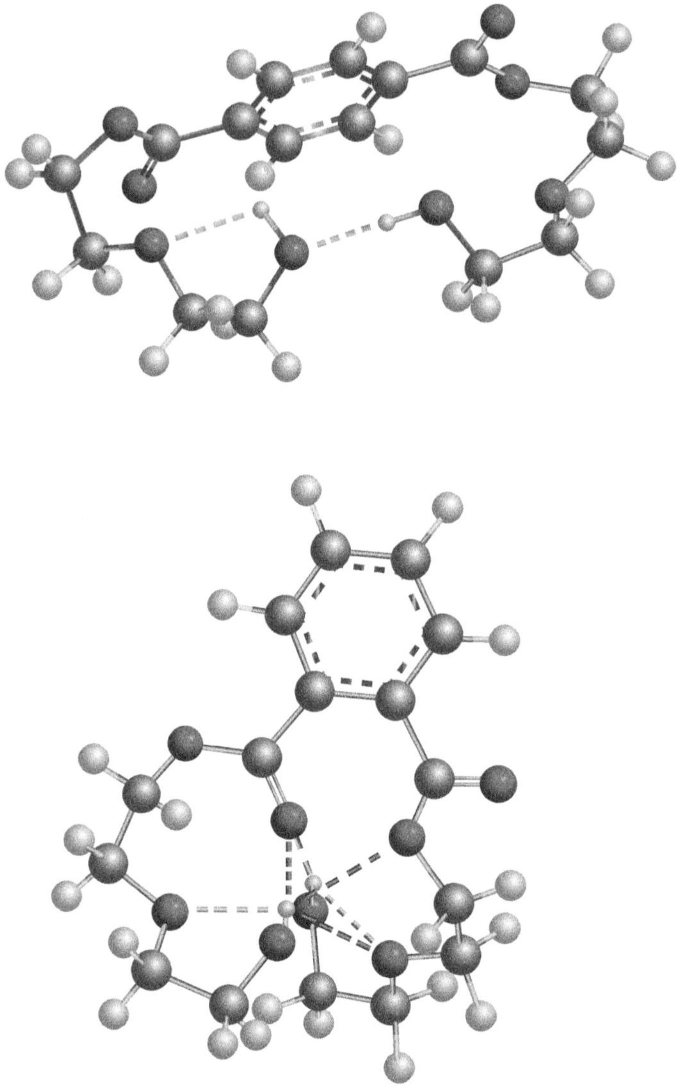

**FIGURE 2.28** Comparison of lowest energy conformations of aromatic polyester polyol from terephthalic acid and orthophthalic acid (or phthalic anhydride) and DEG. The free volume requirement of the orthopositional substitution is higher, which may account for much of the viscosity difference between polyols made with the different comonomers. (*See insert for color representation of the figure.*)

**TABLE 2.8 Properties of rigid polyurethane foams with aromatic polyester polyols having either terephthalic acid or phthalic diacid comonomers reacted with diethylene glycol**

| Diacid | Terephthalic acid | Phthalic anhydride |
|---|---|---|
| Viscosity (cP) | 4000 | 2700 |
| $\lambda$ (mW/mK at 10 °C) | 19.6 | 19.85 |
| Compressive strength (kPa) | 99 | 96 |
| Weight loss upon burning (%) | 27 | 32 |

Viscosity data refer to the viscosity of the polyols used in the formulations. Other foam formulation factors were held constant.
$\lambda$, thermal conductivity.

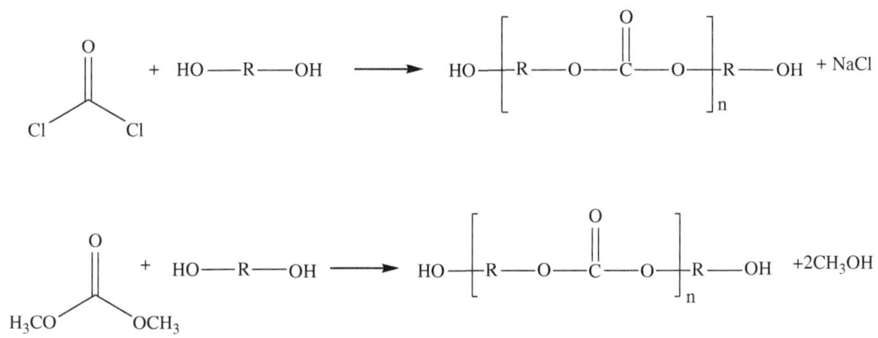


**FIGURE 2.29** Preparation of polycarbonate polyols from phosgene (not recommended) or dimethyl carbonate.

### 2.1.3 Other Polyols

***2.1.3.1 Polycarbonate Polyols*** In application to polyurethanes, polycarbonate polyols are among the highest performance polyol backbones available in the marketplace [65, 66]. They are also among the most expensive and are produced at relatively small volumes. The world capacity for polycarbonate production is 120 million pounds, and growth in consumption is reported to be increasing [5]. Relative to other polyols, the properties most attractive about polycarbonate polyols are thermal, UV, and hydrolytic stability. This suite of attributes has resulted in polycarbonate polyols finding their widest application in very high-performance applications of coatings, elastomers, and adhesives.

*Preparation of Polycarbonate Polyols* The preparation of polycarbonate polyols is straightforward in its simplest form. It is the reaction of phosgene or a carbonate monomer, usually dimethyl carbonate, with a diol monomer or a mixture of diol monomers (Figure 2.29).

In the case of the phosgene route, the liberated hydrochloric acid (HCl) is usually quenched with sodium hydroxide (NaOH) to form the filterable salt. This route is generally discouraged because of the significant environmental health and safety drawbacks associated with large-scale handling of phosgene.

Dimethyl carbonate (DMC) is a relatively more attractive reagent for polycarbonate synthesis because of its low molecular weight, its relatively high unit ratio in the polymerization, and its relatively low cost and commercial availability. One disadvantage of using DMC is the evolution of toxic methanol, which requires safety protocols in production. A more significant issue is the 30/70 DMC/methanol azeotrope that forms, resulting in lost DMC in the condensation product [67, 68]. The practical effect is to make usage of DMC quite inefficient since the methanol/DMC azeotrope is difficult to separate by normal methods [69]. One approach that has worked is to determine the minimum in the DMC/methanol concentration as a function of temperature. The reaction is then run while controlling the reaction vessel to the temperature producing overhead products at the minimum of DMC in the azeotrope. However, this approach can be unworkable from a practical standpoint since the boiling point of dimethyl carbonate is about 90 °C, a temperature at which the reaction rate between the diol and the carbonate is very low. Thus, one efficient procedure for the preparation of polycarbonate polyols using DMC is to run the reaction vessel at the optimum temperature for reaction, which at the same time optimizes the use of DMC by minimizing its content in the azeotrope collected overhead. This can be accomplished by slow feeding ("starve feeding") the DMC to the molten rapidly mixing diol. The optimum overhead temperature is about 58 °C. The exact conditions for this will depend on the geometry of the reaction vessel, the control of reaction vessel heaters, and the mixing of the reaction vessel.

An alternative efficient method for producing polycarbonate polyols is to perform the reaction at elevated pressure. The reaction is performed in an autoclave with all components of the reaction present (Figure 2.30). The temperature and pressure are elevated for the purpose of driving the reaction to a stage at which a large amount of DMC has been converted to the carbonate–diol adduct. Subsequently the pressure and temperature are reduced to drive the reaction to completion by removal of methanol. While still inefficient in DMC, this method provides improved time to a final product, and does so with a smaller production footprint [70].

An alternative polycarbonate polyol is the polyalkylene carbonate formed from the reaction of oxiranes and $CO_2$. A common problem associated with this synthesis has been the formation of five-member cyclic carbonates. There has been recent

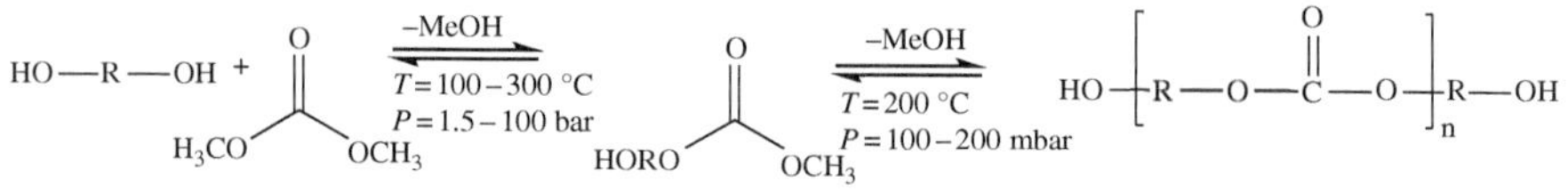


**FIGURE 2.30** Preparation of polycarbonate polyols using an elevated pressure procedure.

activity with new catalysts that have proven themselves able to result in high-molecular-weight polycarbonate polyols (185–250 kDa) and low-molecular-weight polyols useful in polyurethane synthesis (Figure 2.31) [71, 72]. Despite the low cost of the building blocks, pricing of these materials is approximately the same as for those achieved via the condensation techniques described above.

There are numerous catalysts for this polymerization – both heterogeneous and homogeneous – that vary by efficiency, tendency to form the undesired cyclic carbonate, and resulting tacticity of the final polyol product. The tacticity is one of several controls of polyol structure that can be used to design the final properties of polypropylene carbonate polyols (Figure 2.32).

Design of polycarbonate polyol structure is primarily a matter of choice of which diol or mixture of polyols to react with the carbonate linkage. Designed artfully, the properties of the resulting polycarbonate polyol can be varied widely and have considerable effect on a derived polyurethane property. Table 2.9 lists diol monomers polymerized with DMC to make polycarbonate polyols, typical applications, and the physical state of the polyol produced. Hexanediol is a preferred monomer because of its reduced tendency to volatilize during polymerization (compared with 1,4-butanediol). It is also apparent that diol additives that randomize or disrupt regular chain structure help make the polyol liquid.

*2.1.3.2* ***Polyacrylate Polyols*** Acrylic polyols differ in numerous ways from the polyols described above, but they are an important and common occurrence in our day-to-day experience. Unlike the previously described polyols, which we may find ourselves sitting upon, walking upon, wearing, depending upon in numerous but silent functions, acrylic polyols in isocyanate formulations are generally limited to coatings, particularly, but not exclusively, to automotive surface finishes [73–75]. In addition, acrylic polyols find application in wood coatings and anticorrosion finishes. Acrylic polyols are preferred for achieving higher hardness, UV stability, and water and acid stability. In contrast, polyester polyols are preferred in the same applications

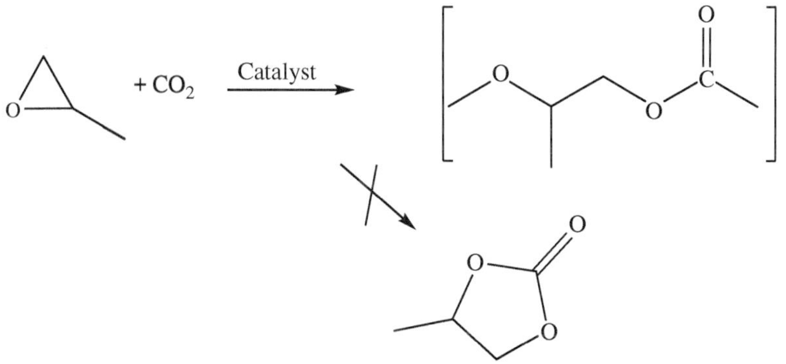


**FIGURE 2.31** Preparation of polycarbonate polyol using an alkoxylate.

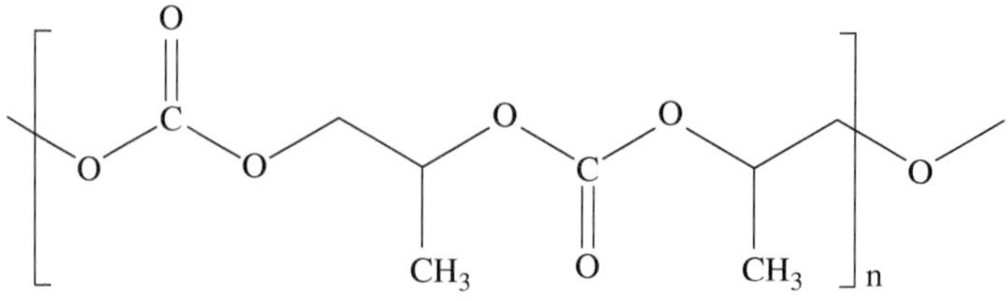


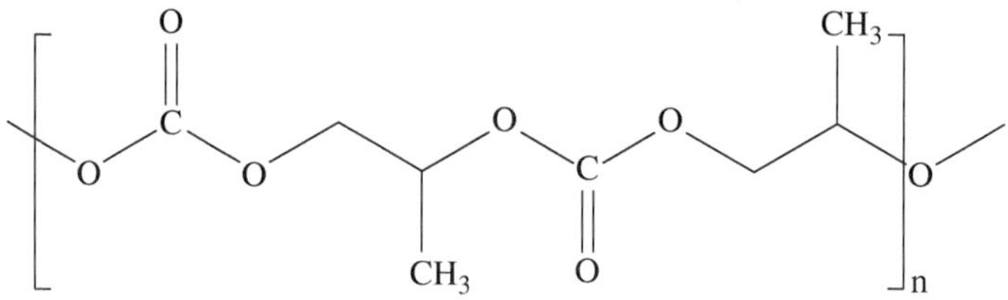


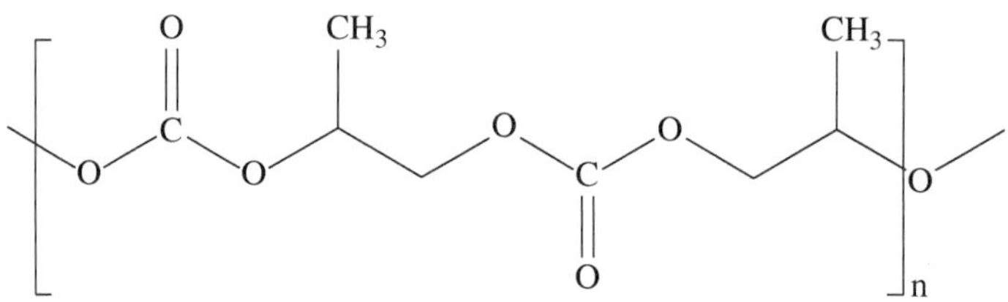


**FIGURE 2.32** Polycarbonate polyols from alkoxylates and the structural variation that can manifest as a result of the addition mechanism.

where solvent resistance, toughness, and flexibility are critical. Polyesters may find usage for the same applications where hydrolytic stability is less important and cost is paramount. As commonly used, almost 1 billion pounds of acrylic polyols were used worldwide in 2018, with an average annual growth rate of over 4%, tracking the overall use of urethane surface coatings [76, 77]. As in all of these numbers, the use and growth rates are not globally homogeneous.

*Preparation of Acrylic Polyols* Acrylic polyols are prepared like many acrylics, via radical initiation. Any acrylic monomer may be used as long as hydroxyl-functionalized monomers such hydroxyethyl methacrylate, hydroxyl butylacrylate, hydroxyethyl acrylate, etc. are present in the backbone [78]. The role of any particular acrylic monomer in the preparation of an acrylic polyol would be the same as in

**TABLE 2.9 Select polycarbonate polyols (based on diol variation) and the difference in properties**

| Diol | Characteristics | Applications |
|---|---|---|
| 1,6-Hexanediol | 750 Da, solid | Coatings, lenses |
| 1,6-Hexanediol | 1000 Da, waxy solid | Coatings |
| 1,6-Hexanediol | 2000 Da, waxy solid | Adhesives, artificial leather |
| 1,6-Hexanediol | 3000 Da, solid | Artificial leather |
| 50/50 1,6-Hexanediol/1,5-pentanediol | 2000 Da, liquid | Adhesives, artificial leather |
| 1,6-Hexanediol/polycaprolactone | 2000 Da, liquid | Adhesives |
| 1,6-Hexanediol/1,4-cyclohexane dimethanol | 1000 Da, solid | Adhesives |
| 2,2-Butyl, ethyl 1,3-propanediol | 1000 Da, liquid | Elastomers, flexible coatings |
| 2,2-Butyl, ethyl 1,3-propanediol | 2000 Da, liquid | Elastomers, flexible coatings |
| 1,3-Propanediol | 2000 Da, solid | |
| Adipate polyester cohexanediol polycarbonate | 2000 Da, liquid | Elastomers, adhesives |

any acrylic resin: to adjust $T_g$, hardness, flexibility, or barrier properties. Styrene is also used commonly in acyclic polyol synthesis to increase hardness. Styrene can also be added to adjust the formulation to lower cost, although this has been reported to lower performance when added in excess. The use of polar nonhydroxy components such as acrylic acid and acrylonitrile can be used to increase interchain polar interactions and so increase the tensile strength and ultimate elongation of a resin at the potential cost of decreasing low-temperature flexibility [79]. Allyl alcohol has been employed as a monomer for modification of molecular weight. Depending on the application, the molecular weight of acrylic polyols can vary from under a thousand to several thousand with hydroxyl equivalent weight from several hundred to several thousand. To maintain control of the polymerization process and maximize conversion of monomer to polymer, the polymerization is usually performed in a solvent such as ethyl acetate or an aromatic such as xylenes. The solvent can then be removed or reused as a solvent for subsequent application of the coating. The carrier solvent aspect of acrylic polyols has become an area for technological innovation as producers try to limit solvent content and reduce regulated volatile organic carbon (VOC) content. Common initiators for acrylic polymerization such as peroxides are also used in this process. An illustrative formulation and polymerization product is shown in Figure 2.33. In an actual formulation the butyl acrylate and styrene could be balanced to provide the desired level of flexibility and hardness for the given application. The amount of hydroxyl-functionalized acrylate would be balanced to provide the desired cross-link density for the coating.

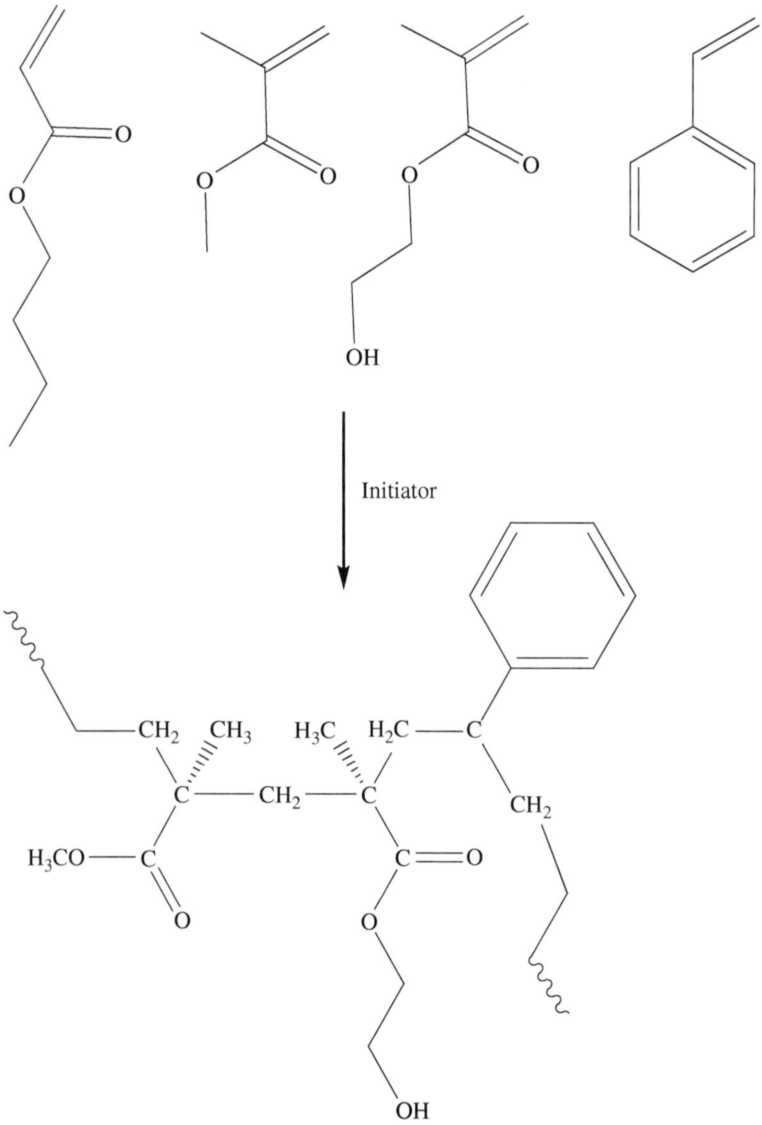


**FIGURE 2.33** Example preparation of an acrylic polyol.

Acrylic polyols have also been designed to optimize compatibility with aqueous systems. This is normally done by including acrylic acid among the monomers [80]. The acid is subsequently neutralized with an amine such as dimethyl ethanolamine and water. The ratio of monomers will control the final properties of the film as well as the differential solubility.

### 2.1.4 Filled Polyols

The low cost and ubiquity of polyols has generated innovation in features and additives to add value to downstream polyurethane processing. One innovation has involved using the polyol as a polymerization medium, and then using those polyol–copolymer mixtures to provide a downstream value. Three so-called "copolymer polyols" have been developed and found significant commercial value. These are (i) styrene–acrylonitrile (SAN)-filled polyols [81], referred to generically as copolymer polyols and prepared via radical polymerization of vinyl monomers; (ii) step growth copolymer polymerizations of isocyanates and amines such as hydrazine, called PHD polyols for *p*oly*h*arnstoff *d*ispersion polyols [82]; and (iii) *p*oly*i*socyanate *p*oly*a*ddition (PIPA) polyols made by reaction of an isocyanate and an alkanolamine such as triethanolamine (TEA) [83]. PHD polyols result in a polyurea filler phase, and the PIPA process results in a urethane dispersed phase. Copolymer polyols are commonly used in comfort applications such as seating, where they are used to provide increased load bearing without having to increase overall foam density. As a result of the improved load bearing the foam will also exhibit less shrinkage upon cooling because of its inherent ability to resist shrinkage stresses. In these roles the combined copolymer polyol market was approximately 800,000 metric tons in 2018, growing at about the rate of the automotive industry. In use, the critical characteristics sought by customers include the percent polymer in the polyol (the solids level), the viscosity of the filled polyol, the filterability or polymer-phase uniformity, and final product characteristics such as color, VOC, and odor minimization. Virtually all producers of polyether polyols sold into comfort seating applications also offer filled polyol options to their customers.

***2.1.4.1 Copolymer Polyols*** The polymerization of styrene and acrylonitrile to form SAN in polyol was first documented by Stamberger in the 1960s [84]. The polymerization is easily accomplished using 2,2′-azodi(2-methylbutyronitrile) (AMBN) or azobisisobutyronitrile (AIBN) as an initiator for the radical polymerization (Figure 2.34).

The limited solubility of SAN in the polyol causes it to phase separate into domains capable of scattering light. This results in the filled polyol being milky white. The phase-separated polyol phase is referred to as the serum because it still contains SAN polymer and oligomers up to the solubility limit. During the polymerization the radical initiators may abstract hydrogen from the polyol, particularly the hydrogen on the tertiary carbon of polypropylene oxide, and provide a radical site on the polyol chain (Figure 2.35). These carbon-centered radicals provide new initiation sites for radical polymerization of SAN polymer and create a graft stabilization mechanism keeping the SAN dispersed in the polyol [85].

The polymer in the serum causes the viscosity of the filled polyol to increase such that it has a useable limit of 20–25% solids before it becomes unusable to foam manufacturers as a result of poor flow and mixing with other foam components with significantly lower viscosity.

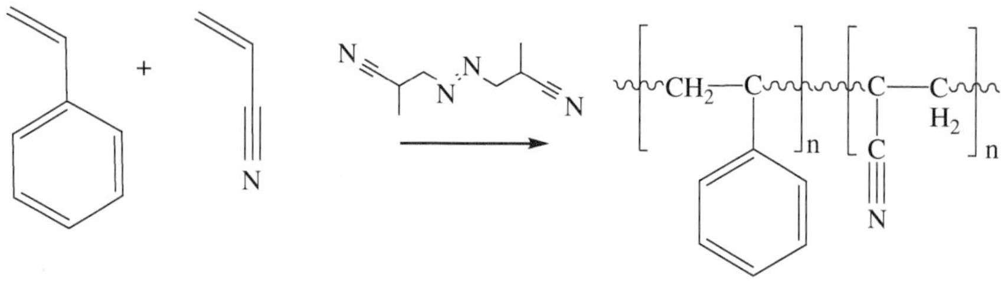

**FIGURE 2.34** Polymerization of SAN using azobisisobutyronitrile (AIBN) as an initiator.

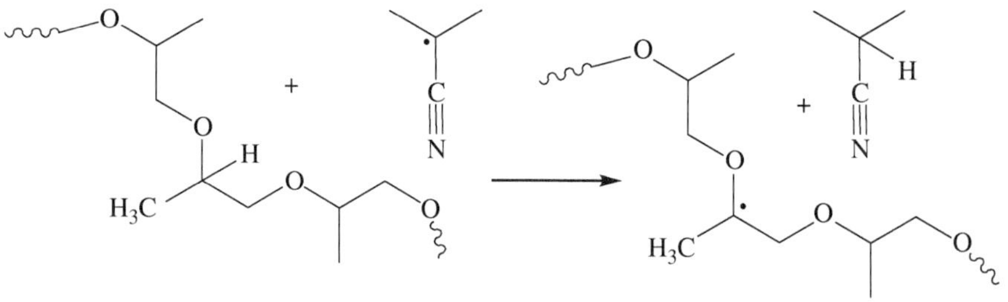

**FIGURE 2.35** Generation of a carbon-centered radical on a propylene oxide polyol. The radical center can serve as a polymerization initiator or chain terminator.

Continuing innovation has focused on achieving ever smaller particles, low color formation, higher solids at lower viscosity, and improved VOC profile. Polyol viscosity, which is directly related to solids content, has been an area of particular innovation and patent activity. Breakthrough performance has been established by the development of stabilizers or "macromers" (macromolecular monomers) [86]. These molecules typically are A–B functional monomers with vinyl and hydroxyl functionality. While reacting readily into the SAN polymer structure they can provide a new stabilization mechanism, preventing agglomeration or settling, without requiring the formation of radical sites on the polyol chain. Macromer components sit on the surface of the SAN particle and act as a buffer to prevent direct collision of the SAN particles with one another. Typical macromer structures are presented in Figure 2.36. The hydroxyl functionalization is related structurally to the host polyol, while the vinyl component is polymerized with a short SAN chain. Artful design of macromer structures, polyol design, and processing have allowed solids contents greater than 50%, with viscosities of c. 5000 mPas and weight average particle sizes under 1 μm (Figure 2.37). Macromer design and optimization continues as a potent area for innovation [87]. Copolymer polyols have an adjusted hydroxyl equivalent weight of between 2500 and 3000 g/OH eq. The polyols usually used for dilution in foam formulation are high-functionality polyols with equivalent weight between 1500 and 2000 g/eq. The polyols are usually EO-capped PO polyols with 15–20% EO content to assure rapid polyurethane formation in molded foam operations.

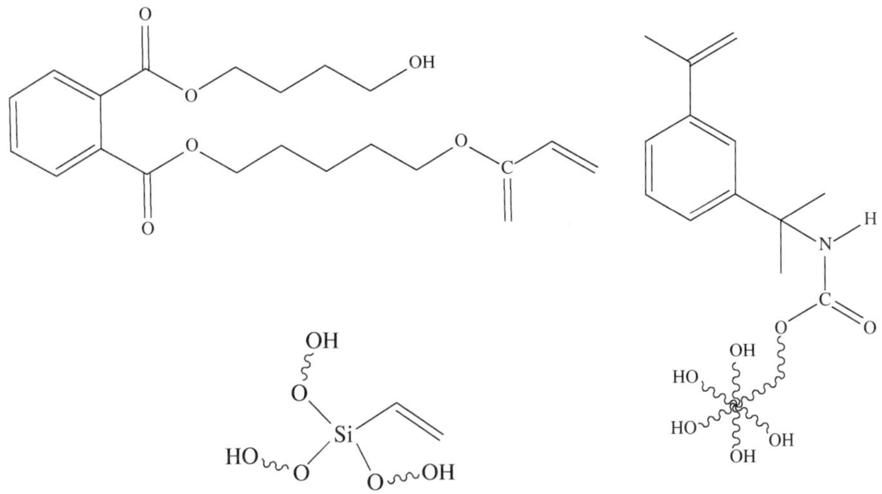

**FIGURE 2.36** Examples of macromer structures.

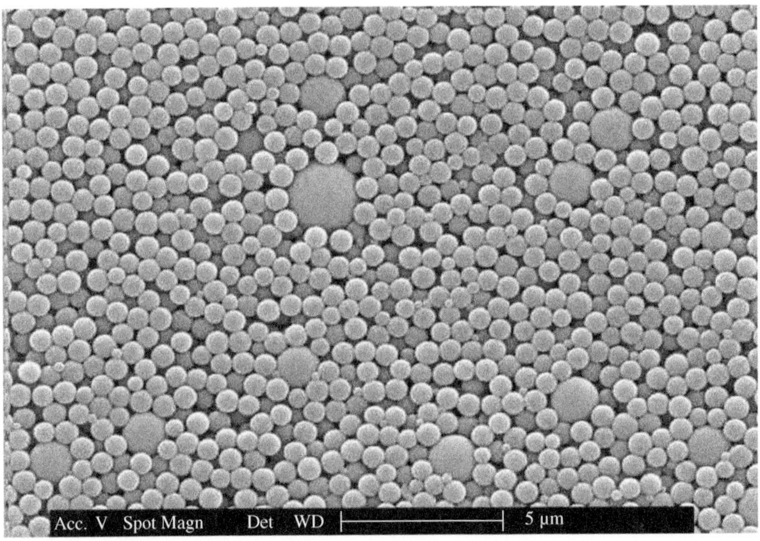


**FIGURE 2.37** Scanning electron microscope image of SAN particles formed in copolymer polyol synthesis. The particle size uniformity points to the exquisite control and optimization of the process. Image by Robert Vastenhout; The Dow Chemical Co.

*2.1.4.2 PHD Polyols* PHD polyols are prepared for the same reason that copolymer polyols are, i.e. to provide a stable filler phase for flexible foams with greater load bearing at low density. PHD polyols are created by rapid reaction of polyisocyanates, usually toluene diisocyanate (TDI), with low-molecular-weight polyamines or hydrazine in a polyol phase. While reaction of isocyanates with amines is much faster than that of isocyanates with the polyol hydroxyls (see Chapter 3), there is sufficient urethane formation with the polyol to create a stabilization mechanism for the formed particles.

A continuous process for making PHD polyols generally involves impingement mixing of the components. Batch descriptions have been described and economic analyses presented [88]. The process involves rapid addition of TDI to hot (80 °C) aqueous hydrazine dispersed in a polyol used for molded foam applications under vigorous mixing conditions. The stoichiometry is such that isocyanate functionality exceeds amine functionality to insure reaction with the polyol phase and minimize the need for system neutralization. The heat of reaction is a major source of control and manufacturing attention. Subsequently the system is dehydrated under heat and vacuum.

The reaction of isocyanates with amines is 1000 times faster than their reaction with primary hydroxyls (see Chapter 3). Thus, the PHD polyol reaction does not promote the level of compatibilization and size limitation available with the copolymer polyol process that employs the macromer stabilizer technology. Additionally, the rapid amine–isocyanate reaction results in poor control of particle nucleation and formation. The high concentration of reactants in the initial stages results in many initial nuclei, but also diffusion-controlled reaction and growth. The result is a relatively broad distribution of particle sizes. The low concentration towards the end of the reaction results in the formation of fewer and smaller particles with an increased incidence of polyol reaction (abetted by the high temperatures of the reaction at the end). This usually results in PHD polyols exhibiting bimodal distribution, or a broad monomodal distribution of particle sizes from hundreds of nanometers to several micrometers. Additionally, viscosities of PHD polyols tend to be somewhat higher than the SAN copolymer polyol. The causes of this effect are not completely understood but probably result from a significant amount of particle polyol colloidal interaction, resulting in less free polyol in the system, and very small particles acting as chain extenders, increasing the effective molecular weight of the fluid phase. The effect is not huge and process improvements have been able to produce 30% solids PHD polyols in the same viscosity realm (~3000–4000 cP) as those from 40% solids SAN copolymers [89].

Like SAN copolymers, most PHD products have a spherical morphology. A less desirable, though still interesting, result can occur when rods are formed as a polyurea-phase dispersion. The less desirable feature is that the high aspect ratio dispersion particles can significantly increase the solution viscosity owing to increased opportunity for particle–particle interactions. On the other hand, higher aspect ratio particles can significantly increase the load-bearing efficiency of PHD polyols that would then be incorporated into the subsequent foam. The different morphology can result by controlling the disposition of isocyanate and amine phases within the polyol reacting medium. In the case that the solid phase forms either interfacially or as a result of two-phase (amine–isocyanate) interdiffusion, the typical and desired (from a low-viscosity standpoint) spherical particle is obtained. In the case that the solvent acts as an intermediating and diffusion rate-controlling medium it becomes possible to obtain rod-like structures.

*2.1.4.3* ***PIPA Polyols*** PIPA polyols are related to PHD polyols except that a polyurethane rather than a polyurea filler phase is produced in a polyol. The process is relatively simple, and several foam producers are known to produce it for captive use rather than purchase alternative filled polyols. The chemistry is relatively straightforward, though optimization requires significant control. The particle is produced through the reaction of a polyisocyanate such as TDI with a short-chain polyol such as TEA, as illustrated in Figure 2.38. The polyol, the TEA, and the catalyst are premixed and the TDI is added rapidly with vigorous mixing [90].

As the particle forms in the polyol medium it can exceed its solubility limit and precipitate. Surface isocyanate groups allow for continued growth of the particle from reaction with TDI and TEA though this occurs at a slower rate. In addition, at all times free isocyanate groups can react with hydroxyl functionality from the polyol. In fact, if the isocyanate reaction with the polyol is not closely controlled, undesirable properties are obtained. For instance, if the reaction is hindered by the presence of too many polyol secondary hydroxyl end groups, the particles will not stabilize in the polyol medium and will develop a size distribution with large average size. If there is too much reaction with the polyol the size distribution will be small but the filled polyol will have a very high viscosity.

In practice this problem is controlled via a technique commonly employed in latex emulsion polymerization, which is to seed the medium prior to isocyanate addition [91]. Thus, to the base polyol–TEA–catalyst, an addition is made of small seed particles in the base polyol. When the TDI is added particle formation begins preferentially on the suspended seeds, resulting in a high density of preformed uniform nucleation sites from which the particles can grow. The result is a significantly more uniform particle distribution centered at smaller sizes. The reaction of base polyol on the seed isocyanate surface also serves to limit the further growth of the particles. Controlling the concentration of seed particles and the size and distribution of seed sizes in the initial state helps obtain an optimum final result.

### 2.1.5 Seed Oil-Derived Polyols

The convergence of advanced industrial biotechnology and advanced chemical manufacturing techniques has opened up an opportunity to develop agriculture-based polyol feedstocks. Whereas most polyols are originally derived from $C_2$ and $C_3$ components of natural gas, the use of seed oils provides an additional source of hydrocarbon either with intrinsic hydroxyl functionality or with an olefin reactive handle to create isocyanate reactive groups. Modern agroscience has at the same time been developing genetic variants that more ideally suit chemical industry requirements for cost-competitive feedstocks. Table 2.10 provides a list of seed oils and their fatty acid

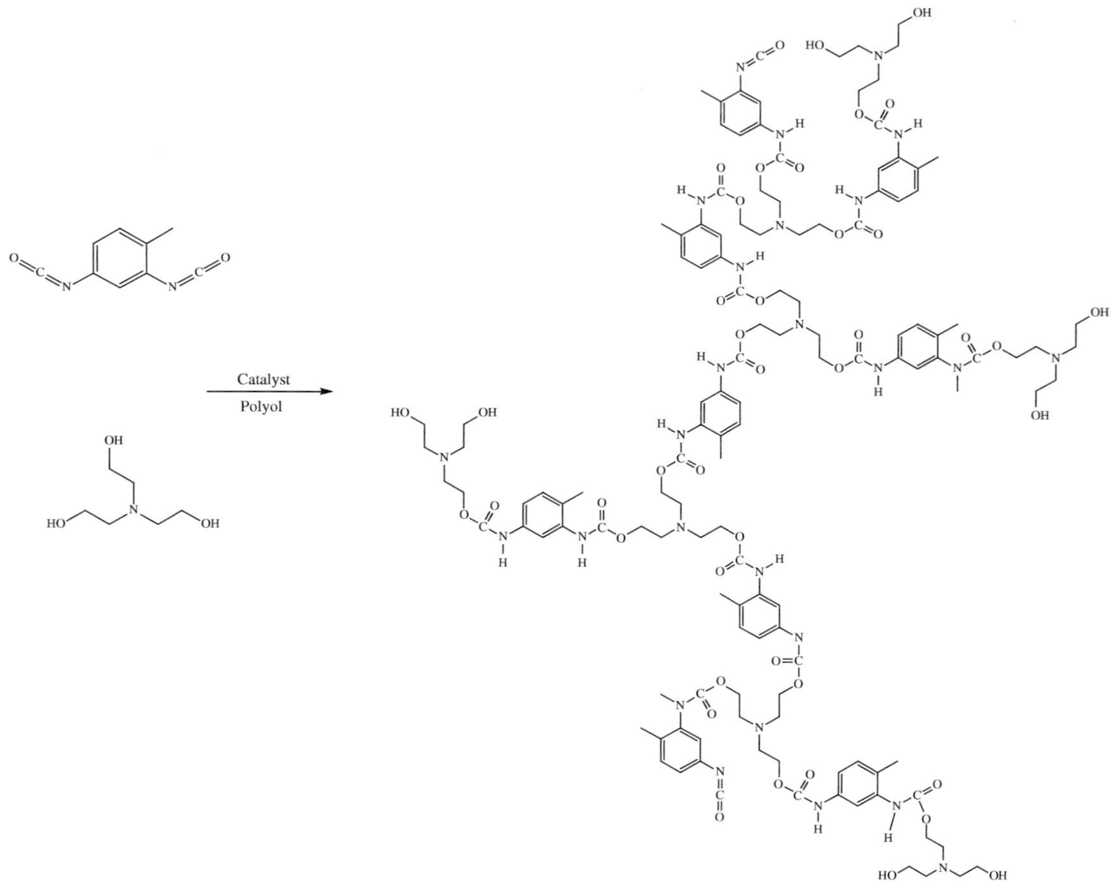


**FIGURE 2.38** Preparation of a PIPA polyol.

**TABLE 2.10 Fatty acid composition of common seed oils used for polyester polyol synthesis**

| Fatty acid composition (%) | Soy | Sunflower | High-oleic sunflower | Canola | Modified canola | Castor |
|---|---|---|---|---|---|---|
| Palmitic | 11 | 6 | 3 | 4 | 4 | 1 |
| Stearic | 5 | 5 | 1 | 2 | 2 | 1 |
| Oleic | 22 | 18 | 93 | 61 | 78 | 5 |
| Linoleic | 54 | 69 | 1 | 21 | 13 | 2 |
| Linolenic | 8 | 2 | 0 | 11 | 2 | 1 |
| Ricinoleic | | | | | | 90 |

composition [92]. The fatty acid compositions can vary around these average values depending on the soil and growing conditions. There are many other seed oils that have been the subject of their potential for use in making polyols based on availability or uniqueness of fatty acid composition. Figure 2.39 shows the structure of the fatty acid components. The seed oil contains the fatty acid components in the form of a triglyceride (Figure 2.40). The triglyceride can be modified for used from its native form by transforming the olefin groups to hydroxyl groups, for instance. Alternatively, the fatty acids can be freed of the glycerine via alcoholysis or hydrolysis to form the ester or acid, respectively. If the native triglyceride is used, the hydroxyl equivalent weight of the polyol will be the molecular weight divided by the average hydroxyl functionality on the triglyceride. The molecular weight between cross-links in the final polyurethane will be very inhomogeneous as some arms of the triglyceride will be stearate or palmitate and have no hydroxyl functionality and some may have as many as three hydroxyls with only a few carbon atoms between them.

Castor oil from the castor bean is a unique substrate since it is the triglyceride of ricinoleic acid, which, as can be seen in Figure 2.39, is already hydroxyl functionalized, albeit a secondary hydroxyl. This allows the unmodified castor oil to be used in some applications with no thermochemical transformation required [93].

### *2.1.5.1 Preparation of Seed Oil-Derived Polyols*

*Epoxidation and Ring Opening* There are numerous ways to create functionality on the seed oil internal olefin bonds for use in polyurethane chemistry. One of the simplest methods is to ring open epoxidized seed oil with ring-opening reagents such as methanol and HCl to produce hydroxy functional esters. Epoxidized soybean oil and linseed oil have been converted this way and tested in rigid foam applications, but performance issues have generally limited their application [94]. A related but more complicated method involves alkoxylation of ring-opened epoxidized seed oil. The epoxides are ring opened with a diol such as ethylene glycol and subsequently alkoxylated to the desired structure (Figure 2.41). These structures have found use by virtue of the reactivity provided by the primary hydroxyl product [95].

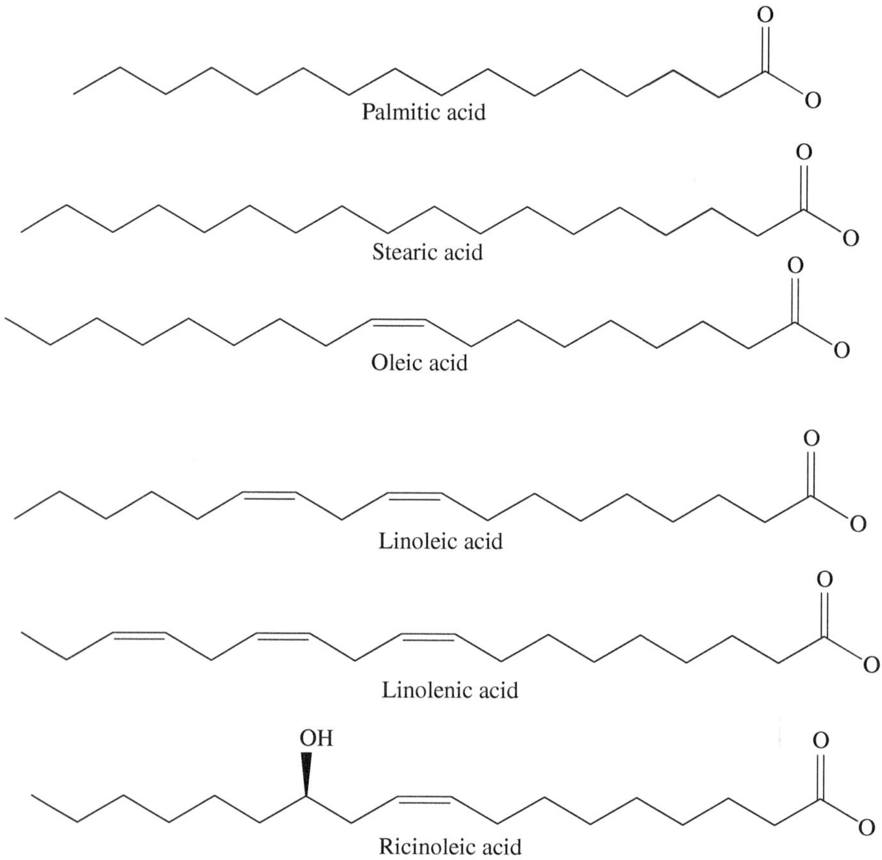


**FIGURE 2.39** Structure of fatty acid components of seed oils.

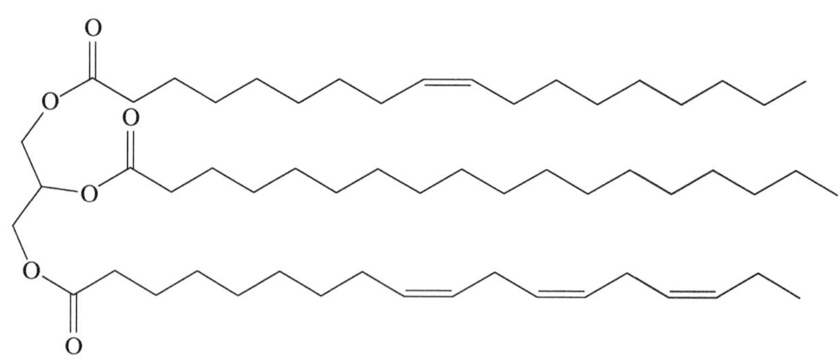


**FIGURE 2.40** Illustrative triglyceride structure showing oleic, stearic, and linolenic esters.

*Ozonolysis* Ozonolysis is a seed oil modification technique that opens avenues to numerous polymerization building blocks. Ozonolysis of the fatty acid double bonds creates ozonide functionalities that can be subsequently reacted to diacids, acid aldehydes, terminal acids, and terminal aldehydes [96]. The difunctional materials can then be modified for convenient polymerization for polyurethanes. While attractive for its versatility and seeming simplicity, the number of steps necessary to get to useable polymerization feedstocks can be a roadblock to economic feasibility. Figure 2.42 shows the pathways available to polymerization building blocks from ozonolysis of oleic acid.

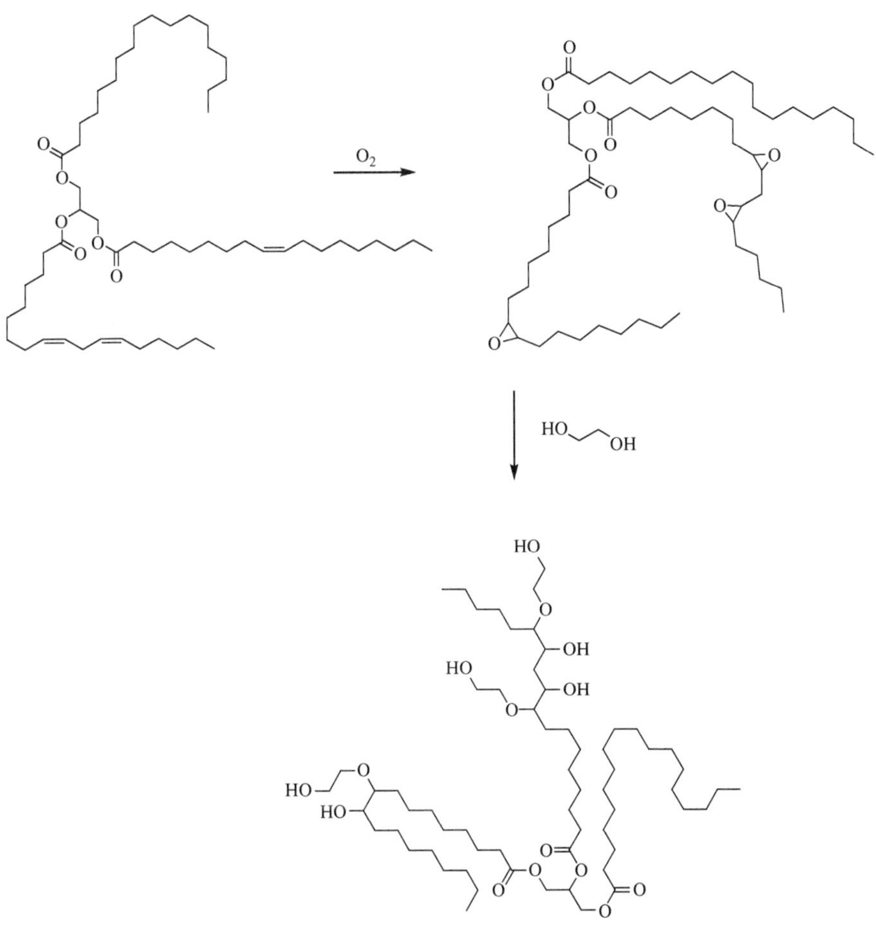


**FIGURE 2.41** Oxidation of an illustrative triglyceride to the epoxidized form followed by alcoholysis to the ester alcohol of the triglyceride. The alcohols can react subsequently with isocyanates to form urethanes.

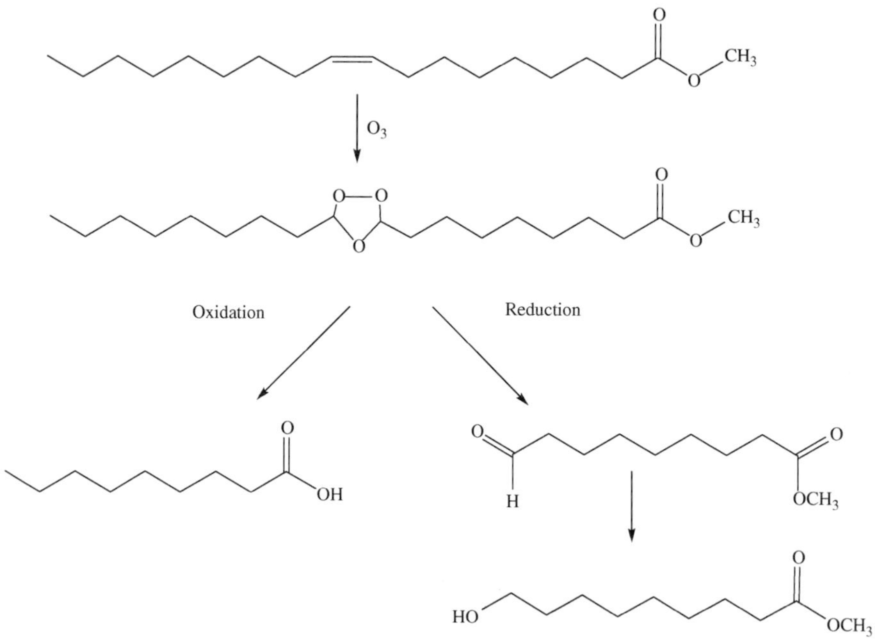


**FIGURE 2.42** Simplified ozonolysis of oleic acid. Other products are also possible. *Source:* Modified from [97].

*Hydroformylation and Reduction* Hydroformylation of seed oil unsaturation forms an aldehyde at each olefin group which can then be reduced to form a primary hydroxy methyl group in place of each aldehyde. This technology has the benefits of quantitatively producing a single primary hydroxyl group for every olefin, resulting in polyester polyol building blocks with uniform reactivity [98]. The hydroformylation of the olefin to the aldehyde is accomplished with very expensive homogeneous rhodium catalysts that must be recovered by quantitative extraction or the product can quickly become very expensive to produce. Reduction of the aldehyde can be accomplished by any number of well-known industrial reduction procedures. When this procedure is applied to the triglyceride the equivalent weight of the polyol is of the order of 300 g/OH group. This equivalent weight is useable in rigid foam formulations, but might require formulation with even lower equivalent weight polyols to provide the necessary cross-link density to result in rapid gelation, high tensile strength, and low gas transmission necessary for most rigid foam applications.

It is also possible to separate the fatty acids from the triglyceride to produce the fatty acid or the methyl ester. Hydroformylation and reduction then results in an A–B monomer having ester and hydroxyl functionality on a single monomer (Figure 2.43) [99].

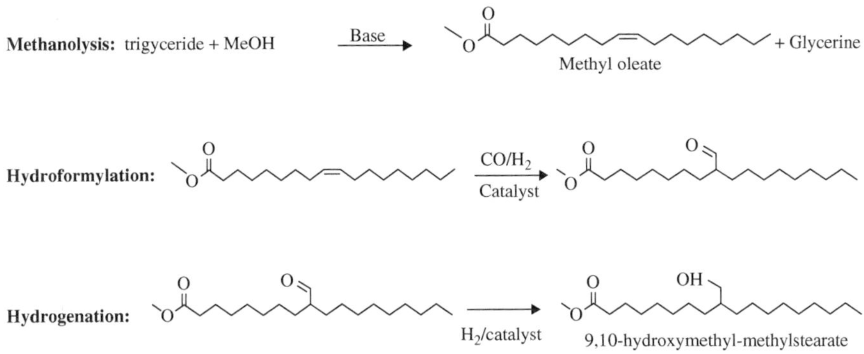


**FIGURE 2.43** Procedure for formation of polymerizable ester alcohols from seed oil triglycerides by a hydroformylation and reduction procedure. From Ref. [100]. © Elsevier.

Using a low-molecular-weight polyol as a polymerization starter or "initiator," it is thus possible to obtain a polyester polyol of nearly any molecular weight or functionality typical for flexible applications. Separation of the different hydroformylation and reduction products resulting in diol and triol esters, for instance, can allow for each product to find its highest value application.

Along with the process requirements associated with catalyst recovery, several issues can pose as obstacles to this elegant procedure. One problem is that, even with substantial optimization, the overall cost can be prohibitive owing to the number of steps necessary to get from seed oil to final building blocks, including separations and recycle. Another potential problem is related to the saturates content found in most seed oils, the presence of which reduces overall asset utilization by occupying volume in the chemical process, but on which no modification occurs [101]. The stearate and palmitate saturates furthermore serve no useful role in polyurethane chemistry, and therefore will either degrade performance or require finding an alternative commercial outlet for their production [99].

*Metathesis* Metathesis is an organic reaction that enables the redistribution of olefin bonds through a scission and recombination mechanism. In the context of seed oil modification for the purpose of creating urethane-capable feedstocks, this modification creates useable products for urethane and polyolefin applications [102]. A pictorial image of the action of cross-metathesis exchange such as envisioned useful in this context is shown in Figure 2.44 [103].

The product ester olefin can then be readily functionalized to a terminal ester alcohol A–B monomer for polymerization by hydroformylation and reduction as described in Section "Hydroformylation and Reduction." The alpha olefin shown can be directly incorporated into polyolefin polymerizations, as is well documented elsewhere.

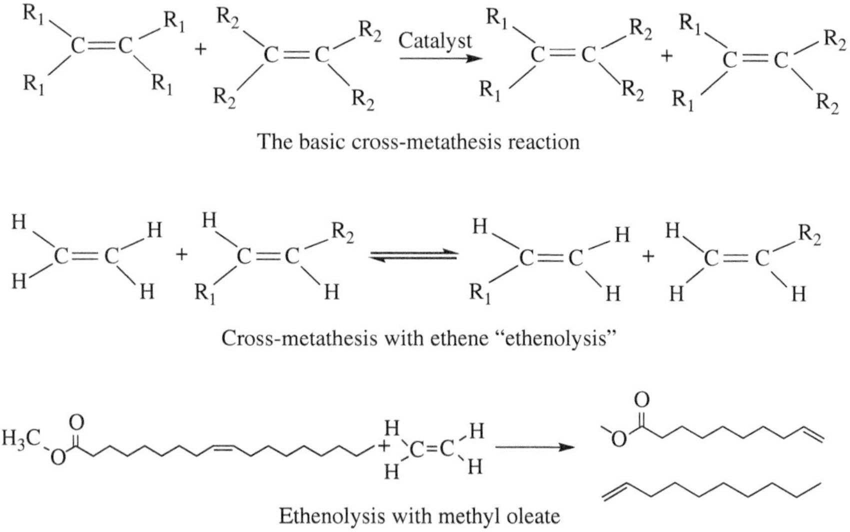


**FIGURE 2.44** Illustrative procedure of olefin metathesis reactions and metathesis applied to the etheneolysis of oleic acid.

New catalysts for metathesis have improved their compatibility with modern industrial processes. The commercialization of these catalysts has resulted in products from metathesis becoming available commercially [104]. Obstacles to fully realizing a large-scale application of this technology are similar to those of the hydroformylation and reduction technology described in Section "Hydroformylation and Reduction associated with separations, and the cost of chemical production requiring large numbers of unit operations.

### 2.1.6 Prepolymers

The term "prepolymer" can be misleading to someone not fully conversant in the patois of polyurethane terminology. A polyurethane prepolymer is one in which all of the polyol hydroxyl end groups have been reacted with isocyanate groups, leaving isocyanate functionality at the termini instead of hydroxyls. In principle a prepolymer molecular weight should be the molecular weight of the polyol plus two times the molecular weight of the diisocyanate capping the polyol. In practice things are rarely that neat [105]. Figure 2.45 shows a simplified view of the reactants to products of prepolymer production.

In principle a prepolymer could be formed by reacting diisocyanates with an excess of polyol to produce a polyol functional prepolymer with urethane functionality at its center, but in practice this is not observed and the term "prepolymer" is always used to refer to polyols terminated with isocyanate functionality. The excess isocyanate is required to prevent molecular weight growth caused by opportunistic chain extension

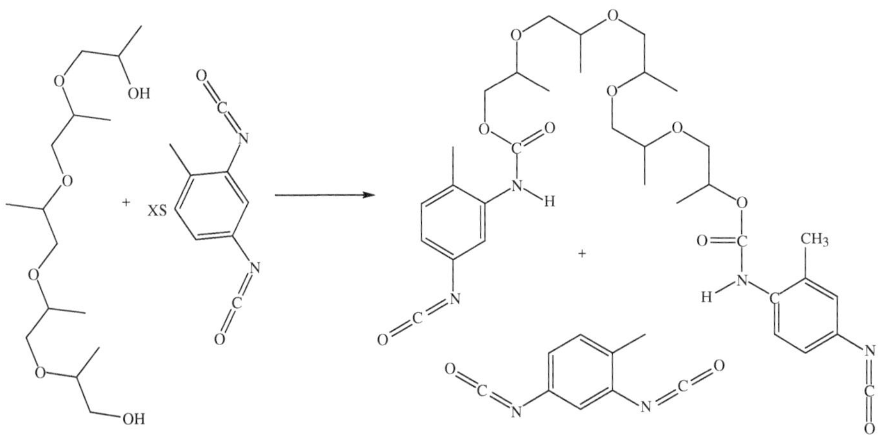


**FIGURE 2.45** General preparation of a TDI–polypropylene oxide prepolymer.

of unreacted polyol reacting with the isocyanate-terminated prepolymer. In practice chain extension always occurs, but can be limited by increasing the ratio of isocyanate to hydroxyl functionality. A true "prepolymer" in this regard would then distill the excess isocyanate from the product. In the case where the excess isocyanate is left in the product this is referred to as a "quasi-prepolymer" [106]. In practice "quasi-prepolymer" refers to systems with relatively high levels of free isocyanate (>12%) while "prepolymer" refer to systems with low levels of free isocyanates (<12%). Over 500 million pounds a year of prepolymer are used commercially for many applications where the advantages of prepolymers can be realized.

Prepolymers find application for several reasons. Among the advantages are that there is improved handling since prepolymers are liquids, they are storage stable (when kept dry), and they provide a source of isocyanate functionality with lower vapor pressure than free isocyanates. Additionally, a prepolymer will usually provide better compatibility between additional polyurethane formulation components since the reacted systems provide lower surface tension to other chemicals within the solubility parameter range. Lastly, it is believed that the prepolymer allows one to "prebuild-in" the final polymer properties, and obtain a more uniform final polyurethane structure [107, 108]. This is particularly true in cast urethane systems where mixing and physiochemical polymer kinetics are more limited, and makes obtaining equilibrium phase structures more difficult (see Chapter 9).

As a practical consideration, prepolymers are usually used and defined on the basis of their percent isocyanate or more commonly as "percent NCO." On reflection, the calculation of this parameter is not as intuitively defined as desirable, and so merits exposition. The definition of %NCO is by equation

$$\frac{\%\text{NCO}}{100} = \frac{(42/Y)\times Y}{X+Y+Z} \tag{2.10}$$

where $Y$ is the equivalent weight of the isocyanate, $X$ is the equivalent weight of the polyol, the weight of NCO is 42 and $X + Z$ equals the number of grams of isocyanate that must be reacted with $Y$ grams of polyol to make an $N$/100% NCO prepolymer or

$$Z = \frac{X + Y}{\left(42/\left(N \times Y\right)\right) - 1} \tag{2.11}$$

As an example, if it is desired to prepare a 12.5% NCO prepolymer with methylene diphenyl diisocyanate (MDI) (equivalent weight = 125 g/NCO eq) with a polyol having equivalent weight = 1633 g/eq, using the equation for $Z$ one would need

$$\frac{1633 + 125}{\left(42/\left(0.125 \times 125\right) - 1\right)} = 1041 \tag{2.12}$$

Thus $N + Z = 125 + 1041 = 1166$ g of isocyanate per 1633 g of polyol (total mass 2799 g) to make the desired 12.5% NCO. Smaller quantities would be prepared by a simple ratio adjustment.

Prepolymers offer an attractive approach that allows one to prepare cast systems, especially elastomers, from entirely liquid components. Using pure MDI, for instance, would require that all ingredients be heated to a temperature greater than the MDI melting point (c. 43 °C), which could shorten the window for convenient processing considerably. As a liquid, it is additionally convenient to begin with a system that is as low viscosity as possible to avoid problems such as entrainment of air during processing. Minimization of viscosity is usually a function of pH control, which, under basic conditions, can result in trimerization of the isocyanate. Other viscosity-building cross-linking reactions occur in making prepolymers such as formation of allophanates and biurets that are more prevalent in NCO-rich environments.

## 2.2 ISOCYANATES

Isocyanates, especially polyisocyanates, are highly identified with polyurethane chemistry. While isocyanates are capable of undergoing many reactions, their significance is usually in the context of their relationship to polyurethane polymerization and structure. Isocyanates represent a class of chemicals that are characterized by high reactivity and versatility. This combination of positive attributes has contributed greatly to the broad application of polyurethane materials, but is part and parcel of the complications associated with isocyanates [109]. The most chemically relevant attribute of isocyanate chemistry is its reactivity with molecules having active hydrogens. Such active hydrogens are typically found on molecules having alcohol and amine functionalities and also water [37]. The basic isocyanate structure is given by Figure 2.46.

**FIGURE 2.46** The isocyanate function.

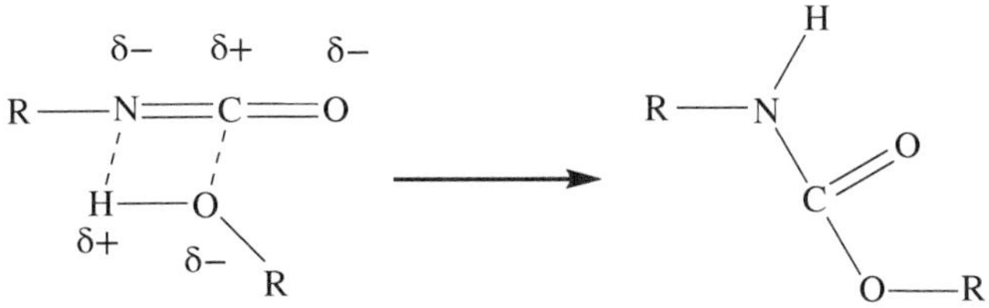

**FIGURE 2.47** Assembling the reactants into a transition state geometry for urethane formation.

Since addition polymerization requires monomers to be able to propagate a chain by undergoing multiple reactions, polymerizable isocyanate monomers have at least two isocyanate functionalities. The growth of polyurethane technology has been associated with the ability to produce polymerizable isocyanate monomers at low cost. The low-cost requirement has driven the chemistry to employ available and low-cost building blocks and industrial production has been optimized to an exceptional extent. The two highest volume isocyanates are based in one case on toluene to make TDI, and in the other case aniline and formaldehyde to make MDI. The preparation of each of these important molecules and their characteristics will be handled separately. Polyurethanes also function in any number of applications requiring good weatherability. Good weatherability can be hindered by aromatic ring structures and entirely aliphatic polyisocyanates have been developed for these uses; the latter will be handled as a class in this chapter.

As discussed in greater detail in Chapter 3, the utility of isocyanates is a result of their electronic structure. As illustrated in Figure 2.47 the unique N=C=O triad results in an electrophilic carbon and a relatively nucleophilic nitrogen [110]. Without going into detail here, this creates an advantageous condition for addition of active hydrogen molecules and formation of urethanes, ureas, allophanates, biurets, and other structures.

The benefits of this seemingly simple reaction have resulted in an industry representing the sixth most common plastic material produced [111]. Much of this is owed to the ability to produce isocyanates from low-cost feedstocks using procedures that, even if complex, could be optimized to an extent that cost is not a barrier to use. As will be discussed at length below, the ability to produce isocyanates with attractive economics is because the pathways to product are well established. The result of this is that many isocyanates can be imagined and have been produced using these processes. Regardless, more than 90% of all isocyanates produced are based on aromatic feedstocks; of these, there are only two substrates for polyisocyanates: one beginning with toluene and the other beginning with aniline [112]. From these substrates are produced TDI and MDI (or pMDI). The remaining isocyanate volume is based on one of several aliphatic substrates [113].

As commodity chemicals, the isocyanate segment of the polyurethane industry is closely associated with the equilibrium between industrial capacity and consumption [114]. Unlike the polyols market, of which there are many varieties and where there are continuous innovations based on structure, the isocyanates market is reasonably defined by TDI and MDI and their component feedstocks. Figure 2.48 describes 2017 estimates for the balance between isocyanate industrial capacity to produce and industrial consumption. The ratio of consumption to capacity was ca. 0.8, suggesting a significant amount of overcapacity in world markets. To an extent this reflects the downturn in Western construction markets and overbuilding of capacity in some countries.

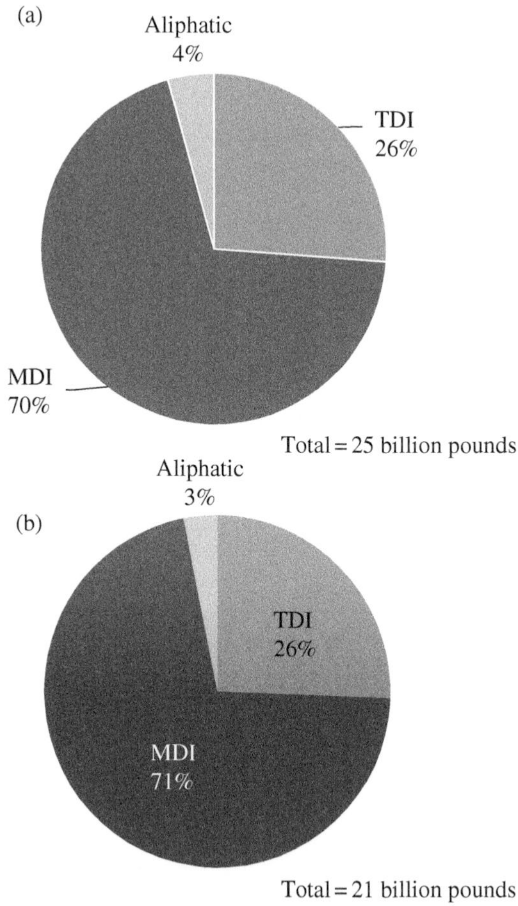


**FIGURE 2.48** World (a) capacity and (b) production of isocyanates for polyurethane polymerization in 2017. *Source:* Based on [112].

Significant global capacity increases have been built under the premise of future economic growth and adoption of rising living standards requiring polyurethane products. The expected annual growth rate is 3.6% for TDI, 4.3% for MDI, and 4.5% for aliphatic isocyanates. These disparate growth rates reflect the differing growth rates for the underlying applications and increasing industrial preference for MDI and its occupational health and safety profile.

### 2.2.1 TDI

TDI is an important industrial chemical because of its central role in polyurethane chemistry. Global production volumes for TDI were 2.6 billion pounds in 1990, 4 billion pounds in 2008, 4 billion pounds in 2010, and 5.0 billion pounds in 2017, and are predicted to be 5.4 billion pounds by 2022 [112, 115] with an implied growth rate of about 3% per year. Demand is driven by the main TDI applications, which are 85% for flexible foams, 10% for coatings, and the remaining 5% spread over numerous small applications. The largest flexible foam applications are furniture, automotive seating, carpet underlay, and mattresses. China was the largest consumer of TDI in 2017 using 34% of available total volume, followed by Western Europe using 18%, and then the USA using about 15% [112].

TDI is produced from toluene. As a basic feedstock it accounts for over 50% of the price of TDI on the market. Interestingly, none of the producers of TDI are back integrated into toluene. The two largest toluene producers are Exxon and Formosa Chemical and Fiber Corporation, together accounting for about 9% of the total. The toluene market is not concentrated; the top 25 producers account for only about 45% of the total produced [116]. Toluene is produced from fuel stocks and the level of supplier diversity and implied competition accounts for the lack of interest by TDI producers to back integrate to toluene. TDI production accounts for only about 5% of toluene consumption; however, as feedstock for chemical synthesis, it is the second largest consumer of value added to toluene (the largest being conversion to benzene/xylenes) (Figure 2.49).

Another key component to making TDI, and isocyanates in general, is phosgene. It is produced by the vapor-phase reaction between carbon monoxide and chlorine. The reaction is exothermic and requires cooling to prevent the reversion of phosgene back to reactants above 200 °C:

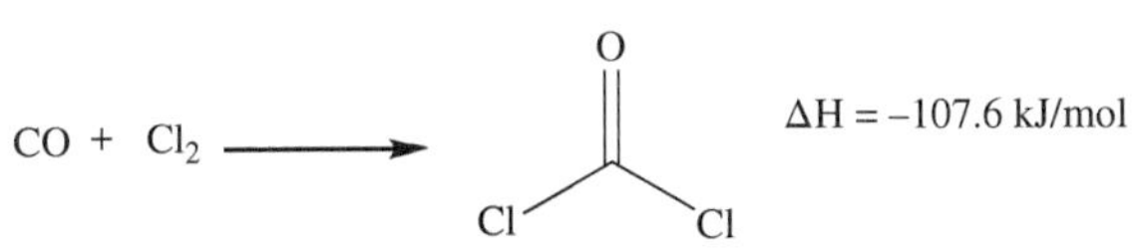


Because of severe health and safety issues associated with exposure to phosgene, this product is almost never shipped and is almost always produced in the plant where its use will occur. A significant amount of engineering optimization is expended on securing phosgene from loss of containment.

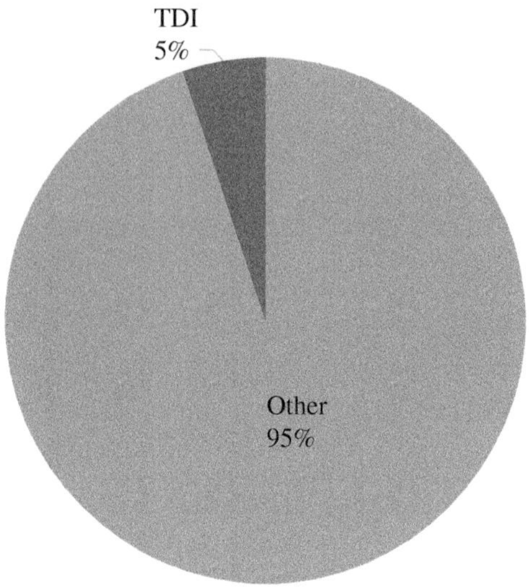


**FIGURE 2.49** World toluene production and portion used for manufacturing TDI in 2018. Total production c. 63 billion pounds. *Source:* Based on [116].

***2.2.1.1 Conventional Production of TDI*** A simplified block diagram for production of TDI is given in Figure 2.50. Most TDI produced in the world is from a three-step process, as shown. It begins with nitration of toluene, which is subsequently hydrogenated to the polyamine. The last step is reaction of the amine with phosgene to make TDI. Innovations have centered on increasing unit efficiency/reduction of waste and minimizing process phosgene [117–119]. Additionally there has been ongoing research into nonphosgene routes to polyisocyanates [120]. In Figure 2.50, blocks associated with waste treatment, intermediate purification steps, hydrogen generation, solvent handling, and handling of bottoms products are not included. The first step of the process is nitration of toluene (Figure 2.51). This simple reaction vastly understates the complexity of pathways leading from reactants to products. Several by-products are formed, including cresole derivatives.

In practice, and as shown in Figure 2.50, nitration is handled in two steps. Formation of the mononitrotoluene is performed with recycled acids from previous production batches at approximately a 30/55 (wt%/wt%) ratio. Along with enhancing overall reagent efficiency, it has the secondary effect of reducing nitric acid in the sulfuric acid prior to sulfuric acid regeneration steps. Formation of dinitrotoluene is performed with nearly the same acid concentrations but at higher temperatures. The yield of the desired 2,4- and 2,6-dinitro positional isomers is approximately 96% and that of other isomers is about 4%. The formation of trinitro isomers is minimized by careful control of acid concentrations and nitration conditions [121]. Figure 2.52 provides explicit pathways of mono- and dinitrotoluene formation. The initial mononitration products consist of 57.5% ortho (substitution at the 2- position), 38.5%

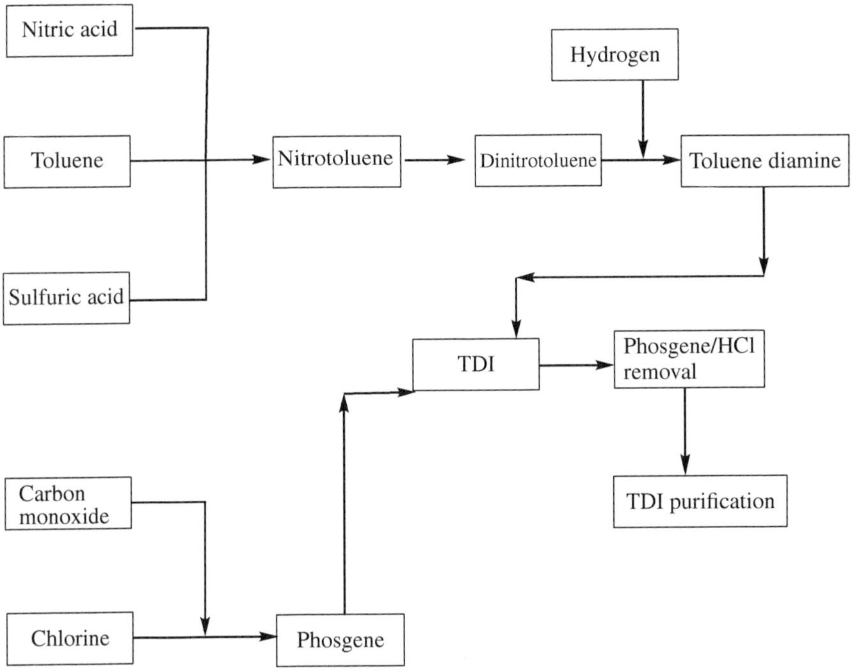


**FIGURE 2.50** Simplified block diagram for TDI production.

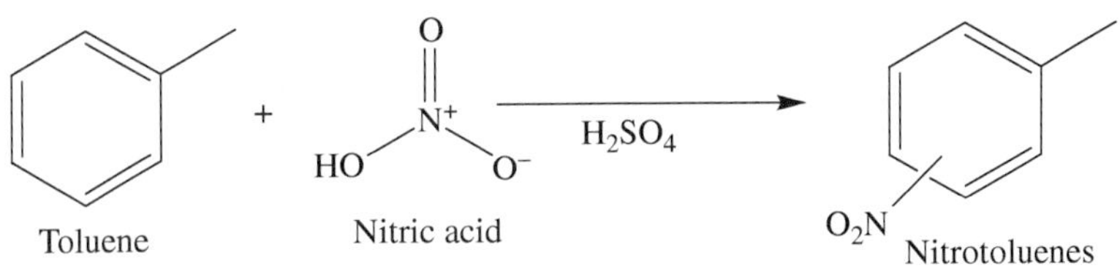


**FIGURE 2.51** Preparation of nitrotoluenes from toluene and nitric acid with a sulfuric acid catalyst [37].

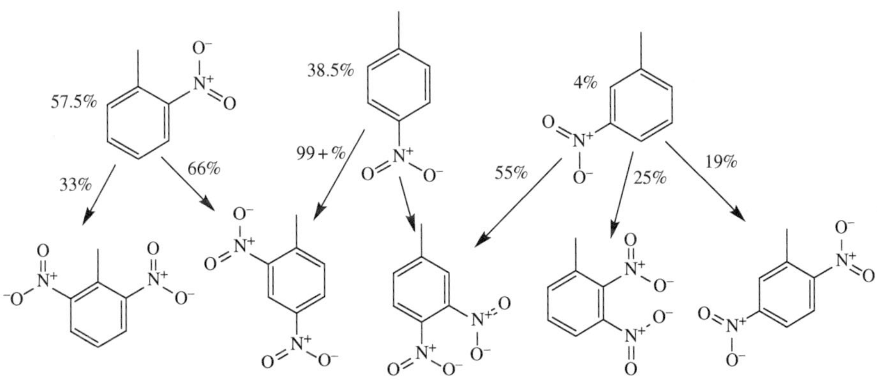


**FIGURE 2.52** Pathways and product distributions of dinitrotoluene from mononitrotoluene [37].

para (substitution at the 4- position), and about 4% meta (substitution at the 3- position). Sulfuric acid is regenerated by removal of water with hot air stripping [122].

The production of nitrotoluenes that avoids the use of sulfuric acid and also the expensive sulfuric acid regeneration step has been reported but is not yet implemented commercially. Problems with the "single acid" nitric acid-only process include the creation of more unwanted by-products, slower rates, difficulty in nitrotoluene recovery because of the solubility of the product in nitric acid, and an expensive nitric acid regeneration procedure [123].

The second step to production of TDI is hydrogenation of dinitrotoluene to toluene diamine (Figure 2.53).

The hydrogenation is accomplished in a continuous process using a palladium–platinum on carbon or other metal heterogeneous catalyst system [124, 125]. The hydrogenation is thus a multiphase system of dinitrotoluene melt, catalyst, and toluene diamines in water and hydrogen gas. Hydrogen pressure can vary from c. 100 psig to over 500 psig. The hydrogenation process is exothermic and temperature is carefully controlled (c. 120 °C). Water and orthoisomers are removed by distillation. As with nitration, the mechanism of hydrogenation to the polyamines is very complex, going potentially through hydroxylamine, azo, and hydrazine intermediates among several others.

The third and final step (apart from distillation) is phosgenation of the toluene diamine isomers to diisocyanates (Figure 2.54).

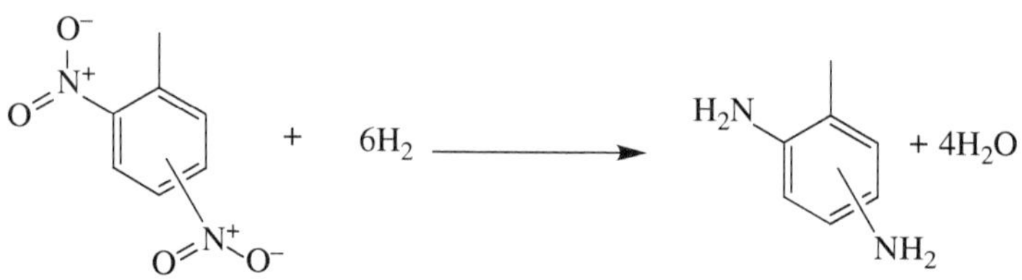


**FIGURE 2.53** Preparation of toluenediamines from nitrotoluenes by catalytic hydrogenation [37].

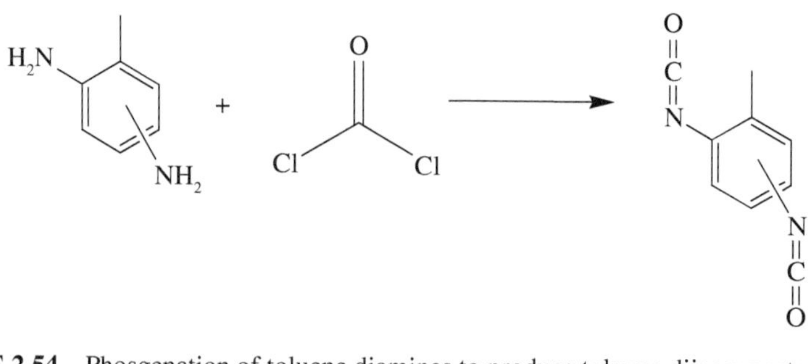


**FIGURE 2.54** Phosgenation of toluene diamines to produce toluene diisocyanates [37].

It is generally accepted that the reaction to isocyanate is a two-step process. First, phosgene reacts with the diamine to form a dicarbamoyl chloride (Figure 2.55).

The dicarbamoyl chloride is not isolated and is directly converted to the diisocyanates at elevated temperature (Figure 2.56) [126].

In a conventional batch process, toluene diamine is dissolved in monochlorobenzene and mixed with liquid phosgene. The mixture is heated to effect reaction, and HCl and excess phosgene purged with nitrogen. The produced amine and HCl form an acid–base pair (sometimes referred to as "slurry"), which must be thermally broken and operationally requires excess phosgene for efficiency. In a continuous process the carbamoyl chloride is formed at ambient conditions and then pumped to a reaction vessel maintained at an elevated temperature (65–80 °C) to form the diisocyanates [127]. A so-called "hot phosgenation" will gradually increase temperature until there is no more HCl evolved, at which point the reaction is considered complete. The hot process has been run up to nearly 90% conversion efficiency of toluene diamine to TDI. Unreacted diamine can be recycled back to the phosgenation step, and partially converted isocyanate can be reverted back to the diamine by hydrolysis

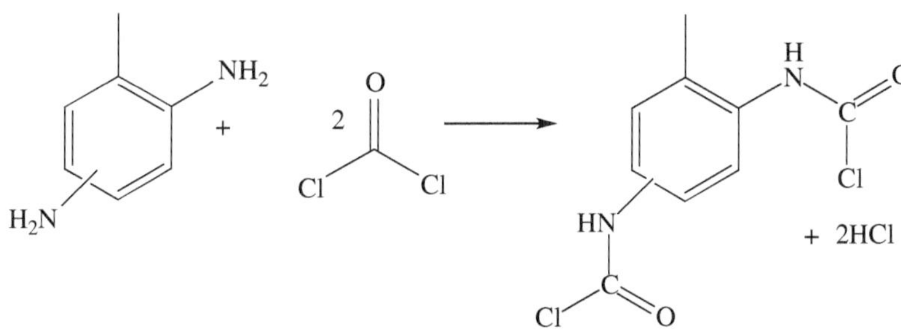


**FIGURE 2.55** First step of the phosgenation process forming the dicarbamoyl chloride. The acid product can complex the free toluene diamine, requiring application of excess phosgene and heat to complete the reaction.

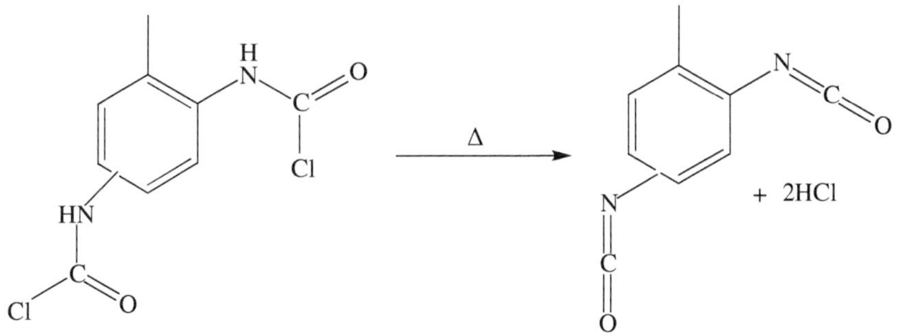


**FIGURE 2.56** Thermolysis of the dicarbamoyl chloride to form the isocyanate and acid coproduct. The acid is recovered by distillation.

and sent back for further phosgenation. Reactions of diisocyanates with amines occurring under concentrated conditions, such as urea, biuret, and carbodiimide formation, are additional routes for loss of final product. These reactions will be covered in detail in Chapter 3.

A significant innovation that has developed commercial reality has been the gas-phase phosgenation process originally developed by Bayer and described in a series of patents [118, 128]. The main innovation allows for the efficient formation of diisocyanates without the need for solvent such as monochlorobenzene. This may not sound like a major breakthrough, but production of commodity chemicals is a very conservative endeavor, especially when there are toxic reagents involved. Although the gas-phase process does incorporate some monochlorobenzene solvent in the phosgenation, the significant reduction decreases the need for handling, heating, and recovering solvents, representing savings in capital as well process expense. Furthermore, the shorter residence time of toluene diamine contact with phosgene results in a reduction of phosgene inventory required to operate the plant reliably – a considerable health and safety outcome.

In gas-phase phosgenation, toluene diamine, phosgene, and a small amount of chlorinated benzene solvent are heated to very high temperatures (>300 °C). The amine and phosgene feeds are mixed in a reactor tube in a manner that maximizes intermolecular interaction [129]. The molar ratio of phosgene to toluene diamine is high to maintain a high reaction rate, reflecting second-order reaction kinetics. Residence times and process are optimized for decomposition of the carbamoyl chloride intermediate, resulting in yields of order of 99% of that expected from theory. Solvent, unreacted phosgene, HCl, and by-products are separated as a vapor phase, and subsequently quenched and separated for recycle or disposal.

The attractive features of this process are resulting in several announcements of new TDI plant construction based on the gas-phase phosgenation technology, and perhaps have contributed to closing of older less profitable assets by other companies. The economic advantage in cost of manufacture using gas-phase technology for a large (660 million pounds per year) TDI plant is of the order of 10% – a large difference for commodity chemicals [130].

***2.2.1.2 Nonphosgene Routes to TDI*** Gas-phase phosgenation technology described in Section 2.2.1.1 results in industrial efficiency, including reduced use of phosgene. Other pathways of TDI production have been explored to forgo phosgene consumption altogether. While these processes are in no way economically competitive with conventional phosgenation, and even less so with gas-phase phosgenation (disadvantaged by about 25% and 35%, respectively), these costs do not reflect the incalculable potential costs associated with a phosgene accident. Three nonphosgene routes to TDI will be covered below. The first two have in common that isocyanate functionality is produced by making urethane, and then thermally decomposing the urethane to isocyanate and alcohols. The differences are in how the initial urethane is achieved. The last, based on the Curtius reaction, is made by decomposition of an intermediate azide.

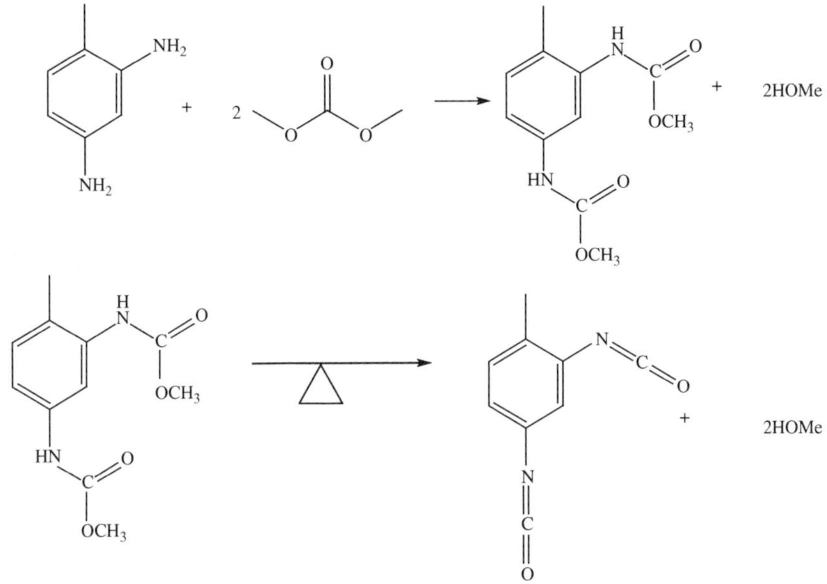


**FIGURE 2.57** Preparation of TDI by reaction of toluene diamines with dimethyl carbonate followed by thermolysis to the diisocyanates.

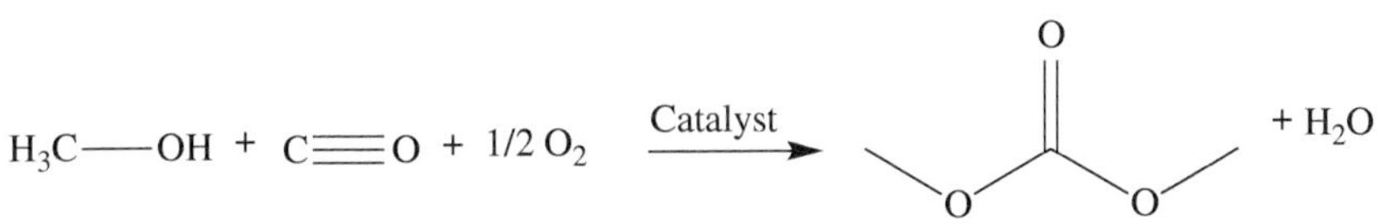


**FIGURE 2.58** Preparation of dimethyl carbonate.

*Thermolysis of Carbamic Acid,N,N′-(4-Methyl-1,3-Phenylene)Bis-,C,C′-Dimethyl Ester made from the Reaction of Toluene Diamine with Methyl Carbonate* The reaction of toluene diamine with dimethyl carbonate (DMC) is known to produce the diurethane of TDI (Figure 2.57) [131]. Dimethyl carbonate is a commodity chemical prepared from the oxidative carbonylation of methanol [132]. The reaction proceeds via progressive heating of DMC and the toluene diamine up to about 170 °C. The diurethane product is subsequently heated to produce the diisocyanates and methanol, which is removed and recycled back to form DMC (Figure 2.58). Yields of TDI are 90% based on input toluene diamine.

*Thermolysis of Carbamic Acid,N,N′-(4-Methyl-1,3-Phenylene)Bis-,C,C′-Dimethyl Ester made from the Reductive Carbonylation of Dinitrotoluene* The catalytic reductive decarbonylation of aromatic nitro compounds to isocyanates has been known since 1967 [133]. Demonstration of this reaction stimulated a significant amount of research on catalysts and cocatalysts to provide ever greater activity and

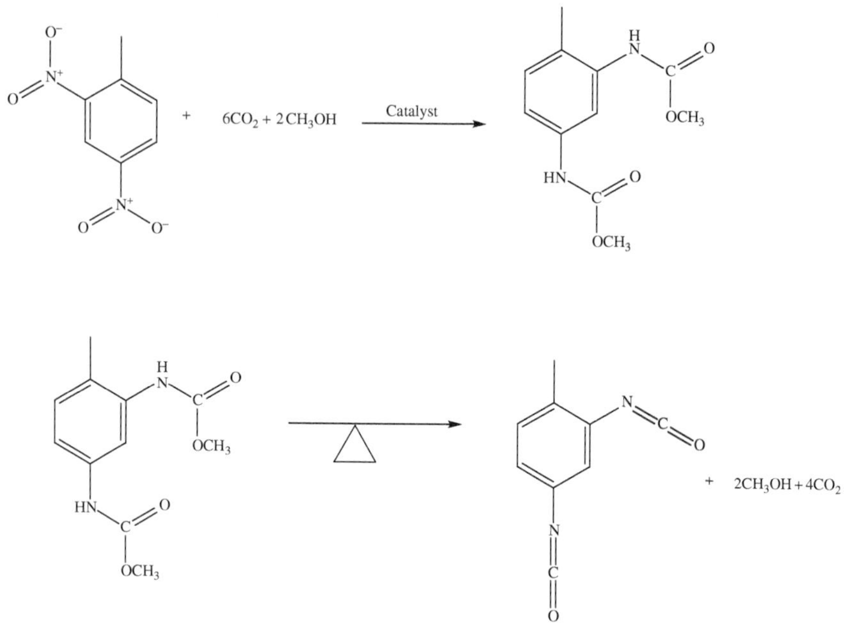


**FIGURE 2.59** Process for reductive decarbonylation for nonphosgene production of TDI [37].

selectivity. It eventually became apparent that, while conversion of dinitrotoluene to TDI could occur, it could not occur with sufficient efficiency to provide a commercial process. It was later observed [134] that with a [Pd(OAc)$_2$ 1,10-phenanthroline] catalyst exceptionally efficient carbonylation of dinitrotoluene to the dicarbamate could be effected in an alcohol solvent. The formation of the carbamate can be conceptualized as the formation of the isocyanate and subsequent stabilization via reaction with the alcohol solvent (Figure 2.59). This concept has been validated by observation of urea formation when the reaction occurs in the presence of amine. The mechanism for such a large and complex transformation is clearly a matter of debate. One simplified mechanism involves a stepwise catalyst coordination, followed by insertion and elimination reactions, ending in the final carbamate product (Figure 2.60) [135, 136]. Less expensive selenium catalysts have also demonstrated high conversion to carbamate. After the carbamate is formed, thermolysis is achieved at about 260 °C with methanol removed overhead. Despite high levels of optimization, this route has not achieved sufficient efficiency to make a viable industrial process.

*Isocyanates by Thermal Decomposition of Acyl Azides: The Curtius Rearrangement* Carboxylic acid chlorides can be converted to isocyanates via prior formation of the carbonyl chloride [138]. The azide can be prepared by initial reaction of the acid

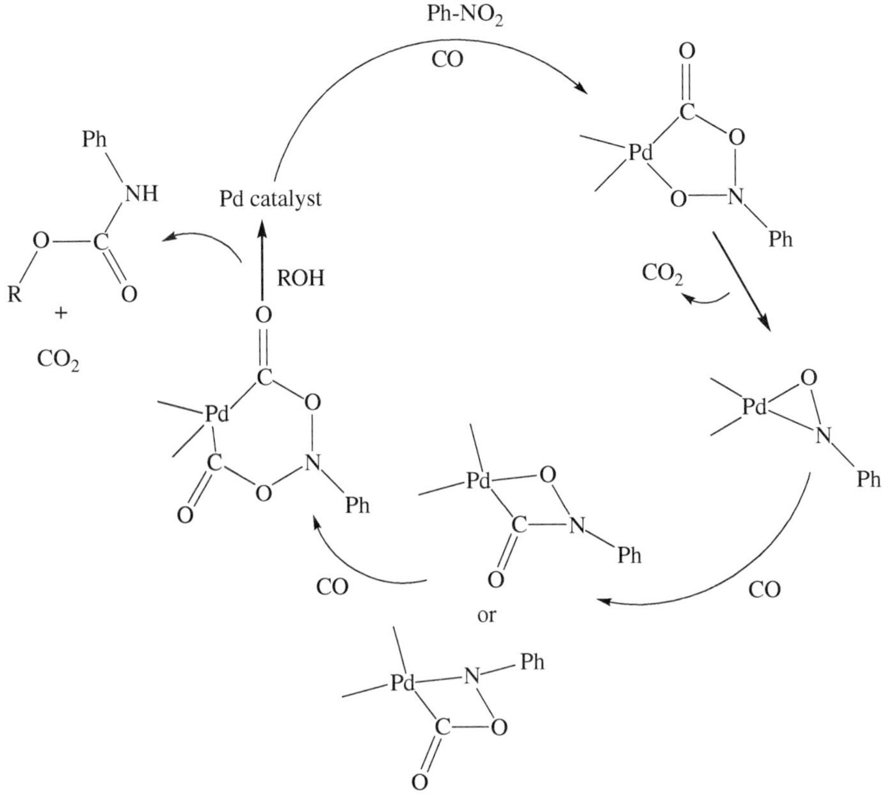


**FIGURE 2.60** Simplified proposed mechanism for Pd-catalyzed reductive carbonylation to carbamate product. *Source:* Based on [137].

chloride with a reagent such as sodium azide or by reaction of the acid hydrazide with nitrous acid. The azide can be converted to the isocyanate by heating to about 100 °C (Figure 2.61) [139]. This reaction has been demonstrated using aliphatic, alicyclic, aromatic, and heterocyclic compounds. The explosion hazards of handling and heating azides make this reaction unworkable from a practical perspective.

### 2.2.2 Diphenylmethane Diisocyanates (MDI)

MDI is the largest volume industrial isocyanate produced (Figure 2.62). Global production of MDI was 4600 thousand metric tons in 2011, was projected in 2011 to be nearly 6000 thousand metric tons in 2016, and actually produced 6700 thousand tons in 2017. This translates to an impressive annual growth rate of about 6%. The largest volume uses for MDI are rigid foams, adhesives, sealants, coatings, elastomers, and flexible polyurethane foam. Rigid foam applications monopolize about 60% of total

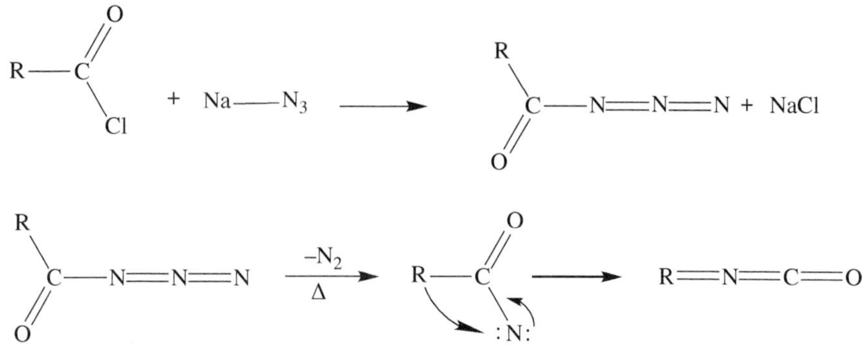

**FIGURE 2.61** Nonphosgene preparation of isocyanates via the Curtius rearrangement.

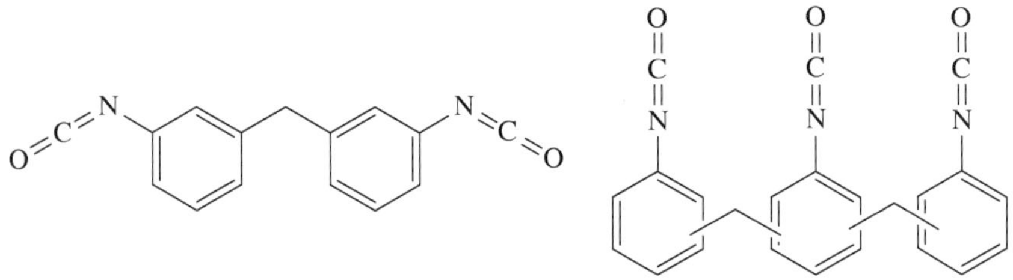

**FIGURE 2.62** Illustrative structures of MDI and pMDI.

MDI volume. China supports the largest MDI production capacity (35%), followed by Europe (33%) and North America (19%). It was estimated in 2011 that China's growth would encompass MDI capacity growth to about 33% of the world total by 2020, taken approximately evenly from the other regions [112]. This early projection has been clearly exceeded.

MDI is produced from aniline and formaldehyde feedstocks. Aniline is produced sequentially from nitration of benzene followed by reduction to aniline by hydrogenation (Figure 2.63). In contrast to TDI being a minor consumer of feedstock toluene (~5%), MDI consumes about 95% of world aniline production (Figure 2.64) [140]. Thus the economics of aniline and MDI production are very tightly correlated and dependent on each other. Furthermore, since most aniline is produced and used captively by MDI producers, MDI manufacturers that are not back integrated to aniline are potentially disadvantaged from the perspectives of supply and price. Potential supply constraints are theoretical since capacity consistently exceeds demand and is projected to for the future. However, MDI manufacturers that purchase rather than make MDI may be between 5% and 10% disadvantaged in overall cost of manufacture. As a commodity chemical, the price of MDI cannot be adjusted for a particular manufacturer's cost positions, and so capital savings associated with not manufacturing aniline result in a reduced profit margin. Capacity for MDI production in 2017

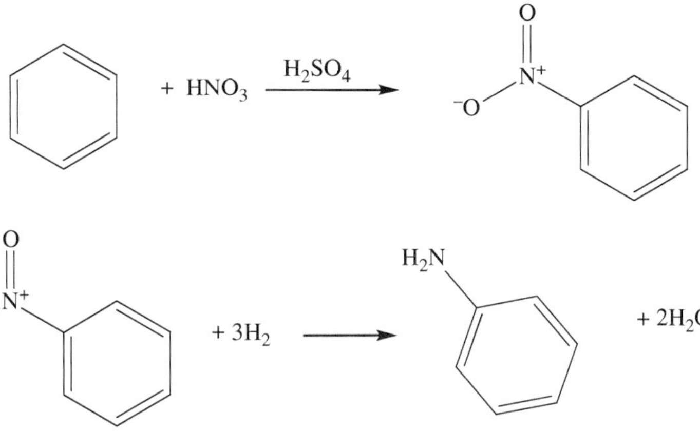


**FIGURE 2.63** Nitration of benzene to nitrobenzene followed by hydrogenation to aniline – a fundamental feedstock for MDI production.

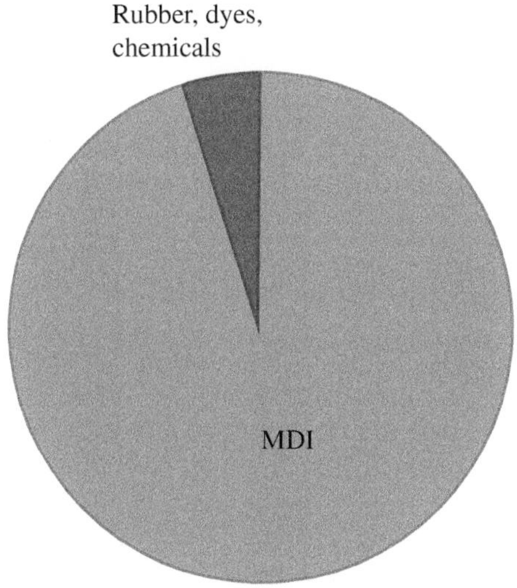


**FIGURE 2.64** Global usage of aniline in 2018. The vast majority is used for making MDI, which tethers their production and costs to each other.

was approximately 7800 thousand metric tons, while consumption was 6700 thousand metric tons, representing about 86% capacity utilization.

Formaldehyde is another feedstock for MDI production. It is a very important commodity feedstock with many applications, particularly in thermoset polymers such as urea, phenol, and melamine formaldehyde resins used particularly in wood bonding. Use in MDI manufacture only accounts for about 5% of total formaldehyde consumption so is not a major determinant of formaldehyde pricing, but may correlate with MDI demand because of overlap in applications with other uses of formaldehyde such as construction [141]. Made from the catalytic oxidation of methanol, Covestro and BASF produce some formaldehyde for captive production of MDI, but most formaldehyde for MDI production is purchased from companies outside of the polyurethane industry.

***2.2.2.1 Production of MDI*** A simplified block diagram for production of MDI is shown in Figure 2.65. There are many similarities to production of TDI with the essential steps of (i) nitration, (ii) reduction, and (iii) phosgenation being the same. Structurally the fusing of the aromatic rings by formaldehyde condensation particularly stands out as a difference and has a pronounced effect on eventual urethane properties and optimal applications. The condensation chemistry also produces isocyanate oligomers. The oligomers of MDI actually have more commercial use than the monomer and this makes it very difficult to treat MDI as a single entity, as generally is the case with TDI. Industrially, 1 kg of MDI is produced from 0.8 kg of aniline and 0.9 kg of phosgene, producing 0.6 kg of waste HCl.

Innovations in MDI process technology center on incremental improvement in product recovery, reagent minimization, distillation versus crystallization of MDI monomer, plant debottlenecking, etc. Heterogeneous catalysis rather than homogeneous acid catalysis would remove the need for handling caustic for reaction neutralization as well product salt. In particular there have been advancements in gas-phase phosgenation processes of methylene dianiline (MDA) to match those

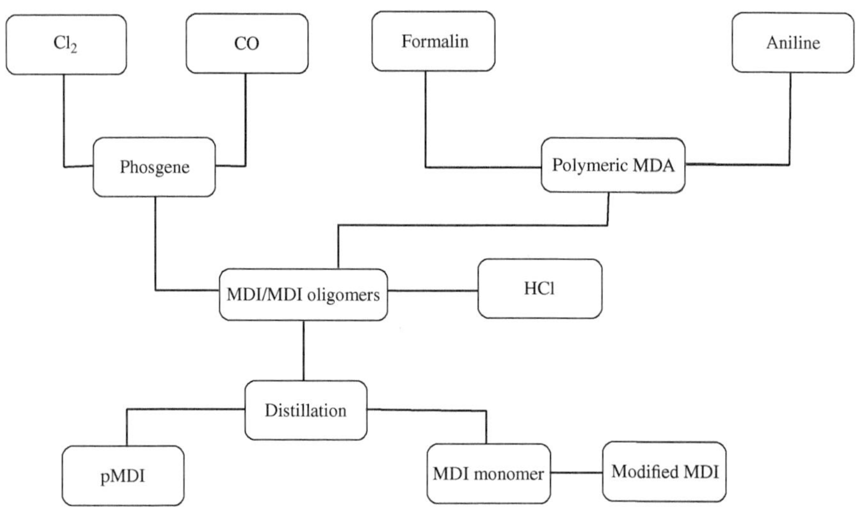


**FIGURE 2.65** Simplified block diagram for the production of monomeric, polymeric, and modified MDI.

practiced with toluene diamine (TDA) [142]. Nonphosgene routes have been explored for the production of MDI, as have been for TDI [143]. As with TDI there is no forecast for nonphosgene routes finding economic viability. Beyond establishment of optimized gas-phase phosgenation processes for making MDI, potential future changes may be to develop plants that flexibly produce monomer MDI, removing the need for distillation steps and serving the higher value monomer markets. Alternatively, construction of plants that exclusively produce pMDI could be envisioned, again removing the need for product separations and maximizing assets to most economically serve the largest volume markets [37].

The diversity of MDI-based products, and the means by which producers distinguish their respective products, originates in the reaction of aniline with formaldehyde to form MDA and its oligomers (Figure 2.66). Details of how this reaction is conducted are tightly controlled process variables and industrial secrets.

The mechanism of acid-catalyzed condensation of aniline and formaldehyde does not lead to a single product, but instead to a mixture of MDA positional isomers and oligomers. Control of the final product distribution depends on process variables such as aniline to formaldehyde ratios, reaction temperatures and times, and the acid catalyst employed. Figure 2.67 is a simplified pathway for aniline and formaldehyde to reach the final products [144, 145]. The initial product of the reaction is *N*-methylolaniline, which loses water rapidly to form the Schiff base. The Schiff base reacts with adventitious aniline to form aminals, one being linear and the other being cyclic. The relative amounts of these forms depend on the ratio of aniline to formaldehyde. The undesired cyclic form is favored when the aniline to formaldehyde ratio is low, while the linear form is favored at aniline to formaldehyde ratios greater than 2:1.

As mentioned, there is a cyclic product (Figure 2.68) that can be formed from the Schiff base that is in principle reversible but in practice is avoided by process control. Furthermore, the products and intermediates after formation of the Schiff base conversion to the aminals are not observed except under acidic conditions. The final step to MDA is not reversible, and the reaction with the carbonium ion intermediate is the basis for modern MDI technology and 75% of commercial volume. It should also be clear that the last step to form MDA is not exclusive to formation of the 4,4′ products and that the 2,4 product can form, and certain process parameters can hinder or enhance its formation.

Process control can influence the product distributions of these reactions. As alluded to earlier, the aniline to formaldehyde ratio is an obvious variable. Ratios of aniline to formaldehyde are typically 2–5 with higher MDA monomer yields

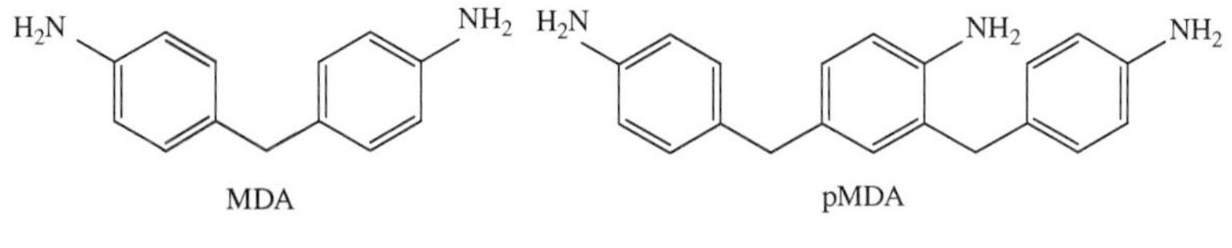


**FIGURE 2.66** Structures of 4,4′-methylene bisphenyldiamine and the MDA polymer.

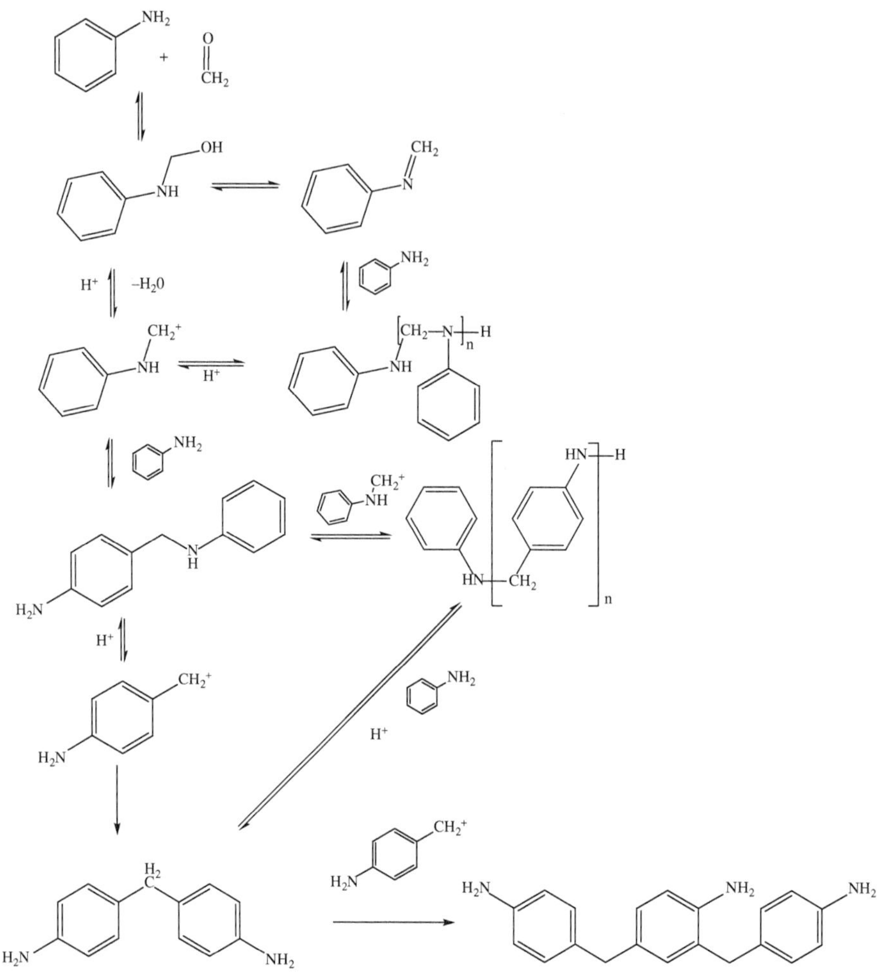

**FIGURE 2.67** Mechanism for the production of polymeric and monomeric MDA by reaction of aniline and formaldehyde. Product structure and distribution is largely determined in this step.

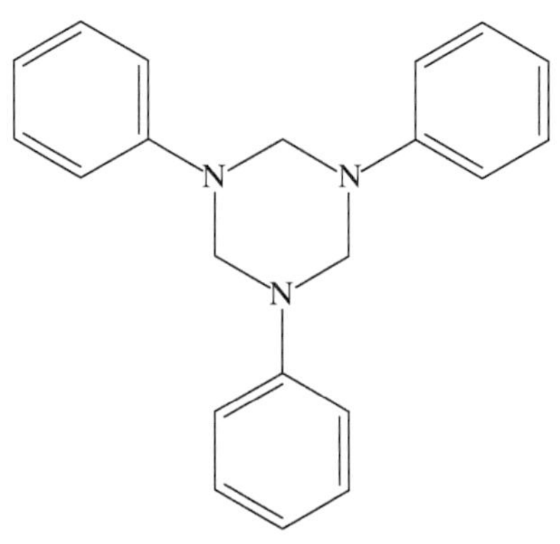

**FIGURE 2.68** 1,3,5-Triphenylhexahydritriazine formed by trimerization of the Schiff base intermediate in the production of MDA

**TABLE 2.11 Effect of process variable ratios on relative production of monomeric and polymeric methylene bisphenyldiamine**

| Process variable | Process direction | MDA | pMDA |
|---|---|---|---|
| Aniline/formaldehyde | Increase | Increase | Decrease |
| Acid/aniline | Increase | Increase | Decrease |
| Temperature | Increase | Decrease | Increase |
| Reaction time | Increase | Decrease | Increase |
| Water content | Increase | Decrease | Increase |

produced at the higher ratios, and MDA oligomer produced at lower ratios [146]. Increasing the acid to aniline ratio is also observed to increase MDA monomer composition in the final product. Higher reaction temperatures result in increased oligomer formation, as do increases in reaction time and water content (Table 2.11).

The phosgenation of MDA and pMDA mixtures follows a similar pathway to that of TDI. A solution of the aromatic anilides is dissolved in a solvent such as monochlorobenzene and then mixed with liquid phosgene to form the carbamyl chloride. The carbamyl chlorides are then heated to drive off HCl, which is stripped along with the excess phosgene for recycle back to the production process (Figure 2.69) [147]. The efficiency of amine conversion to isocyanate in the phosgenation reaction is quite high. About 0.8 kg of MDA produces about 1 kg of MDI (pMDI). The efficiency of the phosgene reaction does not preclude the formation of unwanted side products, some of which are well known and are quantified in product specifications. One such reaction is the diffusion-limited reaction between amines and the freshly produced isocyanates to form urea functionality (Figure 2.70). The by-product of Figure 2.70 is of course fully capable of further additional side reactions. Among the reactions the urea by-product can undergo is the phosgene-stimulated hydrolysis of urea to carbodiimide, which can then react with excess isocyanate to form the imino-uretdione, which in this context is referred to as "APA" (Figure 2.71). The addition of another isocyanate to the four-member ring creates a six-member triazine ring, another side product of the reaction termed "AP6."

Parameters associated with phosgenation and control of the final products are given in Table 2.12 [148].

Industrially, a portion of pure 4,4′ MDI is usually distilled from the pMDI mixture. Historically the polyurethane industry has consumed as much of this 4,4′ component as can be produced, and controlling production has had the effect of maintaining price. The applications for 4,4′ MDI monomer are elastomers, adhesives, and notably spandex fibers. Another reason for manufacturers to limit production is related to the limited shelf life of the pure 4,4′ monomer [149]. In the solid state the 4,4′ monomer undergoes a facile dimerization reaction to form the uretdione (Figure 2.72) [150]. This reaction is intrinsic to the reactivity of isocyanate functionality and the packing of 4,4′ MDI in its crystal state [151]. The maximum in the conversion rate is unfortunately around ambient storage temperatures (Figure 2.73).

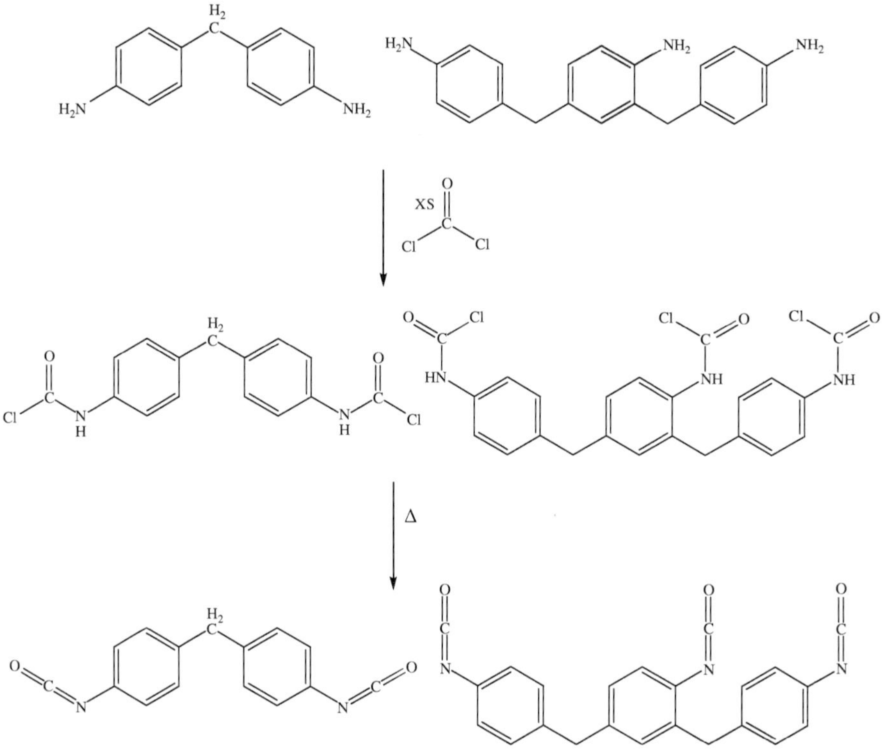

**FIGURE 2.69** Phosgenation of MDA and pMDA to form MDI and pMDI.

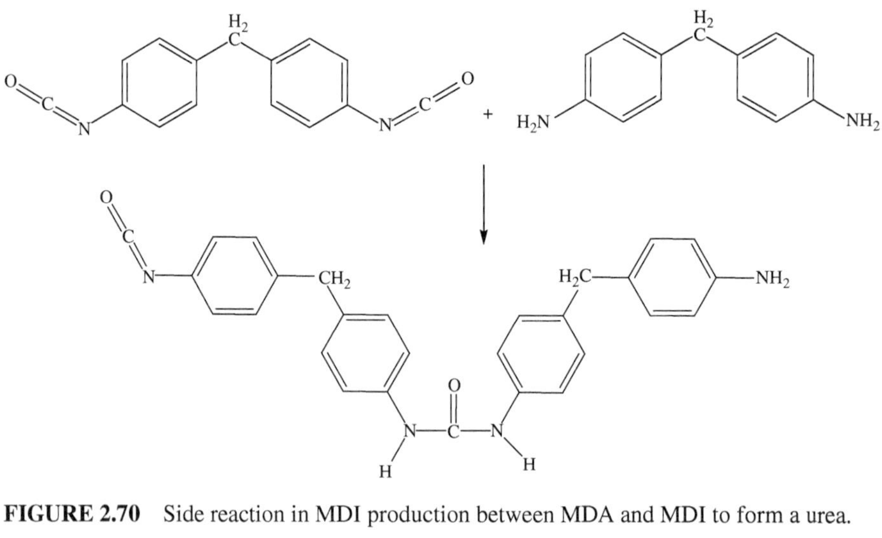

**FIGURE 2.70** Side reaction in MDI production between MDA and MDI to form a urea.

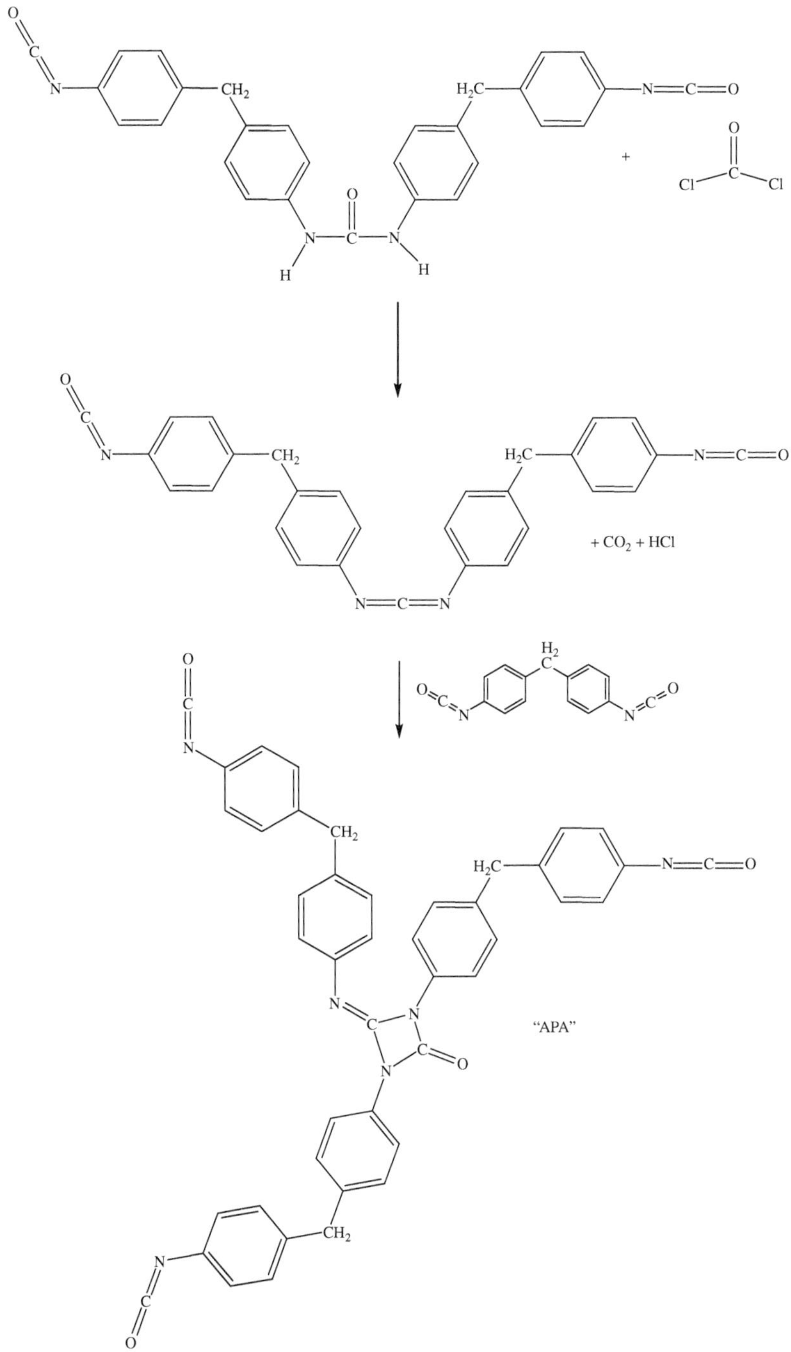


**FIGURE 2.71** Formation of APA – a common impurity in pMDI, usually quantified in a producer's certificate of analysis.

**TABLE 2.12 Effect of process variable ratios on the production of desired MDI and pMDI (the "process") and reactions resulting from numerous potential side reactions**

| Process parameter | Process direction | Side reactions | Notes |
|---|---|---|---|
| Phosgene/pMDA ratio | Increase | Decrease | Ratio usually 6–8 |
| Concentration pMDA in solvent | Decrease | Decrease | ~30–35% |
| Residence time in reactor | Increase | Increase | |
| Reactor temperatures | Increase | Increase | ~200 °C |

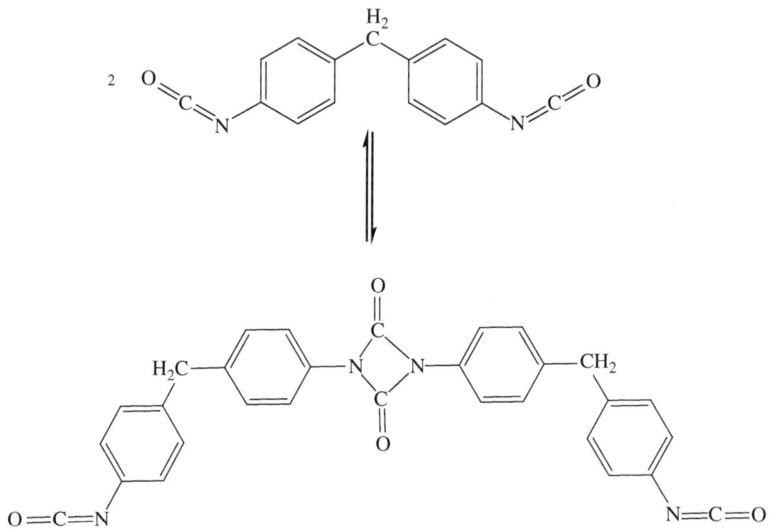


**FIGURE 2.72** Formation of uretdione, a common undesired side reaction resulting from dimerization of the isocyanate function. The reversion from the dimer to the monomers can occur at temperatures over 100 °C.

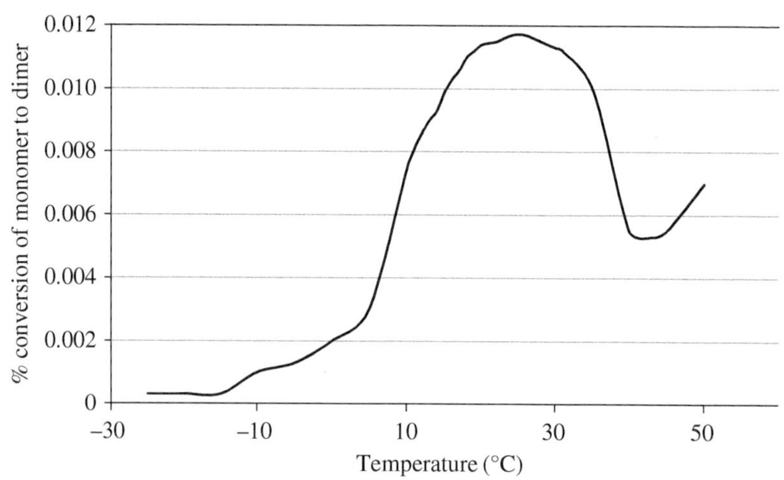


**FIGURE 2.73** Measured percentage molar conversion of MDI monomer to dimer *per day* as a function of temperature.

This warrants normal laboratory storage in freezer conditions. Industrially 4,4′ MDI is maintained in the molten state (but not too hot as the thermally induced dimerization rate increases above 50 °C) to facilitate pumping and minimize dimer formation. Regardless, the monomer does not have a long shelf life and must be used in a relatively short time after production.

### 2.2.3 Aliphatic Isocyanates

About 660 million pounds of aliphatic isocyanates were consumed in 2017, and the expected growth rate for this industrial segment is between 3% and 4% per annum. The vast majority of aliphatic isocyanates are used for coatings. This application particularly takes advantage of the reduced weather-induced coloring of aliphatic-based urethanes versus aromatic-based polyurethanes (Figure 2.74). Additionally, a small amount of aliphatic isocyanates is used in thermoplastic polyurethane films that are also used for weatherable applications. About 45% of aliphatic isocyanates are consumed in Europe, about 35% in Asia, and 20% in North America (Figure 2.75). The growth rate is highest in the developing economies, but does not mirror the difference in overall economic growth, with growth somewhat slower than expected on macroeconomics [112]. In contrast to aromatic isocyanates, aliphatic isocyanates are characterized by superior weatherability, their relatively slow reactivity, and their being substantially more expensive.

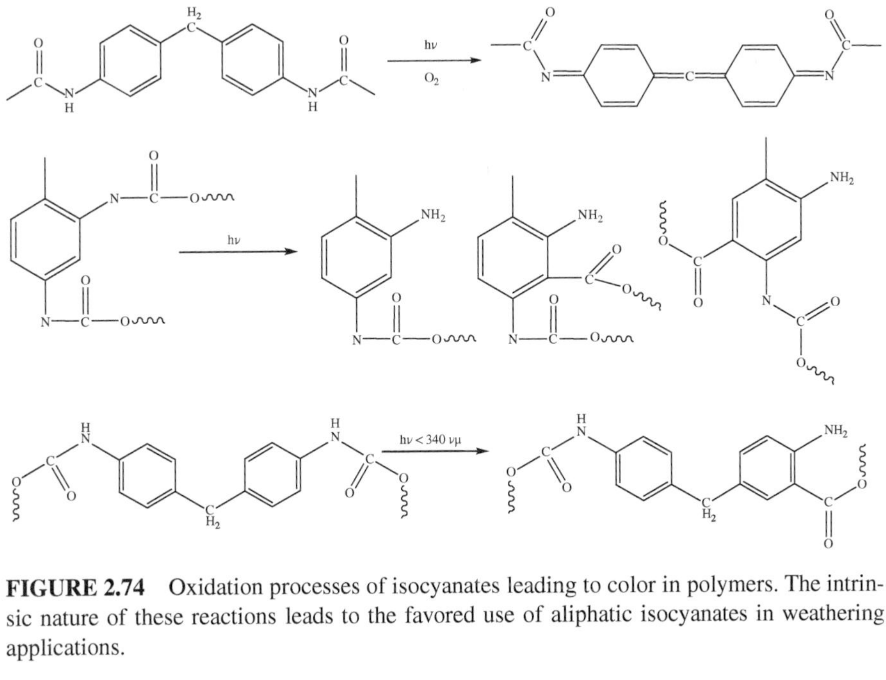

**FIGURE 2.74** Oxidation processes of isocyanates leading to color in polymers. The intrinsic nature of these reactions leads to the favored use of aliphatic isocyanates in weathering applications.

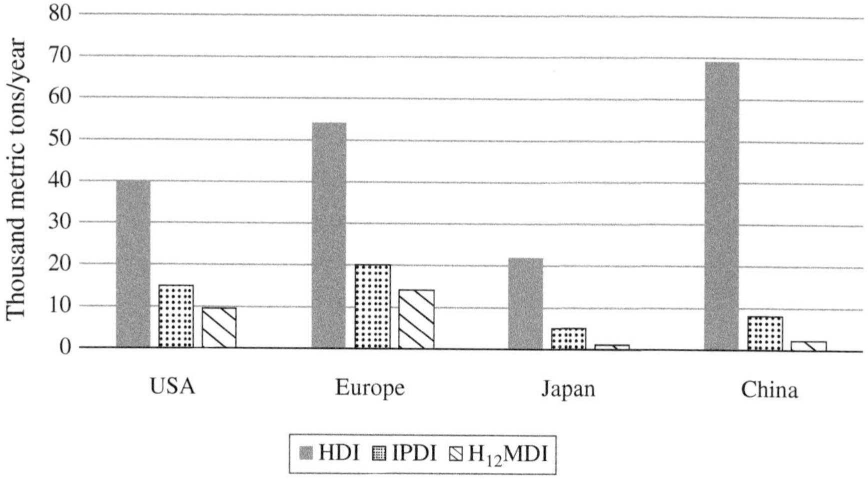


**FIGURE 2.75** Global consumption of major commercial aliphatic isocyanate in 2017. HDI is primarily used in its trimer form.

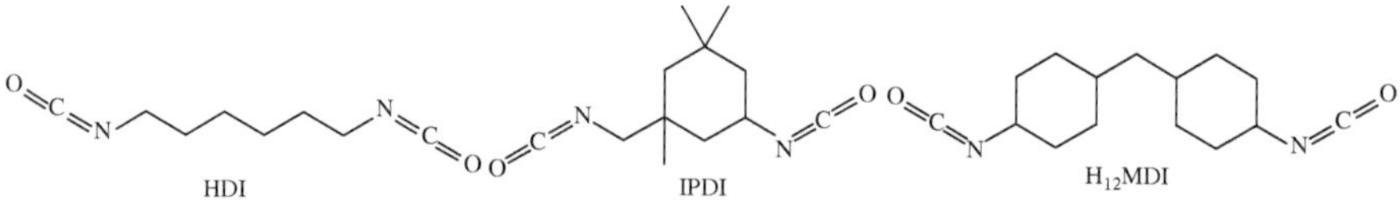


**FIGURE 2.76** Structures of common and commercially available aliphatic isocyanates.

Three aliphatic isocyanates represent more than 95% of the total aliphatic isocyanate world production. They are, in order of size, hexamethylene diisocyanate (HDI), isophorone diisocyanate (IPDI), and the hydrogenated version of MDI, 4,4′-diisocyanatodicyclohexylmethane ($H_{12}$MDI). Approximately 70% of the total used is HDI, 20% is IPDI, and the rest is $H_{12}$MDI. Almost 90% of aliphatic isocyanates are used in coatings and 6% in elastomers. Almost half of the polyurethane coatings that use aliphatic isocyanates are automotive related. The other volumes are spread disparately over many relatively smaller applications. The structures of the three main isocyanates are shown in Figure 2.76.

The need for aliphatic isocyanates stems from the photooxidative instability of aromatic isocyanate structures. The photooxidative instability mechanisms have been well studied and are, of course, complex [152]. MDI-based urethanes tend to show an increased tendency to yellow upon light exposure. It is believed that this is due to formation of the quinone-diimide product shown in Figure 2.74 [153, 154]. It is also shown that, under other circumstances, a photo-Fries-type rearrangement can occur, resulting in anilide formation, which is known to oxidize and discolor upon further oxidation (Figure 2.74). In addition it has been shown that the soft segment can have a significant influence on photodegradation, with increased degradation occurring with increased soft-segment flexibility [155].

#### *2.2.3.1 Production of Aliphatic Isocyanates*

*Hexamethylene Diisocyanate (HDI)* The largest volume aliphatic diisocyanate, HDI, is produced from feedstocks that do not depend on HDI manufacture for their economic structure. HDI is prepared from the phosgenation of hexane diamine in a manner quite analogous to that of aromatic polyisocyanates. Over 1400 thousand metric tons of hexamethylene diamine were produced in 2018 with an expected annual growth rate of about 3%. Most of this volume is prepared for manufacture of nylons (polyamides) and the diamine represents a backward integration into feedstocks [156]. Feedstocks for adiponitrile are butadiene or alternatively acrylonitrile. Once adiponitrile is formed the diamine is formed by reduction with hydrogen (Figure 2.77).

From the diamine the diisocyanate is formed from the acid-catalyzed reaction with phosgene to form the carbamoyl chloride, which converts to the diisocyanates upon heating. As usual the amine is in equilibrium with the hydrochloride salt formed by the evolved HCl, which requires the use of excess phosgene and heat to liberate the free amine (Figure 2.78) [157].

*Isophorone Diisocyanate (IPDI)* Isophorone diisocyanate is prepared by a multi-step process beginning with the base-catalyzed aldol condensation of acetone to form isophorone isomers (Figure 2.79) [158]. The isophorone is subsequently converted to isophorone nitrile by reaction with hydrogen cyanide (Figure 2.80) [159] and converted to the diamine by catalytic reductive amination, followed by formation of the

**FIGURE 2.77** Process to produce hexamethylene diamine from adiponitrile.

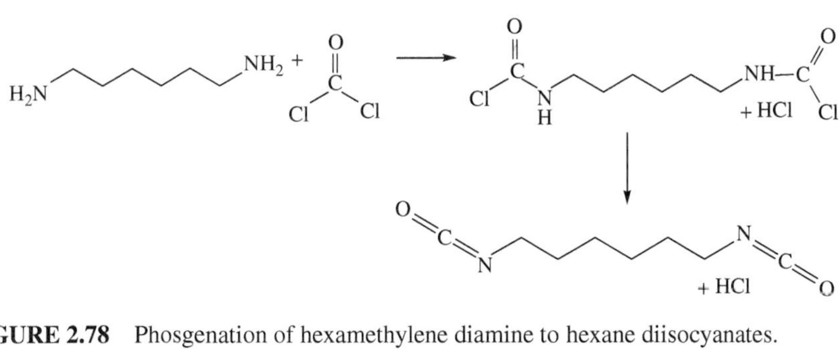


**FIGURE 2.78** Phosgenation of hexamethylene diamine to hexane diisocyanates.

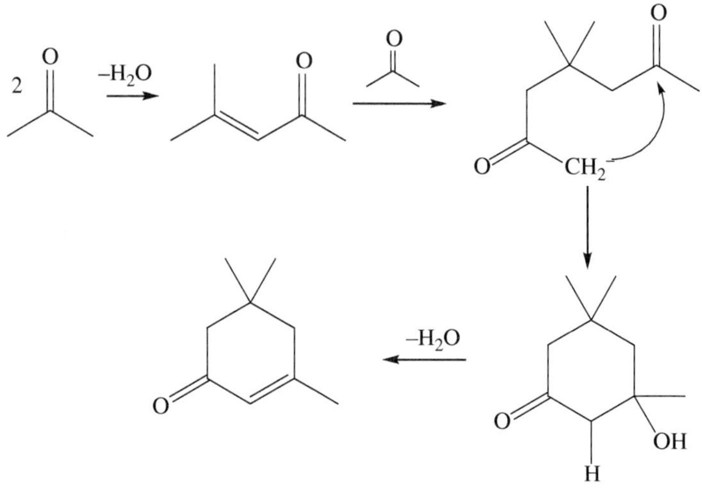


**FIGURE 2.79** Trimerization of acetone to form isophone [161].

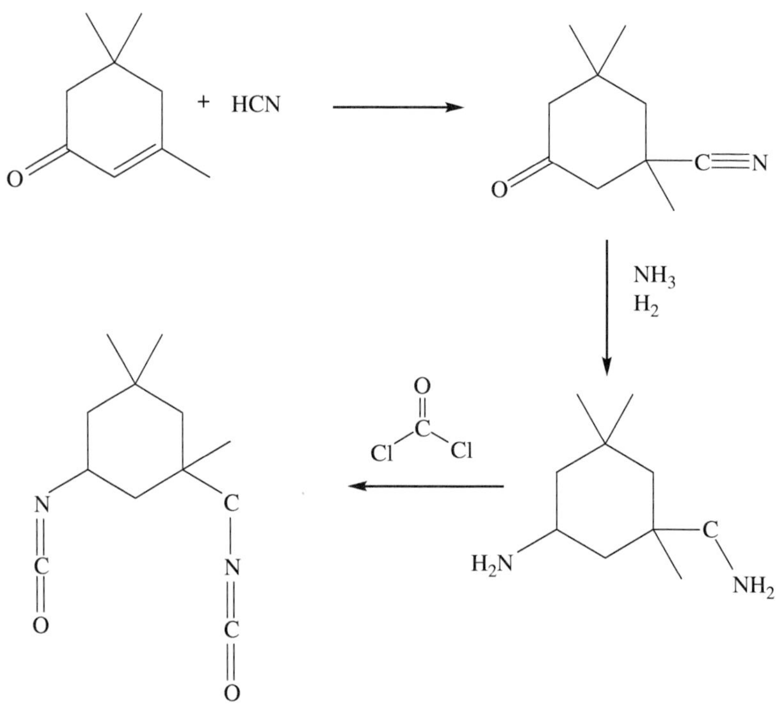


**FIGURE 2.80** Conversion of isophorone to isophorone nitrile, hydrogenation to isophorone diamine, and subsequent phosgenation to isophorone diisocyanate.

isocyanate with phosgene [160]. IPDI is formed in *cis-/trans-isomers* of the reactive centers believed to have equal reactivities. The *nonequivalent isocyanate locations* make for differential reactivity, with the secondary isocyanate being more reactive as a result of steric factors. This reactivity difference is a valued attribute of IPDI since it allows for controlled and lower viscosity prepolymers (Section 2.1.6).

*4,4′-Diisocyanatodicyclohexylmethane ($H_{12}MDI$)* Unlike HDI and IPDI, $H_{12}MDI$ derives from the urethane industrial base – the aniline/formaldehyde process to MDA. The economics and price sensitivities of $H_{12}MDI$ therefore mirror those of MDI and its feedstocks. The process to make $H_{12}MDI$ follows that of MDI up until the phosgenation step. At that point a pure 4,4′ MDA cut is distilled from the crude pMDA and hydrogenated [162].

Along with the predicted and desired reaction shown in Figure 2.81, other reactions can occur. One of the most common is deamination resulting in a monoamine. At lower temperatures and pressures incomplete conversion to aliphatic structures is most common, with one ring aliphatic and the other aromatic. This tendency for incomplete hydrogenation is a significant roadblock to expansion of the hydrogenation to the crude pMDI stream as produced. The hydrogenation is done in alcohol or dioxane medium using rhodium or ruthenium on an alumina catalyst [163, 164]. The use of more common hydrogenation catalysts will often entail the requirement of higher temperatures and the concomitant deamination.

***2.2.3.2 Use of Aliphatic Isocyanates*** The primary uses of aliphatic isocyanates are in coatings applications (see Chapter 10) because of their preferred weathering performance relative to aromatic building blocks [165–167]. However, along with weathering, coatings often must also provide other properties such as optical clarity, a high $T_g$, and a barrier to a variety of chemicals the coating may encounter in use. The ability of isocyanates to form cross-linking structures by self-reaction is made use of in this instance [168, 169]. The mechanism of these cross-linking reactions is covered in detail in Chapter 3. The most common cross-linking applied to aliphatic isocyanates and particularly for coatings applications is their processing to isocyanurates (Figure 2.82) [170–172].

Unlike dimer formation, the trimer is a nonthermoreversible structure. Trimer formation is highly exothermic with a heat of reaction estimated to be –42 kcal/mol of trimer. Since trimerization of difunctional isocyanates leaves free isocyanates to participate in further trimerization, the conditions of the reaction have to be controlled to limit oligomerization. Along with providing increased cross-link density,

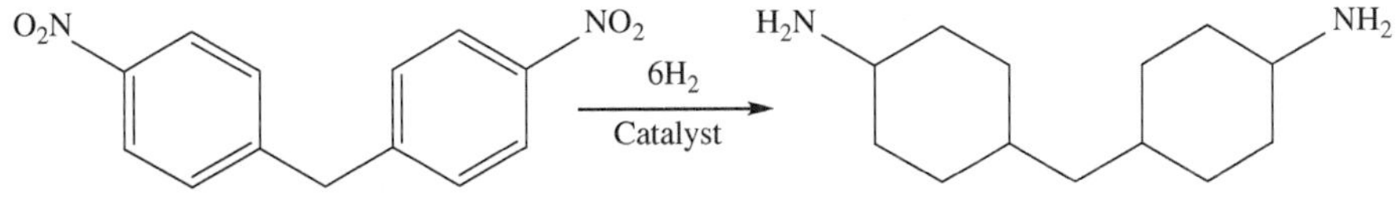


**FIGURE 2.81** Conversion of MDA to hydrogenated MDA.

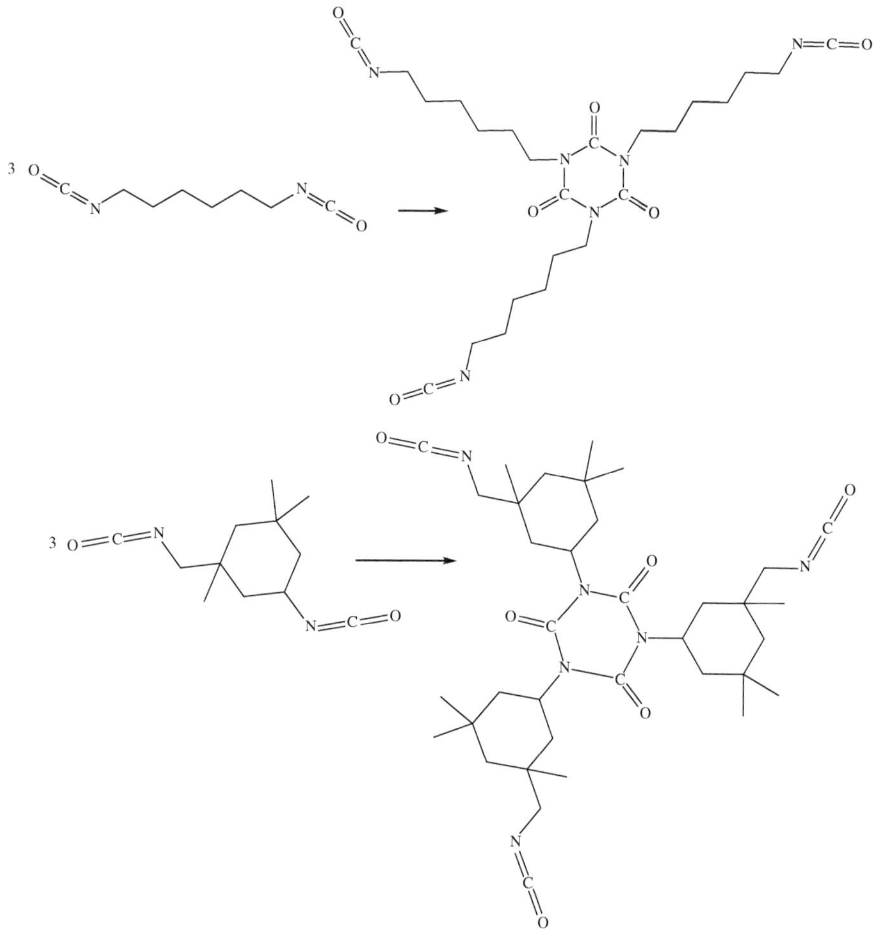


**FIGURE 2.82** Trimerization of aliphatic isocyanates commonly preceding their use in coatings formulations.

trimerization enhances flame retardancy [173], thermal resistance [174], chemical resistance [175], and film forming [176], and improves worker safety [177] by reducing isocyanate vapor pressure. By virtue of the cross-link density, polyurethane phase separation processes are inhibited, and desirable amorphous coatings are promoted.

Distinguishing the monomers based on structures is difficult because HDI is seldom used as a monomer; it is usually used in the trimer form, so its application is substantially cyclic. IPDI is used both as a monomer and in its trimer form, and $H_{12}$MDI is used almost exclusively in its monomer form. As such there is a significant overlap in the resulting polyurethane properties, and customers will often use one or the other based on preference. It is also quite common to mix aliphatic isocyanates for formulation optimization [178]. Forming of mixed trimers has also been reported. In this case, the mixture randomization is achieved by staged addition of isocyanates to compensate for the differences in isocyanate reactivity (e.g. IPDI is somewhat more reactive than HDI).

Although the great majority of aliphatic isocyanates are produced via phosgene chemistry, the higher prices commanded by aliphatic isocyanates has allowed nonphosgene routes to diisocyanates to realize industrial reality. Conversion of isophorone diamine (IPDA) to IPDI via reaction with dimethyl carbonate to form the biscarbamate followed by thermolysis to the isocyanate and methanol (see Section "Thermolysis of Carbamic Acid,*N*,*N*′-(4-Methyl-1,3-Phenylene)Bis-,*C*,*C*′-Dimethyl Ester made from the Reaction of Toluene Diamine with Methyl Carbonate") has been practiced commercially by Daicel Chemical (Japan). Evonik has produced IPDI by reaction of IPDA with urea and alcohol to form the diisocyanates and ammonia by-product (Figure 2.83). The mechanism of this route is potentially complex, with

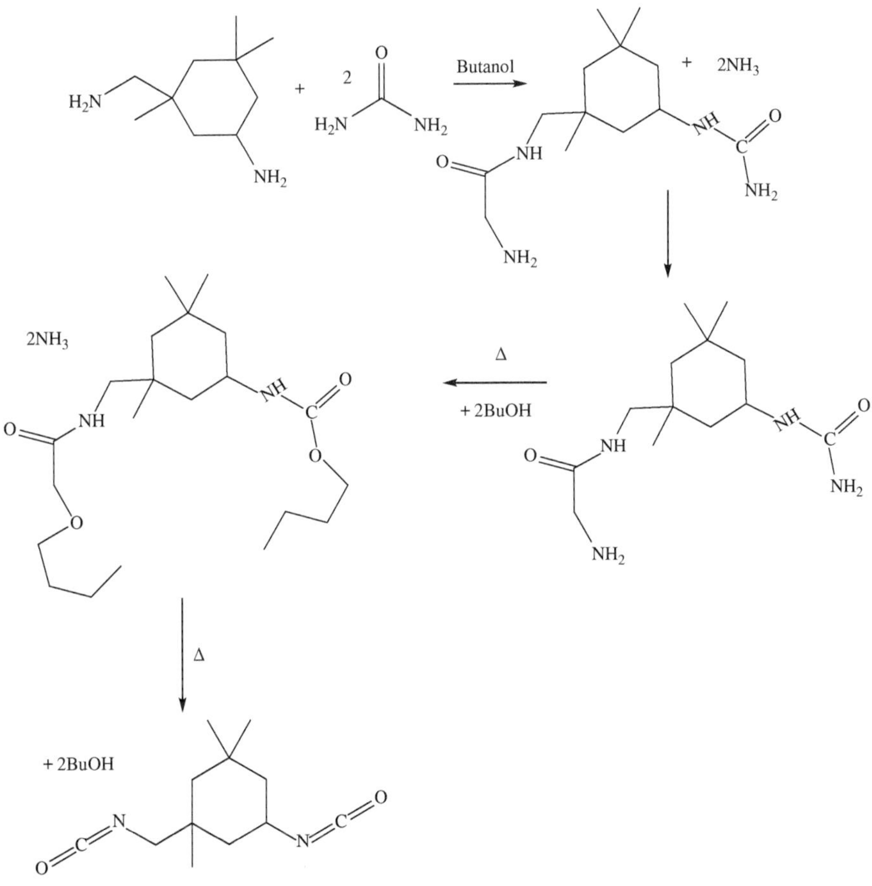


**FIGURE 2.83** Conversion of isophorone diamine to isophorone diisocyanates using urea and aliphatic alcohol reagents to form products. Reaction pathways are speculative.

many possible routes to go from products to reactants. Problems associated with this route have been relatively slow rates at low temperatures and the formation of polyureas at high temperatures. A series of patents document procedures for efficiently producing high yields of the desired diisocyanates [179–181]. Figure 2.83 shows the reaction pathways of reactants to products [182].

An additional route involves the formation of a carbamate by reaction of urea with an alcohol (see Chapter 12) and substitution of the target amine for the unsubstituted carbamate.

## 2.3 CHAIN EXTENDERS

For any given polyurethane, soft segments and isocyanates may make up more than 95% of the polymer volume. Very often a low-molecular-weight component is added that primarily serves as a spacer between neighboring polyisocyanates. Thus, the polymer structure will be, for instance, soft segments of 1000 or 2000 molecular weight and a hard segment consisting of diisocyanates of 250 g/mol (as in the case of 4,4′ MDI) strung together with a short-chain diol or diamine of about 100 g/mol. The low-molecular-weight diol or diamine stringing together diisocyanates is called the "chain extender." Figure 2.84 is a representation of this concept with an ethylene glycol chain extender. These chains of polyisocyanate and chain extender (ethylene glycol here) are termed the "hard segment." The end isocyanate units are implicitly connected to high-molecular-weight polyether or polyester soft-segment polyols. The importance of this structural detail is that, as the addition polymerization occurs, these hard segments and their immiscibility with the soft segment drive the phase structure of the resulting polyurethane (see Chapter 4) [183]. The chain extender then has a significant effect on the kinetics and extent of phase separation in these systems. The chain extender can also have a significant effect on the strength of the hard-segment phase (similar to a crystal-phase melting point) by virtue of the interphase hard-segment hydrogen bonding (Figure 2.85) [184].

Thus, the chain extender structure can have a substantial effect on polyurethane properties by its ability to drive phase separation, its ability to complement or interfere with a regular hard-segment structure, and its ability to promote interhard-segment hydrogen bonding [185–189]. Virtually any low-molecular-weight polyfunctional molecule capable of reacting with isocyanates has been explored for its influence on polyurethane properties. The most prevalent and industrially important chain extender in foams is water, which makes urea hard segments, and 1,4-butanediol, which is used commonly for elastomer hard segments. The urea units can be particularly robust owing to the presence of the bidentate hydrogen bonding available to the interchain structure, as seen in Figure 2.86. While the most common

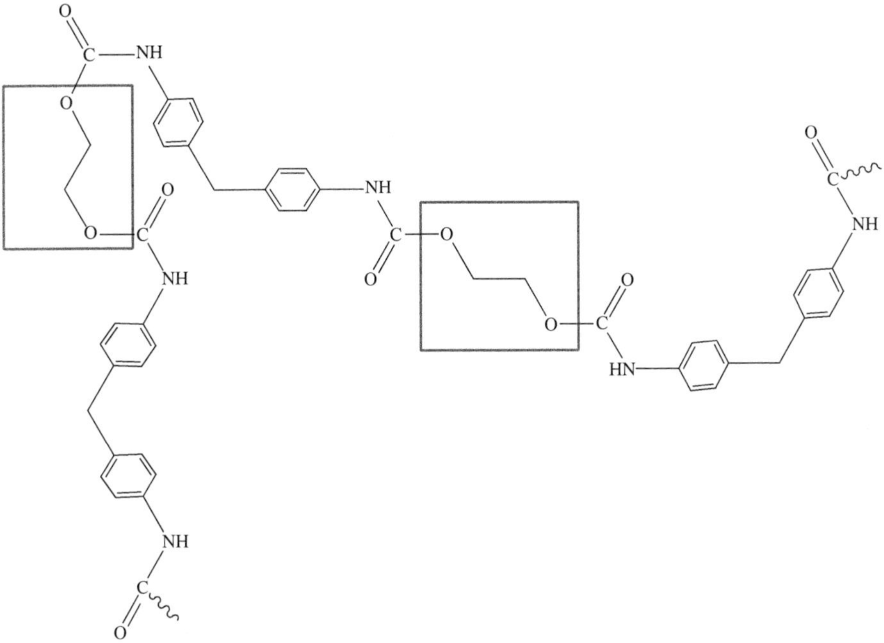


**FIGURE 2.84** Ethylene glycol chain extender (in boxes) linking isocyanates through urethane bonds.

source of urea hard segments is the reaction of isocyanate and water, an alternative source is the reaction of isocyanates and amines. Chain extenders with primary amines have two active hydrogens on each nitrogen, doubling the equivalence of each amine function. The first reaction of a primary amine is to a urea, which subsequently may undergo a slower isocyanate reaction to a biuret. Thus, each primary amine offers a urea reaction product that can enhance hard-segment stability and thermoset formation, but also a cross-link point militating against ordered hard-segment formation and against melt processability. Usually, just the first primary amine reaction is counted in considering chain extender chemistry. Hydroxyl functional chain extenders also serve as cross-linkers. In contrast, secondary amines offer a single isocyanate reaction site and the possibility of maintaining melt processability.

There are a large number of potential chain extenders that can be employed to alter hard-segment structure, polymerization kinetics, and polymer properties. Table 2.13 is a limited list sampling the numerous types and structural parameters of each. The potential effects on properties is covered in Chapter 4 and chapters associated with polymer properties.

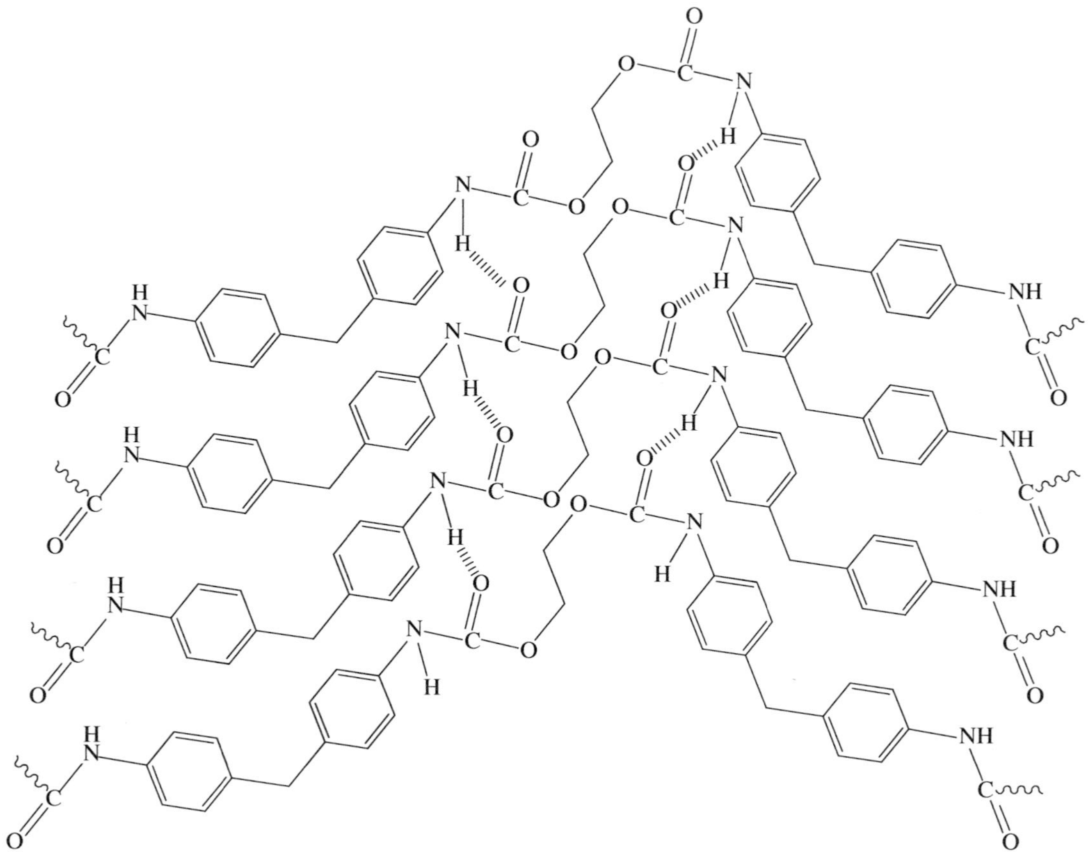

**FIGURE 2.85** Hydrogen bonding between chain-extended urethane hard segments. Cohesive energy between hard-segment chains is influenced by the chain extender via the spacing between H-bonding interactions and by the regularity of the assembled chains into quasi-crystalline structures. Irregularities in hard-segment structure can lead to a more nematic-like liquid-crystal structure.

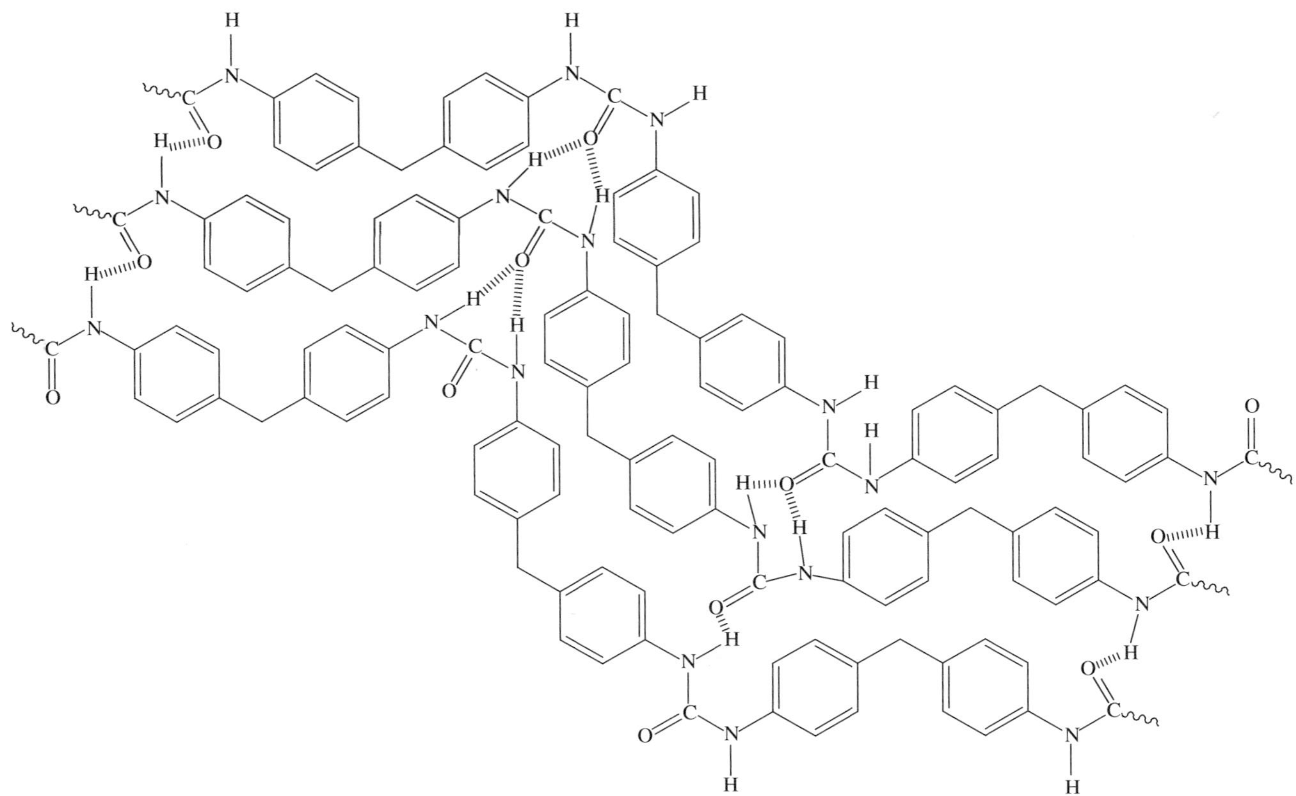

**FIGURE 2.86** Illustration of the origin of the greater stabilization of hard segments afforded by urea chain extension versus the urethane chain extension shown in Figure 2.85.

**TABLE 2.13 Structure and properties of common chain extenders used in urethane chemistry**

| Chain extender/cross-linker | Structure | Mol weight (g/mol) | Equivalence weight (g/eq) | OH/NH number |
|---|---|---|---|---|
| Water | 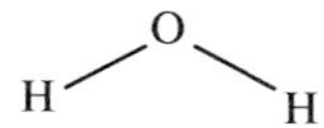 | 18 | 9 | 6233 |
| Ethylene glycol |  | 62 | 31 | 1808 |
| 1,2-propylene glycol | 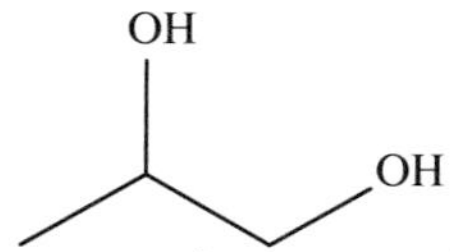 | 76 | 38 | 1475 |
| 1,3-propylene glycol | 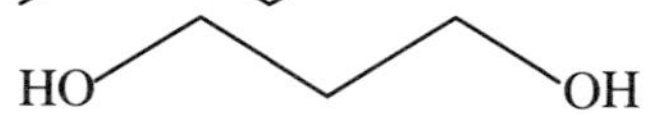 | 76 | 38 | 1475 |
| Diethylene glycol | 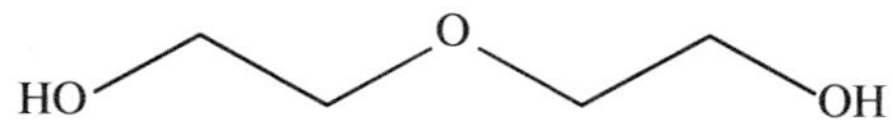 | 106 | 54 | 1057 |
| Neopentyl glycol |  | 104 | 52 | 1078 |
| Butanediol |  | 90 | 45 | 1244 |
| Hexanediol | 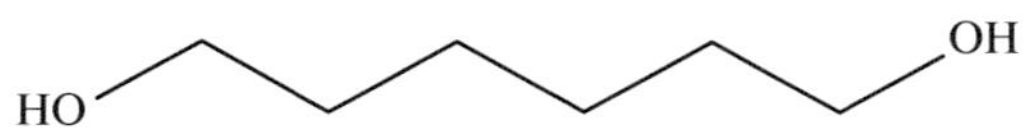 | 118 | 59 | 951 |
| Cyclohexane dimethanol | 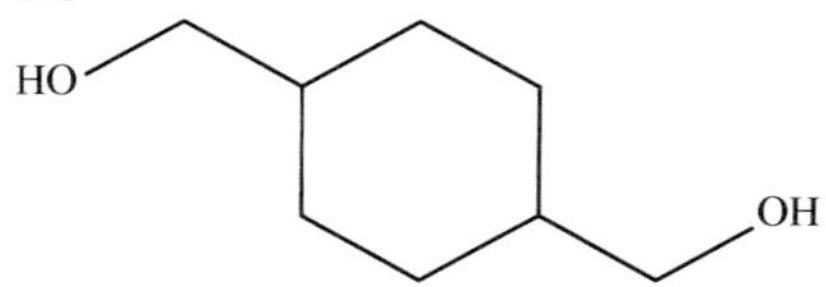 | 144 | 72 | 779 |

| | | | | |
|---|---|---|---|---|
| Hydroquinone bis(2-hydroxy ethyl)ether |  | 198 | 99 | 567 |
| Resorcinol bis(2-hydroxyethyl)ether | 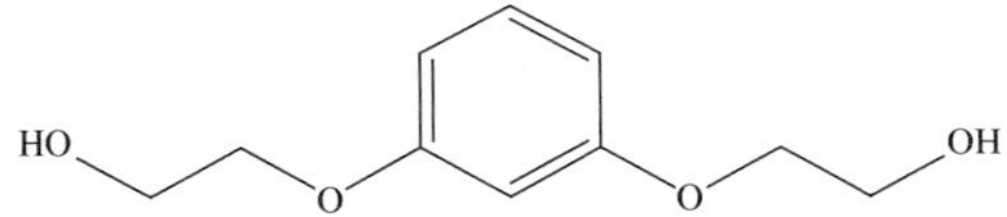 | 198 | 99 | 567 |
| Bisphenol A bis(2hydroxyethyl)ether | 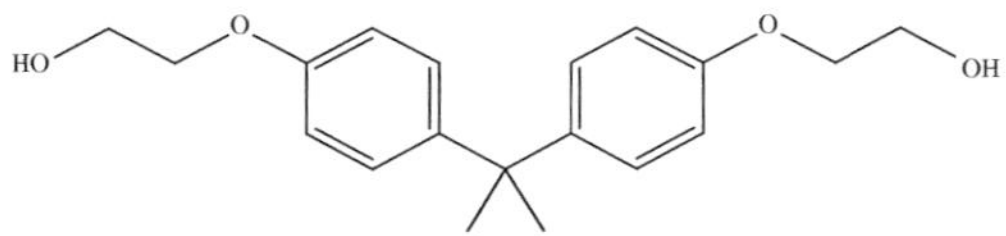 | 316 | 158 | 355 |
| Bisphenol A bis(2,3-dihydroxypropyl)ether | 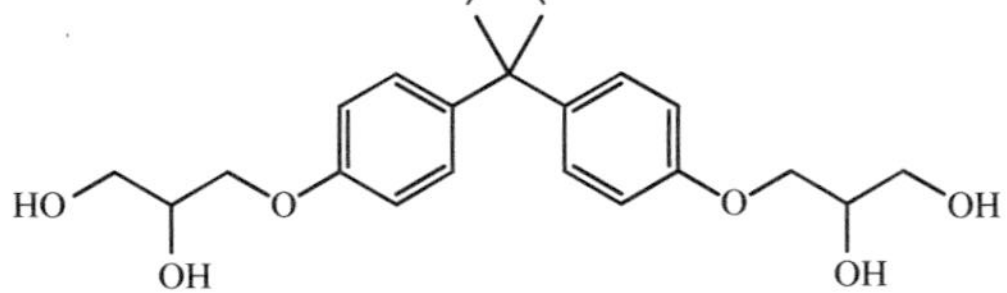 | 376 | 94 | 597 |
| 4,4′-Methylenebis(2-chloroaniline) (MOCA) | 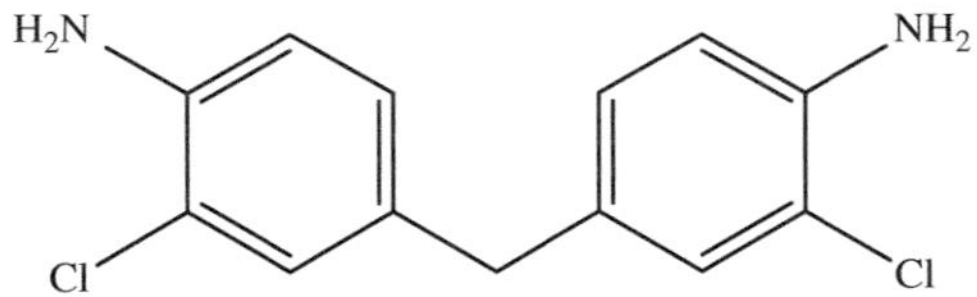 | 267 | 133.5 | 837 |
| Glycerine | 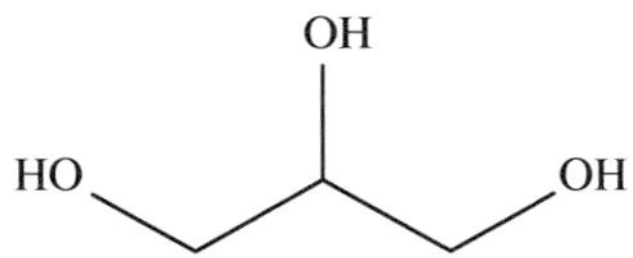 | 92 | 31 | 1828 |

(*Continued*)

**TABLE 2.13 (Continued)**

| Chain extender/cross-linker | Structure | Mol weight (g/mol) | Equivalence weight (g/eq) | OH/NH number |
|---|---|---|---|---|
| Trimethylolpropane | 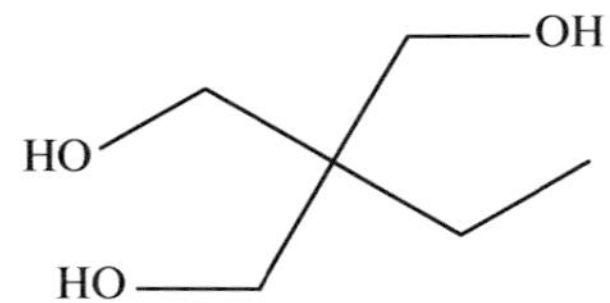 | 134 | 45 | 1247 |
| 1,6-Hexane diamine | 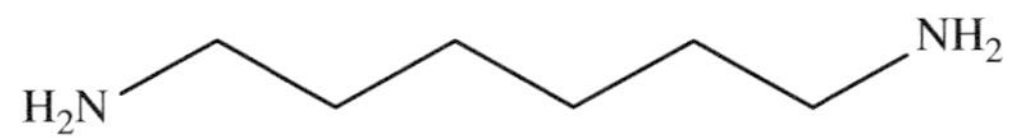 | 116 | 58 | 967 |
| 1,3-Diamino pentane | 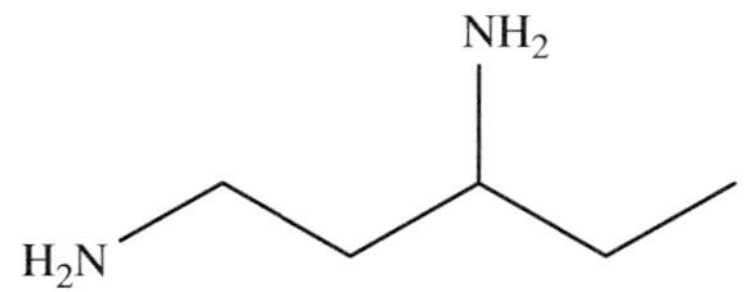 | 102 | 51 | 1100 |
| Diethanolamine | 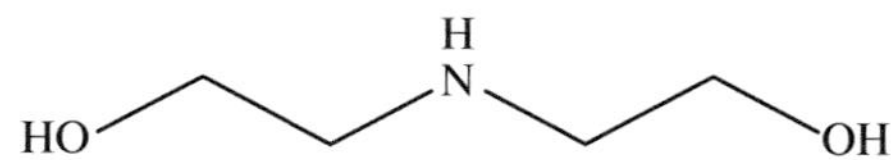 | 105 | 35 | 1602 |
| Triethanolamine | 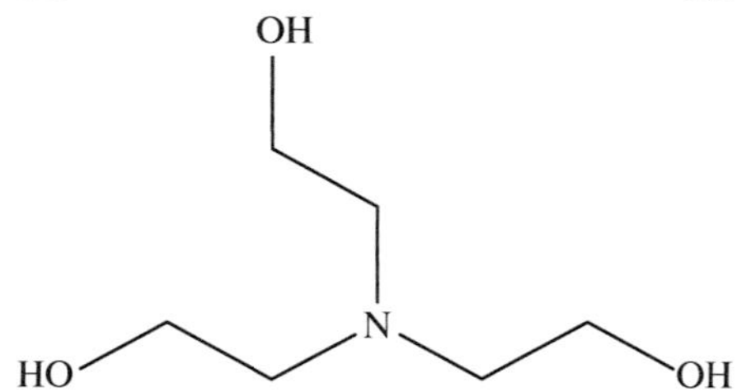 | 149 | 50 | 1122 |

Primary amines may react twice, once to form a urea and subsequently to form a biuret.

## REFERENCES


[1] O Figovsky, L Shapovalov, and O Axenov, *Surface Coatings International Part B*, **87**, (2004), 83–90.

[2] A Cornille, R Auvergne, O Figovsky, B Boutevin, and S Caillol, *European Polymer Journal*, **87**, (2017), 535–552.

[3] M Ghasemlou, F Daver, E Ivanova, and B Adhikari, *European Polymer Journal*, **118**, (2019), 1668–1684

[4] M Yoneyama, E Ormond, H Chinn, U Loechner, and X Xu, *Polyether Polyols for Polyurethanes*, Chemical Economics Handbook, IHS Markit, London, UK, 2018.

[5] M Yoneyama, B Kauer, U Lochner, and X Xu, *Polyester Polyols*, Chemical Economics Handbook, IHS Markit, London, UK, 2019.

[6] T Kumamoto, A Bland, E Greiner, and E Zhang, *Polytetramethylene Glycol*, Chemical Economics Handbook, IHS Markit, London, UK, 2019.

[7] D Nest, T Chung, J Vegh, D Graves, R Bruce, T Lin, R Phaneuf, G Oehrlein, B Long, and C Wilson, *Journal of Physics D: Applied Physics*, **43**, (2010), 1–7.

[8] C Beyler, *Xiaofang Kexue Yu Jishu*, **26**, (2007), 476–484.

[9] M Hirschler, *Polymers for Advanced Technologies*, **19**, (2008), 521–529.

[10] M Hirschler and T Earl, "Is upholstered furniture a flammable solid?," in *Proceedings of the Annual Conference on Advanced Flame Retardancy of Polymeric Materials*, Volume **19**, Mill Valley, CA, 2008, 326–338.

[11] T Rosendahl, T Maeurer, C Dobner, and J Zuehlke, Catalyst for the production of ethylene oxide. Assignee BASF SE. WO 2012140617. 2012.

[12] M Verhaak, Process for the production of ethylene oxide by oxidation of ethylene produced by oxydehydrogenation of ethane. Assignee Shell International. WO 2012101069. 2012.

[13] M Ghanta, D Fahey, D Busch, and B Subramanian, *Sustainable Chemical Engineering* **1**, (2013), 268–277.

[14] S Horstman, H Gardeler, K Fischer, F Koester, and J Gmehling, *Journal of Chemical and Engineering Data*, **46**, (2001), 337–345.

[15] G Blanco-Breva, P deFrutos-Escrig, H Martin, J Campos-Martin, and J Fierro, *Catalysis Today*, **187**, (2012), 168–172.

[16] E Beckman, *Green Chemistry*, **5**, (2003), 332–336.

[17] L Candela, R Cochran, and S Sandler, Ethylbenzene oxidation. Assignee Arco Chemical. US patent 4950794. 1990.

[18] J Yamomoto and J Tsuji, *Studies in Surface Science and Catalysis*, **172**, (2007), 553–554.

[19] P Dimroth and H Pasedach, *Angewandte Chemie*, **72**, (1960), 865.

[20] D Zhang, H Yin, J Xue, C Ge, T Jiang, L Yu, and Y Shen, *Industrial Engineering and Chemistry Research*, **48**, (2009), 11220–11224.

[21] N Pullan, T Liu, and P Topham, *Polymer Chemistry*, **4**, (2013), 2272–2277.

[22] S Schunk, C Baltes, and A Sundermann, *Catalysis Today*, **117**, (2006), 304–310.

[23] U Hong, W Park, J Lee, S Hwang, J Yi, and I Song, *Applied Catalysis A: General*, **415–416**, (2012), 141–148.

[24] T Uryu, H Shimazu, and K Matsuzaki, *Journal of Polymer Science: Polymer Letters Edition*, **11**, (1973), 275–278.

[25] V Sutinska, M Pajtasova, S Lalikova, A Ferjancova, J Palieskova, and S Mojumdar, *Journal of Thermal and Analytical Calorimetry*, **104**, (2011), 923–927.
[26] E Steiner, R Pelletier, and R Trucks, *Journal of the American Chemical Society*, **86**, (1964), 4678–4786.
[27] R Herrington, L Nafziger, K Hock, R Moore, F Casati, and W Lidy, "Basic chemistry," in *Flexible Polyurethane Foams*, 2nd Ed., R Herrington and K Hock, Eds., The Dow Chemical Co., Midland, MI, 1997, 2.8.
[28] *Standard Test Methods for Testing Polyurethane Raw Materials: Determination of Hydroxyl Numbers of Polyols*, ASTM D4274, American Society for Testing Materials, West Conshocken, PA, 2011.
[29] *Standard Test Method for Polyurethane Raw Materials: Determination of Unsaturation of Polyols*, ASTM D4671, American Society for Testing Materials, West Conshocken, PA, 2011.
[30] A deLucas, L Rodgrigues, M Perez-Collado, and P Sanchez, *Polymer International*, **51**, (2002), 1041–1046.
[31] G Wegener, M Brandt, L Duda, J Hofmann, B Klesczewski, D Koch, R Kumpf, H Orzesek, and H Pirkl, *Applied Catalysis, A: General*, **221**, (2001), 303–335.
[32] K Haider and J Pazos, Polyether polyols produced by base-catalyzed alkoxylation in the presence of polyethylene oxide containing compounds. Assignee Bayer MaterialScience. US patent 20070149631. 2007.
[33] I Kim, S Byun, and C Ha, *Journal of Polymer Science, Part A: Polymer Chemistry*, **43**, (2005), 4393–4404.
[34] I Lee, J Ha, C Cao, D Park, C Ha, and I Kim, *Catalysis Today*, **148**, (2009), 389–397.
[35] L Wu, A Yu, M Zhang, B Liu, and L Chen, *Journal of Applied Polymer Science*, **92**, (2004), 1302–1309.
[36] X Zhang, Z Hua, S Chen, F Liu, X Sun, and G Qi, *Applied Catalysis, A: General*, **325**, (2007), 91–98.
[37] M Sonnenschein and W Koonce, "Polyurethanes," in *Encyclopedia of Polymer Technology*, H Mark, Ed., Wiley, Hoboken, NJ, 2012, 1–63.
[38] K Lorenz and J Hofmann, Method for producing polyether polyols by alkoxylation with intermediate neutralization of initial alkali catalysts. Assignee Bayer MaterialScience. WO 2012084762. 2012.
[39] B Klescewski, J Hofmann, K Lorenz, J Reese, and J Pazos, Process for the production of polyether polyols with a high ethylene oxide content. Assignee Bayer MaterialScience. US patent 20110230581. 2011.
[40] M O'Sickey, *Characterization of Structure–Property Relationships of Poly(urethane-urea)s for Fiber Applications*, PhD Thesis, Virginia Technological and State University, (2002).
[41] (a) T Miyajima, K Nishiyama, M Satake, and T Tsuji, *Polymer Journal*, **47**, (2015), 771–778.(b) A Raghuraman, W Heaner, D Babb, and M Paradkar, Polyether–acetal polyol compositions. Assignee Dow Global Technologies. WO 20180273676. 2018.(c) D Chakraborty, A Rodriguez, and E Chen, *Macromolecules*, **36**, (2003), 5470–5481.
[42] S Davis, T Kalin, and T Kumamoto, *Polytetramethylene Ether Glycol*, Chemical Economics Handbook, Market Research Report 688.4000A, IHS Markit, London, UK, 2010.

[43] P Dreyfuss, *Polytetrahydrofuran*, Gordon and Breach, New York, NY, 1982.
[44] D Hollingsworth and J Knifton, Polyether glycol production by polymerization of tetrahydrofuran. Assignee Texaco Chemical. US patent 5262562. 1993.
[45] E Dominguez-Rosado, J Liggat, C Snape, B Eling, and J Pichtel, *Polymer Degradation and Stability*, **78**, (2002), 1–5.
[46] A Duda, and S Penczek, "Mechanisms of aliphatic polyester formation," in *Biopolymers*, Volume **3**, Y Doi and A Steinbuchel, Eds., Wiley-VCH, Weinheim, Germany, 2001, 371–429.
[47] K Weissermel and H Arpe, *Industrial Organic Chemistry*, 4th Ed., Wiley-VCH, Weinheim, Germany, 2003.
[48] A Chauvel and G Lefebvre, *Petrochemical Processes*, Volume **2**, Editions Technip, Paris, France, 1989.
[49] A Rudin and P Choi, *The Elements of Polymer Science and Engineering*, 3rd Ed., Elsevier, Waltham, MA, 2013.
[50] M Sonnenschein, S Guillaudeu, B Landes, and B Wendt, *Polymer*, **51**, (2010), 3685–3692.
[51] K Saunders, *Organic Polymer Chemistry*, 2nd Ed., Chapman and Hall, London, UK, 1988, ch. 16.
[52] A Albertsson and I Varma, *Biomacromolecules*, **4**, (2003), 1466–1486.
[53] K Ding, M Miranda, B Moscato-Goodpaster, N Ajellal, L Breyfogle, E Hermes, C Schaller, S Roe, C Cramer, M Hillmyer, et al., *Macromolecules*, **45**, (2012), 5387–5396.
[54] I Vitkauskiene, R Makuska, U Stirna, and U Cabulis, *Journal of Cellular Plastics*, **47**, (2011), 467–482.
[55] *Polyurethane Chemicals and Products in North and Central America: Rigid Foam Products*, IAL Consultants, London, UK, 2016.
[56] C Cecchini, R Zannetti, and A Stefani, *Journal of Cellular Plastics*, **35**, (1999), 514–530.
[57] E Dominguez-Rosado, J Liggat, C Snape, B Eling, and J Pichtel, *Polymer Degradation and Stability*, **78**, (2002), 1–5.
[58] M Ionescu, X Wan, N Bilic, and Z Petrovic, *Journal of Polymers and the Environment*, **20**, (2012), 647–658.
[59] M Bomberg and J Lstiburek, *Spray Polyurethane Foam in External Envelopes of Buildings*, CRC Press, Boca Raton, FL, 1998.
[60] P Painter and M Coleman, *Essentials of Polymer Science and Engineering*, DEStech, Lancaster, PA, 2009.
[61] P Golini and M Guandalini, Polyurethane rigid foams and formulations and foam preparation. Assignee Dow Global Technologies. WO 2013030101. 2013.
[62] S Crain and J Surma, Spray polyurethane foam for non-covered and non-coated use in structural application. Assignee Dow Global Technologies. US patent 20110086217. 2011.
[63] Q Zhu, C Zhang, S Feng, and J Chen, *Journal of Applied Polymer Science*, **83**, (2002), 1617–1624.
[64] S Hamilton, J Semlyen, and D Haddleton, *Polymer*, **39**, (1998), 3241–3252.
[65] E Foy, J Farrell, and C Higginbotham, *Journal of Applied Polymer Science*, **111**, (2009), 217–227.
[66] S Hofacker, *European Coatings Journal*, **4**, (2006), 22–25.

[67] K Hsu, Y Hsiao, and I Chien, *Industrial & Engineering Chemistry Research*, **49**, (2010), 735–749.

[68] R Francesconi, *International DATA Series, Selected Data on Mixtures, Series A: Thermodynamic Properties of Non-Reacting Binary Systems of Organic Substances*, **26**, (1998), 111.

[69] H Buysch, H Krimm, and H Rudolph, Separation of dimethylcarbonate from its azeotrope with methanol. Assignee Bayer AG. DE 2737265 A1 1. 1979.

[70] J Kuyper, P Lednor, and G Pogany, Process for the preparation of polycarbonates. Assignee Shell International. US patent 4826952. 1987.

[71] S Allen, G Coates, A Cherian, C Simoneau, A Gridnev, and J Farmer, Preparation of polycarbonate polyols from epoxides and carbon dioxide. Assignee Novamer Inc. US patent 8604155. 2010.

[72] D Darensbourg and S Wilson, *Green Chemistry*, **14**, (2012), 2665–2671.

[73] V Gite, P Mahulikar, and DG Hundiwale, *Progress in Organic Coatings*, **68**, (2010), 307–312.

[74] A Kabra and P Mahanwar, *Paintindia*, **58**, (2008), 85–96.

[75] P Vandevoorde and A van Gaans, *European Coatings Journal*, **9**, (2005), 22–29.

[76] E Linak, A Kishi, M Guan, and U Bucholz, *Urethane Surface Coatings*, Chemical Economics Handbook, IHS Markit, London, UK, 2019.

[77] E Linak, A Kishi, M Guanm, and Y Bucholz, *Acrylic Surface Coatings*, Chemical Economics Handbook, Market Research Report, IHS Markit, London, U.K 2018.

[78] P Hoffmann and A Tye, Aqueous, hydroxyl functional acrylic copolymerisate dispersions for two component polyurethane coating compositions. Assignee BASF SE. WO 2012131050 A1. 2012.

[79] A Buijs, J Paulus, W Karzijn, and N Venne, Process for preparing a polymer polyol, by mixing a base polyol, ethylenically unsatd. monomers and a polymn. initiator in a reactor vessel and polymerizing the mixture. Assignee Shell International. US patent 8530610. 2011.

[80] R Subramanian, K Best, C Gambino, C Mebane, M Jeffries, M Shaffer, and M Dvorchak, Direct-to-metal radiation-curable nonaqueous compositions and their application to metal substrates. Assignee Bayer MaterialScience. US patent 20090176907 A1. 2009.

[81] S Guelcher, J England, and R Adkins, Polymer polyols with improved particle properties and a filtering process for their production. Assignee Bayer MaterialScience. EP 2072555. 2009.

[82] K Spitler and J Lindsey, *Journal of Cellular Plastics*, **17**, (1981), 43–50.

[83] P Cookson, R Borella, D Hoehnener, and F Casati, Fine particle, high concentration, polyisocyanate polyaddition/polyurethane-urea polyols. Assignee Dow Global Technologies. WO 2012154831. 2012.

[84] W Kuryla, F Critchfield, L Platt, and P Stamberger, *Journal of Cellular Plastics*, **2**, (1966), 84–96.

[85] R Adkins and S Chauk, Low viscosity polymeric polyols containing vinyl polymers for use in production of polyurethanes. Assignee Bayer MaterialScience. US patent 8383733. 2007.

[86] H Van der Wal, R Dawe, J-M Sonney, and F Casati, Copolymer polyols and a process for the production thereof. Assignee Dow Global Technologies. WO 2008005708. 2008.

[87] R Borella, S Claussens, J-P Masy, and S DeLoos, Macromer and process for making polymer polyols. Assignee Dow Global Technologies. US patent 9994701 B2. 2018.

[88] Y Yen and T Tsao, *SRI Process Economics Program Report 45B Polyols for Polyurethanes*, IHS Group, Englewood, CO, 1995, 25.

[89] M Koshute, Stable dispersions and their use in the preparation of polyurethane foams. Assignee Miles Inc. US patent 5332763. 1994.

[90] G Verhelst and A Bruyninckx, PIPA polyol dispersions and their production. Assignee Huntsman ICI Chemicals. WO 2000073364. 2000.

[91] P Cookson, F Casati, R Borella, D Hochener, and I Amici-Kroutilova, Polyisocyanate polyaddition manufacturing process and product. Assignee Dow Global Technologies. US patent 9840578 B2. 2017.

[92] F Gunstone, "Fatty acids and lipid structure," in *Lipid Technologies and Applications*, F Gunstone and F Padley, Eds., Marcel Dekker Inc., New York, NY, 1997, ch. 1.

[93] H Caupin, "Products from castor oil: past present and future," in *Lipid Technologies and Applications*, F Gunstone and F Padley, Eds., Marcel Dekker Inc., New York, NY, 1997, 787–796.

[94] F Firdaus, *Journal of Chemistry and Chemical Engineering*, **6**, (2012), 96–100.

[95] L Zhang, H Jeon, J Malsam, R Herrington, and C Macosko, *Polymer*, **48** (2007), 6656–6667.

[96] S Narine, X Kong, L Bouzidi, and P Sporns, *Journal of the American Oil Chemists' Society*, **84**, (2007), 65–72.

[97] P Tran, D Graiver, and R Narayan, *Journal of the American Oil Chemists' Society*, **82**, (2005), 653–659.

[98] M Sonnenschein, M Greaves, A Sanders, Z Lysenko, G Spilman, G Frycek, J Phillips, A Schrock, and S Martin, Capped polyester polyol lubricant composition. Assignee Dow Global Technologies. US patent 201000087350. 2008.

[99] M Sonnenschein, V Ginzburg, K Schiller, and B Wendt, *Polymer*, **54**, (2013), 1350–1360.

[100] M Sonnenschein and B Wendt, *Polymer*, **54** (2013), 2511–2520.

[101] G Bohlman, *Process Economics Program Biobased Polyols Report 217*, IHS Group, Englewood, CO, 2007.

[102] Y Schrodi, R Pederson, H Kaido, and M Tupy, Synthesis of terminal alkenes from internal alkenes via olefin metathesis. Assignee Materia Inc. US patent 8501973. 2008.

[103] A Chatterjee, S Eliasson, and V Jensen, *Catalysis Science and Technology*, **8**, (2018), 1487–1499.

[104] M von Czapiewski, O Kreye, H Mutlu, and M Meier, *European Journal of Lipid Science and Technology*, **115**, (2013), 76–85.

[105] Y He, X Zhang, X Zhang, H Huang, J Chang, and H Chen, *Journal of Applied Polymer Science*, **127**, (2013), 2176–2183.

[106] A Daniel-da-Silva, J Bordado, and J Martin-Martinez, *Journal of Applied Polymer Science*, **107**, (2008), 700–709.

[107] R Fuest, "Polyurethane elastomers," in *Rubber Technology*, JS Dick, Ed., Hanser, Cincinnati, OH, 2001, 238–263.

[108] H Janik, M Debowski, and A Balas, "Microscopy in quality control of different polyurethanes," in *The International Polyurethane Industry Conference UTECH 2000*, March 28–30, The Hague, Netherlands, 2000, F4/1–F4/3.

[109] E Delebecq, J-P Pascault, JB Boutevin, and F Ganachaud, *Chemical Reviews*, **113**, (2013), 80–118.

[110] G Raspoet and G Nguyen, *Journal of Organic Chemistry*, **63**, (1998), 6867–6877.
[111] PRWeb News Center, Global Foamed Plastics (Polyurethane) Market to Reach 9.6 Million Tons by 2015, Global Industry Analysts Inc., San Jose, CA. Vocus PRWeb Information available at: http://www.prweb.com/releases/foamed_plastics/polyurethanes/prweb8128151.htm (accessed February 11, 2011).
[112] H Chinn, M Yoneyama, U Lochner, and X Xu, *Diisocyanates and Polyisocyanates*, Chemical Economics Handbook, Marketing Research Report, IHS Markit, London, UK, 2018.
[113] J Miron, A Brief, S Epstein, and C Robinson, *Polyurethanes VI, Skeist Incorporated*, Market Research Report, Skeist Publishers, Whippany, NJ 2006.
[114] J Greenwood and Z Hercowitz, *Investment, Capacity Utilization and the Real Business Cycle*, Institute for International Economic Studies, Paper No. 357, University of Stockholm, Sweden, June 1986, 1–27.
[115] *Polyurethane Chemicals – Raw Materials*, IAL Consultants, London, UK, 2016.
[116] M Monconduit and C Carr, *Toluene*, Chemical Economics Handbook, IHS Markit, London, UK, 2019.
[117] J Baldyga, E Molga, S Szarlik, W Wojcik, P Machniewski, L Rudniak, S Piechota, J Lachmiajer, L Ruczynski, L Wojcik, et al., Producing toluene diisocyanate (tdi) in the process of the toluene diamine (tda) phosgenation reaction in the gaseous phase. Assignee Zaklady Chemisczne Zachem Poland. EP 2463272. 2012.
[118] W Lorenz, M Boehm, R Adamson, F Steffens, and A Hielscher, Process for the preparation of isocyanates. Assignee Bayer MaterialScience. US patent 7645900. 2009.
[119] R Smith, *TDI Process Update Toluene*, PEP Review, IHS Markit, London, UK, 2017.
[120] H van den Berg, L van der Ham, H Gutierrez, S Odu, T Roelofs, and J de Weerdt, *Chemical Engineering Journal*, **207–208**, (2012), 254–257.
[121] R Russell, R Switzer, and R Longmore, *Journal of Chemical Education*, **67**, (1990), 68–69.
[122] H Beckhaus, J Munnig, and D Grenner, Process for purifying and concentrating spent sulfuric acid. Assignee Bayer AG. US patent 6156288. 2000.
[123] A Villanti and G Ravetta, A continuous process and catalysts for the dinitration of aromatic substrates. Assignee Finchimica S.p.A. CA 2221811. 1998.
[124] G Neri, G Rizzo, C Milone, S Galvagno, M Musolino, and G Capannelli, *Applied Catalysis, A: General*, **249**, (2003), 303–311.
[125] A Raichle, T Coelho, S Neto, U Penzel, S Oehlenschlaeger, M Zoellinger, H Braunsberg, S Haase, and J Buettner, Process for preparing toluenediamine by hydrogenation of dinitrotoluene in the presence of catalysts. Assignee BASF SE. US patent 20110295039. 2011.
[126] T Mattke and G Olbert, Process for preparing isocyanates by reacting amines with phosgene in reaction plant. Assignee BASF SE. US patent 20130060062. 2013.
[127] H Ulrich, R Richter, P Whitman, A Sayigh, and W Rabourn, *Journal of Organic Chemistry*, **39**, (1974), 2897–2899.
[128] H Bruemmer, M Eichmann, V Haverkamp, B Sojka, J Laue, and J Sanders, Gas-phase phosgenation process. Assignee Bayer MaterialScience. US patent 8692016. 2007.
[129] J Meyn and H Stutz, Phosgenation process and tubular reactor for the preparation of tri- and diisocyanates. Assignee Bayer MaterialScience. US patent 6930199. 2005.
[130] P Pavlechko, *Toluene Diisocyanate From Toluene Diamine by Bayer Gas Phase Phosgenation Preliminary Evaluation*, Process Economics Program, SRI Consulting, IHS Group, Englewood, CO, 2008.

[131] Y Pei, H Li, H Liu, and Y Zhang, *Catalysis Today*, **148**, (2009), 373–377.
[132] F Rivetti, *Comptes Rendus de l'Academie des Sciences, Serie IIc: Chimie*, **3**, (2000), 497–503.
[133] W Hardy and R Bennett, *Tetrahedron Letters*, **9**, (1967), 961–962.
[134] Q Yang, A Robertson, and H Alper, *Organic Letters*, **10**, (2008), 5079–5082.
[135] M Gasperini, F Ragaini, C Cazzaniga, and S Cenini, *Advanced Synthesis and Catalysis*, **347**, (2005), 105–120.
[136] A Tafesh and J Weiguny, *Chemical Reviews*, **96**, (1996), 2035–2052.
[137] F Paul, *Coordination Chemistry Reviews*, **203**, (2000), 269–323.
[138] E Scriven and K Turnbull, *Chemical Reviews*, **88**, (1988), 297–368.
[139] S Brase and K Banert, *Organic Azides: Syntheses and Applications*, Wiley, Chichester, UK, 2010, 322.
[140] M Pampell, *Aniline*, Chemical Economics Handbook, Marketing Research Report, IHS Markit, London, UK, 2019.
[141] P Sriram, M Alvarado, V Baghdian, and X Ma, *Formaldehyde*, Chemical Economics Handbook, Marketing Research Report, IHS Markit, London, UK, 2017.
[142] H Schelling, U Penzel, E Stroefer, M Eiermann, J Spier, K Thiele, and M Bock, Process for the production of isocyanates, preferably diisocyanates and polyisocyanates with solvent recirculation. Assignee BASF SE. US patent 8759569B2. 2014.
[143] Y Pei, H Li, H Liu, and Y Zhang, *Catalysis Today*, **148**, (2009), 373–377.
[144] G Verardo, S Cauci, and A Giumanini, *Journal of the Chemical Society, Chemical Communications*, **24**, (1985), 1787–1788.
[145] C Wang, H Li, L Wang, Y Cao, H Liu, and Y Zhang, *Chinese Chemical Letters*, **23**, (2012), 1254–1258.
[146] C Mitchell, C Corma, R Carr, and A Botella, Process for production of methylene-bridged polyphenyl polyamines. Assignee Huntsman LLC. WO 2010072504. 2010.
[147] R Smith and R Carr, Process for production of polyisocyanates with low impurities. Assignee Huntsman LLC. WO 2006022641. 2006.
[148] H Schelling, H-J Pallasch, E Stroefer, J Speier, and B Kim, Process for the preparation of light-colored isocyanates of the diphenylmethanediisocyanate series. Assignee BASF SE. WO 2011006970. 2011.
[149] A da Silva, J Martin-Martinez, and B Moura, *International Journal of Adhesion and Adhesives*, **28**, (2007), 29–37.
[150] M Szycher, *M. Szycher's Handbook of Polyurethanes*, 2nd Ed., CRC Press, Boca Raton, FL, 2013, 98.
[151] H Ulrich, *Cumulenes in Click Reactions*, Wiley, Chichester, UK, 2009, 82.
[152] C Hoyle, Y No, K Malone, S Thames, and D Creed, *Macromolecules*, **21**, (1988), 2727–2730.
[153] D Thorpe, "Isocyanates," in *The Polyurethane Book*, D Randall and S Lee, Eds., Huntsman International/Wiley, Everberg/Hoboken, NJ, 2002.
[154] C Hoyle, Y No, and G Nelson, *Polymer Preprints (American Chemical Society, Division of Polymer Chemistry)*, **28**, (1987), 415–416.
[155] C Hoyle and K Kim, *Journal of Polymer Science, Part A: Polymer Chemistry*, **25**, (1987), 2631–2642.
[156] M Monconduit, B Merril, and M Woo, *Hexamethylene diamine/adiponitrile*, Chemical Economics Handbook, Product Review, IHS Markit, London, UK, 2019.

[157] R Daussin, P Cameron, L Chen, J Jimenez, and J Gehrke, Process for production of aliphatic isocyanates. by rapid dispersion of primary amines or their salts in phosgene used as partial solvent. Assignee Dow Global Technologies. US patent 20110275854. 2011.
[158] Y Liu, K Sun, H Ma, X Xu, and X Wang, *Catalysis Communications*, **11**, (2010), 880–883.
[159] J Thunberg and W Begonis, Process for the preparation of 3-cyano-3,5,5-trimethylcyclohexanone by addition of hydrogen cyanide to isophorone. Assignee W.R. Grace. US patent 5011968. 1991.
[160] D Thorpe and R Smith, Eur. Pat. Appl. Process for making aliphatic and cycloaliphatic polyisocyanates. Assignee ICI Inc. EP 327231. 1989.
[161] R Teissier and G Martino-Gauchi, Continuous process for the preparation of 3,5,5-trimethylcyclohexa-3-en-1-one, β-isophorone. Assignee Atofina. US patent 6274772. 2000.
[162] E Powers, Diisocyanate for the preparation of polyurethane plastics. Assignee Mobay Chem Co. US patent 3590002. 1969.
[163] R Yamamoto, Y Hirai, A Takagi, and Z Tashima, Alicyclic isocyanates. Assignee Mitsui Toatsu Chem Inc. US patent 4299766. 1981.
[164] T Nagata, M Wada, and H Mizuta, Preparation of aliphatic isocyanates by reaction of aliphatic amines with phosgene in esters as solvents. Assignee Mitsui Toatsu Chem, Inc. US patent 5136086. 1992.
[165] B Erdem, D Bhattacharjee, and L-J Chen, "A detailed understanding of polyurethane dispersions, their process and applications," in *Proceedings of the 31st International Waterborne, High-Solids, and Powder Coatings Symposium*, February 18–20, Hattiesburg, MS, 2004, 497–513.
[166] S Grace, H Mertes, F Richter, and M Sonntag, "Isocyanates and isomerism – new developments in aliphatic polyisocyanate research," in *Proceedings of the 27th International Waterborne, High-Solids, and Powder Coatings Symposium*, March 1–3, Hattiesburg, MS, 2000, 169–184.
[167] R Saxon, R Dexter, and G Hewitt, *Cellular Polymers*, **4**, (1985), 17–49.
[168] W Slack and H Kemp, Trimer catalyst system for aliphatic and aromatic isocyanates and isocyanurate trimer preparation. Assignee Bayer Corp. US patent 5955609. 1999.
[169] H Sehovic, A Sendijarevic, V Sendijarevic, and K Frisch, *Journal of Polymer Science, Part A: Polymer Chemistry*, **27**, (1989), 1287–1299.
[170] N Nodelman, D Steppan, and M Super, RIM elastomers based on isocyanurates of isophorone diisocyanate and prepolymers thereof. Assignee Bayer MaterialScience. US patent 20070142601. 2007.
[171] F Moghaddam, M Dekamin, and G Koozehgari, *Letters in Organic Chemistry*, **2**, (2005), 734–738.
[172] H Schaefer and H Binder, PCT Int. Appl. Method for producing colorless, isocyanurate group-containing polyisocyanates of (cyclo)aliphatic diisocyanates. Assignee BASF SE. US patent 8373004. 2008.
[173] C Hilado and E Olcomendy, *Journal of Thermal Insulation*, **2**, (1979), 87–89.
[174] M Pavlova and M Draganova, *Polymer Communications*, **32**, (1991), 443–445.
[175] K Best, C McClurg, and L Boretzky, Two-component water-reducible chemical-resistant polyurethane coating. Assignee Miles Inc. CA 2107957. 1994.
[176] A Kaminski and M Urban, *Journal of Coatings Technology*, **69**, (1997), 113–121.

[177] S Allmaras, *Applied Occupational and Environmental Hygiene*, **18**, (2003), 151–153.

[178] J Argypropoulas, A Xu, D Bhattacharjee, A Yang, D Schall, and L Luo, Two-component polyurethane coating compositions comprising isocyanurate compositions from bis (isocyanatomethyl) cyclohexane and from aliphatic diisocyanates. Assignee Dow Global Technologies. US patent 20120130016. 2012.

[179] G Bohmholdt, J Disteldorf, P Kirchner, and H Michalczak, Recycle process for the preparation of (cyclo)aliphatic diisocyanates from their corresponding diamines. Assignee Huels AG. US patent 5087739. 1992.

[180] A Otterbach, H Schwarz, F Merger, W Schwarz, E Brandt, P Magnussen, and O Mattner, Multistep continuous preparation of organic polyisocyanates. Assignee BASF AG. US patent 5386053. 1995.

[181] G Bohmholdt, W Heitmann, P Kirchner, and H Michalczak, Continuous, multistep preparation of (cyclo)aliphatic diisocyanates. Assignee Huels AG. US patent 5360931. 1994.

[182] R Smith, Alkyl carbamates from urea and aliphatic alcohols. Assignee West Point Pepperell. US patent 4443622. 1984.

[183] W Li, A Ryan, and I Meier, *Macromolecules*, **35**, (2002), 5034–5042.

[184] M Sonnenschein, N Rondan, B Wendt, and J Cox, *Journal of Polymer Science Part A: Polymer Chemistry*, **42**, (2004), 271–278.

[185] YV Savelyev, ER Akhranovich, AP Grekov, EG Privalko, VV Korskanov, VI Shtompel, VP Privalko, P Pissis, A Kanapitsas, *Polymer*, **39**, (1998), 3425–3429.

[186] J Bae, D Chung, J An, and D Shin, *Journal of Materials Science*, **34**, (1999), 2523–2527.

[187] H Han, F Wei, and J Brown, *Journal of Testing and Evaluation*, **39**, (2011), 463–467.

[188] S Oprea, *Journal of Materials Science*, **46**, (2011), 2251–2258.

[189] K Gisselfaelt and B Helgee, *Macromolecular Materials and Engineering*, **288**, (2003), 265–271.

# 3

# INTRODUCTION TO POLYURETHANE CHEMISTRY

## 3.1 INTRODUCTION

In principle this chapter should be a simple matter of discussing means of achieving the urethane structure (Figure 3.1); however, given the reactivity of polyurethane chemical building blocks and the kinetics of the urethane-forming reaction, many other structures form and these can have significant effects on the final materials. In some cases the experimental approach is to control variables to a level that optimizes the final product and its structure. In other cases the reaction is allowed to follow its course and find equilibrium, with the object of simply achieving a desired level of reproducibility. This chapter will cover the mechanism and catalysis to polyurethane structure, the side reactions that can take place, and the results these variances can have on the final product. Also covered are the means of minimizing some routes and maximizing others and the properties achieved in such control. Only isocyanate-based reactions will be covered in this chapter; other routes to urethane utilizing cyclic carbonates and aminal reactions will be covered in Chapter 12.

Although isocyanates are very reactive species, the observed room temperature reaction of isocyanate with hydroxyl is often rather slow. To some extent, the slow speed is a reflection of phase incompatibility of the relatively nonpolar and denser isocyanate phase and the relatively polar and less dense polyol phase (hydroxyl components). However, even when phase compatibilized with a surfactant, the reaction is slow at room temperature. In general, when simply allowed to sit in each other's

*Polyurethanes: Science, Technology, Markets, and Trends*, Second Edition. Mark F. Sonnenschein.

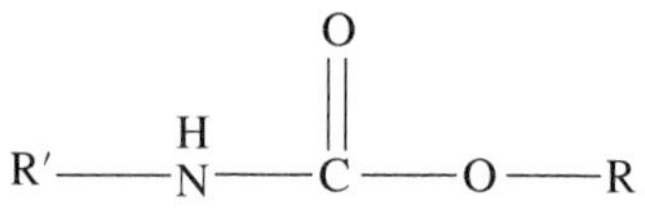

Urethane

**FIGURE 3.1** Urethane structure.

presence a hydroxyl unit and an isocyanate will slowly form urethane or urea (in the case of reaction with water) at the interface, forming a crystalline barrier, which slows the observed reaction rate further. For this reason it is commonplace and industrially essential to utilize catalysts and provide some method of phase mixing to enhance the reaction rate between hydroxyls and isocyanates. Even in the presence of catalysts, isocyanate reactions can show sharp temperature threshold behavior. Observations such as these have created numerous theories about the action of catalysts in these systems. In the absence of a catalyst it is believed that urethane formation proceeds through a six-member ring transition state [1, 2] (Figure 3.2). The simplest transition state would be a four-member ring, unmediated by an additional third body (i.e. second alcohol) [3, 4]. Several points argue against this. The first is that Fourier transform infrared (FTIR) analysis shows that polyols are predominantly in a hydrogen-bonded state, suggesting that the availability of lone OH groups may be a limiting factor. Figure 3.3 shows the FTIR spectrum for a 1000 equivalent weight commercial polyol. The spectrum focuses on the relevant OH stretch region. The breadth of the peak centered at 3500 $cm^{-1}$ is indicative of the extensive hydrogen bonding these molecules engage in. An OH group that was uninvolved in intermolecular interactions would reveal itself at about 3650 $cm^{-1}$ [5] and would increase its intensity as a function of temperature. The absence or very slight population indicated by the peak intensity and the temperature independence of the response (25–85 °C) indicate very little free OH under these conditions. Additionally, quantum mechanical calculations predict that the expected rate of reaction in the absence of bond polarization (as shown in Figure 3.2) is a much slower reaction than observed. However, when mediating influences such as shown in Figure 3.2 are introduced, the observed and calculated reaction rates are in good agreement [6].

Along with self-associations there are a large number of structural and experimental factors that can influence the rate of reaction. One of the main factors is the isocyanate structure since it was first shown that phenyl isocyanate was nearly a factor of 50 more reactive with an alcohol than ethyl isocyanate with the same alcohol [7–9]. This has inevitable consequences for the application of aliphatic isocyanates in weatherable formulations. In addition, it has been proposed that the reaction rate might depend on the dielectric strength of a solvent environment. The rate of reaction can decrease with increasing solvent dielectric strength [10–12]; it can also be strongly affected by the solvent's ability to associate with hydroxyls through polar or hydrogen-bonding interactions [13, 14], thereby decreasing the availability of hydroxyl functionality to react with the isocyanate.

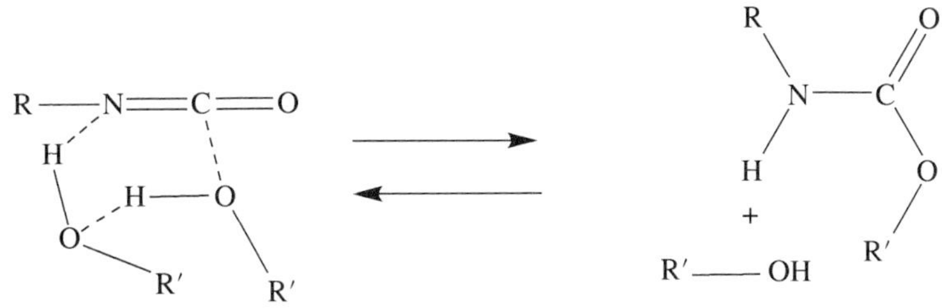


**FIGURE 3.2** Generally accepted transition state for the uncatalyzed formation of urethane.

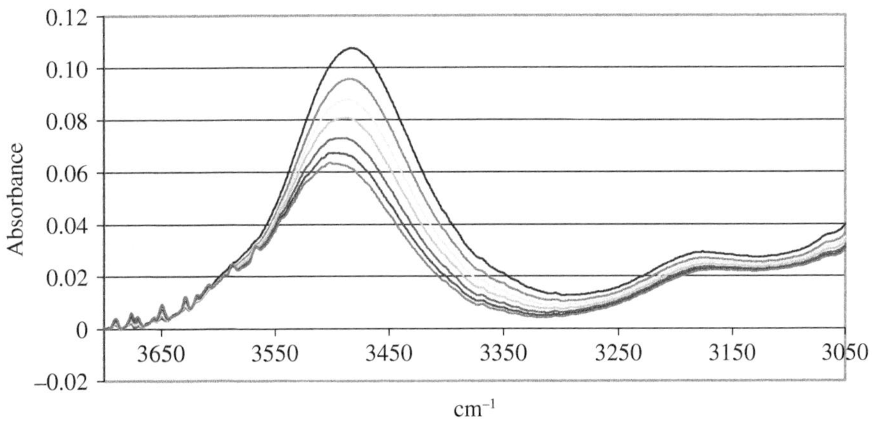


**FIGURE 3.3** FTIR spectroscopy of a polyetherpolyol focusing on the OH stretch region. The nested spectra are the evolution of the signal from 25 °C (top curve) to 85 °C (bottom curve) in 10 °C increments.

## 3.2 MECHANISM AND CATALYSIS OF URETHANE FORMATION

There are a large number of molecules and unique structures that catalyze urethane reactions. Examples are shown in Figure 3.4, along with their commercial names. Many others are available as well [15]. The mechanism of their action is a subject of debate. The structures in Figure 3.4 can be conveniently broken into Lewis acid transition metals and tertiary amines. The mechanisms of action of the two classes of catalyst are postulated to proceed by different routes. Basic models usually have the catalyst polarizing either the hydroxyl or the isocyanate moieties via simple polar interactions (Figure 3.5). One can expect that catalyst removal of electron density from the oxygen of the NCO group will in turn enhance the electrophilic (electron deficient) character of the isocyanate carbon. This effect not only will increase the rate of urethane formation consistent with the structure of Figure 3.2 but may also influence the myriad side reactions isocyanates undergo, a topic that will be covered at greater length later in this

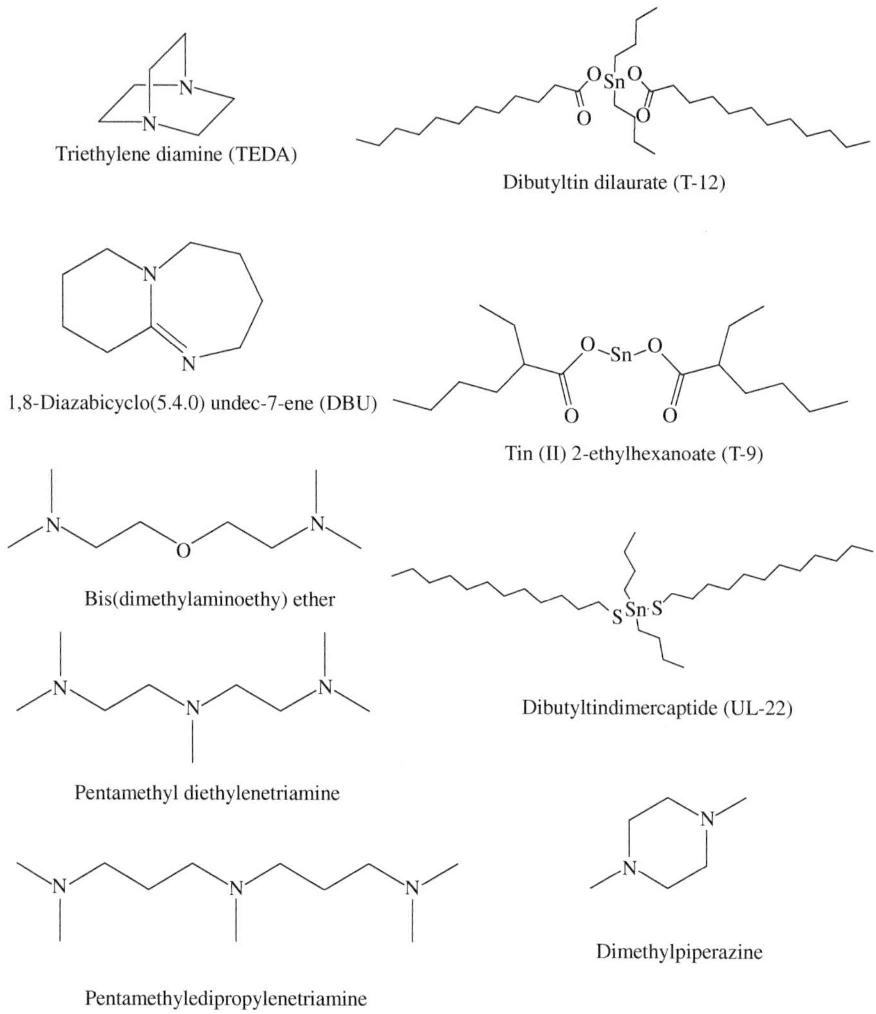


**FIGURE 3.4** Common catalysts for urethane and urea formation. Trade names are in parentheses.

chapter. Alternative mechanisms of catalyst action have also been proposed [15–20]. In many of these studies the mechanism of action was probed by controlling catalyst concentration or catalyst activity (e.g. by varying acidity). These mechanisms can sometimes be quite elaborate and still be compelling based on presented evidence. A mechanism of Lewis acid metal catalyst activity has been presented by several independent workers (Figure 3.6) [16, 17]. This mechanism has a unique attribute in that its initial assumption is based on the metal's ability to increase its complexation number and increase reactivity of the associated ligands. It also presents opportunities to probe its validity by varying structures and concentrations of the reactants to see if the

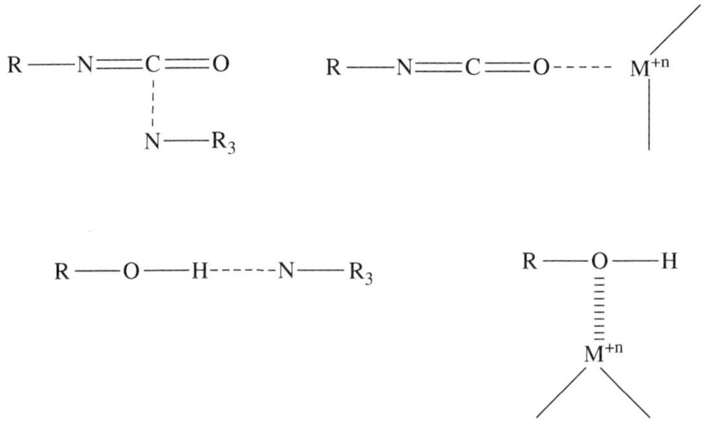


**FIGURE 3.5** Examples of bond polarization mechanisms of catalyst action.

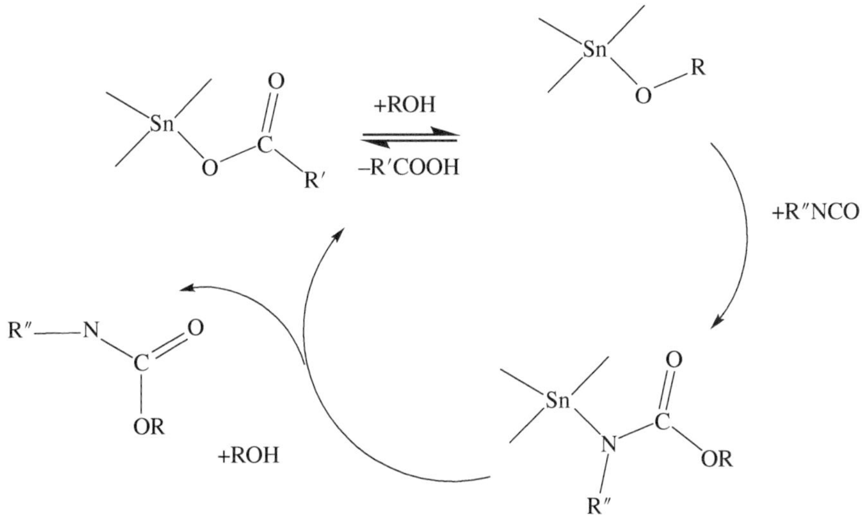


**FIGURE 3.6** Proposed mechanism of Lewis acid catalysis of urethane formation.

expected influence is realized. This deductive approach has had mixed mechanistic success since the kinetics have been observed independent of the ligand acid strength; it is also independent of the presence of total system carboxyl or carboxylate strength. Thus the validity of the mechanism rests on the formation of the alkoxide form of the catalyst. The thermodynamics of this transformation have been evaluated and nuclear magnetic resonance data based on significant downfield OH proton shifts and line broadening in the presence of bismuth carboxylate catalysts suggest that the alcohol is both coordinated and undergoing proton exchange in the presence of catalyst. Such data suggest the mechanism illustrated in Figure 3.7 to explain this result [18].

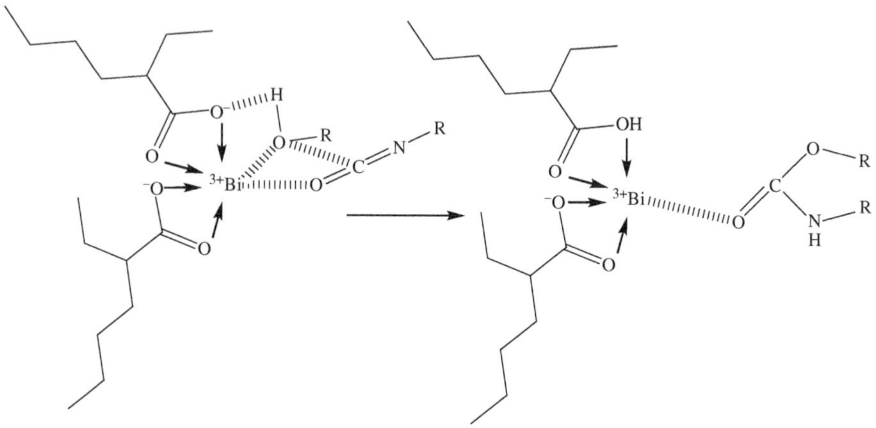


**FIGURE 3.7** Proposed mechanism of Bismuth carboxylate urethane catalysis. Coordination of the metal is simplified for illustrative purposes.

Because of the difficulties associated with (i) making highly controlled measurements, (ii) the catalytic (albeit weak) action of the product urethane groups formed, (iii) the difficulty quantifying the effects of varying associations of reactants and products affecting the rate constants, and (iv) the overall thermal sensitivity of the reactions and the catalysis, it is likely that a defined Lewis acid catalyst mechanism enjoying universal endorsement will remain elusive.

The mechanism of tertiary amine catalyst action is also unsettled and further complicated by the tendency of tertiary amine catalysts to drive reactions other than urethane formation [21–24]. In fact almost all tertiary amines are capable of promoting the formation of urethane, urea (reaction of water with isocyanate), and numerous cross-linking side reactions of isocyanates with reaction products. The catalytic activity of tertiary amines to form urethane bonds is conventionally referred to as the "gel" reaction since, in polyurethane chemistry, the urethane reaction is most tied to the rapid growth of molecular weight and concomitant increase in viscoelasticity, resulting in gelation. The catalyzed reaction of isocyanates with water is referred to as the "blow" reaction because of the formation of carbon dioxide, which contributes to the formation of foam or frothed morphologies [25]. As shown in Figure 3.5, the ability of catalytic tertiary amines to interact strongly with both hydroxyl and isocyanate groups accounts for their versatility in polyurethane chemistry. Amines can also be designed to preferentially interact with water, accounting for the ability of some amines to drive the blow reaction with much higher efficiency than the gel reaction (see Chapter 6) (Figure 3.8).

It is also easy to visualize how a polyamine catalyst can interact simultaneously with a water (or polyol hydroxyl) and an isocyanate in a manner to rapidly promote their reaction with one another; however, it is difficult to predict a priori what the behavior of a given amine will be based on measurables such as basicity or even steric hindrance [24, 26, 27]. Once an amine is measured for overall activity and blow/gel reaction ratio, the connection is usually then asserted.

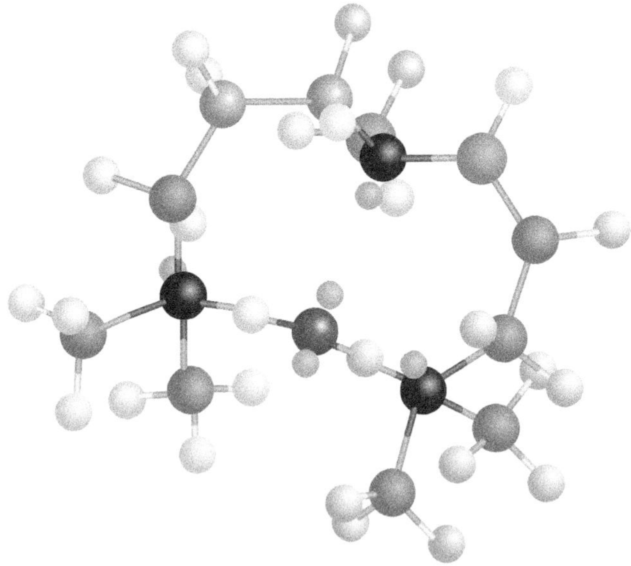

**FIGURE 3.8** Energy-minimized interaction between water and pentamethylene dipropylenetriamine demonstrating the favorable water interactions that result in good blow catalyst properties (light red surfaces indicate nonbonding electron pairs; red is oxygen and blue is nitrogen). (*See insert for color representation of the figure.*)

At the same time that amine catalysts are able to promote both blow and gel reactions, they are also capable of promoting numerous side reactions. Most of these side reactions involve the reaction of isocyanates with already formed reaction products, and their efficiency for any given side reaction will depend on isocyanate concentration, coreactant concentration, temperature, catalyst concentration, etc. Isocyanate side reactions will be covered in greater detail later in this chapter. One isocyanate reaction that has great industrial relevance, and uses wholly different catalysts than amines or heavy metals, is the trimerization of isocyanates to isocyanurates (Figure 3.9). Isocyanurates are particularly useful in the manufacture of construction foams because of their contribution to network cross-link formation, their thermal resistance, and their relative inflammability.

While some amines can catalyze isocyanate trimerization, the temperature at which they are efficient can be high enough that many other reactions are also operating. In general, excellent trimerization catalysts are highly basic and nucleophilic [28–30]. One hypothesis is that they operate through a stepwise addition beginning with nucleophilic attack of the catalyst on the carbon of a first isocyanate. This increases the nucleophilicity of the neighboring nitrogen to then react with a carbon on a second isocyanate and culminates with the rapid ring closure to the six-member isocyanurate ring.

The final step of the sequence is irreversible and drives the final product formation. Recent work has obtained nearly quantitative conversion at low temperatures using carbenes [31] and proazaphosphatrane catalysts [32]. The mechanism of Figure 3.10 suggests that the reaction should be first order in the catalyst; this has in

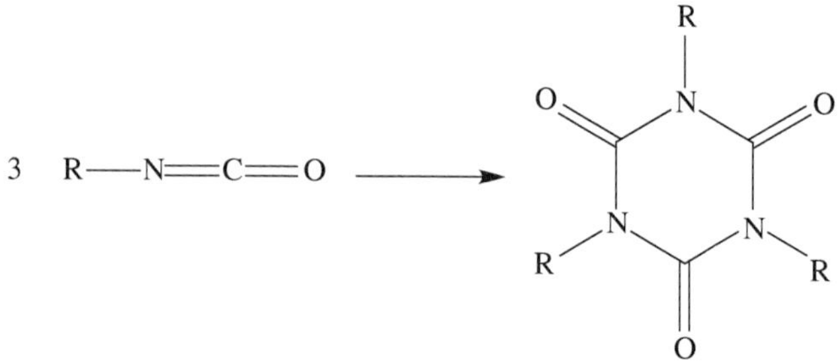


Isocyanurate

**FIGURE 3.9** Trimerization of isocyanate to isocyanurate.

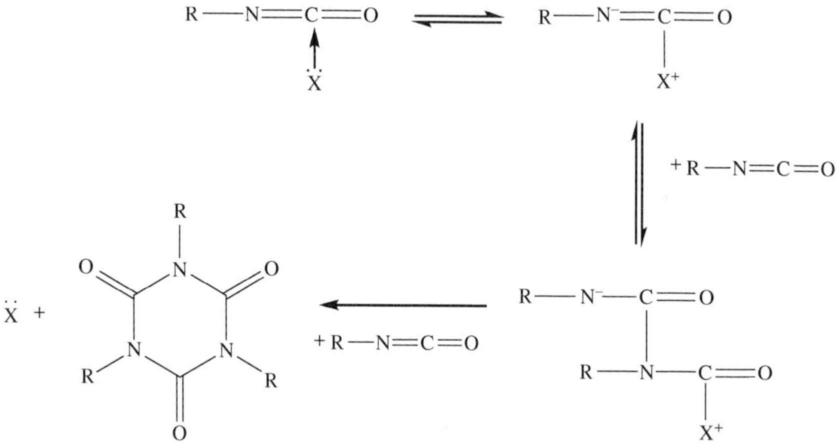


**FIGURE 3.10** Proposed mechanism of isocyanurate formation using highly nucleophilic catalysts.

fact been observed in several studies of trimerization kinetics, though reports of kinetics exhibiting second order in the catalyst have also been published. The heat of reaction is measured to be –42 kcal/mol trimer formed with a low activation energy consistent with a third-power dependence on isocyanate concentration, a linear dependence on the catalyst, and only a weak dependence on temperature. Many industrial processes use carboxylate salts and specific amines because of their availability and cost, and compensate for their relative inadequacies through process optimization. Other six-member ring structures can be formed by unintended side reactions of carbodiimides and will be covered later in this chapter.

## 3.3 REACTIONS OF ISOCYANATES WITH ACTIVE HYDROGEN COMPOUNDS

Polyurethanes are the product of mixing isocyanates with hydroxyls to form the structure depicted in Figure 3.1. As shown in the previous section, this is a relatively straightforward addition reaction; however, side reactions can occur that have profound effects on polyurethane properties. There is no condensation elimination product in the formation of polyurethane, and the product is in a strongly temperature-dependent equilibrium with reactants (isocyanates and hydroxyl compounds) [33, 34]. Literature references report a heat of formation ranging from −16 to −34 kcal/mol. Most of the data range from −22 to −25 kcal/mol [35] and are only slightly dependent on the alcohol (primary or secondary) and structural aspects of the isocyanate (Table 3.1) [36, 37]. However, as shown in Figure 3.3, reactants may exist in varying states of self-association and the heat of dissociation and hydrogen bonding will vary in measurement and calculations as well.

The urethane bond is able to undergo reversion back to isocyanate and hydroxyl functionality at elevated temperatures [38–40]. The temperature at which this reversion commences is a function of the stability of the urethane structure from which reversion is occurring. Table 3.2 provides detail of the structural aspects contributing to urethane thermal stability. The greater the aliphatic content of the structure, the more thermally stable, since aromatic delocalization can stabilize the activated states from which reversion results [9]. It has been observed that methylene diphenyl diisocyanate (MDI)-based urethanes have higher thermal stability relative to toluene diisocyanate (TDI). Urethanes from aromatic alcohols exhibit particularly low thermal stability; for this reason, phenol has been widely used as a blocking agent for protecting isocyanate groups and creating their availability at elevated temperatures (>100 °C) [42, 43]. In the absence of a change in stoichiometry (e.g. because of selective volatilization or degradation of one of the reactants), upon cooling the formation of urethane is highly favored and there is little or no change in the overall molecular weight [44]. Figure 3.11 shows a temperature-dependent dynamic mechanical analysis demonstrating the loss of modulus and the increasingly liquid-like behavior as reversion reactions become increasingly prevalent.

**TABLE 3.1 Enthalpy of urethane formation (kcal/mol) of various isocyanates with varying alcohol order**

| | *n*-Butyl | Isobutyl | Sec-butyl |
|---|---|---|---|
| Phenyl isocyanate | 25.1 | 24.1 | 23.4 |
| *o*-Tolyl isocyanate | 23.5 | 22.1 | 21.7 |
| *m*-Tolyl isocyanate | 19.5 | 19.1 | 18.5 |
| *p*-Tolyl isocyanate | 24.7 | 23.9 | 23.6 |
| 2,4-Toluene diisocyanate | 22 | 21.3 | 20.6 |

**TABLE 3.2 Approximate reversion temperatures of urethane product back to constituent isocyanates and polyols**[a]

| Urethane structure | Approximate limit of thermal stability (°C) |
|---|---|
| Alkyl—N(H)—C(=O)—O—Alkyl | 250 |
| Aryl—N(H)—C(=O)—O—Alkyl | 200 |
| Alkyl—N(H)—C(=O)—O—Aryl | 180 |
| Aryl—N(H)—C(=O)—O—Aryl | 120 |

[a] *Source:* Based on [41].

It is generally accepted that the reactivity of an active hydrogen compound will increase as its nucleophilicity increases (and so reactivity with the isocyanate carbon increases). The general trend follows the sequence [10]:

$$\text{primary amine} > \text{aromatic amine}(\text{aniline}) > \text{primary alcohol}$$
$$> \text{aromatic alcohol} > \text{thiol}$$

The low reactivity of thiols with isocyanates perhaps reflects the low dipole moment of the S–H bond instead of the nucleophilicity. Within each grouping important steric influences can occur. For instance, the measured reactivity of alcohols follows the sequence:

$$\text{primary} > \text{secondary} > \text{tertiary}$$

with a ratio of 200:60:1 [45]. This ratio has been determined by measurement of reaction rates of isocyanate with butanol variants.

The effect of isocyanate structure on reactivity can be understood in terms of the relationship of electronic and steric factors affecting the NCO groups. Quantum mechanical calculation on hydrogen isocyanate (isocyanic acid) H–N=C=O resulted in relative charge values of H, +0.29; N, –0.17; C, 0.51; O, –0.34 [4]. In general, electronic factors that decrease the electron density on the carbon will increase the isocyanate reactivity. In addition, from the relative charge values it is

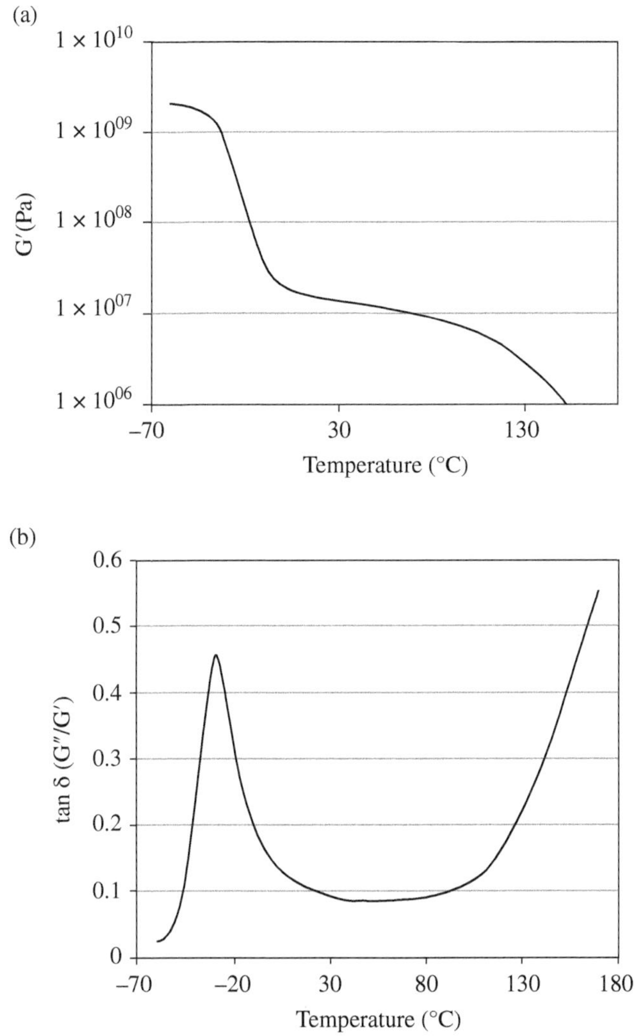


**FIGURE 3.11** Dynamic mechanical analysis of an aromatic polyurethane showing the loss of modulus (a) and increase in viscous (liquid like) behavior (b) at temperatures above 120 °C, reflecting the reversion reaction.

reasonable to conclude that increasing the negative charge density on the isocyanate oxygen will increase reactivity by electrophilic attack. While that will be true, it has also been calculated that the isocyanate nitrogen is actually the favored position for attack because of the greater spatial dispersion of electronic charge around the oxygen. If steric factors are held constant, any electron-withdrawing group attached to the nitrogen enhances the electrophilic character of the isocyanate carbon, resulting in the relative reactivity ratios shown in Table 3.3. TDIs are not shown in Table 3.3 because of the significant reactivity difference of the second isocyanate upon reaction of the first isocyanate, and the significant difference in reactivity of the isocyanate group depending on the ring position [47]. This is further complicated by the practical fact that most commercially available TDIs are mixtures of 2,4 and 2,6 diisocyanate isomers. The relative reactivity of these isomers and of the reactivity of the second ring isocyanate following reaction of the first is given in Table 3.4. This deactivation results from the replacement of a strongly activating isocyanate group (electron withdrawing increasing the reactivity of the isocyanate carbon) with a weakly activating urethane group. The effect of the first isocyanate reaction on the second for MDI is not as important since they are on different aromatic rings separated by a nonconjugated methylene linkage, but it has been reported that the second reaction may be slowed down by as much as a factor of 2, though the cause of this retardation is not known.

As alluded to earlier, the effect of solvent on urethane reactions has been noted in the literature [45, 48, 49]. Particular notice was taken when a 71-fold increase in rate was observed in acetonitrile versus benzene [45]. While it was initially thought that this was a result of the solvent dielectric strength it was subsequently shown that the activation energy for urethane formation was the same in toluene, dibutyl ether, and acetonitrile. Theoretical calculations have also suggested that dielectric strength plays a minor role in urethane formation [2]. It is now thought that the major effect of solvent on urethane formation is the effect of solvent on the hydroxyl phase self-association and the concentration of free or dimeric alcohol available for reaction. Additional studies have been carried out into the effect of catalyst and activation parameters in different solvents at varying conversions [50].

**TABLE 3.3 Reactivity of several common diisocyanates relative to MDI**

| Diisocyanate | Reactivity |
|---|---|
| Paraphenylenediisocyanate | 1.85 |
| MDI | 1 |
| Isophorone diisocyanate | 0.15 |
| 1,6-Hexamethylene diisocyanate | 0.13 |

**TABLE 3.4 Reactivity of the isocyanates of TDI isomers relative to the para position (fourth position) of the 2,4 isomer**

| Isocyanate | $k_1$ | $k_2$ |
|---|---|---|
| 2,4-TDI | 100 | 12 |
| 2,6-TDI | 56 | 17 |

### 3.3.1 Urea Formation

The reaction of isocyanates and water to form urea is essential to many polyurethane industrial applications. The reaction with water creates the essential stability and heat of reaction to make the kinetics of foam formation practical [51]. In addition, it creates a gas reaction product ($CO_2$) that provides an expanding volume, helping to create the porous structure of foams (Figure 3.12). The initial reaction of water and isocyanate forms an unstable carbamic acid intermediate which immediately decomposes to amine and $CO_2$ [52].

The reaction of amine with isocyanate is diffusion controlled, forming the urea and a significant enthalpy of reaction (Figure 3.13) (c. 22 kcal/mol) [53]. It is reported that the use of a catalyst has no kinetic effect on the rate of amine and isocyanate reaction and that the reaction of each NCO with water liberates a similar amount of heat to the urethane reaction. Disparities in the literature may depend on experimental factors affecting the self-associations that promote or hinder the formation of the new urea bond.

The reactivity of water with isocyanate relative to that of hydroxyls is generally regarded as following a path similar to that of alcohol hydroxyl, requiring the concerted interaction of two water molecules across the N=C bond. Quantum mechanical calculations have suggested that a third water molecule would actually provide an enhanced geometry for the reaction across the isocyanate [46]. Certainly, the insolubility of most isocyanates with water will have a significant effect on the observed rate of reaction and can conceivably account for much disparity of the measured rates found in the literature [54].

The kinetics of urea formation from the water reaction are of course a convolution of the initial kinetics of isocyanate and water, which is slow unless catalyzed, followed by the reaction of the subsequent amine with isocyanate to urea, which is diffusion

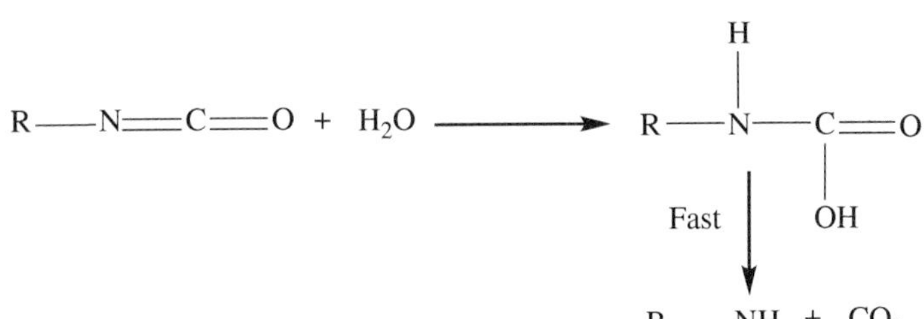


**FIGURE 3.12** The reaction of isocyanate and water, forming an amine.

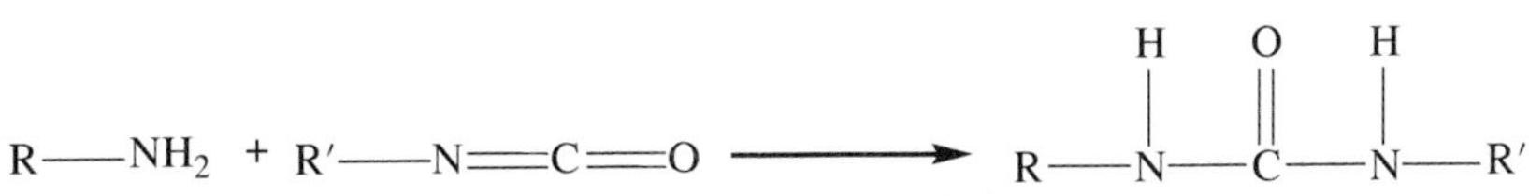


Urea

**FIGURE 3.13** The reaction of amine with an isocyanate to form a urea.

controlled [48]. An interesting phenomenon is the apparent first-order kinetic dependence on isocyanate concentration and second-order dependence on the amine. This second-order dependence on the amine is realized for both aniline and aliphatic amines. This high-order dependence coupled with the relatively low activation energy (~17.5 kJ/mol) account for the very high initial reactivity of these systems [48].

### 3.3.2 Allophanate Formation

The reactions described above are those based on reactants and intended by the chemist. However, because of the reactivity of the isocyanate group, the adiabatic heat buildup in the reacting mass, and local inhomogeneities in reactant concentration, other reactions can and do occur. The reaction of an isocyanate and a urethane to form an allophanate is one such reaction (Figure 3.14) [55]. The active H on the urethane nitrogen is quite a bit less active than those available on water or alcohol, resulting in relatively unfavorable kinetics of formation (Table 3.5). Allophanate formation can occur under conditions of elevated temperature and the presence of excess isocyanate in the immediate vicinity of the urethane linkage. However, like the urethane bond, at temperatures above 100–150 °C the allophanate linkage will revert back to the urethane and the free isocyanate, reflecting the stability of the parent urethane [56]. It is seemingly inevitable that allophanates form in the practice of making urethanes. The industrial approach has, in this case, made the best of it by formulating reactant concentrations to optimize the number of allophanate linkages, which provides cross-linking to obtain the desired final polymer properties [57, 58]. The gradual increase of polyurethane tensile properties upon cooling over a period of days is sometimes considered a result of slow formation of allophanates (among other possibilities, including maturation of phase morphology). Because of the thermal reversibility of the allophanate bond, these cross-links do not have to negatively affect the processability of the polymer [59].

### 3.3.3 Formation of Biurets

Just as isocyanate in the presence of urethane at elevated temperatures can create allophanates, isocyanate in the presence of urea can create a biuret linkage (Figure 3.15).

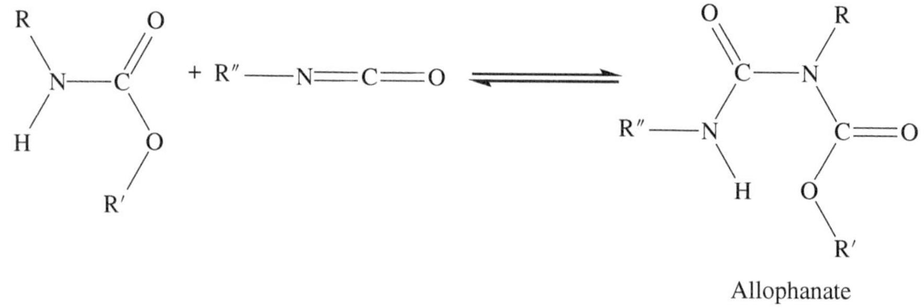


**FIGURE 3.14** Formation of an allophanate by reaction of a urethane with an isocyanate. The R group can constitute an additional isocyanate moiety, resulting in cross-linking.

**TABLE 3.5 Relative reaction rates of isocyanates with varying reaction partners**

| Active H compound | Structure | Relative uncatalyzed reaction rate at 25 °C |
|---|---|---|
| Primary aliphatic amine | R–N**H**$_2$ | 100,000 |
| Primary secondary amine | R$_2$N–**H** | 20,000–50,000 |
| Primary aromatic amine | Ar–N**H**$_2$ | 200–300 |
| Primary hydroxyl | R–CH$_2$–O**H** | 100 |
| Water | **H**$_2$O | 100 |
| Carboxylic acid | R–COO**H** | 40 |
| Secondary hydroxyl | R$_2$CHO**H** | 30 |
| Urea | R–N**H**–CO–NH–R | 15 |
| Tertiary alcohol | R$_3$COH | 0.5 |
| Urethane | R–N**H**–COOR | 0.3 |
| Amide | R–CO–N**H**$_2$ | 0.1 |

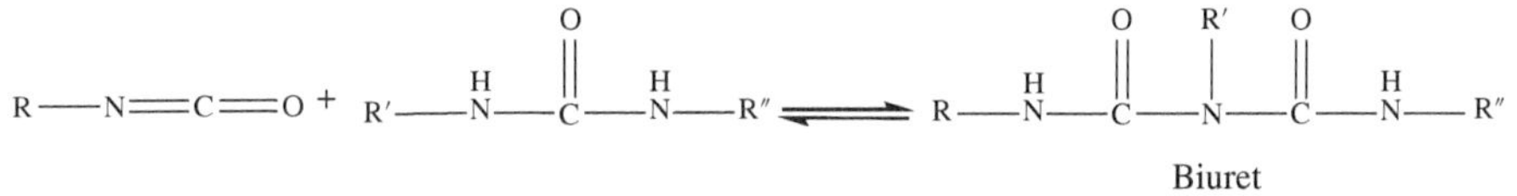


**FIGURE 3.15** Formation of a biuret linkage by reaction of urea with isocyanate. The R group can constitute an additional isocyanate moiety, resulting in cross-linking.

The hydrogen on the urea nitrogen is slightly more reactive than that of the urethane, resulting in the biuret linkage having slightly more thermal stability that the allophanate (Table 3.5). Regardless, the biuret linkage is thermoreversible and has temperature-dependent effects on overall polymer properties [60, 61].

### 3.3.4 Formation of Uretdione (Isocyanate Dimer)

The dimerization of isocyanate to form a uretdione is a slow reversible reaction (Figure 3.16) [62]. Uretdione structures have found industrial use as a cross-linking structure within urethane powder coatings [63, 64], but are otherwise most notable as a problem related to MDI storage. Because of fortuitous packing of MDI in the crystal, the formation of MDI dimers occurs more rapidly in the solid state than in the melt [65]. In fact, the rate of formation of MDI dimer is most rapid at normal room temperatures, converting approximately 0.01% (wt/day) at 21 °C. For this reason, it is common practice to store MDI either under refrigeration or in the melted state to increase shelf life, and to filter the molten MDI prior to use [66]. The dimer has a nearly zero heat of formation and dissociation occurs at relatively low temperatures (80–100 °C). This permits the uretdione to serve as a protected isocyanate group in a manner similar to that described previously with phenols. This is particularly useful for 2,4-TDI protection since there is a substantial difference in reactivity of the isocyanates, and formation of trimer does not present itself as a competitive reaction [10].

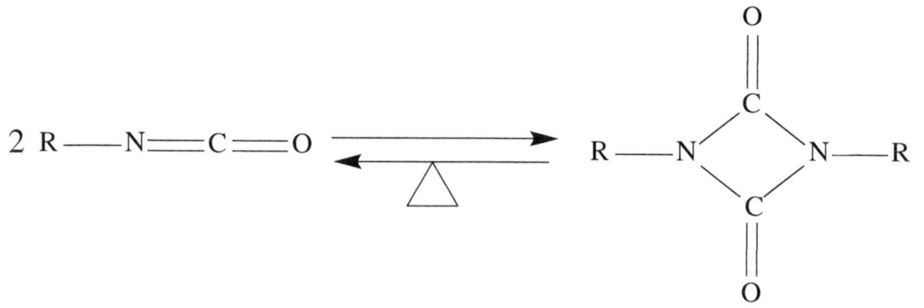

**FIGURE 3.16** Dimerization of isocyanate to form the uretdione.

### 3.3.5 Formation of Carbodiimide

Carbodiimide is formed by a catalyzed irreversible dimerization reaction of isocyanate [67, 68]. Carbodiimide has significant practical utility since it can be used to oligomerize 4,4″-MDI, which is useful for numerous applications, including the manufacture of cast elastomers and foams and for making composite laminates. In its simplest expression its formation is described by Figure 3.17. Polymers formed from carbodiimides can exhibit greater flexibility and less friability than analogous polymers synthesized from conventional diisocyanates.

Various descriptions and catalytic mechanisms have been put forward to explain the evolution of reactants to products (Figure 3.18). Significant progress was made with the advent of phospholene oxide catalysts, which are highly efficient at forming carbodiimide while minimizing the formation of other products. The carbodiimide is detectable by infrared spectroscopy by the characteristic N=C=N absorption at 2140 $cm^{-1}$ but may exhibit partial overlap with the isocyanate resonance at 2270 $cm^{-1}$. Another practical aspect of the formation of carbodiimide results from its unintentional formation in a closed volume (i.e. in a drum for instance), with the concomitant evolution of $CO_2$ resulting in an explosion hazard from pressure buildup.

Carbodiimide chemistry is also notable as part of urethane degradation pathways [71, 72]. When urethanes are heated to high temperatures (c. 240–270 °C) for extended periods of time in an FTIR cell, the characteristic carbodiimide resonance will appear at 2140 $cm^{-1}$. It was subsequently noticed that the carbodiimide concentration initially increased and subsequently reached an equilibrium concentration. Evolution of isocyanate was also observed but declined with time, as did the urethane-associated carbonyls between 1715 and 1730 $cm^{-1}$ with the growth of carbonyl resonance between 1700 and 1710 $cm^{-1}$. The overlap between the trimer resonance and the urethane resonance makes assignment of carbodiimide trimerization somewhat imprecise, and there was unfortunately no mass spectral data. However, the cross-linked behavior of the heated material was noted. It has been postulated that carbodiimide (which clearly forms) is capable of trimerizing according to the reaction of Figure 3.19. Additionally, the reaction of two isocyanates and one carbodiimide was postulated (Figure 3.20).

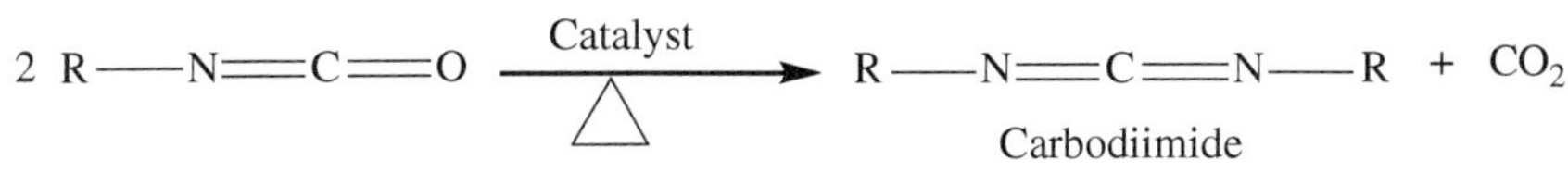


**FIGURE 3.17** Catalyzed dimerization of isocyanate to form carbodiimide.

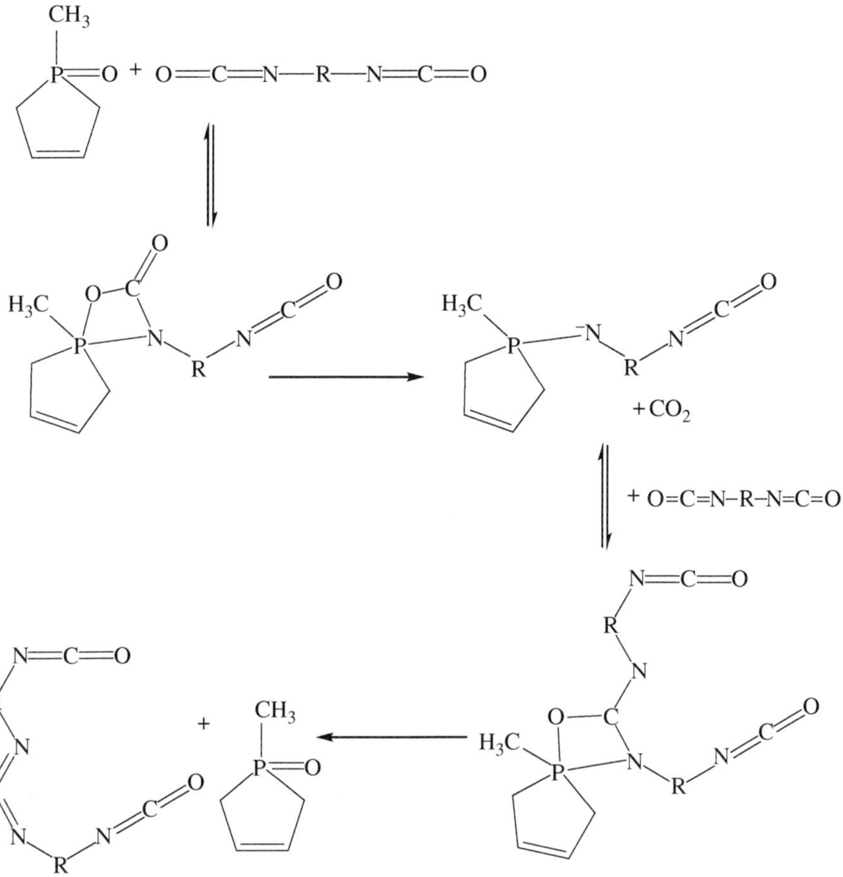


**FIGURE 3.18** Phospholene-catalyzed formation of carbodiimide. *Source:* Modified from [69, 70].

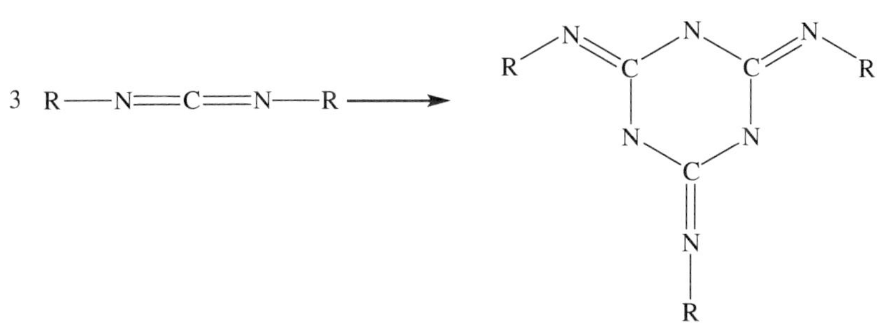


**FIGURE 3.19** Proposed trimerization of carbodiimide occurring during high-temperature decomposition.

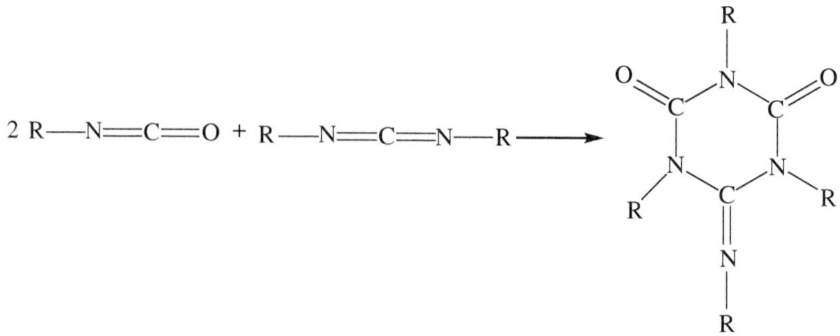


**FIGURE 3.20** Proposed reaction of carbodiimide to form a six-member ring during high-temperature decomposition.

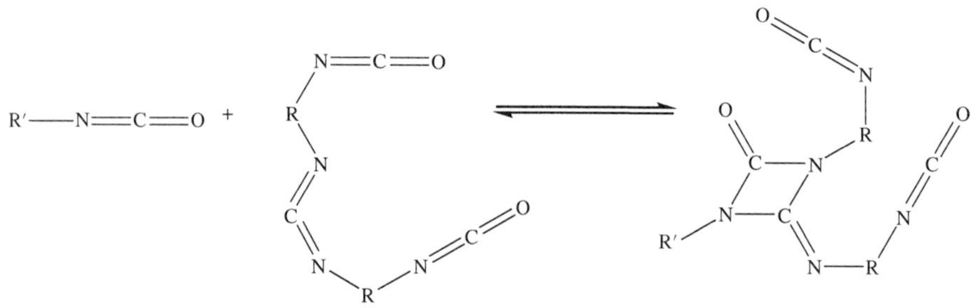

**FIGURE 3.21** Reaction of carbodiimide and isocyanate to form the industrially important uretonimine.

### 3.3.6 Formation of Uretonimine

In the presence of excess isocyanate a carbodiimide can react reversibly to form the uretonimine [73]. The uretonimine adduct is of practical importance within polyurethane technology because it results in a low-viscosity room temperature liquid [74]. The modification is accomplished by initially forming the carbodiimide with a catalyst at approximately 100 °C. The phospholene catalyst is subsequently neutralized (e.g. with an organic acid chloride). The uretonimine forms upon cooling (Figure 3.21). Subsequent heating to approximately 200 °C dissociates the uretonimine back to the carbodiimide and the isocyanate. At 200 °C the dissociation/formation rate constant ratio has been reported to be 1.63 [75].

### 3.3.7 Formation of Amides

The reaction of isocyanates and carboxylic acids has been reported in the literature, but its practical application has not yet been appreciated [76–80]. Commodity aromatic isocyanates such as MDI and TDI do not form amides with carboxylic acids; instead, they reportedly form urea bonds. Aromatic isocyanates with strong electron-withdrawing groups such as cyano-, nitro-, and trifluoromethyl can result in

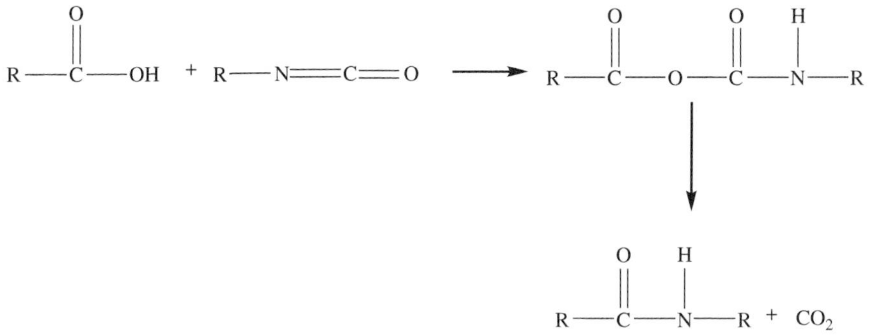

**FIGURE 3.22** Reaction of aliphatic isocyanate and carboxylic acid to form the intermediate anhydride and the final amide.

the formation of amides in reaction with carboxylic acids. Alternatively, aliphatic isocyanates react readily with carboxylic acids in solvent to form amides according to Figure 3.22. In some cases, the anhydride is isolated; in others, the anhydride proceeds immediately to the amide. The reaction of carboxylic acid with a dimethyl-pyrazole-blocked isocyanate and a magnesium triflate catalyst reportedly results in 100% amide yield in chloroform. While no mechanism for the isocyanate/carboxylic acid reaction has been reported, it has been determined that the carbon dioxide carbon originates from the isocyanate carbon.

## REFERENCES


[1] G Raspoet, M Nguyen, M McGarraghy, and A Hegarty, *Journal of Organic Chemistry*, **63**, (1998), 6878–6885.

[2] A Caraculau and S Coseri, *Progress in Polymer Science*, **26**, (2001), 799–851.

[3] L Phillips and D Barker, *Polyurethane Chemistry: Technology and Properties*, Life Books Ltd, London, UK, 1964.

[4] M Tang and X Fu, *International Journal of Quantum Chemistry*, **42**, (1992), 403–409.

[5] JB Lambert, HF Shurvell, LR Verbit, RG Cooks, and GH Stout, *Organic Structural Analysis*, McMillan Publishing, New York, NY, 1978, 274.

[6] LP Smirnov and EV Deyun, "Formation of spatial dissipative structures during synthesis of polyurethanes," in *Heterophase Network Polymers*, BA Rozenberg and GM Sigalov, Eds., Taylor & Francis Ltd, London, UK, 2002, 139–150.

[7] M Sato, *Journal of Organic Chemistry*, **27**, (1962), 819.

[8] G Oertel *Polyurethane Handbook*, Hanser Publishers, New York, NY, 1993.

[9] M Szycher, *M. Szycher's Handbook of Polyurethanes*, 2nd Ed., CRC Press, Boca Raton, FL, 2013.

[10] R Arnold, J Nelson, and J Verbanc, *Chemical Reviews*, **57**, (1957), 47.

[11] S Wong and K Frisch, *Journal of Polymer Science Part A: Polymer Chemistry*, **24**, (1986), 2877–2890.

[12] K Schwetlick and R Noack, *Journal of the Chemical Society, Perkin Transactions*, **2**, (1995), 295–402.

[13] U Shinji, O Tomeyoshi, T Tanaka, M Suzuki, H Shirai, *Macromolecular Chemistry and Physics*, **200**, (1999), 2645–2650.
[14] YM Lee and LJ Lee, *Polymer*, **28**, (1987), 2304–2309.
[15] R Herrington and K Hock, *Flexible Polyurethane Foam Reference Book*, 2nd Ed., Dow Chemical Co., Midland, MI, 1992.
[16] A Draye and J Tondeur, *Journal of Molecular Catalysis*, **238**, (1999), 135–144.
[17] R Devendra, N Edmonds, and T Sohnel, *RSC Advances*, 60, (2015), 48935–48945.
[18] J Arnevir, "Bismuth carboxylates for polyurethane catalysis," in *Proceedings of the SPI Annual Technical/Marketing Conference*, October 1–4, San Francisco, CA, 1989, 623–627.
[19] L Thiele and R Becker, *Advances in Urethane Science and Technology*, **12**, (1993), 59–85.
[20] K Wongkamolsesh and JEK Kresta, *Polymeric Materials Science and Engineering*, **49**, (1983), 465–468.
[21] DN Tarasov, RP Tiger, Y Chirkov, N Yu, SG Entelis, and J-J Tondeur, *Kinetics and Catalysis (Translation of Kinetika i Kataliz)* **41**, (2000), 355–358.
[22] R Pandey, R Dagade, P Sharda, MK Dongare, and P Kumar, *Synthetic Communications*, **33**, (2003), 4019–4027.
[23] G Tillet, B Boutevin, and B Ameduri, *Progress in Polymer Science*, **36**, (2010), 191–217.
[24] R Van Maris, Y Tomano, H Yoshimura, and K Gay, *Journal of Cellular Plastics*, **41**, (2005), 305–322.
[25] L Huber, "Flexible cellular plastics," in *Reaction Polymers*, W Gum, W Riese, and H Ulrich, Eds., Hanser Publisher, New York, NY, 1992, 297–305.
[26] P Chaffanjon, RA Grisgby, EL Rister, and RL Zimmerman, *Journal of Cellular Plastics*, **39**, (2003), 187–210.
[27] R Sojecki and S Trzcinski, *European Polymer Journal*, **36**, (2000), 1511–1517.
[28] M Roman, B Andrioletti, L Bruno, MB Lemaire, J-M Schwartz, and JP Barbeau, *Tetrahedron*, **67**, (2011), 1506–1510.
[29] M Spirkova, M Kubin, P Spacek, I Krakovsky, and K Dusek, *Journal of Applied Polymer Science*, **52**, (1994), 895–904.
[30] R Romero, R Grigsby, E Riser, J Pratt, and D Ridgway, *Journal of Cellular Plastics*, **41**, (2005), 339–359.
[31] H Duong, M Cross, and J Louie, *Organic Letters*, **6**, (2004), 4679–4681.
[32] S Raders and J Verkade, *Journal of Organic Chemistry*, **75**, (2010), 5308–5311.
[33] WW van Pelt and JG Goossens, *Analytica Chimica Acta*, **604**, (2007), 69–75.
[34] G Suhel, R Giesen, and K Lucas, *Journal of Chemical and Engineering Data*, **49**, (2004), 826–831.
[35] M Coban, F Konuklar, and S Aylin. *Computational & Theoretical Chemistry*, **963**, (2011), 168–175.
[36] K Kulesza, K Pielichowski, and K German, *Journal of Analytical and Applied Pyrolysis*, **76**, (2006), 243–248.
[37] S Karpov, V Lodygina, V Komratova, A Dzhalukhanova, G Malka, and E Badasmshina, *Kinetics and Catalysis*, **57**, (2016), 422–428.
[38] L Okrasa, P Czech, G Boiteux, F Mechin, and J Ulanski, *Polymer*, **49**, (2008), 2662–2668.
[39] I Mohammed and G Sankar, *High Performance Polymers*, **23**, (2011), 535–541.

[40] P Yang, X Zhu, J Li, Y Xia, M Yog, and T Li, *Journal of Applied Polymer Science*, **117**, (2010), 1095–1099.
[41] B Dombrow, *Polyurethanes*, 2nd Ed., Reinhold Publishers, New York, NY, 1957.
[42] H Kothandaraman, A Nasar, and R Lakshmi, *Journal of Applied Polymer Science*, **53**, (1994), 31–38.
[43] D Wicks and Z Wicks, *Progress in Organic Coatings*, **36**, (1999), 148
[44] H Sui, X Liu, K Cheng, Y Luo, and F Zhong, *Polymer Degradation and Stability*, **101**, (2014), 109–113.
[45] AC Draye and J Tondeur, *Reaction Kinetics and Catalysis Letters*, **66**, (1999), 319–324.
[46] G Raspoet and G Nguyen, *Journal of Organic Chemistry*, **63**, (1998), 6867–6877.
[47] RA Martin, K Hoy, and R Peterson, *Industrial and Engineering Chemistry. Product Research Development*, **6**, (1967), 218.
[48] MC Pannone and C Macosko, *Journal of Applied Polymer Science*, **34**, (1987), 2409.
[49] V Guzikhin, M Zaiac, and D Liamkin, Chimishkian.Kauchuck and Rubber (Paper in Russian, translated by Andrei Merenov), (2001), 4.
[50] F Hostettler and E Cox, *Industrial and Engineering Chemistry*, **52**, (1960), 609–610.
[51] M Sonnenschein and W Koonce, "Polyurethanes," in *Encyclopedia of Polymer Science and Technology*, 4th Ed., H Marks, Ed., John Wiley & Sons, Inc., Hoboken, NJ, 2012, 1–63.
[52] K Saunders, Ed., "Polyurethanes," in *Organic Polymer Chemistry*, 2nd Ed., Chapman and Hall, London, UK, 1988.
[53] Y Zhao and G Suppes, *Polymer Engineering & Science*, **55**, (2015), 1420–1428.
[54] R Tabor, K Hinze, R Priester, and R Turner, "The compatibility of water with polyols," in *Proceedings of the SPI 34th Annual Technical/Marketing Conference*, October 21–24, New Orleans, LA, 1992, 514–528.
[55] A Lapprand, F Boisson, F Delolme, F Mechin, and J-P Pascault, *Polymer Degradation and Stability*, **90**, (2005), 363–373.
[56] K Wagner, *Angewandte Chemie*, **74**, (1962), 799.
[57] I Kogon, *Journal of Organic Chemistry*, **24**, (1959), 83–86.
[58] A Sebenik, U Osredkar, and I Vizovisek, *Journal of Macromolecular Science. Chemistry*, **A23**, (1986), 369–379.
[59] S Dassin, M Dumon, F Mechin, and J-P Pascault, *Polymer Engineering & Science*, **42**, (2002), 1724–1739.
[60] W Seo, J Park, Y Sung, D Hwang, W Kim, and H Lee, *Journal of Applied Polymer Science*, **93**, (2004), 2334–2342.
[61] K Dusek, M Spirkova, and M Ilavsky, *Makromolekulare Chemie, Macromolecular Symposia*, **45**, (1991), 87–95.
[62] M Edelmann and G Heinrich, *Thermochimica Acta*, **499** (1–2), (2010), 160–165.
[63] E Spyrou, H Metternich, and R Franke, *Progress in Organic Coatings*, **48**, (2003), 201–206.
[64] F Schmitt, A Wenning, and J-V Weiss, *Progress in Organic Coatings*, **34**, (1988), 227–235.
[65] RB Wilson, YS Chen, IC Paul, DY Curtin, *Journal of the American Chemical Society*, **105**, (1983), 1672–1674.
[66] N Shah, Method of storing 4,4′-methylene diphenylisocyanate. Assignee E.I. Du Pont deNemours Co. US patent 2999106. 1961.
[67] W Darwish, E Seikel, R Kaesmarker, K Harms, and J Sundermeyer, *Dalton Transactions*, **40**, (2011), 1787–1794.
[68] J Anderson and R Bou-Moreno, *Tetrahedron*, **66**, (2010), 9182–9186.

[69] J Monagle, *Journal of Organic Chemistry*, **27**, (1962), 3851.
[70] T Cambell, and K Smeltz, *Journal of Organic Chemistry*, **28**, (1963), 2069.
[71] D Joel and A Hauser, *Angewandte Makromolekulare Chemie*, **217**, (1994), 191–199.
[72] R Aguirresarobe, L Irusta, and M Fernandez-Berridi, *Polymer Degradation and Stability*, **97**, (2012), 1671–1679.
[73] N Luo, D-N Wang, and S-K Ying, *Journal of Applied Polymer Science*, **61**, (1996), 367–370.
[74] A Mayr, W Cook, G Edward, and G Murray, *Polymer International*, **49**, (2000), 293–301.
[75] R Turner, C Macdonald, J Blackwell, J Quay, and C Lee, *ACS Symposium Series*, **270**, (1985), 53–64.
[76] K Sasaki and D Crich, *Organic Letters*, **13**, (2011), 2256–2259.
[77] N Leventis, C Chidambareswarapattar, D Mohite, Z Larimore, H Lu, and C Sotiriou-Leventis, *Journal of Materials Chemistry*, **21**, (2011), 11981–11986.
[78] R Gertzman and C Gurtler, *Tetrahedron Letters*, **46**, (2005), 6659–6662.
[79] D Crich and K Sasaki, *Organic Letters*, **11**, (2009), 3514–3517.
[80] I Blagbrough, N Mackenzie, C Ortiz, and A Scott, *Tetrahedron Letters*, **27**, (1986), 1251–1254.

# 4

# THEORETICAL CONCEPTS AND TECHNIQUES IN POLYURETHANE SCIENCE

There is very good physical understanding behind polyurethane structure, properties, and the chemical processes leading to material properties. Perhaps unique among modern polymers, theoretical understanding of polyurethanes has had to await developments in different fields of materials science. Theoreticians then applied those developments to polyurethanes, almost as an afterthought. Thus, modern polyurethane science has dined well at the table serving other materials such as composites, polymer blends, and block copolymers. Even as our *qualitative* appreciation of polyurethanes has grown very rapidly, so has appreciation that some aspects of *quantitative* understanding are limited because of potentially uncontrolled parameters of polyurethane polymerization. These uncontrolled variables can have very significant effects on final polymer structure and properties. Thus, theoretical foundations of polyurethane science are sound and predictive; however, they do not as yet allow one to know ahead of an experiment if one of myriad influences on polymerization will invalidate a predicted result. Current theory allows one to look back after materials analysis and make sense of the result, but not predict ahead of time if that particular result will occur.

The purpose of this chapter will not be to *derive* the equations that are currently well applied to polyurethane science. Instead the basis for those equations and their physical meaning and limitations will be presented. An intuitive understanding of the equations and their extension to predicting polyurethane performance will also be provided. The best understanding of polyurethane structure comes from a simultaneous appreciation of

*Polyurethanes: Science, Technology, Markets, and Trends*, Second Edition. Mark F. Sonnenschein.

the thermodynamic concepts underpinning the drive from reactants to final polymer structure coupled with the micromechanical models that allow estimation of properties from that structure. Tensile properties of elastomers can be predicted by applying micromechanical results to calculations with assumptions of polymer extensibility. Bulk polymer structure can also be propagated into foam properties assuming well-known relationships associated with properties as a function of bulk density and cell structure.

## 4.1 FORMATION OF POLYURETHANE STRUCTURE

The chemical reactions that lead from reactants (isocyanates, chain extenders, polyols) to polyurethane polymer have been covered in detail in Chapter 3. The hard-segment blocks are a result of isocyanate–chain extender reactions. The chain extender can be low-molecular-weight polyols, polyamines, or water.

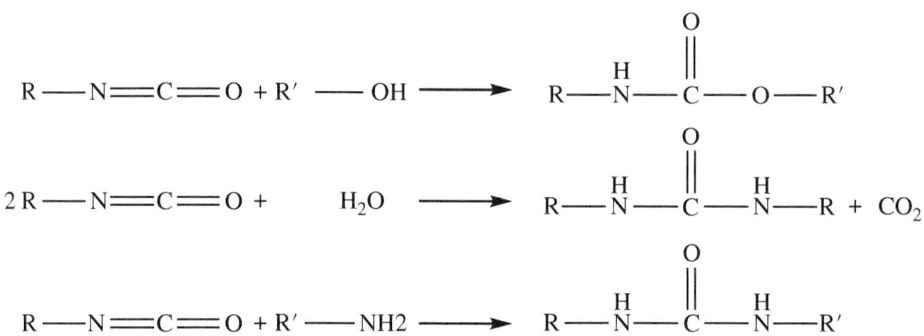

The first and third reactions are first order in each reactant while the second reaction is first order in water and second order in isocyanate. Each reaction has a rate constant characteristic of the particular reactants and mutual solubilities that evolve as the reaction proceeds and the heat of reaction is released. Soft segments are typically low-polydispersity polyalkylene glycols (Mw/Mn ~ 1) or polyesters of polydispersity of about 2.0 made separately from the polyurethane polymerization.

It has been proposed that polyurethane structure from these reactions results in a block copolymer (Figure 4.1) that can be treated using models for block copolymers [1–3] derived for well-characterized or ideal block copolymer structures [4–7]. Structures developed from a basis of ideal conditions usually begin from assumptions that the structure under consideration is fully phase separated and at a point of lowest free energy. Furthermore, these thermodynamic structures often comprise polymers with ideal polydispersity (i.e. Mw/Mn = 1) in their molecular weight as well in block size.

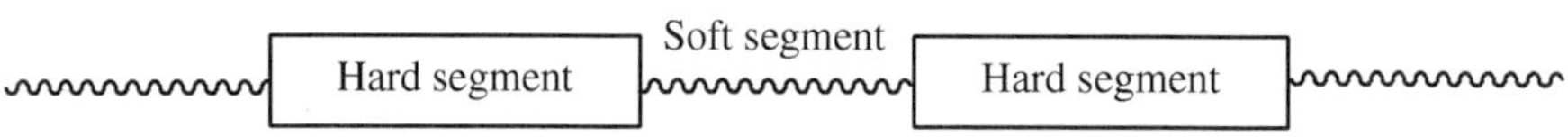


**FIGURE 4.1** Hard and soft polyurethane segments make for a multiblock copolymer.

Addition polymers such as polyurethanes can never have these attributes but, because of their chain growth process, they will have polydispersities from about 2.0 to the undefined dispersity of a thermoset. This issue of nonideal polydispersity of polyurethanes versus that employed to simplify theoretical calculations should not deter us, however, since the effect of polydispersity can be anticipated and should not deny us the insights such an analysis can provide. An additional problem is the equilibrium assumption, which means that there is the expectation that the theoretical analysis is performed on a structure that is in the lowest energy state. Actual polyurethane structure reflects the thermodynamics of the polymer chains coming to an equilibrium structure convoluted with the kinetic limitations of a system that may be coming to a nonequilibrium state as a result of a cooling rate that is not infinitely slow (Figure 4.2). Again, such considerations will apply to all real block copolymers that phase separate upon either cooling from a melt or assembling from a dilute condition. These variances of polyurethane reality from model assumptions should not prevent us from obtaining a good qualitative understanding of our systems in terms of both how close they lie to ideal expectations and how far they lie from them.

Figure 4.2 expresses final conditions in which hard segment is separated or partially phase separated from the soft segment. However, the path from the initial conditions to a final state more or less represented by Figure 4.2 is dependent on several factors, including the thermodynamic incompatibility of the hard and soft phases. The analysis of phase separation in block copolymer systems can be treated in limits of very strong phase segregation [5] or weaker phase separation [4, 6, 7]. The strong and weak limits are essentially indications of how much the phases mix even when they have achieved a phase-separated state such that each phase has more (weak segregation) or less (strong segregation) solubility in the other phase at equilibrium. Closer to the case for polyurethanes, Benoit and Hadziioannou [7] covered the case

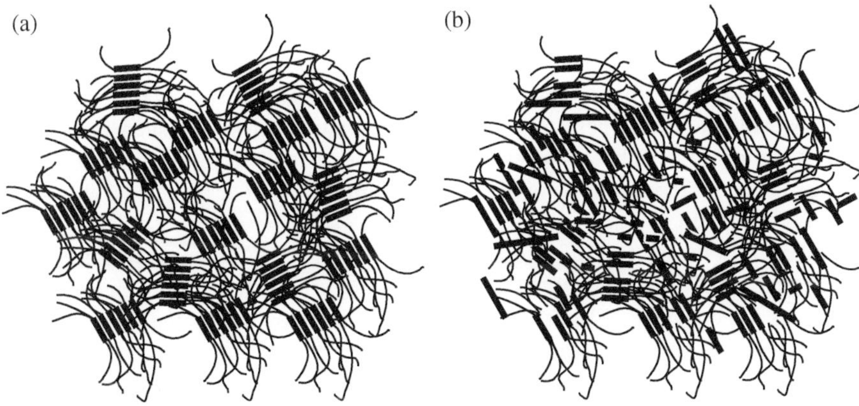


**FIGURE 4.2** Representations of polyurethane final structures. (a) An idealized structure showing monodisperse hard-segment lengths and block sizes isotropically dispersed in a monodisperse soft segment. (b) A more realistic representation showing polydisperse hard-segment lengths and block sizes, along with unassociated hard segments dissolved in the soft segment.

**FIGURE 4.3** Scanning electron micrograph of a 20%/80% (w/w) blend of cross-linked polydimethylsiloxane in polymethylmethacrylate showing macrophase separation and minimization of interfacial surface area. Full scale = 100 μm.

for block copolymers that are multiblock rather than simple diblock or triblock copolymers, for instance. This matches polyurethanes in that over the course of a typical polymer chain there may be many runs of hard and soft segments in sequence. Also cogent to polyurethane discussion is the work of Fredrickson and Helfand [6], who included the polydispersity of molecular weights. A more recent treatment by Ginzburg et al. [8] considered the structures created under moderate phase separation relevant to polyurethanes. In a situation of strong phase separation the interfacial thickness between phases (see Section 5.2.3.2) is much less than the spacing between phase periodicity.

The force driving phase separation is phase incompatibility [9]. The usual case is for two blended polymers to phase separate into macrophase-separated structures. A typical macrophase-separated structure between two immiscible polymers (in this case, polydimethylsiloxane and polymethylmethacrylate) is shown in Figure 4.3.

However, in the case of block copolymers such as polyurethanes, such a macrophase separation does not usually occur since the chemically incompatible sequences are confined by their covalent connections, as illustrated in Figure 4.1. The result is that phase separation is on a much finer scale and interphase adhesion is complete. Structures that form in the block copolymer case still reflect the phase incompatibility, but at the same time are highly influenced by the forced intimacy of the phases [10]. This can result in a new menu of structures such as the one shown in Figure 4.4.

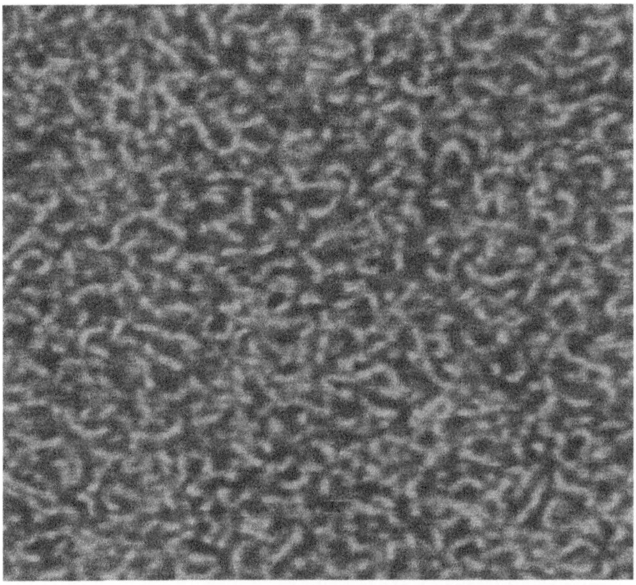

**FIGURE 4.4** Tapping-mode atomic force microscopy of a thermoplastic polyurethane comprising 40% hard 4,4′-methylene diphenyl diisocyanate hydroquinone dietherethanol (MDI-HQEE) and soft polycaprolactone. Structure shows a cylindrical structure with no obvious orientational ordering. Full scale = 500 nm.

The incompatibility is reflected in the Flory interaction parameter (conventionally denoted by the Greek letter $\chi$ [chi]):

$$\chi_{AB} = \nu_{ref} \frac{(\delta_A - \delta_B)^2}{RT} \tag{4.1}$$

In Equation 4.1 $\nu_{ref}$ is a reference volume that can represent a monomer unit such as that for propylene oxide (PO) (58 $cm^3/mol$), while $\delta$ is a solubility parameter (sometimes called the Hildebrand solubility parameter or the van Krevelen solubility parameter). The solubility parameter is a measure of a material's ability to interact with another. Two materials with similar solubility parameters will be able to interact or mix, while those with different solubility parameters will phase separate. The $\delta$ values are calculated from the cohesive energy density, which is a measure of the energy needed to remove a unit of molecules from the molecules surrounding them. Equation 4.1 can then be rewritten as:

$$\chi_{AB} = \nu_{ref} \frac{\left(\sqrt{E_{coh_A}} - \sqrt{E_{coh_B}}\right)^2}{RT} \tag{4.2}$$

Cohesive energy densities can be simply calculated using group additivity relationships provided in several well-known references [11, 12]. Polymer phase separation is strongly driven by the repulsion of unlike sequences even when the solubility parameters of the monomers of each sequence are relatively close. This is a result of the very unfavorable enthalpy of mixing different polymer phases resulting in the polymer or block phase separation being driven by $\chi^N$, where $N$ is the number of repeat units in the polymer chain or in the block sequence. Despite the block repulsion, which in the case of polyurethanes can begin at quite low $N$ values [3, 13], the effect of the forced intimacy of the blocks resulting from their covalent attachments is to potentially produce periodic structures such as that shown in Figure 4.4. The structures that form will be a result of the block incompatibility, the volume fraction of the blocks, and often processing conditions, which can greatly complicate a priori quantitative prediction of the final structure [14]. When homopolymers phase separate they are usually observed to minimize their interfacial surface area, as seen in Figure 4.3. When block copolymers phase separate they may also form spherical structures, but they may also form alternative structures such as elongated cylinders, as shown in Figure 4.4; spherical structures, such as observed in Figure 5.6b; or lamellar structures, such as in Figure 4.5 and Figure 5.6a. In limited cases a gyroid-like structure, such as in Figure 5.9, may be observed (Table 4.1).

Figures 4.4 and 4.5 point out the very important role that the relative phase volumes play in the final phase-separated polyurethane structure.

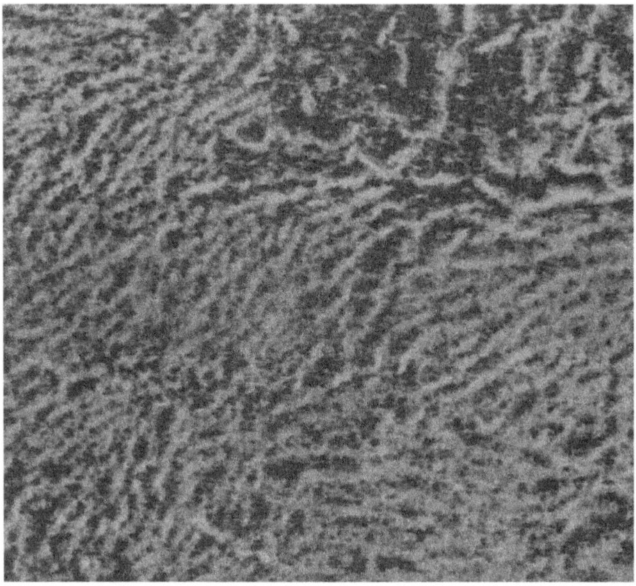

**FIGURE 4.5** Polyurethane elastomer of the same composition as the polyurethane in Figure 4.4, except with 45% hard segment. Full scale = 1 μm.

**TABLE 4.1 Examples of solubility parameters calculated by the method of Bicerano[a]**

| Polymer | Solubility parameter $\frac{\text{cal}^{1/2}}{\text{cm}^{3/2}}$ |
|---|---|
| Polyethylene oxide | 8.8 |
| Polypropylene oxide | 8.5 |
| PTMEG | 8.8 |
| Polycaprolactone | 8.3 |
| Seed oil-derived polyol | 7.6 |
| MDI-BDO | 11.2 |
| IPDI-BDO | 10.4 |
| MDI-diethyltoluenediamine | 12.2 |
| TDI-water | 14.1 |
| MDI-water | 10.7 |

Conversion to $Pa^{1/2}$ multiply by 2.045.
BDO, 1,4-butanediol; IPDI, isophorone diisocyanate; MDI, 4,4′-methylene diphenyl diisocyanate; PTMEG, polytetramethylene glycol; TDI, toluene diisocyanate.
[a] *Source:* Modified from [12].

With the Flory interaction parameter (Equation 4.1) $\chi$ being defined and the solubility parameters calculated from group additivity principles, it is possible to make a rough estimation of the amount of hard segment in soft segment that will immediately result in phase separation. Equation 4.2 allows one to predict the block size of a block copolymer that will spontaneously phase separate. With $N$ values equaling the number of "mer" repeat units within a block, if the right-hand side of Equation 4.2 is less than the Flory interaction parameter, i.e. if $\chi_{HS} > \chi_C$, the block structure will be predicted to phase separate at 298 K:

$$\chi_c = \frac{1}{2}\left(\frac{1}{\sqrt{N_A}} + \frac{1}{\sqrt{N_B}}\right)^2 \tag{4.3}$$

For a polyurethane with an MDI hard block molecular weight of about 1000 the $N$ value is approximately 4. For a PO soft-segment molecular weight of about 2000 the $N$ value is approximately 34, which predicts a $\chi_c$ of approximately 0.2. With a calculated Flory hard segment–soft segment interaction parameter of approximately 0.46 clearly exceeding the $\chi_c$, this material is predicted to phase separate, and this is in fact observed. For the block copolymer system to phase mix the requirement for this formalism is that $\chi_{HS} < \chi_C$. Calculation of the appropriate $N$ values in Equation 4.3 produces useful predictions.

Theoretical treatments of microphase separation transitions such as those illustrated by Figures 4.4 and 4.5 result in phase diagrams typified by Figure 4.6. The boundaries indicate the combined effects of incompatibility and molecular weight (i.e. $\chi^N$) on phase separation and structures. The equation by which the phase boundaries are computed is given by Equation 4.4.

$$\chi N = \left(\chi^N\right)_o \frac{1}{1 - \dfrac{0.25 - f_i\left(1 - f_i\right)}{0.25 - \alpha_i}} \tag{4.4}$$

The details of this equation are reasonably straightforward but mask a very sophisticated derivation and line of reasoning [8]. The value $(\chi^N)_o$ is the value of $\chi^N$ below which there will be no phase separation. In Figure 4.6 it is approximately 18. This value is particular for polyurethanes and is the limit below which all phase boundaries converge to a disordered phase. The value of 10.495 has been calculated for $\chi^N$, representing infinite molecular weight diblock copolymers. The effect of the finite and polydisperse molecular weight of polyurethanes and their multiblock structure is to increase $(\chi^N)_o$ [6, 7]. This is equivalent to a suggestion that for multiblock, polydisperse block, and cross-linked copolymers, such as a polyurethane, a higher level of phase incompatibility is required to obtain structured phases. Alternatively, it could be interpreted that for a block copolymer such as a polyurethane a disordered phase can be obtained upon heating to a lower temperature than an ideal diblock of infinite molecular weight. In other words, one must cool from a melt to a lower temperature to obtain microphase separation. The constant 0.25 is calculated from the expression $f_i(1 - f_i)$ for the value above which one would expect a phase inversion. In the illustrative case depicted in Figure 4.6, phase inversion occurs at a hard-segment fraction ($f$) greater than 0.5. It can also be intuitively thought of as the ratio of the two blocks in any single chain. The term $\alpha_i$ represents the evaluation (calculated, measured, or guessed) for $f_i(1 - f_i)$ resulting in a phase boundary for a particular morphology i. Thus, in Figure 4.6, the phase boundary for the disordered/spherical transition is calculated with $\alpha_i = 0$, for the spherical to cylindrical phase transition with $\alpha_i = 0.098$, and for the cylindrical to lamellar transition with $\alpha_i = 0.206$. Conditions in which

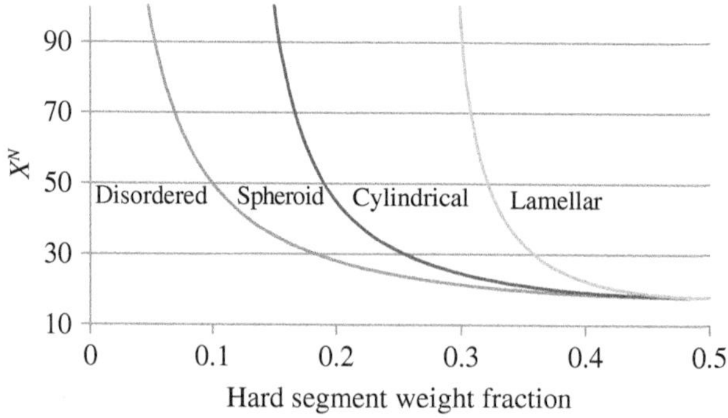


**FIGURE 4.6** Illustrative phase diagram of a polyurethane capable of exhibiting different phase structures as a function of hard-segment volume or molecular weight.

multiple phases can be simultaneously observed have been treated as a special case for the common occurrence where mixing of reactants is nonideal or the hard-segment fractions are very close to the phase boundaries and statistical fluctuations may result in phase diversity. The means by which these phases form from the melt and affect properties is discussed below.

The phase separation mechanism by which phase morphologies (Figure 4.6) can occur has been reviewed and validated experimentally for polyurethanes [15–18]. Since polyurethane is simultaneously forming, phase separating, and potentially creating a gaseous coproduct, the path from high-temperature melt to final product is quite a bit more complicated than that for an ideal block copolymer. For example, considerations of *ideal* block copolymers are that, as the melt cools (quenches), the overall system free energy is at a minimum at each temperature, arriving at a final equilibrium structure at the final state. The depth of quench is sometime referred to as the difference between the upper critical solution temperature and the system temperature [19]. In reactive systems such as polyurethanes the system is evolving both chemically and rheologically, changing the definition of equilibrium conditions continuously (i.e. there is no reference state). Furthermore, as the reactions proceed and oligomers of hard segment coupled to soft segment form, they can act as surfactants with the resulting change in the distribution of $\chi$ values and the interfacial energy of components in the mixture. In this case the quench is potentially determined by temperature and simultaneously by the degree of conversion to polymer. As the polymer is forming and the block repulsions are initiating phase separation, the phases may separate by either nucleation and growth or by spinodal decomposition.

The difference between the nucleation-and-growth and spinodal phase-separation mechanisms is critical to understanding final polymer morphologies and controlling properties via formulation and processing [17, 18]. In phase separation by nucleation and growth, the sample is quenched to a state where the solubility of each pure phase of the block exceeds the capacity of the melt. The phase separation begins as the homogeneous mixture enters this unstable region of the phase diagram (Figure 4.7). Nucleation and growth is the expected mechanism when the system proceeds slowly in temperature to the binodal curve [20]. The composition of the phase-separated region proceeds along the arms of the binodal (or coexistence) curves to a composition of two major phases, each one solubilizing the minor component at the solubility limit of the other phase [21]. Nucleation occurs since, in this mechanism, dispersed nuclei of each phase will form in the melt as the activation energy for nucleus formation exceeds the free energy of surface formation [22, 23]. Radiation scattering such as small-angle x-ray scattering (Section 5.2.3.2) from a system phase separating by a nucleation-and-growth mechanism will exhibit a linear decrease with scattering angle, and increase with time at a fixed angle according to Equation 4.5.

$$I = K\left(t - \tau\right)^n \tag{4.5}$$

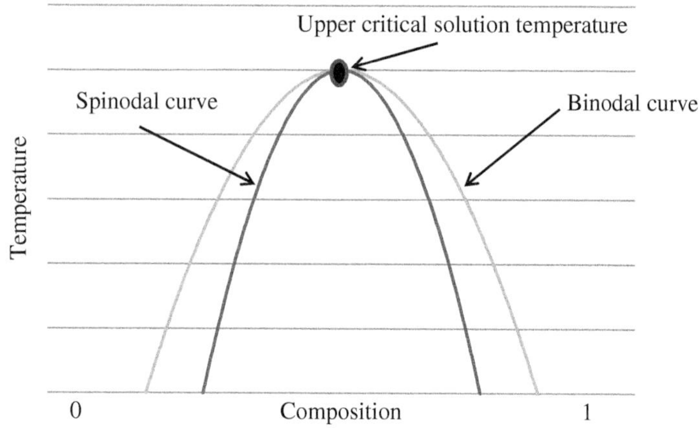


**FIGURE 4.7** Illustrative phase diagram for polyurethane phase decomposition through the upper critical solution temperature upon temperature decrease or molecular weight increase.

Time lag prior to the inception of nucleation, $\tau$, is often arbitrarily made zero. The value $n$ for nucleation and growth is well established to be between 3 and 4, such that a plot of log intensity increase with log time at a fixed angle will have a slope of between 3 and 4. This behavior is strong evidence for a nucleation-and-growth mechanism of phase separation.

In the case of spinodal phase decomposition the system is rapidly quenched (i.e. cooled rapidly or made unstable by rapid molecular weight growth) and the system will occupy an unstable position bounded by the spinodal curve of Figure 4.7. In this case phase separation will initiate through the existing concentration fluctuations with increasing amplitude and with a characteristic interphase spacing. The separating phases will form two cocontinuous and sometimes interpenetrating structures [19, 24–26]. In some cases, the final morphology may assume the appearance of a structure developed by nucleation and growth because of phase coalescence and coarsening at long times. However; the difference in mechanism as measured by radiation scattering is diagnostic. In a spinodal decomposition the system may exhibit a time lag in the development of structure; however, once phase separation begins, intensity will grow at a characteristic angle, which can be translated via the Bragg equation (Equation 5.6) into a characteristic spacing. The scattering intensity will have a finite breadth reflecting inhomogeneity of the periodicity within the sampled volume, but the maximum will be at a constant angle and will exhibit exponential growth according to Equation 4.6, where $R(q)$ is the amplitude growth rate of fluctuations of wavenumber $q$.

$$I(q,t) = I(q,0)e^{[2R(q)t]} \tag{4.6}$$

Thus a plot of ln $I(q,t)$ versus time ($t$) should yield a straight line during the early stages of phase separation. While there is no question that the reaction-induced phase separation that occurs during polyurethane polymerization is different from the thermally induced phase separation upon cooling the melt of an ideal block copolymer, the behavior of polyurethane foam during foam formation exhibits scattering behavior consistent with a spinodal mechanism for phase separation. In the case of thermoplastic elastomers the resulting structure can be consistent with either mechanism of phase separation. Specific elastomer processes can certainly exhibit the appearance of a phase separation that has developed via nucleation and growth (Figure 4.8) using a slow cooling process, and also scattering signatures not confirming spinodal decomposition using a fast reactive injection-molding process [17]. Many other thermoplastic polyurethane elastomers will exhibit finely divided cocontinuous gross morphologies typical of spinodal decomposition [16].

The significant differences in appearance observed between a morphology from spinodal and nucleation-and-growth mechanisms is a result of the difference in details of the mechanism. In both mechanisms the individual blocks of an A–B block copolymer will possess some level of mutual insolubility and insolubility in the melt. As the blocks diffuse the local concentrations of components will exhibit increasing fluctuations. Fick's law stipulates that flux of an ideal mixture will be proportional to a concentration gradient.

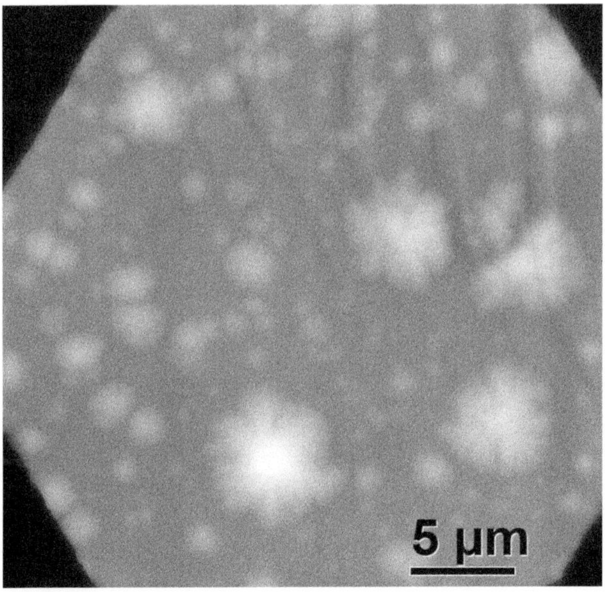


**FIGURE 4.8** Transmission electron microscopy of a cast polyurethane elastomer with a 35% 4,4′-methylene diphenyl diisocyanate 1,4-butanediol (MDI-BDO) hard segment with the spherulitic appearance typical of a nucleation-and-growth phase separation. Polymer was formed at 85 °C and sat quiescently at temperature for 4 hours.

$$J_{\mathrm{A,B}} = -\frac{D_{A,B}C_{A,B}}{RT}\frac{\partial \mu_{\mathrm{A,B}}}{\partial C_{\mathrm{A,B}}} \tag{4.7}$$

For nonideal mixtures or solutions the driving force is the gradient of chemical potential.

$$J_{\mathrm{A,B}} = -\frac{D_{\mathrm{A,B}}C_{\mathrm{A,B}}}{RT}\frac{\partial \mu_{\mathrm{A,B}}}{\partial x_{\mathrm{A,B}}} \tag{4.8}$$

In Equations 4.7 and 4.8, $D$ is the diffusion coefficient proportional to the squared velocity of the diffusing particles depending on temperature, viscosity of the surrounding medium, and the size and shape of the diffusing block, and $C$ is the concentration of the individual species.

If the value of $\partial\mu_{(\mathrm{A,B})}/\partial C_{(\mathrm{A,B})}$ is positive then the diffusion coefficient is positive and diffusion occurs in the direction from high concentration to lower concentration, as is usually observed. When $\partial\mu_{(\mathrm{A,B})}/\partial C_{(\mathrm{A,B})}$ is negative then the direction of diffusion is from lower concentration to higher concentration. The former case is typical of diffusion occurring in a nucleation and growth mechanism [27, 28]. This is because the interface between two phases growing along the binodal curve is "infinitely" sharp and so the concentration just outside this phase is quite low, and away from the growing separating phase the concentration of species is relatively higher. The latter case is typical of phase growth by a spinodal mechanism. This is because the phase formed via spinodal separation is not sharp and the concentration between concentration fluctuation waves is relatively lower between the concentration wave peaks yet still diffuses in the direction of the higher concentration peaks. In some cases and at long times, the morphologies created by spinodal decomposition and nucleation-and-growth mechanisms can be indistinguishable because of coarsening ("Ostwald ripening") of the spinodal phase structure with time.

## 4.2 PROPERTIES OF POLYURETHANES

Modern understanding of polyurethane structure–property relationships is conventionally first associated with the work of Cooper and Tobolsky in 1966 [1]. Their recognition that modulus values of thermoplastic polyurethanes could not be modeled using standard equations for filler reinforcement led to detailed understanding of these materials. Greater theoretical insight awaited further experimental progress. Koberstein and Stein [3] suggested that phase separation of a reactive block copolymer system such as polyurethane occurs only after the attainment of a critical hard-segment length (or average number of repeat units). Consistent with the conclusion of Cooper and Tobolsky they ruled out a random two-phase morphology on the basis of their x-ray scattering curves. They estimated a critical hard-segment length of between 2 and 3 depending on numerous hard-segment and processing factors. Subsequent work by Ryan and coworkers [9] suggested a minimum critical length slightly lower than this. These results are consistent with calculation of the Flory interaction parameter discussed earlier.

With experimental description of the polyurethane structure at many length scales, and the theoretical formalisms of polymer superstructure from block copolymer theories, composite models were developed that could account for the properties of polyurethane foams and elastomers. Block copolymer theory (see Section 4.1) had suggested that block copolymers could arrange themselves into spherical, cylindrical, lamellar, and gyroid structures, for instance, and various microscopy techniques were enlisted to detect morphologies that were forming during reactive polyurethane processing. In fact many block copolymer structures that can be observed have indeed been observed in polyurethanes. Spherical (Figure 4.8), cylindrical (Figure 4.4), and lamellar (Figure 4.5) morphologies have been observed by the author and others [14, 29]. Gyroid-like morphologies are also observable (see Figure 5.9). What has not been observed is ordered packing of the hard-segment phases, such as is observed in ideal block copolymers. Thus cylindrical hard-segment phases can be observed, but hexagonal packing of those cylinders, for instance, is not.

Regardless of the phase-separation temperature or molecular weight, the calculation of properties based on that structure is significantly affected by the assumptions made in the composite model. Simple equations can result from sophisticated analyses and subtle assumptions.

### 4.2.1 Models and Calculations for Polymer Modulus

An example of a relatively simple equation for a two-phase composite model that has been employed to model polyurethane properties is the Kerner equation [30]. Like many such equations, the derivation of the final equation assumes a macroscopically homogeneous and isotropic solid with properties defined by the properties and concentrations of the components in the bulk. One component is assumed to be on average spherical and randomly distributed within a uniform matrix. Key considerations in this model are that the composite inclusions are bonded to the matrix and that they are isolated from each other – thus, they do not form a cocontinuous structure within the composite. The approach used by Kerner is termed a generalized self-consistent field approach. Kerner assumed that the spherical particle is embedded in a spherical shell of matrix surrounded by an "infinite" body with the average composite properties. A basis for the equation is also that the adhesion between the filler and the matrix is perfect, so that displacements across boundaries are continuous and invariant. The spatially averaged properties of the embedded particle must match the properties of the matrix. Analysis of the cell and its properties and distributions are iterated until there is a match between the modeled volume element and the bulk reference [31]. The resulting Kerner equation is given by Equation 4.9, where $G_c$ is the shear modulus of the composite, $G_p$ is the shear modulus of the matrix phase, $G_f$ is the shear modulus of the filler phase, $v_p$ and $v_f$ are the volume fraction of polymer and filler, respectively, and $\nu$ is Poisson's ratio.

$$\frac{G_c}{G_p} = \frac{\dfrac{G_f v_f}{\left[(7-5v)G_p+(8-10v)G_f\right]}+\dfrac{v_p}{\left[15(1-v)\right]}}{\dfrac{G_p v_f}{\left[(7-5v)G_p+(8-10v)G_f\right]}+\dfrac{v_p}{\left[15(1-v)\right]}} \tag{4.9}$$

The form of Equation 4.9 is that for the derived Mossotti–Clausius equation for defining the properties of two-component systems, with each component having definable and unique properties [32]. The terms $G_i(7-5v)/(8-10v)$ are derived for the purpose of setting upper and lower bounds on the bulk moduli [33]. The value of Poisson's ratio, $\nu$, will have a quantitative effect on the result, but not a qualitative effect. In principle $\nu$ should be a function of the filler volume fraction for the usual case that the matrix and the filler have different Poisson's ratio values, but in practice this is usually not a significant effect.

In contrast to the isolated particle assumption of Kerner's equation (Equation 4.9), the Davies equation (Equation 4.10) assumes that the filler phase and the matrix phases are cocontinuous and adhesion between the phases is perfect [34]. Other than the cocontinuity assumption, the physical foundation of Davies' model is similar to that of the Kerner equation [35]

$$G^{1/5} = \varnothing_1 G_1^{1/5} + \varnothing_2 G_2^{1/5} \tag{4.10}$$

In Equation 4.10 $\varnothing_i$ represents the fraction of component $i$ while $G_i$ represents the respective shear modulus. The Davies equation has been used successfully to model the composition dependence of the moduli of numerous two-phase systems, including semicrystalline polymers [36], block copolymers [37], and block copolymers with an interstitially polymerized third polymer phase [31, 38].

The simplicity and success of the Davies equation has inspired modifications to take account of physical phenomena associated with the phase-separation process such as phase percolation associated with the transition from the spherical to the cylindrical phase transition (Figure 4.6). An example might be typified by Equation 4.11 [37], which has demonstrated additional utility in evaluating polyurethane elastomer data.

$$G^{1/5} = \varnothing_p G_H^{1/5} + \left(1-\varnothing_p\right) G_S^{1/5} \tag{4.11}$$

$$\varnothing_p = \left[\varnothing_H - \varnothing_{SC}\right]$$

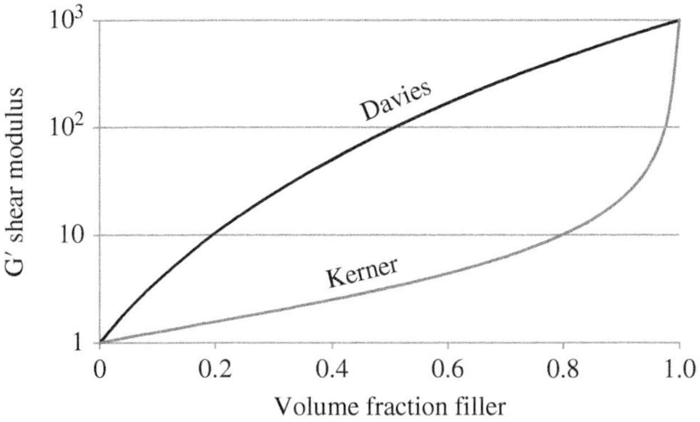


**FIGURE 4.9** Graphical representations of the Kerner and Davies equations using the assumption of a $10^3$ difference between filler and polymer moduli (relevant to polyurethane systems).

In Equation 4.11 $\varnothing_p$ is the hard-segment weight fraction that is reinforcing and $\varnothing_{SC}$ is the hard-segment fraction at which the phase morphology transitions from a spherical to a cylindrical phase morphology, which can be interpreted as a percolation threshold.

A visual comparison of the Davies and Kerner equations is provided in Figure 4.9. The rapid and continuous increase of modulus in the Davies equation is reflective of the percolation assumption. The slow increase in modulus observed for systems modeled by Kerner's equation reflects the noninteraction of particles and their noncontinuous phase status.

Both the Davies and Kerner equations lack recognition that at some hard-segment fraction there can occur a phase inversion in which the hard segment is the matrix phase and the soft segment is in effect the filler. Budiansky [39] proposed an alternative two-phase composite model that would provide a means for such an inversion to occur. The Budiansky model differs fundamentally in that this model embeds the filler particle (hard segment in a polyurethane) directly into the effective medium rather than in the matrix polymer. In this manner the physics and the equation are symmetric with replacement of the matrix and filler fractions (Equation 4.12).

$$\varnothing_2 = \frac{G_1 - G}{G(G_1 - G_2)} \times \left( \frac{3}{5}G + \frac{2}{5}G_2 \right) \tag{4.12}$$

In Equation 4.12, $\varnothing_2$ is the variable volume fraction, $G$ is the composite modulus, and $G_1$ and $G_2$ are the filler and matrix moduli, respectively. A graphical representation of the Budiansky equation is shown in Figure 4.10. This graph shows the realistic case for systems intermediate to the Kerner and Davies cases, in that the filler may at some range of volume fraction exist in isolated domains. In the Budiansky representation, the modulus function begins to show deviation from an initial linear increase at approximately 30% volume percentage filler. At a volume fraction of about 60% the function again assumes a linear increase, suggesting a phase inversion has occurred. While a phase inversion is not a requirement of a specific composite system in this range, and the

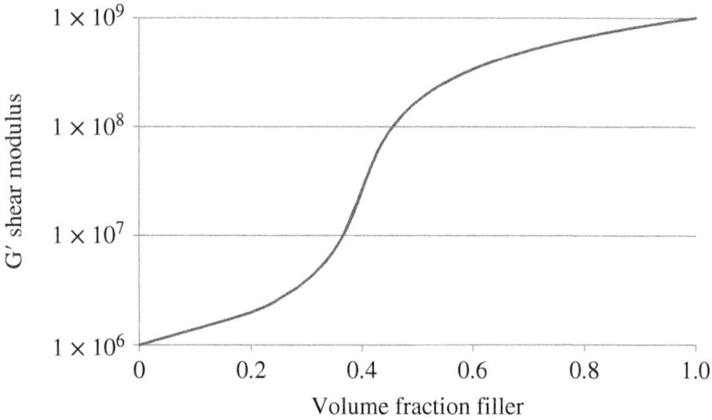


**FIGURE 4.10** Graphical representation of the Budiansky equation using a soft-phase modulus of 1 MPa and a hard-phase modulus of 1 GPa as relevant to polyurethane systems.

specific details of the derivative in any specific range may be inexact, the basic form of the Budiansky equation is quite similar to that observed in actual polyurethanes as a function of hard-segment volume. The actual values obtained in the moderate to high hard-segment volumes are often better approximated by the Davies equation [31, 40].

Another commonly invoked equation to predict composite modulus is the Halpin–Tsai equation (including the numerous variations and refinements that have been developed). This equation is unique in that it allows for the distinction of filler shape. It has been well applied to many composite materials and materials that can be modeled as composites, extending from semicrystalline polymers to nanocomposite-filled thermosets [38, 41–44]. For polyurethanes it is possible to use a simple form of the equation such as Equation 4.13.

$$E_{\text{composite}} = \frac{\left\{E_{\text{filler}} + 2A\left[E_{\text{filler}} + (1-\varnothing)E_{\text{matrix}}\right]\right\}}{\left[(2A+\varnothing)E_{\text{matrix}} + (1-\varnothing)E_{\text{filler}}\right]} \tag{4.13}$$

where $A$ is the filler aspect ratio, $\varnothing$ is the filler volume fraction, and the other symbols are the respective moduli. The aspect ratio is the ratio of the filler's length to width. Thus a spherical particle would have an aspect ratio of 1, while a fiber might have an aspect ratio of 100 or higher. An assumption of Equation 4.13 is that anisotropic fillers are perfectly aligned. The modulus for the direction perpendicular to fiber orientation can be calculated with the alternative assumption of $A = 1$ and the modulus of the overall random composite calculated by taking the average. A graphical representation of the Halpin–Tsai expression is given in Figure 4.11, which shows the significant orientation effect of high-aspect-ratio fillers. The similarity between the Kerner result (Figure 4.9) and the Halpin–Tsai result for $A = 1$ reflects the parallel initial assumptions.

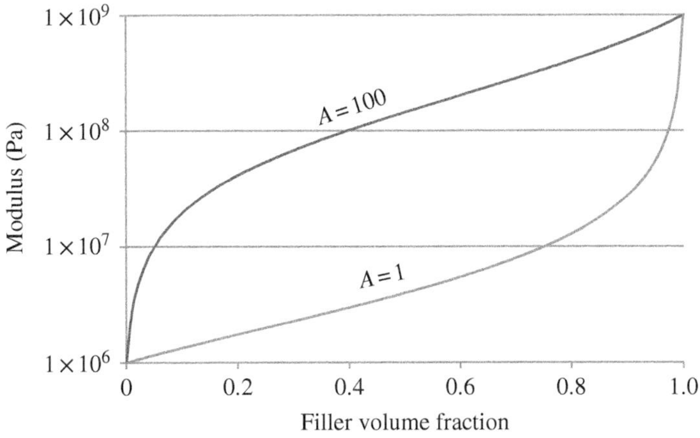


**FIGURE 4.11** Graphical representation of the Halpin–Tsai equation (Equation 4.12) using a soft-phase modulus of 1 MPa, a hard-phase modulus of 1 GPa, and with aspect ratio (*A*) of either 100 or 1.

An approximate expression (Equation 4.14) that can also be applied is based on the expressions of Kolarik [45] and Bicerano [46] and on a simple percolation equation.

$$E_{\text{composite}} = E_{\text{filler}}\left(\frac{\left(\varnothing_{\text{filler}} - \varnothing_{\text{perc}}\right)}{1-\varnothing_{\text{perc}}}\right)^{\beta} \tag{4.14}$$

where $\varnothing_{\text{perc}}$ is the volume fraction at which the filler percolates as a cocontinuous phase. For polyurethanes the percolation threshold can be construed at the spherical to cylindrical phase transition or even at the cylindrical to lamellar phase transition (Figure 4.6). The exponent $\beta$ is sometimes called the percolation exponent and is an adjustable variable of the order of 2. An obvious simplification in this calculation is the invitation to neglect the soft segment since its modulus is much smaller than that of the hard segment. The equation does not have a defined result below the percolation threshold but can be refined to compensate if desired. Results are qualitatively similar to the Davies equation (Figure 4.12).

### 4.2.2 Models for Elastomer Stress–Strain Properties

As discussed in Section 4.2.1, models predicting the modulus of polyurethanes have been proposed. Based on realistic models of polyurethane structure, several have been shown to reproduce measured properties with good fidelity. Part of the success of these models is that they occur in the low-strain–linear response range for the materials such that there are no irreversible plastic deformations that occur during the test. Modeling

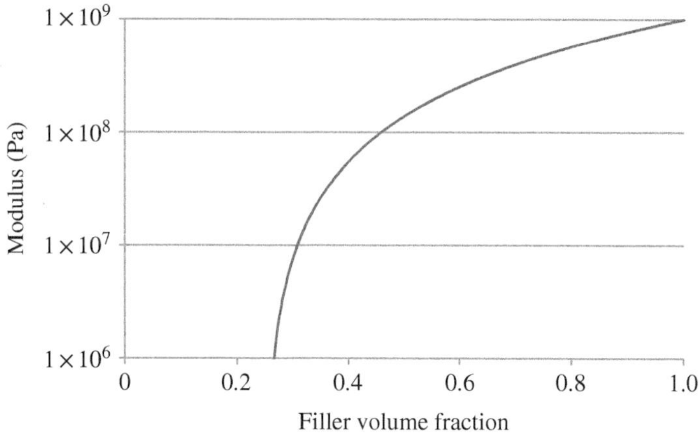


**FIGURE 4.12** Graphical representation of the modulus predictions based on a percolation equation (Equation 4.14) with a filler modulus of 1 GPa and a percolation threshold of 0.25 filler volume fraction.

the whole stress–strain curve is a much larger challenge. High-strain experiments on polyurethanes enter the domain of nonlinear viscoelasticity and irreversible processes that occur to the soft segment (i.e. disentanglement, bond rupture), hard segment (shear yielding and orientation), and physical processes that may be reversible, but only partially so on the time scale of the measurement (i.e. hysteresis and elastic set). To some extent polyurethane stress–strain behavior can be anticipated based on composition, phase structure, and reasonable assumptions about how polymers behave.

***Factors that Affect Polyurethane Stress–Strain Behavior*** For any given polyurethane sample undergoing mechanical testing there are four scenarios that one should consider when trying to anticipate large strain behavior. These scenarios are shown schematically in Figure 4.13.

Scenario 1: Hard and soft phase are below their respective glass transition temperatures ($T_g$), exhibiting properties of a hard glassy polymer. The plastic exhibits linear elastic behavior up until large-scale flow is initiated at the yield point, accompanied by a clear necking as the material exhibits Poisson narrowing initiating at some nonuniformity in the material. The material elongates plastically up until failure, which is a function of strain rate, temperature, defectivity, and details of the material structure. An example of a polyurethane sample tested within this range is shown in Figure 4.14.

Scenario 2: Hard phase below its $T_g$ and soft phase above its $T_g$. This is the typical situation for a polyurethane elastomer in application. In this case the material strain properties will be sensitive to the microphase separation such that if the material is not microphase separated it will elongate like a cross-linked rubber (Figure 4.15a). If the material has microphase separated but is in a spherical phase, or is a different phase below its percolation threshold, it may elongate like a filled rubber (Figure 14.5b). If the material is phase separated and the hard phase is above its percolation threshold the material will behave as a composite (Figure 14.5c) with a high initial modulus that drops rapidly at elongations above a poorly defined yield point.

| 1 | 2 |
|---|---|
| **Hard phase below $T_g$**<br>**Soft phase below $T_g$** | **Hard phase below $T_g$**<br>**Soft phase above $T_g$** |
| **3** | **4** |
| **Hard phase above $T_g$**<br>**Soft phase below $T_g$** | **Hard phase above $T_g$**<br>**Soft phase above $T_g$** |

**FIGURE 4.13** Diagram of the categories of phase-separated polyurethanes that affect stress–strain properties apart from molecular weight.

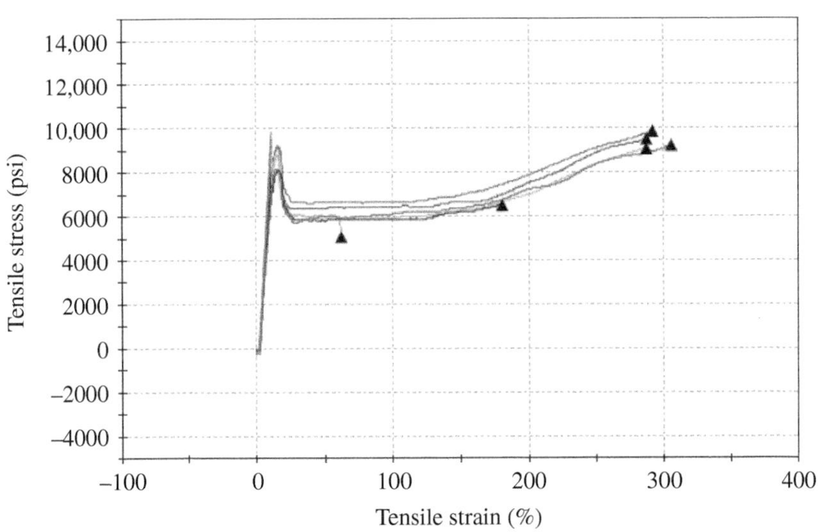


**FIGURE 4.14** Illustrative stress–strain curve for a polyurethane elastomer characterized by Scenario 1 with both soft and hard segments below their $T_g$.

Scenario 3: This is a special case that is rarely observed in practice. Such a case is conceivable in the instance where the soft segment is an aromatic polyester phase and the hard segment comprises an aliphatic isocyanate and a branched diol. Depending on their volume fractions the stress–strain curves may appear as a special case of those observed in Figure 4.15.

Scenario 4: In the case where the soft segment and the hard segment are both above their respective $T_g$ one should expect that the material will extend as a weakly cross-linked elastic network, elongating at relatively low stress and breaking at relatively high extensions and low stress (Figure 14.6).

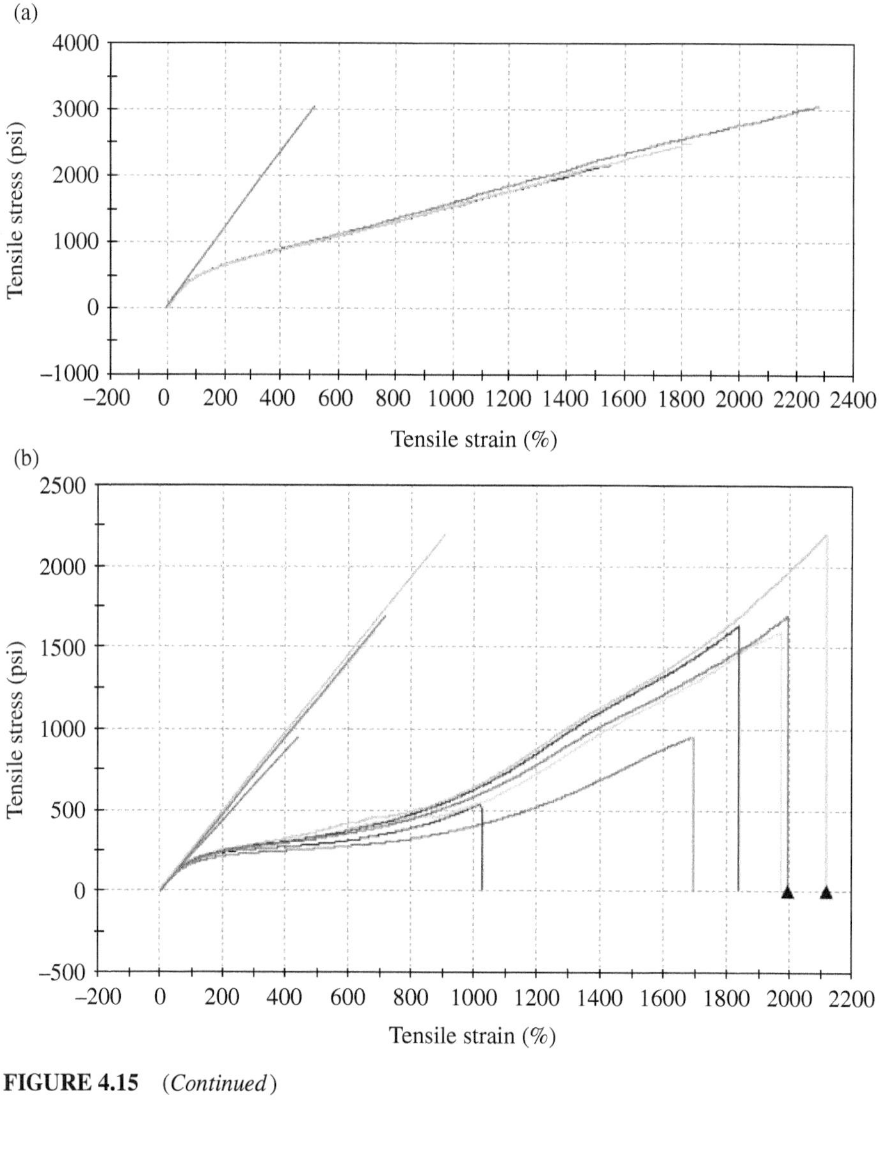


**FIGURE 4.15** (*Continued*)

(c)

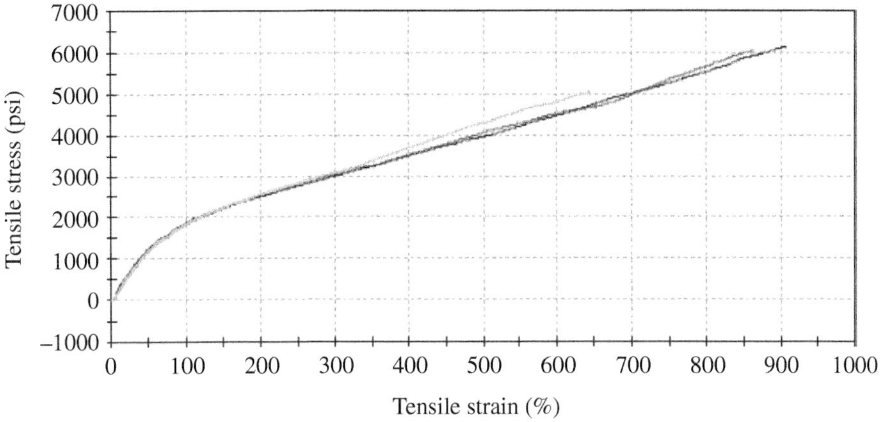


**FIGURE 4.15** Illustrative stress–strain curves for polyurethane elastomers of scenario 2. (a) The hard phase is below the microphase separation limit. (b) The hard phase is microphase separated but not cocontinuous. (c) The hard segment is phase separated and above the percolation threshold.

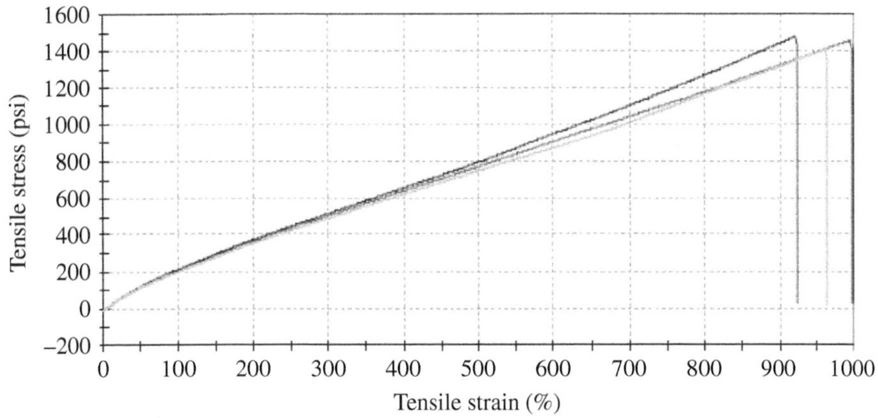


**FIGURE 4.16** Illustrative stress–strain curve for a polyurethane elastomer with both the hard and soft segments above their respective glass transitions.

Attempts to model stress–strain curves have usually followed an "equivalent box model" such as employed by Kolarik [45, 47] and others [48]. This model separates the total stress into the sum of stress contained by the hard segment and a soft segment (Equation 4.15) that may have dissolved components of hard segment in it (Figure 4.17).

$$\sigma = \sigma_{HS} + \sigma_{SS} \tag{4.15}$$

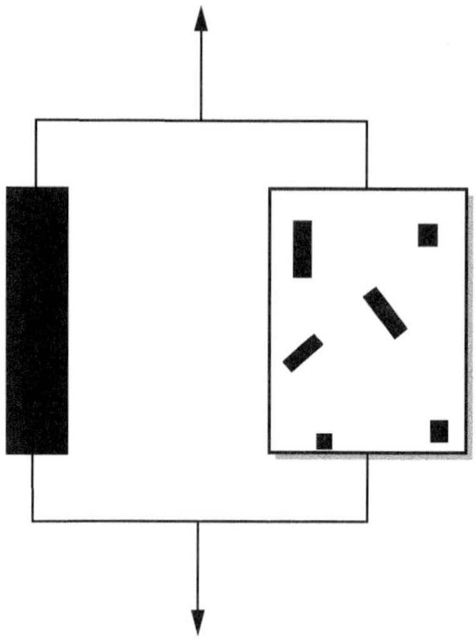


**FIGURE 4.17** The equivalent box model for calculating large strain properties of polyurethanes.

A representation of the functional dependence of stress with strain that allows a separable and additive calculation would possibly be represented by Equations 4.16. In Equations 4.16 $\varphi_h$ is the volume fraction of the hard phase, ε is the strain, $\varepsilon_y$ is the yield strain, $E_h$ is the modulus of the hard phase (estimated from the modulus of a conventional thermoplastic below its $T_g$), $E_s$ is the modulus of the soft segment, and $P(\lambda)$ is a complex function reflecting the stress amplification of the dissolved hard phase within the soft segment as it reinforces the soft segment as a function of the soft-segment strain. This phenomenon of strain reinforcement was first observed as a deviation of measurements on polyurethanes from the behavior predicted by the Mooney–Rivlin equation [49]. The Mooney–Rivlin equation has been used with good success to model stress–strain behavior to moderate strain levels, at which point the limited or constrained extensibility of chains becomes significant [50–52].

$$\sigma_h = \varphi_h E_h \varepsilon_y \left( 1 - e^{\left[\frac{\varepsilon}{\varepsilon_y}\right]} \right) \tag{4.16}$$

$$\sigma_s = (1 - \varphi_h) E_s P(\lambda)$$

As strain is increased the stress amplification of dissolved hard segments decreases since some of the hard-segment clusters yield. At the limit of maximum extension, when all the soft-segment chains have fully extended, all or most of the hard-segment clusters would have yielded prior to soft-segment chains breaking, and the stress amplification of dissolved hard segment is low (assuming the dissolved cluster cohesive energy is less than the bond energy of the soft-segment chain) [53, 54].

Calculation of ultimate elongation and toughness (the integral under the stress–strain curve) for an unfilled Gaussian network is usually related to the contour length of the chains when they are completely stretched, and incremental strain increases will result in chain breakage. For polyurethanes within case 2 (Figure 4.13), the situation is substantially more complex. Elongation of these systems is known to depend on unquantifiable factors associated with sample history and processing minutiae. Reasonable attempts have been made to model the stress–strain behavior of cross-linked glassy materials such as epoxies. Predictions from semiempirical equations (i.e. the Martin, Roth, and Stiehler equation) and an analogous equation replacing stress and strain variables with normalized failure stress and strain (Equations 4.17) and experimental validation have been obtained [55, 56].

$$\sigma = E\left(\frac{\varepsilon}{\lambda^2}\right)e^{\left[A\left(\lambda-\frac{1}{\lambda}\right)\right]} \tag{4.17}$$

$$\left(\frac{\sigma_b}{\nu_e}\frac{273}{T}\right) = 3RT\left(\frac{\varepsilon_b}{\lambda_b^2}\right)e^{\left[A\left(\lambda_b-\frac{1}{\lambda_b}\right)\right]}$$

In Equation 4.17 $E$ is the tensile modulus, ε is the strain normalized to gauge length, $\lambda$ is the extension ratio (equal to ε + 1), $\varepsilon_b$ is the elongation at break, $\nu_e$ is the cross-link density, and A is an adjustable constant of the order of 0.5. A plot of log $\left(\frac{\sigma_b}{\nu_e}\frac{273}{T}\right)$ versus log $\varepsilon_b$ results in a consistent envelope of failure stress and failure strain for a variety of materials when normalized for their molecular weight between cross-links A graphical representation of Equation 4.17 for elastomers with different molecular weights between cross-links as a function of elongation at break is shown in Figure 4.18.

If one assumes that the ultimate strength of the elastomer is dictated by the ultimate strength of the soft-segment chains, it is expected that, as the hard-segment volume increases, the molecular weight between (physical) cross-links will go down and elongation at break will decrease. This is in qualitative agreement with observations. As the hard segment traverses the phase morphology space to a cocontinuous structure the effect of increasing cross-link density on strength and elongation to break becomes more obvious, as seen, for example, in Figures 4.15 and 4.16.

***Calculating Foam Properties*** The properties of polyurethane foams are a convolution of the polymer physical properties with the mechanical properties of foam structure. Almost any solid can be foamed and so the concepts that qualitatively and quantitatively describe generic foam mechanical properties can be applied to polyurethane foams, at least as a starting point. In fact, it has been shown in numerous places that polyurethane foam properties are often well described by a picture of a polyurethane elastomer at lower density [57–59].

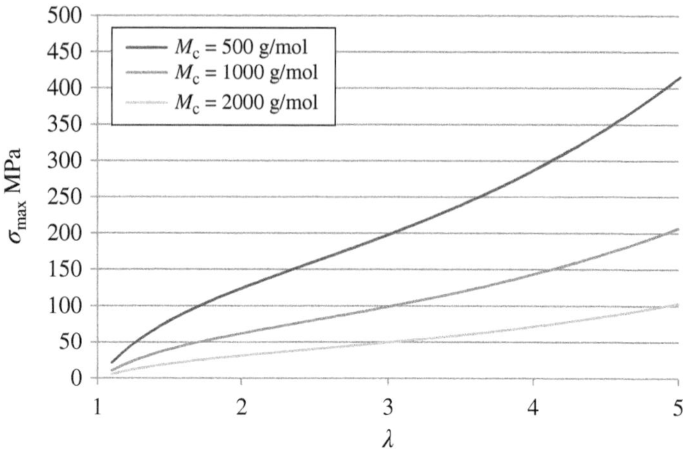


**FIGURE 4.18** Graphical representation of Equation 4.17 calculating the maximum stress as a function of maximum elongation and molecular weight between cross-links ($M_c$). The maximum elongation at break is arbitrarily, though reasonably, set at 400%. (*See insert for color representation of the figure.*)

Observations of properties translating from an elastomer to a foamed elastomer are accurate, but only in a limited sense. The fact is that morphological factors that distinguish polyurethane from nonmicrophase separating materials can have significant effects on the foaming properties. The chemical and physical processes that occur during the foam process can be quite different from those that occur when making a polyurethane elastomer, and they can modify the foam properties by resulting in greater or lesser phase separation and differing levels of network connectivity and molecular weight between cross-links. All the physical concepts that create a specific polyurethane elastomer translate to the polymer morphology of the polyurethane foam, but processing conditions can result in unexpected variance. Furthermore, foam properties are highly influenced by differing choices for catalysis, and surfactancy, which can affect physical properties, and cannot easily be accounted for in a predictive calculation [60].

For open cell foams the equation relating the Young's and shear moduli of foam to the modulus of unfoamed material is given by Equations 4.18. These equations are useful for a wide variety of polymers, densities, and polymer thermal properties [61]

$$\frac{E_{\text{foam}}}{E_{\text{solid}}} \sim \left(\frac{\rho_{\text{foam}}}{\rho_{\text{solid}}}\right)^2 \tag{4.18}$$

$$\frac{G_{\text{foam}}}{E_{\text{solid}}} \sim \frac{3}{8}\left(\frac{\rho_{\text{foam}}}{\rho_{\text{solid}}}\right)^2$$

where $E_i$ are the Young's moduli, $G$ is the shear modulus, and $\rho$ is the density. Measured variances from these simple relationships can be significant and can result from imperfections in foam cell struts, and the presence of closed cells or partially closed cells within the foam. Deviations from expected properties might well be expected to be sensitive to structural imperfections since the foams are usually tested in compression and the flexural modulus $E_b$, bending stress $\sigma_b$, and maximum strain at the bottom surface of a bending beam $\varepsilon_b$ are exceptionally sensitive to geometry (Equations 4.19)

$$E_{\text{bending}} = \frac{mL^3}{4bh^3} \tag{4.19}$$

$$\sigma_{\text{bend}} = \frac{3PL}{2bh^2}$$

$$\varepsilon_{\text{bend}} = \frac{6dh}{L^2}$$

In Equations 4.19 $L$ is the support span, $b$ is the width of the beam, $h$ is the height of the beam, $m$ is the slope of the linear elastic portion of the load deflection curve, $d$ is the deflection at the applied load, and $P$ is a given point on the load deflection curve.

Closed cell foams are more complicated than open cell foams from a fundamental point of view, if only slightly more difficult from a computational perspective. Polyurethane foams are made from an initial liquid state. As the chemical and physical processes proceed to form a foam, material will be drawn from the expanding cell bubble into the edges. In open cell foam, cells matriculate as closed cells up until the thinning cell face membrane ruptures and open cells percolate a continuous open channel from one end of the foam to the other. In the case of closed cell foams, the internal gas pressure that builds within the individual cell bubbles is insufficient to break the thinning cell face membrane prior to the system vitrifying [62]. In this case a substantial amount of the foam tensile properties will be the result of the foamed material contained within the cell faces.

An additional contribution to cell tensile properties will be made by the gas that is trapped within the closed cells. If the sample is compressed by given strain ε, the decrease in the volume of the foam is given by Equation 4.20, where $\nu$ is Poisson's ratio.

$$\frac{V}{V_o} = 1 - \varepsilon;(1 - 2\nu) \tag{4.20}$$

However, the gas occupies only that space not occupied by the foam bubble edges and faces, so that the gas volume decreases by

$$\frac{V_g}{V^0{}_g} = \frac{1-\varepsilon;(1-2v)-\frac{\rho}{\rho_s}}{1-\frac{\rho}{\rho_s}} \tag{4.21}$$

where $\rho/\rho_s$ is the ratio of the foam density to the solid density. The change in the cell internal gas pressure upon compression is given by Boyle's law.

$$pV_g = p_o V^o{}_g \tag{4.22}$$

The pressure resisting compression and contributing to the foam stiffness is the difference between the calculated $p$ and the internal gas pressure under no strain ($p'$). The contribution of the gas trapped within closed cells to the foam modulus is given by Equation 4.23.

$$E_g = \frac{dp'}{d\varepsilon} = \frac{p_{o(1-2v)}}{\left(1-\frac{\rho}{\rho_s}\right)} \tag{4.23}$$

Thus the modulus of the closed cell foam can be calculated from a simple additivity assumption as equal to the sum of moduli from the cell edges, cell faces, and internal gas pressure (Equation 4.24) [61].

$$\frac{E_{ccf}}{E_s} = \varphi^2\left(\frac{\rho}{\rho_s}\right)^2 + (1-\varphi)\left(\frac{\rho}{\rho_s}\right)^2 + \frac{p_{o(1-2v)}}{\left(1-\frac{\rho}{\rho_s}\right)} \tag{4.24}$$

$$\frac{G}{E_s} = \frac{3}{8}\left[\varphi^2\left(\frac{\rho}{\rho_s}\right)^2 + (1-\varphi)\left(\frac{\rho}{\rho_s}\right)^2\right]$$

where $\varphi$ is the fraction of solid in the bubble edges and $1 - \varphi$ is the fraction in the cell faces. There is no shear modulus contribution from the cell internal gas pressure so it does not appear in that equation. Figure 4.19 compares the modulus ratio of the foam to the unfoamed polymer as a function of the ratio of the foam to the solid density using Equations 4.18 and 4.24. For reference, a rigid (closed cell) polyurethane foam has been reported to have a $\varphi$ value of 0.8 [63].

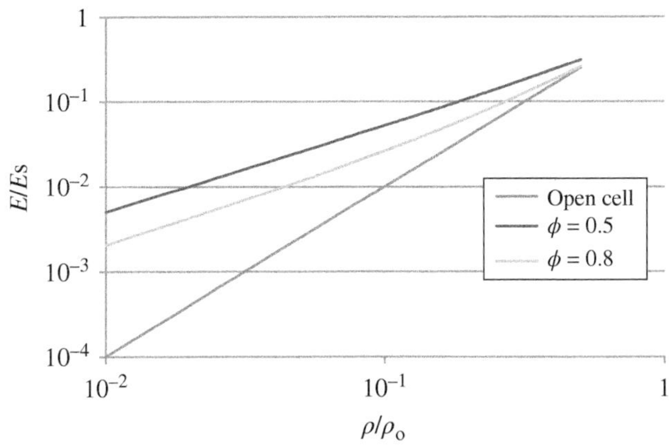


**FIGURE 4.19** Graphical representations of Equations 4.18 and 4.24 showing the modulus ratio of foam to its solid polymer as a function of density and the fraction of polymer in the cell edges ($\phi$). *(See insert for color representation of the figure.)*

### 4.2.3 The Polyurethane Glass Transition Temperature

Figure 4.13 defines the tensile properties of polyurethanes in relation to the phase-separated hard and soft segment $T_g$ values. The $T_g$ is certainly one of the more difficult physical parameters to treat from a theoretical perspective because of its "nonunique" nature [64]. This nonunique distinction is manifested in the value being a function of how it is measured (such as the rate of measurement), whether the transition is approached from low or high temperatures, and also by the instrumental technique. It is also problematic that the glass transition is not well defined by a single temperature but by a temperature region, with a finite and sometimes very substantial breadth (e.g. Figure 5.24). In such cases the $T_g$ will be treated as a specific point in an instrumental response curve, but this is often more by convention than actual utility [65].

Theoretical approaches to the polymer $T_g$ typically treat measured $T_g$ values as kinetic manifestations of underlying thermodynamic interactions among polymer segments, end groups, and additives. The specifics of these interactions can vary widely. Furthermore, disagreement on the thermodynamic definition of $T_g$ relative to discontinuities in thermodynamic quantities further complicates first-principle theoretical treatment. Nevertheless, there are numerous equations that are used in practice and in the scientific literature to estimate the $T_g$ of polyurethanes. Many of these equations are semiempirical, requiring input of measured data obtained from pure materials to make calculations based on interpolation. Others use measured reference values combined with fitting parameters that are defined in elaboration of fundamental principles. The choice of which equation a practitioner uses will depend on the individual's preferences and access to data that might be required for refined prediction of values. Another method has utilized group additivity principles, which can result in relatively good predictions if enough group factors are incorporated. Specific methods and values using group additivity principles have been presented [66].

One of the most common equations used for predicting the $T_g$ of binary systems, of which polyurethanes might be considered, is the Fox equation (Equation 4.25) [67]. The exceptionally simple form and the sole need to obtain the $T_g$ values of the pure components (which can normally be measured or even estimated with good precision) make it widely used. The simplicity of the equation is an expression of the numerous underlying assumptions. These assumptions are that the pure material $T_g$ values are constant with blend composition; that the contributions of the polymer components are additive (a small molecule plasticizer or a polymer blended with a polymer have equivalent effects) and symmetrical in their influence on the $T_g$; that the two polymers are miscible and amorphous; and that the heat capacity changes at the phase transition are the same for the two materials.

$$\frac{1}{T_g} = \frac{w_1}{T_{g1}} + \frac{w_2}{T_{g2}} \tag{4.25}$$

In Equation 4.25 $T_g$ represents the glass transition of the miscible blend, while $w_i$ represents the weight fractions of the respective polymer components and $T_{gi}$ the glass transition of the pure polymer component. The graphical representation of Equation 4.25 is given in Figure 4.20. It shows the weak $T_g$ depression that results from mixing two materials due to the slight increase in free volume and entropy in the absence of specific interactions. This is visualized by comparison with the graphed function, in which the blend $T_g$ is represented by a function in which the $T_g$ values are simply added in a mass weighted linear combination.

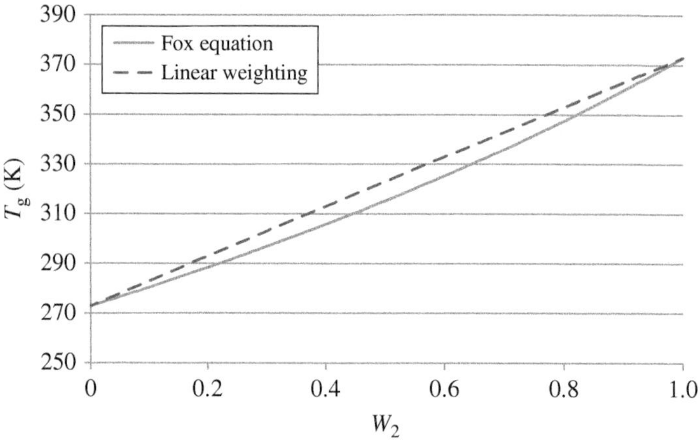


**FIGURE 4.20** Graphical representation of the Fox equation (Equation 4.25) relative to a simple mass-weighted linear combination of pure material $T_g$ values. For illustration $T_{g1}$ was chosen to be 273 K and $T_{g2}$ was chosen to be 373 K – relevant to many polyurethane systems. (*See insert for color representation of the figure.*)

The Fox equation is in fact a special case of the Gordon–Taylor equation (Equation 4.26) [68]. This equation, which is very widely and successfully used, contains the assumptions of the Fox equation but introduces a parameter ($k$). $k$ depends on the specific volume of the components ($V$) and the thermal expansion coefficients $\alpha$. In practice, $k$ is usually treated as a fitting parameter but should be of the order of 1. Practically this parameter allows for asymmetry between the blend component contributions to the blend $T_g$. This would allow for a physical situation in which the contribution of blend component 1 to the blend $T_g$ is weighted disproportionately to that of component 2, resulting in a stronger deviation (Figure 4.21) than predicted by the Fox equation.

$$T_g = \frac{w_1 T_{g1} + k w_2 T_{g2}}{w_1 + k w_2} \tag{4.26}$$

$$k = \frac{(\rho_1 T_{g1})}{(\rho_2 T_{g2})} \text{(The Simha – Boyer rule)}$$

The Fox equation can be derived from the Gordon–Taylor expression as a limiting case of the Simha–Boyer rule for $k$ in Equation 4.26, where specific volumes of the components are treated as identical, i.e. $k \sim \frac{T_{g1}}{T_{g2}}$ [69].

A special case of the Gordon–Taylor equation offered by Ginzburg and Sonnenschein [37] was developed specifically for polyurethanes reflecting microstructural details, as described by Figures 4.6 and 4.17. Equation 4.27 specifies the $T_g$ for the soft segment depending on the amount of dissolved hard segment in the soft segment ($w_H$). The dissolved hard segment is a function of the microphase structure defined by Figure 4.6 and quantitatively in Equation 4.4 by variable $(\chi^N)_o$.

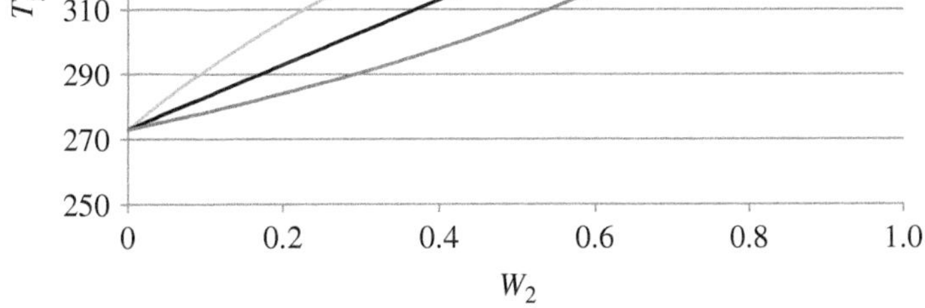


**FIGURE 4.21** Graphical representation of the Gordon–Taylor equation for varying values of $k$, yielding either positive or negative deviations from a mass-weighted linear combination of pure material $T_g$ values. (*See insert for color representation of the figure.*)

$$T_g = \frac{w_H T_{gH} + k(1-w_H)T_{gs}}{w_{1H} + k(1-w_H)} \tag{4.27}$$

$$w_H = \text{HSWF for } f_{(DS)} > \text{HSWF}$$

$$= \frac{f_{DS}}{1-[\text{HSWF} - f_{DS}]} \text{for } f_{(DS)} < \text{HSWF}$$

In this variation of the Gordon–Taylor formalism, if the hard-segment weight fraction (HSWF in Equation 4.27) is below the order–disorder transition (the D–S isomorph of Figure 4.6) the fraction hard segment in the soft phase is the total hard-segment mass in the polymer. However, if the polymer has a phase-separated microstructure, the amount of hard segment in the soft phase reflects the total hard segment minus the amount of hard segment that has precipitated into a separate phase.

An occurrence sometimes encountered in polymer blends is that two polymers interact through a specific and relatively strong interaction. Such an interaction would be, for example, strong hydrogen bonding between blend components. In this instance significant deviations from the previous equations may be observed. For this case the Kwei equation (Equation 4.28) has been utilized for systems exhibiting both negative and positive deviations from the linear combination profile (Figure 4.22) and "S"-shaped profiles [70].

$$T_{g=} \frac{w_1 T_{g1} + k w_2 T_{g2}}{w_1 + k w_2} + q w_1 w_2 \tag{4.28}$$

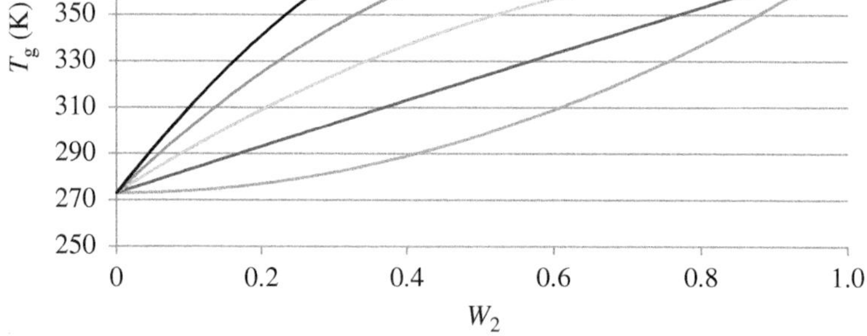


**FIGURE 4.22** Graphical representation of the Kwei equation, keeping $k$ at or near 1 (as in the Gordon–Taylor formalism) and varying $q$ to demonstrate its effect on the functional form. (*See insert for color representation of the figure.*)

The functional versatility of the equation results from the quadratic function in the second half of the equation. While $k$ and $q$ are usually treated as fitting parameters, there has been significant investment in creating a fundamental justification and elaboration for the $q$ variable as it relates to the intermolecular interactions responsible for the observed $T_g$ behavior [71, 72].

Another equation employed in the case of significant deviations from a simple weighted linear combination of $T_g$ values was offered by Lu and Weiss [73]. It is also a modification of the Gordon–Taylor equation (as is the Kwei equation) and incorporates a quadratic dependence, as does the Kwei equation. However, the $q$ value is elaborated in terms of the Flory $\chi$ value, which can be obtained by a best fit routine (Equation 4.29). In Equation 4.29 $b$ is the ratio of amorphous densities of polymer 2 and polymer 1, $\Delta C_{p1}$ is the change in specific heat of polymer 1 at $T_{g1}$, and the other variables are as previously defined for the Gordon–Taylor equation. In cases where the blend polymer interactions are quite strong this equation has shown utility and resulted in reasonable values of $\chi$. In the case of weak or moderate interactions, the Gordon–Taylor equation has provided equally predictive results to the Lu and Weiss equation [74].

$$T_g = \frac{w_1 T_{g1} + k w_2 T_{g2}}{w_1 + k w_2} - \frac{\chi R \left(T_{g2} - T_{g1}\right) b w_1 w_2}{\Delta C_{p1}\left(w_1 + k w_2\right)\left(w_1 + b w_2\right)^2} \tag{4.29}$$

While all of these equations have been applied to polyurethanes, there are quite a few other approaches which have been tried or could be applied [61, 75–79].

## REFERENCES


[1] S Cooper and A Tobolsky, *Journal of Applied Polymer Science*, **10**, (1966), 1839–1844.

[2] J Work, *Macromolecules*, **9**, (1976), 759–763.

[3] J Koberstein and R Stein, *Journal of Polymer Science: Polymer Physics Edition*, **21**, (1983), 1439–1472.

[4] L Liebler, *Macromolecules*, **13**, (1980), 1602–1617.

[5] T Ohta and K Kawasaki, *Macromolecules*, **19**, (1986), 2621–2632.

[6] G Fredrickson and E Helfand, *Journal of Chemical Physics*, **87**, (1987), 697–705.

[7] H Benoit and G Hadziioannou, *Macromolecules*, **21**, (1988), 1449–1464.

[8] V Ginzburg, J Bicerano, C Christenson, A Schrock, and A Patashinski, *Journal of Polymer Science Part B: Polymer Physics*, **45**, (2007), 2123–2135.

[9] M Elwell, S Mortimer, A Ryan, and W Bras, *Nuclear Instruments and Methods in Physics Research B*, **97**, (1995), 261–264.

[10] B Chu, T Gao, Y Li, J Wang, C Desper, and C Byrne, *Macromolecules*, **25**, (1992), 5724–5729.

[11] D van Krevelen, "Cohesive properties and solubility," *Properties of Polymers*, 3rd Ed., Elsevier, Amsterdam, the Netherlands, 1997.

[12] J Bicerano, "Cohesive energy and solubility parameter," in *Prediction of Polymer Properties*, 3rd Ed., CRC Press, Boca Raton, FL, 2002.

[13] M Elwell, A Ryan, H Grunbauer, and H van Lieshout, "Polymer structure development during reactive processing: in-situ studies of a cellular, multiphase block copolymer," in

*Polymer Foams, ACD Symposium Series 669*, K Khemani, Ed., CRC Press, Boca Raton, FL, 1997.

[14] J Garrett, C Siedlecki, and J Runt, *Macromolecules*, **34**, (2001), 1066–7070.

[15] W Yang and C Macosko, *Macromoleculare Chemie, Macromolecular Symposia*, **25**, (1989), 23–44.

[16] A Ryan, *Polymer*, **31**, (1990), 707–712.

[17] A Ryan, W Wilkomm, T Bergstrom, C Macoscko, J Koberstein, C Yu, and T Russell, *Macromolecules*, **24**, (1991), 2883–2889.

[18] W Li, A Ryan, and I Meier, *Macromolecules*, **35**, (2002), 5034–5042.

[19] T Ohnaga, W Chen, and T Inoue, *Polymer*, **35**, (1994), 3774–3781.

[20] C Fernandez, M Bermudez, R Versteegen, E Meijer, G Vansco, and S Munoz-Guerra, *European Polymer Journal*, **46**, (2010), 2089–2098.

[21] S Nunes and T Inoue, *Journal of Membrane Science*, **111**, (1996), 93–103.

[22] Y Li, J Liu, H Yang, D Ma, and B Chu, *Journal of Polymer Science: Part B: Polymer Physics*, **31**, (1993), 853–867.

[23] H Lee, Y Wang, W Macknight, and H Hsu, *Macromolecules*, **21**, (1988), 270–273.

[24] T Hashimoto, "Structure formation in polymer mixtures by spinodal decomposition," in *Current Topics in Polymer Science, Volume 2*, RM Ottenbrite, LA Utracki, and S Inoue, Eds., Carl Hanser Verlag, Munich, Germany, 1988, 199–242.

[25] I Hamley, J Stannford, A Wilkenson, M Elwell, and A Ryan, *Polymer*, **41**, (1999), 2569–2576.

[26] M Junker, I Allig, G Fleisher, and M Schulz, *Macromolecules*, **30**, (1997), 2085–2091.

[27] J Bicerano, R Daussin, M Elwell, H van der Wal, P Berthevas, M Brown, F Casati, W Farrissey, J Fosnaugh, R de Genova, et al., "Flexible polyurethane foams: a review of the state-of-the-art," in *Polymeric Foams*, S Lee and N Ramesh, Eds., CRC Press, Boca Raton, FL, 2004, 173–251.

[28] C Han and A Akcasu, "Dynamics and kinetics of phase separation in polymer systems," in *Scattering and Dynamics of Polymers: Seeking Order in Disordered Systems*, Wiley, Singapore, 2011.

[29] M O'Sickey, B Lawrey, and G Wilkes, *Journal of Applied Polymer Science*, **84**, (2002), 229–243.

[30] E Kerner, *Proceedings of the Physical Society, London*, **69B**, (1956), 808–813.

[31] G Allen, M Bowden, S Todd, D Blundell, G Jeffs, and W Davies, *Polymer*, **15**, (1974), 28–32.

[32] J Partington, *An Advanced Treatise on Physical Chemistry, Volume IV*, Longmans, Green and Co., New York, NY, 1953, 538–645.

[33] Z Hashin and S Shtrikman, *Journal of Mechanics and Physics Solids*, **11**, (1963), 127–140.

[34] W Davies, *Journal of Physics D: Applied Physics*, **4**, (1971), 318–328.

[35] K Wakabayashi and R Register, *Polymer*, **46**, (2005), 8838–8845.

[36] J Janzen, *Polymer Engineering and Science*, **32**, (1992), 1242–1254.

[37] M Sonnenschein, V Ginzburg, K Schiller, and B Wendt, *Polymer*, **54**, (2013), 1350–1360.

[38] M Sonnenschein, D Brune, and B Wendt, *Journal of Applied Polymer Science*, **113**, (2009), 1739–1744.

[39] B Budiansky, *Journal of Mechanics and Physics of Solids*, **13**, (1965), 223–227.

[40] I Soos, W Lenggoro, and H Kodaka, *Journal of Polymer Engineering*, **16**, (1996), 73–86.
[41] J Halpin and J Kardos, *Polymer Engineering and Science*, **16**, (1976), 344–352.
[42] L Sorrentino, M Aurilia, and S Iannace, *Journal of Applied Polymer Science*, **119**, (2011), 1239–1247.
[43] D Brune and J Bicerano, *Polymer*, **43**, (2001), 369–387.
[44] M Sonnenschein, Z Lysenko, D Brune, B Wendt, and A Schrock, *Polymer*, **46**, (2005), 10158–10166.
[45] J Kolarik, *European Polymer Journal*, **34**, 585–590, (1998).
[46] J Bicerano, Ed., "Shear viscosity," in *Prediction of Polymer Properties*, 3rd Ed., CRC Press, Boca Raton, FL, 2002.
[47] J Kolarik, *Polymer Engineering and Science*, **36**, (1996), 2518–2524.
[48] L Robeson and R Berner, *Journal of Polymer Science, Part B: Polymer Physics*, **39**, (2001), 1093–1106.
[49] G Spathis, *Journal of Applied Polymer Science*, **43**, (1991), 613–620.
[50] L Treloar, *Journal of Polymer Science: Polymer Symposium*, **48**, (1974), 107–123.
[51] E Arruda and M Boyce, *Journal of the Mechanics and Physics of Solids*, **41**, (1993), 389–412.
[52] M Boyce and E Arruda, *Rubber Chemistry and Technology*, **73**, (2000), 504–523.
[53] G Spathis, *Journal of Applied Polymer Science*, **43**, (1991), 613–620.
[54] O Thomas, R Priester, K Hinze, and D Latham, *Journal of Polymer Science B: Polymer Physics*, **32**, (1994), 2155–2169.
[55] L Shan, N Verghese, C Robertson, and K Reifsnider, *Journal of Polymer Science Part B: Polymer Physics*, **37**, (1999), 2815–2819.
[56] P Santangelo and C Roland, *Rubber Chemistry and Technology*, **67**, (1994), 359–365.
[57] J Armistead, G Wilkes, and R Turner, *Journal of Applied Polymer Science*, **35**, (1988), 601–629.
[58] M vander Schuur, E vander Heide, J Feijen, and R Gaymans, *Polymer*, **45**, (2004), 2721–2727.
[59] M Sonnenschein and B Wendt, *Polymer*, **54**, (2013), 2511–2520.
[60] R Herrington and K Hock, Eds., *Flexible Polyurethane Foams*, 2nd Ed., The Dow Chemical, Midland, MI, 1997.
[61] L Gibson and M Ashby, "The mechanics of foams: basic results," *Cellular Solids*, 2nd Ed., Cambridge University Press, Cambridge, UK, 1997.
[62] H Gruenbauer, J Bicerano, P Clavel, R Daussin, H de Vos, M Elwell, H Kawabata, H Kramer, D Latham, C Martin, et al., "Rigid polyurethane foams: a review of the state-of-the-art," in *Polymeric Foams*, S Lee and N Ramesh, Eds., CRC Press, Boca Raton, FL, 2004, 253–310.
[63] D Reitz, M Schnetz, and L Glicksman, *Journal of Cellular Plastics*, **20**, (1984), 104.
[64] I Kalogeras and W Brostow, *Journal of Polymer Science: Part B: Polymer Physics*, **4**, (2009)**7**, 80–95.
[65] W Brostow, R Chiu, I Kalogeras, and A Vasilikou-Dova, *Materials Letters*, **62**, (2008), 3152–3155.
[66] J Bicerano, "Transition and relaxation temperatures," in *Prediction of Polymer Properties*, 3rd Ed., CRC Press, Boca Raton, FL, 2002.
[67] T Fox, *Bulletin of the American Physics Society*, **1**, (1956), 123.

[68] M Gordon and J Taylor, *Journal of Applied Chemistry*, **2**, (1952), 493–500.
[69] R Simha and R Boyer, *Journal of Chemical Physics*, **37**, (1962), 1003–1007.
[70] T Kwei, *Journal of Polymer Science: Polymer Letters*, **22**, (1984), 307–313.
[71] H Schneider, *Journal of Research of National Institutes of Technology*, **102**, 229–248, (1997).
[72] A Lin, T Kwei, and A Reiser, *Macromolecules*, **22**, (1989), 4112–4119.
[73] X Lu and R Weiss, *Macromolecules*, **25**, (1992), 3242–3246.
[74] L An, D He, J Jing, Z Wang, D Yu, and B Jiang, *European Polymer Journal*, **33**, (1997), 1523–1528.
[75] R Pinal, *Entropy*, **10**, (2008), 207–223.
[76] E Dimarzio, *Polymer*, **31**, (1990), 2294–2298.
[77] P Couchman and F Karasz, *Macromolecules*, **11**, (1978), 117–119.
[78] E Jenkel and R Heusch, *Colloid and Polymer Science*, **130**, (1953), 89–105.
[79] G Kanig, *Journal of Polymer Science, Polymer Symposia*, **16**, (1967), 1957–1967.

# 5

# ANALYTICAL CHARACTERIZATION OF POLYURETHANES

Advances in instrumentation coupled with computerization and information technology have greatly advanced our understanding of urethane chemistry and polyurethane materials. This increased knowledge base has resulted in greater predictability in polyurethane use and increased proliferation of polyurethane materials across the spectrum of applications. Modern methods of analysis have permitted nearly every kind of time- and frequency-resolved analysis. Morphology can be characterized down to angstrom levels, allowing confident distinction between subtle differences in phase structure and mixing between materials. Prediction of properties, making of materials, and proper application of polyurethanes now requires that a practitioner be familiar with the operation of numerous analytical instruments. Even more crucial is understanding of the data produced by analyses, and the ability to perform root cause analyses of material or application failure based on those data. This chapter will provide an introduction to selected experimental techniques and more in-depth discussion of the connection between data and their use. Although information is almost always useful, gathering information can be wasteful if the measurements do not apply directly to the problems to be solved. This chapter will also provide guidance on the most efficient methods of determining the root cause of material or application failure.

Before a material can be tested or applied it must be made, and the incorrect making of a polyurethane is one of the most common causes for disappointing performance. The properties of a polyurethane are critically dependent on the proper ratio

*Polyurethanes: Science, Technology, Markets, and Trends*, Second Edition. Mark F. Sonnenschein.

of reactants being combined, and determining the correct ratios requires accurate measurement of reactant functionalities. As discussed in Chapters 2 and 3, there are numerous unintended side reactions that can occur in the preparation of polyols and isocyanates that can significantly affect the amount of hydroxyl or isocyanate functionality in a given weight of material. Incorrect measurement of reactant functionalities can have many negative effects on final physical properties, primarily by resulting in reduced final polymer molecular weight or an incomplete network structure.

It is nearly impossible to employ perfect stoichiometric ratios when making a polyurethane; however, it is possible to be close enough that a high final molecular weight polymer network is formed that molecular weight is not the limiting factor driving polyurethane performance. The first and most critical initial steps to making high-performance polyurethane materials is accurate measurement of the hydroxyl and isocyanate functionality per unit mass. ASTM (formerly the American Society for Testing and Materials) is an organization that provides industry-defined, accepted, and explicit methods for performing these analyses. They provide a basis for confidence that all manufacturers of polyurethane building blocks are producing materials that can be used to make a desired material using normal laboratory techniques. ASTM also provides guidance on alternative ways of obtaining the same piece of information using different analyses, which can substitute if a particular technique is not accessible or if independent methods of measurement of the same quantity are desired. Furthermore, ASTM methods provide a gauge of measurement reliability and, thus, a means of determining if a measurement is being made with the proper level of precision to meet an industrial level of expectation.

## 5.1 ANALYSIS OF REAGENTS FOR MAKING POLYURETHANES

### 5.1.1 Analysis of Polyols

There are several analyses that accompany virtually every batch of industrially produced polyol. These are indicated in Table 5.1 along with the relevant ASTM method developed to measure the property.

The application of these standard techniques will invariably result in a number whose precision can be well characterized by replicate samples. However, the ability to obtain optimal polymerization results will depend on the accuracy of the result. There is no a priori means of knowing if the analytical result a practitioner is handed is useful. The result of a poor analysis is a polyurethane with poor properties, and it will be ambiguous if the problem is a result of analysis or polymer design [1, 2]. The practitioner should never simply accept the result of a single experiment, and should always be suspicious of the polyol analysis, particularly the hydroxyl value. It may be necessary to bracket the polymerization assuming the measurement of the property is wrong by as much as 10%. This may require a new polymerization attempt at

**TABLE 5.1 Conventional polyol specifications and ASTM procedures for measurement**

| Property | ASTM method |
|---|---|
| Hydroxyl number | D4274 measurement by derivatization and titration |
| Hydroxyl number | D6342 measurement by near-infrared spectroscopy |
| Viscosity | D4878 measurement by Brookfield viscometry |
| Controlled polymerization rate (CPR) | D6437 measures alkalinity remaining from glycol polymerization, which can affect subsequent isocyanate polymerization |
| Unsaturation | D4671 Titration method uses mercuric acetate in a methanol solution |
| Cloud point | D2024 standard test method for nonionic surfactants |
| Water content | D4672 – Karl Fischer coulometric titration technique for alcohols or E203 general Karl Fischer titration method |

**TABLE 5.2 Critical evaluation of the ASTM method-derived property value in Table 5.1**

| Property | Reliability |
|---|---|
| Hydroxyl number derivatization method | Below average |
| Hydroxyl number near-infrared spectroscopy method | Above average |
| Viscosity | Above average |
| Controlled polymerization rate (CPR) | Average |
| Unsaturation | Average |
| Cloud point | Average |
| Water content | Below average |

Manufacturer numbers on well-established products are usually reliable. New materials made by new processes may be far less so.

the analyzed value, and additional attempts at a ratio 10% greater and 10% less than the measured value. Improved polymer properties in either direction may require additional empirical iteration. Table 5.2 lists the polyol quality tests of Table 5.1 and a relative evaluation of the reliability of the test based on statistical analysis within laboratories and between laboratories (this information is usually provided within the ASTM standard). The discrepancy is often a result of systematic operator errors, differences in execution of the ASTM tests (i.e. catalysts, solvents, etc.), and the use of autotitration [1] versus manual titration techniques. The author has observed that, unless an autotitrator is employed by a very experienced and methodical technician, manual titration is more likely to provide a reproducible result.

Each of the tests in Table 5.1 can be potentially important; however, the hydroxyl number and the controlled polymerization rate (CPR) value are the variables that can greatly affect the final result of the polyol use.

***5.1.1.1 Hydroxyl Number*** Measurement of hydroxyl number is usually obtained by the derivatization and titration technique. The general procedure is to carry out an esterification reaction of the alcohol with an anhydride, usually phthalic or acetic (Figure 5.1) [3]. The need to react all of the alcohol groups necessitates the use of an excess of anhydride and a strong base catalyst (i.e. standard 0.5 N $NaOH_{aq}$ or imidazole). The reaction is often run in pyridine solvent and carried out at elevated temperatures (c. 100 °C) for up to 2 hours. A blank (i.e. no polyol) is run side by side and the amount of acid is subsequently titrated using an indicator such as phenolphthalein for colorimetric end-point determination. The procedure is sufficiently exacting that operator experience and exactitude is a must to obtain useable results.

The hydroxyl number in units of milligrams KOH/gram of sample is calculated from Equation 5.1:

$$\left[(A-B)\mathrm{N}\times 56.1\right]/W \tag{5.1}$$

where *A* is the milliliters NaOH required for titration of the blank, *B* is the milliliters NaOH required for titration of the sample, N is the normality of the NaOH solution, 56.1 is the equivalent weight (g/eq) of KOH, and *W* is the weight of the sample used. The weight percentage OH in the sample is given by Equation 5.2, where 17 is the equivalent weight of the OH functionality [4]:

$$\frac{[(A-B)\mathrm{N}\times 17\times 100]}{\mathrm{mg\,sample}} \tag{5.2}$$

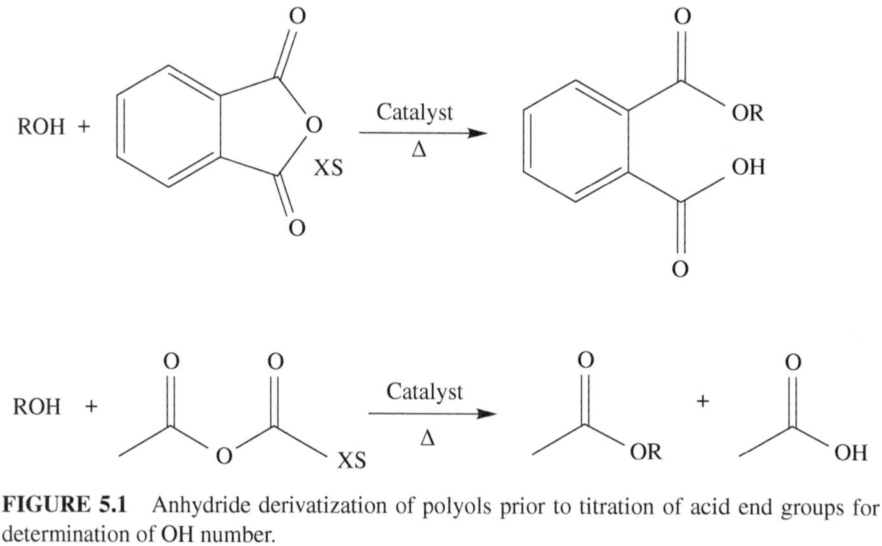


**FIGURE 5.1** Anhydride derivatization of polyols prior to titration of acid end groups for determination of OH number.

The application of an inaccurate hydroxyl number will result in materials with low polymer molecular weight and poor network formation, resulting at the minimum in degraded break elongations and, at worst, a viscous mess.

When performed correctly, the precision of the measurement at the 95% confidence level is about 1% and is of similar reliability. It can be used for almost all polyols (polyether, polyester, polycarbonate, and amine initiated). Many common additives such as most fire retardants, antioxidants, and catalysts do not interfere with the precision or accuracy of the measurement. Excessive water (capable of hydrolyzing the anhydride and biasing the OH number to lower values) and primary or secondary amines also able to react with anhydrides do have the potential to interfere with the titrimetric evaluation of polyols. Lastly, if the sample contains intrinsic acidity or alkalinity prior to derivatization the results must be corrected. The ASTM standard provides an explicit method for the correction of a sample having intrinsic acidity (biasing the result toward lower hydroxyl numbers) or alkalinity (biasing the result toward higher hydroxyl numbers), which can interfere with the titration of derivatized acid end groups.

An alternative means of measuring the percentage hydroxyl in a sample is by infrared (IR) spectroscopy. ASTM specifies the method and use of commercial near-IR instruments for the measurement. This technique is dependent on calibration curves of characteristic OH absorbance in the 5000–9000 $cm^{-1}$ region. Absorption in this region is relatively weak and there are relatively fewer resonances [5, 6]. While mid-IR spectra can also be used [7, 8], the weak absorption of the near-IR is an advantage in this case since it allows thicker samples for transmission. As a secondary method its reliability depends on the calibration curve. However, if a reliable calibration curve can be obtained and the instrumental application can be standardized, near-IR techniques can result in more reliable determination of hydroxyl values than titration techniques since human systematic error is minimized [9]. The greater potential reliability of the near-IR technique has been documented in industrial round-robin testing [10].

Modern IR techniques, especially attenuated total reflectance (ATR) spectroscopy, open up the potential for very rapid and convenient measurement of hydroxyl number and change of hydroxyl number during a reaction. The ATR technique eliminates problems associated with high absorbance causing instrumental response to no longer function in the detector's linear range. The experiment requires that the sample be dry and not have competing resonances due to N–H bonds. When these conditions are fulfilled the O–H stretch resonance (between 3100 and 3600) can be isolated with baseline resolution (Figure 5.2) from other absorption peaks, integrated (usually with instrument supplied software), and compared with an appropriate calibration curve constructed from structurally related polyols (Figure 5.3).

***5.1.1.2 CPR*** Another standard polyol analysis is the CPR measurement, which provides an indication of the trace basicity left in the polyol following production. This is a particularly important variable for individuals involved in preparation of

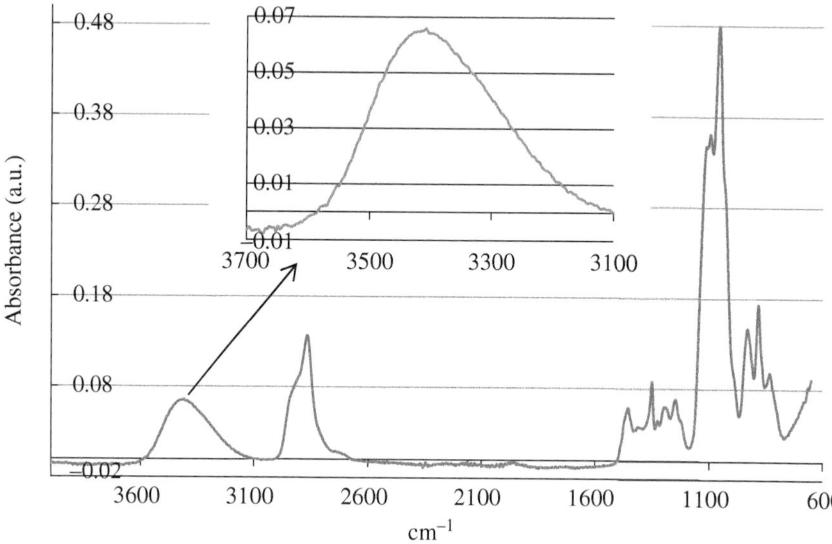


**FIGURE 5.2** ATR Fourier transform infrared (FTIR) spectroscopy of polyethylene glycol of molecular weight 200. The inset is a magnification of the hydrogen-bonded OH absorbance, which can be area integrated as part of a calibration curve.

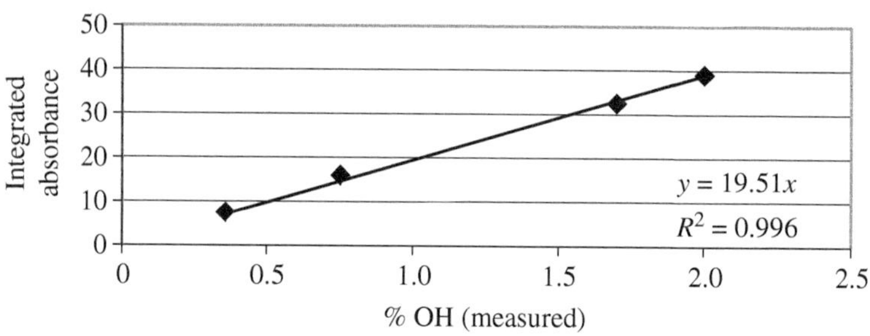


**FIGURE 5.3** Integrated OH absorbance versus percentage OH measured. Example calibration curve generated for FTIR determination of OH content of polyols.

prepolymers (isocyanate-terminated polyols; see Section 2.1.6). The importance of CPR reflects the fact that the vast majority of polyols are produced using base catalysis. Although modern manufacturing practice has numerous means of nearly eliminating the free base left after production, there is normally a trace of basic activity left in the polyol sample being used for polyurethane or prepolymer preparation. This leftover base activity can have negative influences on prepolymer shelf stability because of the ability of base components to catalyze isocyanate–isocyanate reactions. The actual basic components present are not identified but are simply the total weak acid salts left in the sample. Since the amount of base active elements left after manufacture and processing is usually small, a large sample of polyol must be employed for the test. The reported number is in units of milliequivalents of KOH per 30 kg of sample, or often 10 times that number. The typical CPR values are from about 0.1 to 1 (or from 1 to 10). The net CPR reflects that, when mixed with an isocyanate, net acidity from the manufacture of isocyanate (in the form of HCl; see Chapter 2) can in part counterbalance the basicity of the polyol. Thus, the net CPR is given by Equation 5.3:

$$\text{Net CPR} = \text{CPR}_{\text{polyol}} + \text{AHC}_{\text{isocyanate}} \tag{5.3}$$

where $\text{AHC}_{\text{isocyanate}}$ is the active hydrolyzable chloride in the isocyanate. A high net CPR will result in poor prepolymer shelf life and poor polyurethane formation owing to uncontrolled catalysis. The CPR test itself is a titration to an indicated end point or readout of a potentiometric electrode. The calculation is based on Equation 5.4:

$$\text{CPR} = \left(V_{\text{sample}} - V_{\text{blank}}\right) \times \text{N} \times 30{,}000 / W \tag{5.4}$$

where $V_{\text{sample}}$ is the volume of titrant (0.01 N HCl) to the last end point or potentiometric break, $V_{\text{blank}}$ is the volume of titrant to the last break in blank titration, N is the normality of HCl, 30,000 is the conversion factor for 30 kg, and $W$ is the weight of the sample. As indicated in Table 5.2, the precision of this test (±20 % ) is average, at best, based on round-robin industry testing and depends strongly on the skill of the practitioner. Thus, CPR measurement reliability should always be suspected if prepolymer shelf life stability is an issue.

### 5.1.2 Analysis of Isocyanates

Isocyanate functionality and the quality of the isocyanate used for polyurethane polymerization is as critical to the final material result as the polyols. However, the characterization of isocyanates is in some ways easier since, with the exception of polymeric 4,4′-methylene diphenyl diisocyanate (pMDI), they are monomeric and can be distilled to a pure state or possess an isomeric distribution that is relatively easy to characterize using modern analytical techniques. Another distinction from polyols is that isocyanates are somewhat easier to titrate than alcohol groups since their reaction with amines is diffusion limited. Just as with polyols, ASTM provides several explicit standard methods for characterizing an isocyanate sample, and, given the advantages of characterizing the isocyanate monomers and functionality, the tests exhibit superior precision and reliability to those for polyols. However, the reactivity of isocyanates does not permit the practitioner to trust the functionality per unit mass. Given the number of isocyanate–isocyanate (see Chapter 3) reactions and the facile reaction of isocyanate with opportunistic water, the quality of an isocyanate should always be regarded with suspicion if a reaction is not proceeding as anticipated.

**TABLE 5.3 Conventional isocyanate specifications and ASTM procedures for measurement**

| Property | ASTM method |
|---|---|
| Percentage hydrolyzable chloride | D5523 potentiometric method following reaction of sample with isopropanol to form HCl |
| Percent MDI monomer in pMDI and monomer–isomer distribution | D7252 method of HPLC separation resulting in baseline resolution of species and well-determined elution times |
| Determination of isocyanate content per unit mass | D5155 Titration of isocyanate with a secondary amine (dibutylamine) followed by titration of excess amine |
| Standard specification for TDI | D1786 methods and expectations for TDI based on purity, color, total chlorine, hydrolyzable chlorine, isomer distribution |
| Isomer content of TDI | D4660 employs IR spectroscopy using standard materials and internal standard for quantification |

HPLC, high-performance liquid chromatography; TDI, toluene diisocyanate.

**TABLE 5.4 Critical evaluation of the ASTM method-derived property value in Table 5.1**

| Property | Reliability |
|---|---|
| Percentage hydrolyzable chloride | Unknown but expected to be average |
| Percent MDI monomer in pMDI and monomer–isomer distribution | Very good |
| Determination of isocyanate content per unit mass | Very good |
| Standard specification for TDI | NA |
| Isomer content of TDI | Very good |

Manufacturer numbers on well-established products are usually reliable. New materials made by new processes may be far less so.
NA, not applicable.

There are a number of ASTM techniques applicable to isocyanates. Many of them relate to safety and air monitoring for isocyanates. Table 5.3 is a partial list of ASTM methods related to the use of polyisocyanates, the values of which might be included on a certificate of analysis from a manufacturer and relate to the reliability in use.

There are several other standards available associated with, for instance, measurement of viscosity, color, and density, but these are related more to manufacturing control than reliability in use. As alluded to, the reliability of these tests can be expected to be good as enumerated in Table 5.4.

Since the tests for isocyanate quality are expected to be relatively reliable, there is a reasonable expectation that the values attending a particular product were trustworthy at the time they were made. Problems in polyurethane properties that may result from poor isocyanate quality are generally a reflection of poor storage conditions or unoptimized handling during the polymerization. It is often the case that a slight excess of isocyanate is added to the polymerization to account for loss of unit mass content of isocyanate. This addition can be from about a 1% to 5% excess isocyanate equivalence. Experience is the best guide here.

In many applications using pMDI it is found that the performance of a particular manufacturer's product will consistently vary from that of another. While this could be a reflection of systematic handling differences, it is often the case that the composition of the pMDI is sufficiently different that the resulting polyurethane performance is significantly affected. Compositional differences can reflect numerous manufacturing process variables as well the amount of 4,4′ MDI monomer that is distilled from the crude product. Thus, the practitioner should understand that when using pMDI it may be necessary to analyze according to ASTM 7252 for composition to understand the results, and perhaps develop a more useful patent strategy when involved in innovation and patent application filing.

***5.1.2.1 Analysis of pMDI Composition*** ASTM 7252 is an explicit method to determine the composition of pMDI using high-performance liquid chromatography (HPLC). The method depends on the use of acetanilide as an internal standard and employs calibration curves based on pure standard samples obtained from well-known chemical suppliers. The precision of this method based on round-robin testing is about 1%. Example chromatograms are found in the standard. The percentage of a given isomer based on the chromatographic data using an internal standard based on integrated peak areas is given by Equation 5.5:

$$\text{Isomer}_{\text{i}}\left(\%\right)=\frac{A_{\text{i}}^{\text{sample}}A_{\text{IS}}^{\text{std}}W_{\text{i}}^{\text{std}}\times 100}{A_{\text{i}}^{\text{std}}A_{\text{IS}}^{\text{sample}}\times W_{\text{sample}}} \tag{5.5}$$

where $A_{\text{i}}^{\text{sample}}$ is the area of the *I* isomer peak in the chromatogram, $A_{\text{IS}}^{std}$ is the area of the internal standard in the standard chromatogram, $A_{\text{i}}^{std}$ is the area of the *I* isomer in the standard chromatogram, $A_{\text{IS}}^{\text{sample}}$ is the area of the internal standard peak in the sample chromatogram, $W_{\text{i}}^{std}$ is the weight of the *I* isomer in the standard solution, and $W_{\text{sample}}$ is the weight of the unknown sample.

Many laboratories use gas chromatography (GC) for this analysis when HPLC is not convenient, employing a diphenylmethane internal standard, but this is not suggested in the ASTM method. GC analysis is considered to be reliable to a precision less than 1% standard deviation for these measurements [11]. Size exclusion chromatography has also been demonstrated and found to be reasonably reliable with a precision of about 1%, becoming slightly less reliable for quantitative determination of the higher molecular weight oligomers (pentamers and higher) [12].

## 5.2 INSTRUMENTAL ANALYSIS OF POLYURETHANES

Modern technology has developed the means to acquire data at relevant length scales such that many aspects of polyurethane morphology and function can be readily understood. Many techniques offer corroborating as well as extending information. For instance, it is often the case that characteristic phase-separation lengths visibly discernable by atomic force microscopy (AFM) can be validated by small-angle x-ray scattering (SAXS). The SAXS data, in addition, can be further interpreted for information about the boundary between the hard and soft phases. An elastomer that exhibits slow elastic response by touch can have its room temperature glass transition temperature ($T_g$) corroborated and quantified by dynamic mechanical analysis (DMA). Modern instrumental techniques have become so disseminated and accessible to experimentalists that it is now expected that data obtained by a single technique leave too many questions unanswered to be compelling. Experimentalists now routinely apply numerous techniques to a single material [13–18]. The use of these techniques, apart and in unison, has enabled scientists to have unprecedented understanding of polyurethane morphology and supported theoretical analyses that further deepen our understanding and predictive capabilities.

### 5.2.1 Microscopy

Microscopy of polyurethanes is of great interest because its evaluation can span length scales from millimeters to nanometers. Since important polyurethane structures express themselves across these lengths, the information obtained at one scale may often be corroborated at another length scale [19]. For instance, a polyurethane specimen may be prepared and be either transparent or opaque. The cause of this optical distinction can usually be ascribed to phase characteristics of the urethane blocks [20], composite additives, which when of sufficient size will scatter visible light, or artifactual air bubbles. The samples can subsequently be observed by scanning electron microscopy (SEM) on a scale of microns to hundreds of microns, which may confirm the presence of phase-separated copolymer particles (see Chapter 2), soft-segment crystals, composite additives, or gas bubbles generated by the inadvertent reaction of isocyanate and water to form $CO_2$. Alternatively, the SEM may reveal no feature of sufficient contrast applied to an optically transparent specimen. Transmission electron microscopy (TEM) (tens to thousands of nanometers) may be employed with specialized stains to increase contrast between phases in order to reveal phase-separated hard segments that are spherulitic or lamellar. Lastly, AFM (1 to hundreds of nanometers) may be applied to the sample to reveal phase-separated spherical structures developed as a result of nucleation and growth (opaque), the lacy structures developed from spinodal-phase decomposition, or the stunted-phase structures associated with incomplete phase separation resulting from polymer vitrification [21].

*5.2.1.1* ***Optical Microscopy*** Although one of the most readily accessible tools in modern analysis, optical microscopy only finds occasional use in modern polyurethane research. Modern techniques of optical microscopy are well covered in various books and reviews devoted to the subject [22–24]. Use of optical microscopy for the study of polyurethanes often exploits the ability to polarize the light from the light source and so illuminate crystal structures that may evolve as a result of soft-segment crystallization or formation of mesophase (liquid crystalline) structures [25, 26]. The use of a hot stage is well applied in this work since it allows the slow melting/recrystallization of a material to be observed in real time [27]. Optical techniques allow for a descriptive evaluation of the phase structure (i.e. spherulitic, lamellar, etc.), often accompanied by quantitative thermodynamic measurements obtainable by differential scanning calorimetry [28]. Measurements such as these can be compelling, especially when applied to highly systematic structural studies that can track the size of crystallites as a function of some molecular parameter [29, 30].

Classical optical microscopy can also find uses for gross descriptions of polymer wear, aging, and forensic analysis. Other uses of optical microscopy can be frustrated by the mismatch of relevant scales of interest, poor contrast between phases, interference by light scattering off a top surface, poor transmission of light through a polymer film, and light scattering off the myriad surfaces offered by a foam specimen. Modern techniques in optical microscopy, such as optical phase contrast imaging, can provide more information than other optical techniques by increasing contrast between phases (Figure 5.4) [31]; however, other techniques such as TEM and AFM can provide a higher degree of resolution and are more widely used.

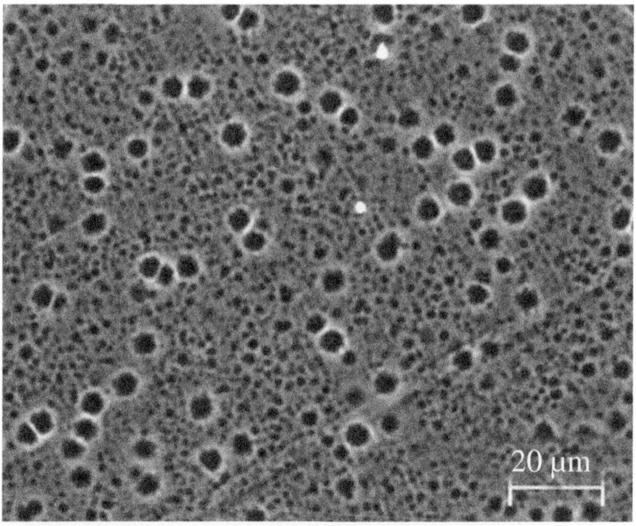


**FIGURE 5.4** Cast polyurethane elastomer: dark areas are hard segment. See Figure 5.6 for comparative image quality and information content versus TEM. Image courtesy of Robert Cieslinski and Justin Virgili.

*5.2.1.2* ***SEM*** This is a broad field of analysis that has improved dramatically over the years [32]. Resolution of electron microscopy rivals that of TEM and AFM. An electron beam from a heated filament is focused and scans the surface of a sample in a raster pattern. The electrons interact with the materials of the surface and are absorbed, scattered, or reflected. Reflected or secondary electrons generated from the specimen as a result of the electron beam collision, or electromagnetic radiation, can be subsequently detected depending on the microscope hardware. A representation of the surface is generated after electron surface collisions are detected and correlated with an $x,y$ coordinate on a display. The resulting images can have tremendous verisimilitude and produce three-dimensional depth of high fidelity.

SEM is very commonly used in polyurethane research; however, because the signal is primarily a representation of surface features, there are limitations. SEM is often used to produce gross images of polyurethane structures such as foam cell structure, which can be more difficult with optical techniques because of incident light scattering. Figure 5.5 shows an SEM image of an open cell foam. Notice that the magnification is not particularly high, but the same picture by optical microscopy would typically not contain the observed level of detail. SEM is also commonly utilized to visualize polyurethane two-phase composites such as fiber- and mineral-reinforced structures and their interfacial features. Such an image of a urethane composite is shown in Figure 5.5b, which demonstrates high levels of adhesion of urethane to the embedded glass fiber.

Hard segments and soft segments do not scatter electrons with sufficient distinction to be observable since polyurethane samples are normally coated with a conductive layer such as gold to conduct electrons away from the surface and reduce de-resolution of the image caused by electron scattering from a negatively charged insulating polyurethane sample. Gross changes in fracture surfaces at different hard-segment contents have been rationalized as demonstrating the effect of hard segment, but do not reveal the hard-segment structure observed by other techniques.

*5.2.1.3* ***TEM*** This is an electron microscopy that has established itself as a very useful tool in polyurethane materials science [33, 34]. Like SEM, the sample depends on the interaction of the sample with a thermally generated or field-generated electron beam. Unlike SEM, which bounces electrons off a sample surface, TEM passes electrons through the sample. Since electrons readily interact with materials, the sample must be very thin, of the order of tens of nanometers thick, to allow sufficient electrons to pass through the sample that a signal can be collected. The image transmitted is a function of the interaction of the electrons with different components of the sample. Although staining is not always required, image contrast can be improved if one phase or another has affinity for a heavy metal contrast agent such as osmium tetroxide or uranyl (uranium) acetate (Figure 5.6).

The many applications and varieties of use for TEM can be found in resources dedicated to the subject. The most critical aspects of the process to obtain a TEM of polyurethane include sample preparation [35, 36] and sampling bias [37]. The need

(a)

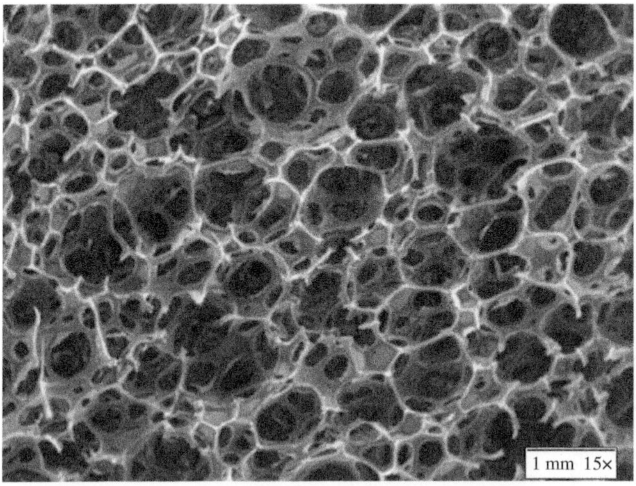


(b)

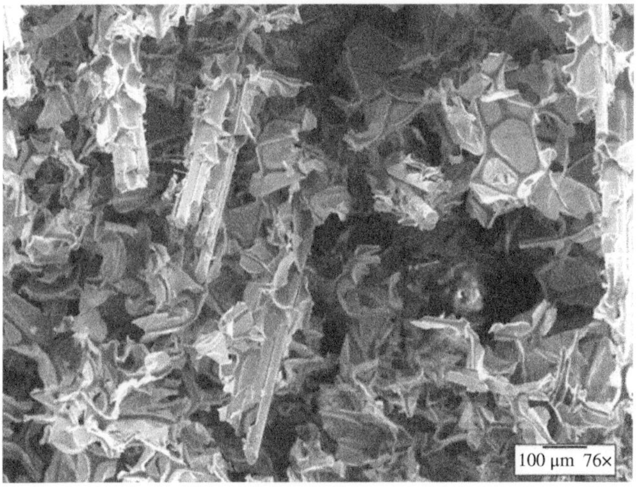


**FIGURE 5.5** Illustrative scanning electron micrographs of (a) open cell polyurethane foam and (b) glass mat polyurethane foam composite.

to prepare exceptionally thin samples is complicated by the fact that many urethanes are relatively soft materials, resulting in deformation during thin sectioning (microtoming). This may result in the need to perform the microtome at very cold temperatures. However, the utility of TEM for understanding polyurethanes (and, for that matter, many polymers) has resulted in standardization and commercial instruments that can be used to minimize the complications. Another problem is one of

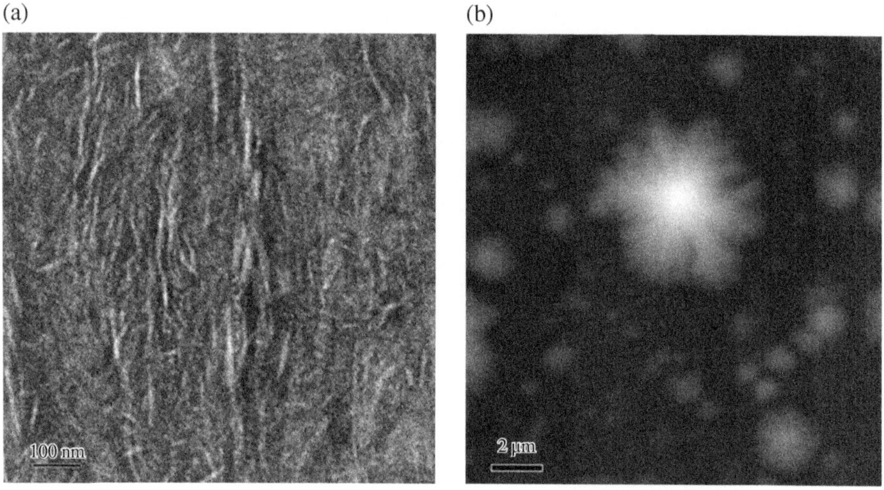


**FIGURE 5.6** Illustrative TEM images of polyurethanes. (a) Oriented rod-like hard segments – an artifact of the compression molding process. (b) A TEM image of the elastomer shown as an optical image in Figure 5.4. Images courtesy of Robert Cieslinski and Justin Virgili. Reprinted with permission from Ref. [34]. © John Wiley & Sons, Inc.

systematic error related to the very small areal coverage of the measurement. This can lead to the search for interesting but rare structures within a sample. It is not uncommon for the rarity of a structure to be lost in the novelty. It is of course the role of the experimentalist to search for representative *and* novel structures. A last issue is the possibility of localized heating of the sample by the beam altering the morphology. This is often visible to the experimentalist and, of course, should not be ignored.

***5.2.1.4 AFM*** This is a relatively recent technology development for imaging materials and has found wide applicability to the study of polyurethanes [38, 39]. Unlike electron microscopies, AFM depends on the interaction of a sample's surface with a scanning tip, which can provide an analog signal to recording electronics or be detected remotely by laser detection of the tip about its coordinate position and its deflection in the *z*-position. This obviates the need for large vacuum chambers (although atomic-level detection may still employ high vacuum to prevent gas adsorption). These advantages mean that AFM can be well employed in ambient environments, solvent laden systems, or even on biological specimens. Numerous textbooks and research monographs [40, 41] are available with detailed information on instruments and techniques

When an AFM tip is placed in direct contact with a surface (in essence, a measure of the repulsive force) the technique is called "contact mode." Contact mode can be useful for investigating the surface of hard materials. It has the advantage of providing a direct interrogation of the surface, but can provide misinformation on surface topology if the surface is deformable by the scanning tip and can also cause tip wear over time, which can affect measurement fidelity. In practice, the tip is scanned over the surface with a constant downward force and the tip position monitored by sensors.

In "noncontact" mode the tip does not directly contact the surface [42]. Instead, the tip position is carefully controlled to hover just above the sample surface. In contrast to the repulsive forces monitored by contact mode, noncontact mode monitors attractive forces between the tip and the surface. The tip is not abraded by wearing on a surface and the sample surface is not affected by contact with the high-modulus (silicon or silicon nitride) tip. The results of contact and noncontact modes applied to rigid samples will usually be very similar. However, soft materials, or materials with loosely bound surface layers, will be better understood using noncontact techniques. The scanning of the tip over the surface can be operated in a frequency modulation mode, where intertip surface forces modulate at a resonant frequency, or can be operated in an amplitude oscillation modulation mode, where amplitude modulation or phase perturbation can be monitored. These modulations of a controlled frequency or amplitude can be correlated with the characteristics of the probed material.

Tapping-mode AFM [43] is, in some ways, a compromise between contact and noncontact methods of tip operation. The tip is oscillated above the surface of the sample but it approaches the surface much more closely, such that it intermittently samples the repulsive or contact regions of the tip–surface interactions. As the material taps the surface it probes directly the surface topography but does not drag or distort the surface as a result of lateral surface interactions. Furthermore, the interactions incurred during the taps are readily related to material properties through the sinusoidal phase change induced in the oscillation. Figure 5.7 shows a method of tip amplitude control and sample positional control used by one particular manufacturer of tapping-mode AFMs.

Figure 5.8 illustrates the portion of the attraction/repulsion potentials sampled by each of these techniques. As the probe is brought into contact with the surface the force measured by the probe is greater than zero. This allows a direct measurement of surface topography with position, but may lose information about subtle material properties. It also may suffer some resolution insufficiency if the probe is too blunt to follow acute surface features. In tapping mode, the probe oscillates above, but very close to, the sample surface. As the probe approaches the surface near the minimum in the force, the attractive forces will bring the probe into physical contact with the surface and bring the measured force into the repulsive region of the potential. This allows the probe not only to sample surface topography but also to directly probe material properties such as adhesion forces and modulus. Noncontact mode samples the region away from the sample surface and samples the positional change in the attractive forces of the material for the probe. As the probe samples the relative height of the sample it will register a change in attractive forces.

AFM has brought a new appreciation for the underlying microstructure responsible for the properties of polyurethanes [44–47]. It has furthermore elevated the theoretical basis for understanding the mechanism for polyurethane phase separation and

(a)

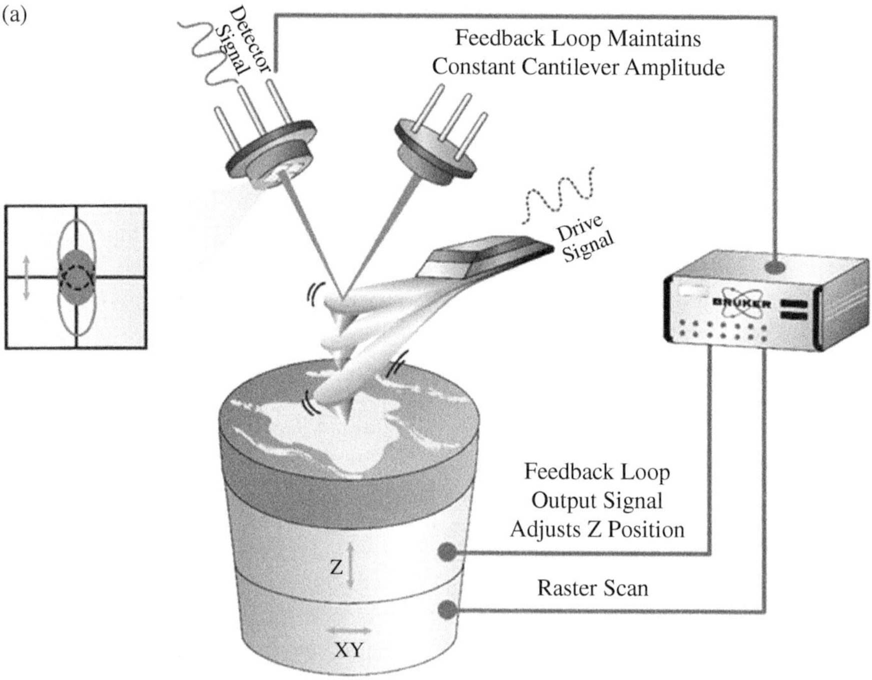


(b)

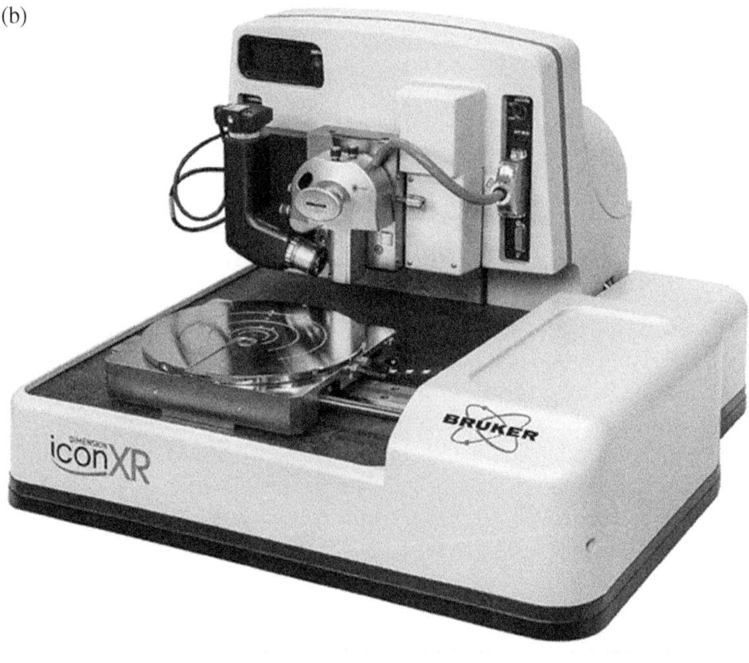


**FIGURE 5.7** Simplified illustration showing sample and signal detection controls for a tapping-mode atomic force microscope and the appearance of a working instrument. Images courtesy of Bruker Nano Surfaces. (*See insert for color representation of the figure.*)

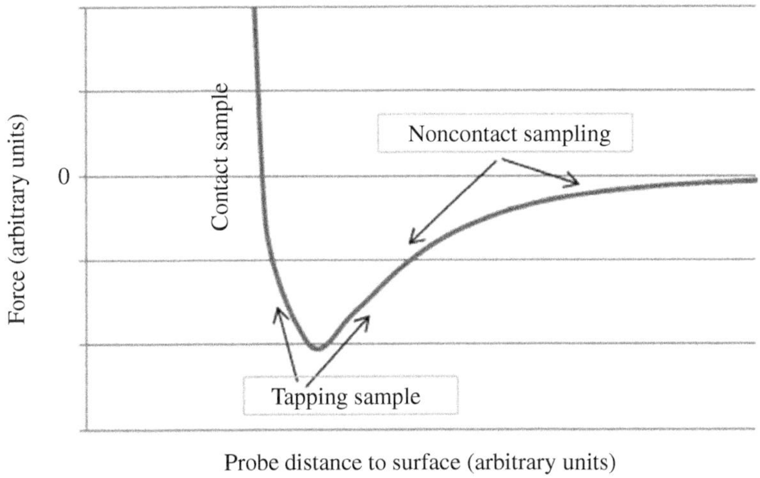


**FIGURE 5.8** Potential diagram of attractive and repulsive forces influencing an AFM tip as it approaches a surface in contact, noncontact, and tapping modes.

the position polyurethanes occupy as a class of block copolymers [48]. Figure 5.9 is an example of information available from, in this case, tapping-mode AFM (TMAFM). The figure shows the height image garnered from the potential forces experienced by the probe as it oscillates over the surface; this is termed the "height." The phase image shows the information obtained by the probe in contact with the surface and as it is preferentially repulsed by the very high modulus hard segment (white in the image) and the relatively adhesive soft segment (dark). The image is very consistent with proposed theories (see Chapter 4) related to mechanisms of phase separation and block copolymer equilibrium structures. Experimental details are provided in the figure caption to impress upon the reader that even a relatively straightforward experimental technique must be performed with precision and attention to small details to obtain meaningful and useful results. Beyond that, the membrane/lacy structure of the hard segment is remarkable for its aesthetic value as well.

### 5.2.2 IR Spectrometry

IR spectroscopy is one of the most widely and successfully applied analytical techniques to polyurethane chemistry and structure [49–52]. The chemistry of polyurethanes is the story of isocyanate and active hydrogen conversion to urethane, urea, and isocyanate–isocyanate reactions. During polymerization, strongly IR-absorbing functional groups are consumed, and other strongly IR-absorbing functional groups are formed. This makes IR spectroscopy particularly useful for following reaction kinetics and perturbations as a function of composition. Table 5.5 is a reference compendium of structures commonly encountered in polyurethane chemistry and

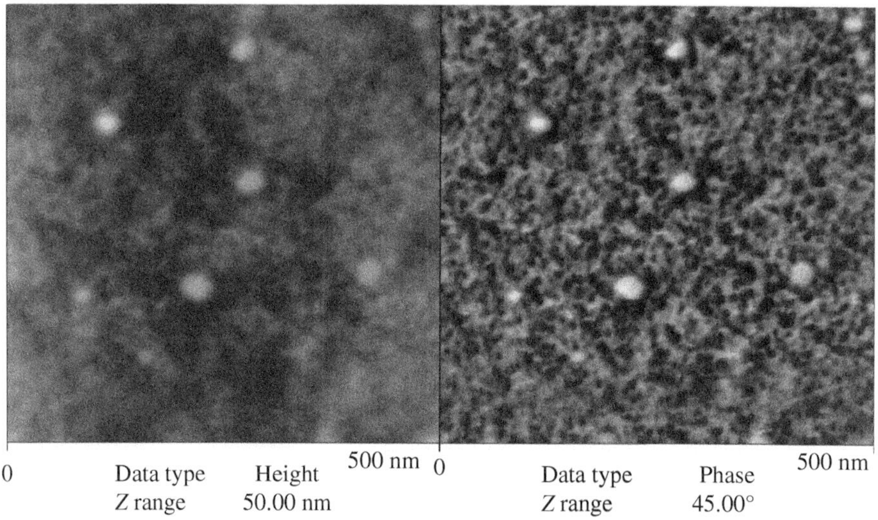


**FIGURE 5.9** A piece of the foam was embedded in epoxy and then cured. The cured block was cryopolished at –110 °C. The resulting face was exposed to $RuO_4$ vapor and then the block was repolished at room temperature. The face on which the TMAFM was done probably had regions that retained the Ru stain. TMAFM was obtained on a Digital Instruments MultiMode using a Nanoscope IV controller (software v. 5.12r3). Silicon cantilevers and tips were used (Nanosensor, LTESP, wafer number 1795L262, frequency = 158 kHz). Typical tapping conditions were Ao ~ 1.0 V, Asp ~ 0.6 V, and rsp ~ 0.60, with the tip–surface interaction repulsive in nature. Image courtesy of Greg Myers. Reprinted with permission from Ref. [44]. © Elsevier Pub.

associated resonance frequencies. At the same time, the final product structures can exist in various states of interchain interaction that modify the IR resonances in highly predictable ways. Furthermore, since polyurethanes involve active hydrogen reactants, and produce hydrogens in IR-active functional groups, experiments using deuterium isotope effects can be contemplated for information about reaction mechanisms and removing ambiguity in final structures [53]. The systems are also amenable to study by model compounds, which can be designed to isolate certain chemical and structural aspects of interest and mitigate confounding or complicating features [54–57]. Lastly, IR measurements are adaptable to other analytical techniques, such as x-ray analyses, which allow simultaneous measurements of structure formation and provide understanding of the order of events, especially phase separation, during reaction [58]. Thus, by combining x-ray and IR analyses, polyurethane reaction dynamics can be combined with structural details of the final product.

The power of IR spectroscopy applied to polyurethane chemistry and structure should not imply that reliable data using this technique are easily obtained. Many aspects of polyurethane chemistry hinder obtaining quantitatively useful data. In the

**TABLE 5.5 Reference for resonance frequencies for functional groups commonly encountered in the analysis of polyurethanes**

| Frequency ($cm^{-1}$) | Resonant functional group | Structure |
|---|---|---|
| 1100 | Ether | R——O——R |
| 1250, 1070 | Ester | R′——O——R |
| 1413, 1710 | Isocyanurate | 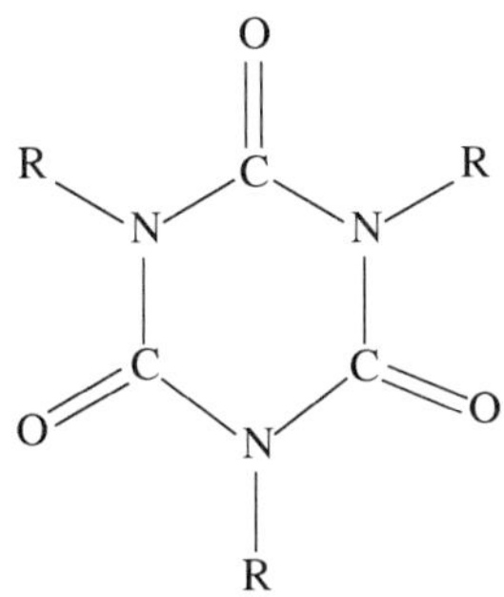 |
| 1598 | C=C (TDI ring) | 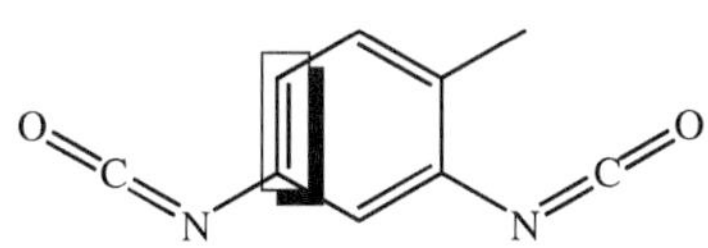 |
| 1640–1645 | H-bonded urea in domains | 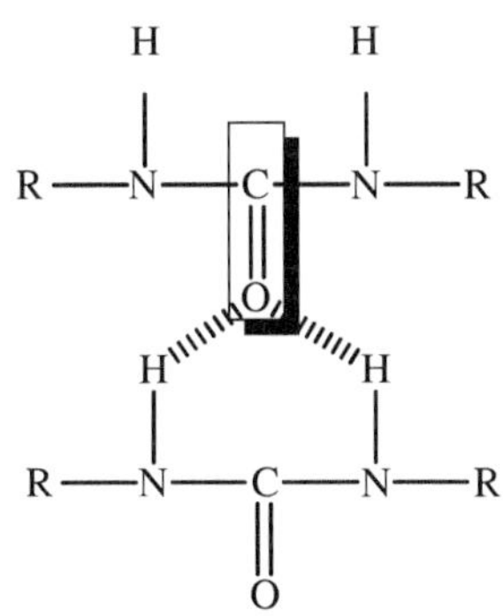 |
| 1695–1700 | Free urea | 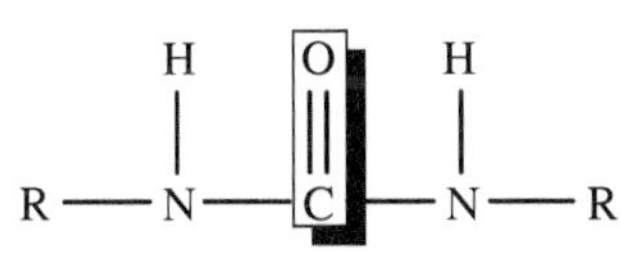 |

(*Continued*)

**TABLE 5.5 (Continued)**

| Frequency ($cm^{-1}$) | Resonant functional group | Structure |
|---|---|---|
| 1700–1710 | H-bonded urethane | 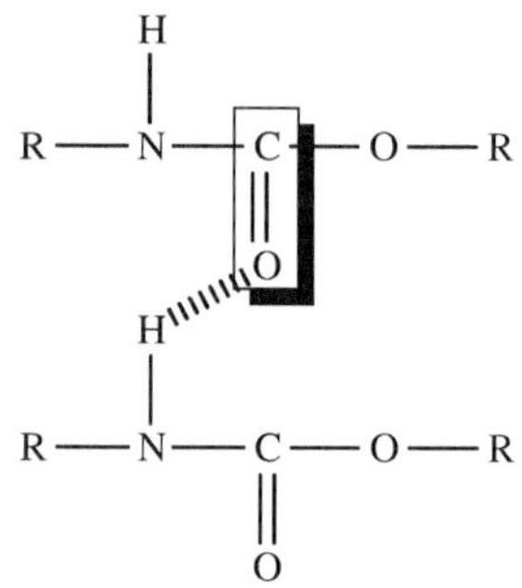 |
| 1662 | H-bonded urea disordered | |
| 1683 | Acid | 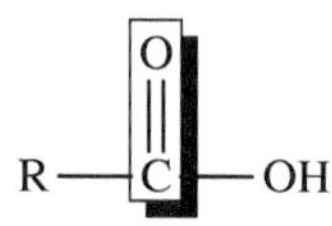 |
| 1737 | Ester | 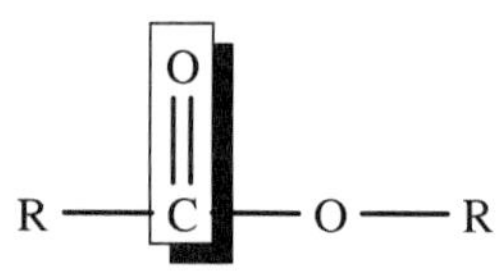 |
| 1730 | Free urethane | 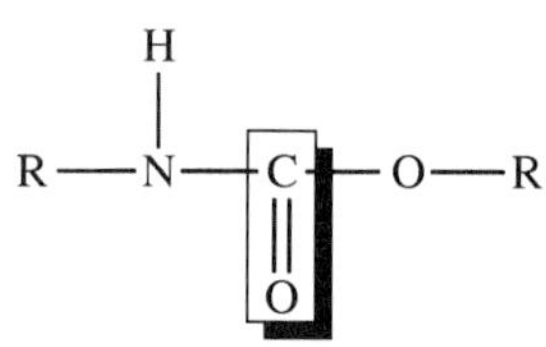 |
| 1710–1740 | Monodentate H-bonded urea | 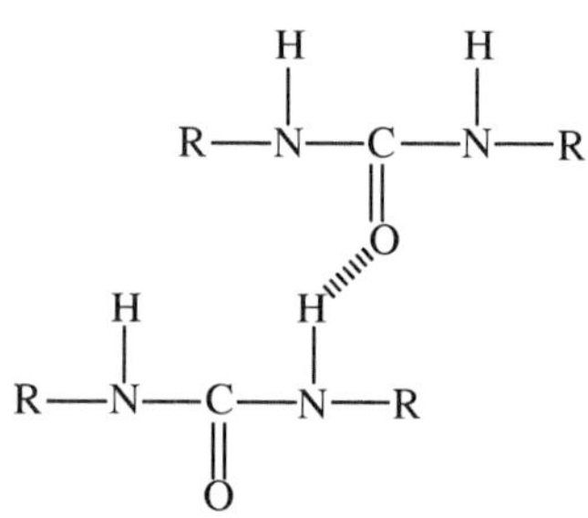 |
| 2100 | Carbodiimide | R—N=C=N—R |

**TABLE 5.5 (Continued)**

| Frequency ($cm^{-1}$) | Resonant functional group | Structure |
|---|---|---|
| 2270 | Isocyanate | R—N=C=O |
| 1660–1680 | Biuret | 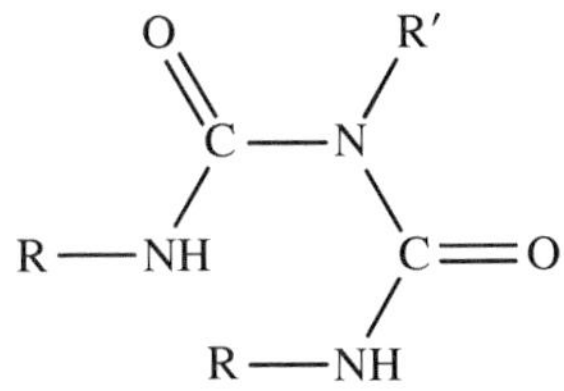 |
| 1710 | Allophanate | 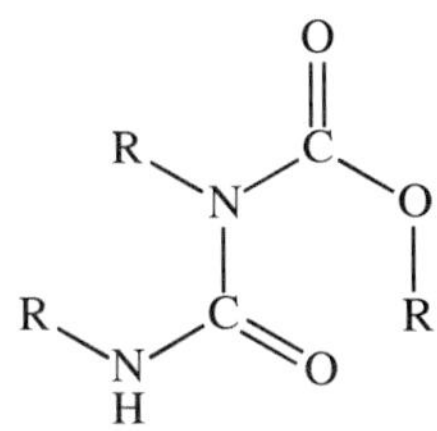 |
| 1785 | Uretdione (dimer TDI) | 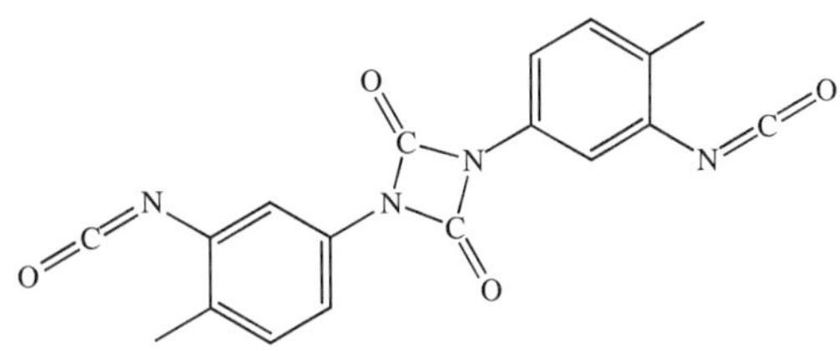 |
| 2970 | $CH_3$ | 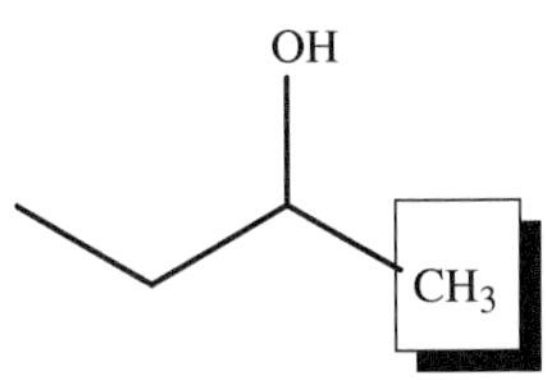 |
| 3200–3500 | N–H | |
| 3100–3700 | OH | |

case of polyurethane foaming and foams, the chemistry is very rapid, is exothermic, and involves very large changes in sample density. The product itself is opaque and so is best studied by ATR spectroscopy using Fourier transform instruments. Figures 5.10–5.12 are example foam spectra illustrating interesting resonance features and a complication associated with operating in a transmission versus a surface reflectance mode.

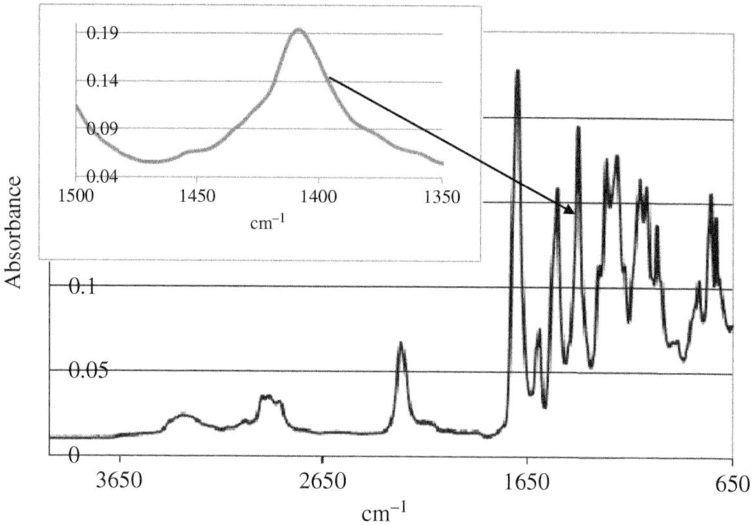


**FIGURE 5.10** ATR-FTIR spectroscopy of isocyanurate foam. These foams are made with a significant excess in isocyanate to assure that a large amount of trimer is formed. Isocyanate functionality is clearly visible as an isolated peak at 2270 cm$^{-1}$. The insert details the IR resonance associated with the isocyanurate ring at 1410 cm$^{-1}$.

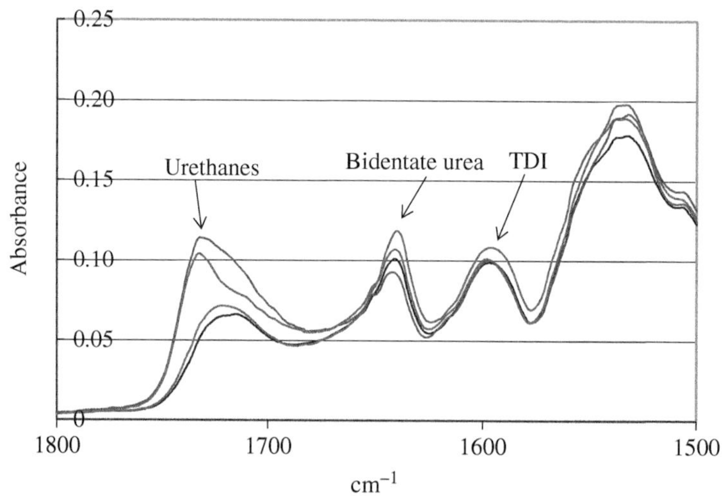


**FIGURE 5.11** ATR-FTIR spectroscopy of flexible foams showing the sensitivity of urea/urethane populations of foaming formulation details. The TDI peaks show nearly identical levels of TDI in all four foams.

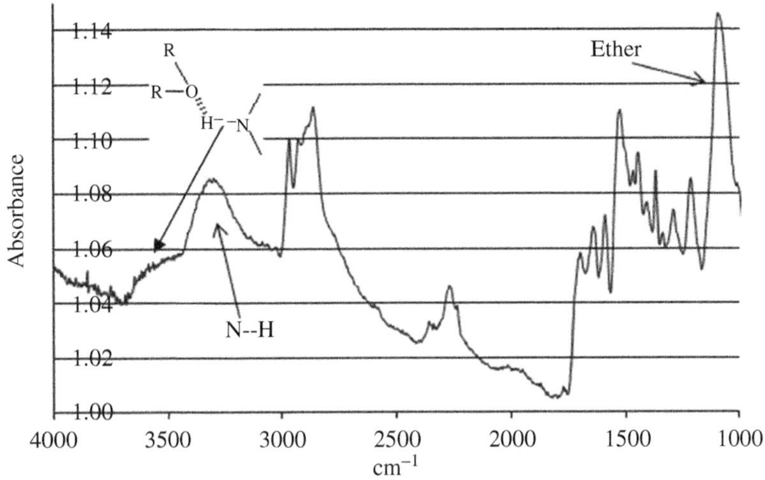


**FIGURE 5.12** Transmission FTIR spectroscopy of TDI–polyurethane flexible foam. Notice the difference in absorbance values and baseline offset due to complications associated with working in transmission mode.

### 5.2.3 X-Ray Analyses

X-ray analysis has become a standard tool for developing in-depth understanding of polyurethane microstructure. It is particularly finding adherents in academic studies of structure–property relationships since scattering techniques can provide unique information on the angstrom to nanometer scale [27, 59, 60]. At this scale a researcher can learn about packing within the hard segment, the ratio of crystal (or quasicrystalline) to amorphous phase volume, the periodicity of phase spacing, and the characteristics of the interface between the polyurethane soft segment and the hard segment. While x-ray methods are surely complementary to other less instrumentally demanding techniques (such as calorimetry, IR spectroscopy, AFM, and mechanical analyses), x-ray analysis, when done with proper care and control, provides a measurement that requires less inductive reasoning than other techniques applied for the same information. The most basic information provided by x-ray analysis is derived by the very simple Bragg equation (Equation 5.6) for determining the interdomain spacing $\lambda$, regardless of the intra- or interphase nature of the scattering center:

$$\lambda = 2\pi / q_m \tag{5.6}$$

where

$$q_m = 4\pi / \lambda \sin(\Theta / 2) \tag{5.7}$$

$q_m$ is referred to as the scattering vector, $\lambda$ is the wavelength of the incident x-ray radiation (often 1.54 Å for a Cu $K_\alpha$ source), and $\Theta$ is the scattering angle. The wealth of additional information contained within scattering diagrams depends on highly involved geometrical reasoning, and obtaining information depends critically on correct application of precise experimental techniques.

***5.2.3.1 Wide-Angle X-Ray Scattering*** Wide-angle x-ray scattering (WAXS) is an x-ray diffraction technique used to characterize the crystallinity and characteristic spatial relationships between atoms within a crystalline phase [61]. WAXS spectroscopy is performed with a sample between the x-ray source and the detector. While scattering is strictly from constituent electron density, this is usually well correlated to local density, making it a sensitive probe of local structural variations. WAXS is differentiated from SAXS by positioning the detector closer to the sample, so that scattering at wide angles may still be captured. The detector may be a film or digitizing two-dimensional (2D) detector, such as shown in Figure 5.13. Alternatively, the source may scan an angular range and the detector scan the diffracted intensity as a function of detector angle to provide the so-called one-dimensional (1D) spectrum. Such a 1D spectrum would be the equivalent of recording the intensity along a single radius extending away from the center of the film to the film's edge.

Applied to many polyurethane structures, WAXS is often not particularly enlightening [13, 35, 52, 62, 63]. Since polyurethane hard segment is disordered, sometimes compared to a nematic liquid crystalline structure (like matchsticks in a box), it often results in diffuse WAXS. Thermoplastic polyurethanes (TPUs) are particularly diffuse in this regard; however, many TPUs are prepared with glycol soft segments capable of crystallizing. Among these are polybutylene adipate, polybutylene succinate, and polycaprolactone (see Chapter 2). WAXS of TPUs prepared with crystallizable soft segments may show up convoluted with the diffuse polyurethane hard segment, as shown in Figure 5.14. Figure 5.14a shows broad diffuse scattering with a well-known peak centered at about $2\Theta=20°$. A 2D image of this pattern would show up as a diffuse halo. Application of the Bragg equation indicates a characteristic scattering reflection of about 4.5 Å. Figure 5.14 also shows that, as design of the

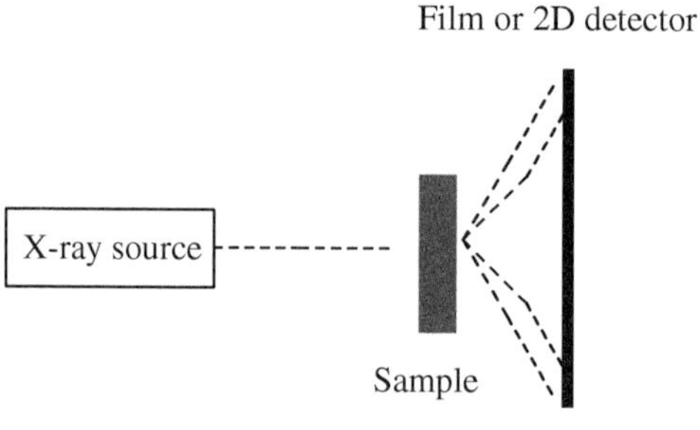


**FIGURE 5.13** Relationship of the sample to the detector for WAXS analysis.

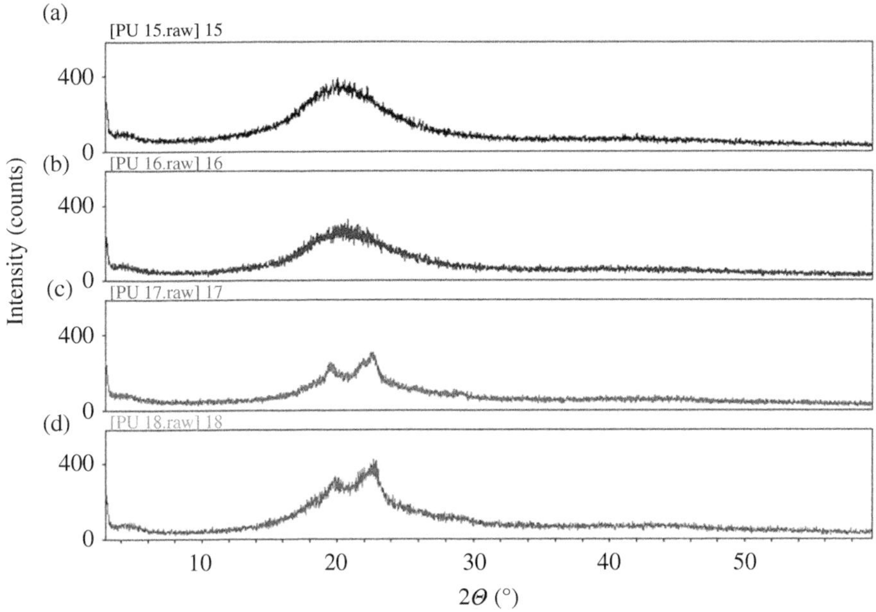


**FIGURE 5.14** WAXS data for polyurethane elastomers with polybutylene succinate soft segments and 4,4′-methylene diphenyl diisocyanate 1,4-butanediol (MDI-BDO) hard segments. As conditions are optimized for soft-segment crystallization the polybutylene succinate diffraction becomes more prominent in the data. Data courtesy of Brian Landes. Reprinted with permission from Ref. [64]. © Elsevier Pub.

TPU is adjusted to encourage soft-segment crystallization (i.e. higher molecular weight soft-segment and less hard-segment volume), the characteristic scattering from, in this case, polybutylene succinate diol appears and begins to dominate the diffraction pattern.

The polyurea hard-segment scattering emanating from polyurethane foams can show a greater amount of ordering and more reflections than observed for TPUs. This might be expected given the higher ordering power of the bidentate hydrogen bonding within the polyurea hard segment. Figure 5.15 shows a representative WAXS from a water-blown foam. In this figure the characteristic 2θ =20° is prominent as well as a shoulder at about 2θ = 10° and 50°. A 2D image of this pattern would show a less diffuse scattering intensity than observed in Figure 5.15 with sharper intensity maxima (still showing circular symmetry in the image) at the indicated 2*θ* distances. The exact morphological nature associated with the scattering is not established but is believed to be characteristic of the hard-segment hydrogen-bonding network inferred from the commonality of many of the reflections independent of whether the isocyanate is MDI, pMDI, or TDI [35]. In principle, processing variations such as compression molding, injection molding, annealing, etc. could affect WAXS

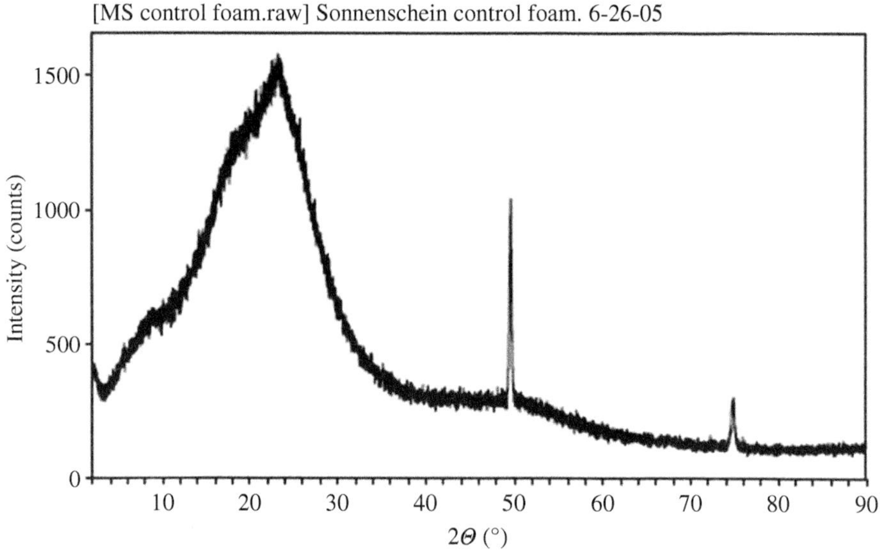


**FIGURE 5.15** WAXS from a polyurethane foam with a polyurea hard segment. Reprinted with permission from Ref. [13]. © Elsevier Pub.

patterns, but in practice this is rarely observed. Soaking polyurethane samples in strong solvents is known to decrease or even eliminate some of the WAXS, but this is certainly unsurprising. Large-scale sample deformation has been observed not to change peak maxima, but rather, in some cases, to change the overall peak shape [65].

***5.2.3.2 SAXS*** Application of SAXS alone and in concert with other techniques, such as WAXS, rheology, Fourier transform infrared (FTIR), and others, has provided powerful insights into polyurethane formation and structure [66–71]. The broad concepts related to SAXS are quite similar to those of WAXS [72]. The detector is moved quite a bit farther from the sample than in a WAXS experiment, of the order of 0.5 meters to more than 3 meters when employing synchrotron radiation sources [63]. Wide-angle scattering signal is not detected or is faintly registered by the detector. Instead, only the scattering at low angular dispersion from the main beam intensity is recorded. Low-angle scattering detection results in the need to have good removal of the main beam (and other background sources, as will be discussed later) to obtain the most useful and interesting information. A means of removing this direct beam intensity is shown in Figure 5.16, where a physical beam stop is employed. The passed scattered intensity will fall upon the detector, representing the amplitude of the Fourier transform of the shape in the beam [73]. Polydispersity of the size and shape of randomly oriented scattering objects, as usually exist in polyurethanes, results in a superposition of scattering patterns, leading to a smooth function such as that shown in Figure 5.17. Alignment of nonspherical scattering

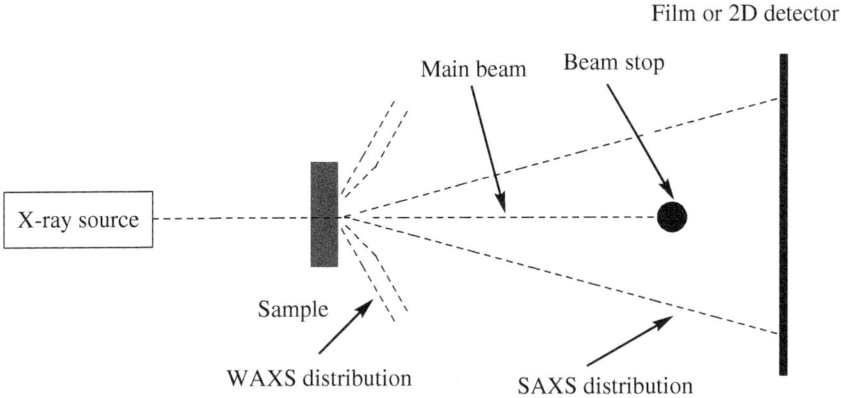


**FIGURE 5.16** Relationship of sample to detector for SAXS analysis. Compare with Figure 5.12 for WAXS analysis.

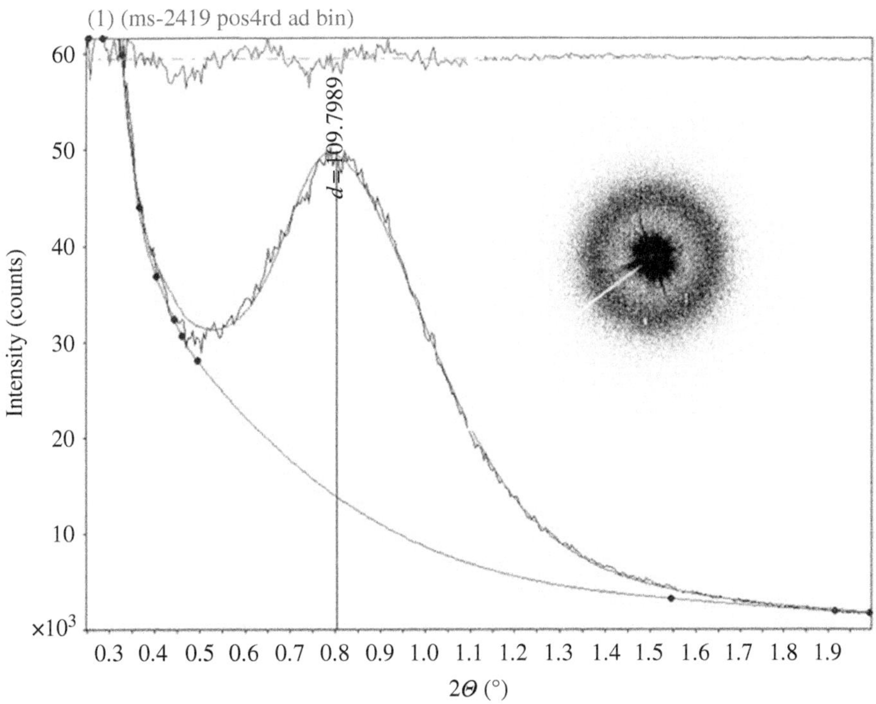


**FIGURE 5.17** SAXS data for a polyurethane foam. Inset are the 2D data.

objects can lead to anisotropic scattering patterns. Interpretation of scattering using classical analyses involves fitting the measured scattering to models based on polydisperse known shapes such as spheres [74]. The modeling and interpretation become more challenging as complex shapes must be accounted for. From the fit of the data to models, insightful information on the spatial distribution of particles in the sample as well as their potential agglomeration, their interfacial characteristics, and their shapes can be obtained [75].

The center of the scattering peak in Figure 5.17 is indicated, and the calculated d-spacing shown is based on the Bragg equation given by Equation 5.6. The d-space quantity is of interest by itself in polyurethane science since it is a quantitative measure of the interhard-segment spacing, which can then be correlated to various formulation and processing variables. It should be pointed out that d-spacing based on picking out the center of the first peak can sometimes be fairly easy, as in the pattern of Figure 5.17. Other times, as will be shown below, picking the peak intensity will be imprecise enough that a d-space calculation will be best characterized as an estimate. Data analysis techniques can be employed to improve the accuracy and precision of the d-space calculation. One such correction is the Lorentz correction for removal of vestigial main beam intensity. This technique is appropriate for lamellar scattering structures, which many, but certainly not all, polyurethane hard segments are. Systematic control of variables and careful analysis of data can then allow for improved understanding and prediction of polyurethane properties. Furthermore, it has been demonstrated that simultaneous measurement of the growth of SAXS with rheological, spectroscopic, and thermal properties can provide new insights into polyurethane structure and property development [51].

The fact that scattering is a function of geometry, and that ideal shapes scatter x-rays in well understood ways, has provided additional opportunity to understand polyurethanes for the frequent instances when ideal response models do not perfectly fit. The variance with expectations can be modeled, and the closeness of fit to data can be used to provide additional structural insights that can be tied to polymer properties. Among the most useful models is that based on the theoretical analysis by Porod on deviations from ideal scattering behavior owing to the real interfacial thickness of phase-separated systems [76–78]. These interfacial thicknesses are also referred to as boundary widths or diffuse boundary thicknesses.

It is vital to note that the Porod analysis requires that, for a two-phase system with a perfectly sharp interface, the long $q$-values must decrease according to an inverse fourth power:

$$\lim_{q\to\infty}\left[I(q)\right]=\frac{\mathrm{K_p}}{q^4} \tag{5.8}$$

where $\mathrm{K_p}$ is a constant defined by Porod as

$$K_p = \frac{Q}{2\pi^3 I_p} \quad (5.9)$$

In Equation 5.9 $I_p$ is the Porod inhomogeneity length and $Q$ is a term described variously as the "invariant" or the "Porod invariant," defined explicitly by

$$Q = 4\pi \int_0^\infty q^2 I(q)dq = V\varnothing_1\varnothing_2(\rho_1 - \rho_2) \quad (5.10)$$

$K_p$ can also be calculated as the limiting value reached by the product of $I(q)q^4$. In Equation 5.10 $\varnothing_1$and$\varnothing_2$ are the volume fractions of the respective two phases, and $\rho_1$ and $\rho_2$ are the respective electron densities.

It is in the variation of the data from the expected ideal values that the interpretation of polyurethane structure can be understood and properties systematically refined. In the cases where $I(q)q^4$ does not reach a limiting value, the inferences can be particularly illuminating. In polyurethane research, the usual case is for a graph of $I(q)q^4$ versus $q^2$ to have a negative slope. This result can be interpreted as a result of diffuse phase boundaries (as opposed to a very sharp phase boundary) and phase mixing [79]. The result of phase mixing is a depletion of scattering intensity at high scattering angles, which are more closely associated with the region closest to each scattering center and less with the regions between the scattering centers [80, 81].

As seen in Figures 5.17 and 5.18a the scattering intensity at high values of $q$ is small and declines rapidly (at or near the fourth power), as predicted by the Porod analysis. Thus, the means of subtraction for background intensity, beam collimation, and even vestigial wide-angle scattering is critical to obtaining unbiased reliable results. In a critical review of correction techniques, it has been shown by Koberstein et al. [75] that representation of the background correction as a constant will always lead to overestimation of interfacial mixing. This is because of overestimation of the decline in background intensity, which leads to an exaggerated negative deviation from ideal expectations based on the Porod analysis.

Obtaining the correlation function of the scattering requires calculation of the Fourier transform of the 1D SAXS curve [82, 83]. This yields a curve of $\gamma_1$, termed the correlation function, which is the spatial correlation of the electron density correlations (the probability of finding a scattering center at distance $r$ from an origin point). A useful mental image is to imagine the function $\gamma_1(r)$ as the probability of scattering from a rod of length $r$ with equal electron densities at either end. A peak in the resulting correlation function will then occur when there is a frequently occurring spacing. An example of SAXS and the resulting correlation function is given in Figure 5.18. The first peak in the correlation function represents the d-space distance, which would normally be estimated from the peak in the scattering intensity. In the provided example, picking out a maximum may not be a particular challenge, but there are many examples where defining a d-space could be somewhat more

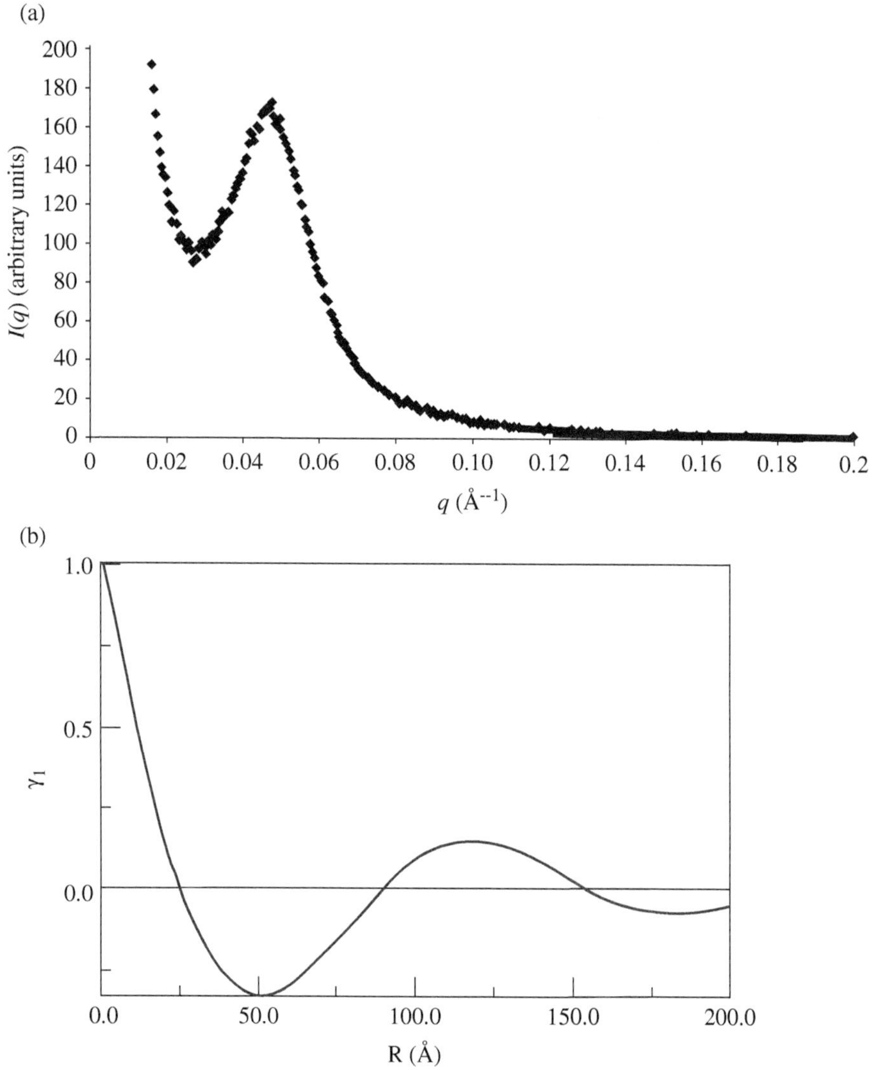


**FIGURE 5.18** (a) SAXS data for a molded polyurethane foam and (b) the derived correlation function. Reprinted with permission from Refs. [13, 44]. © Elsevier Pub.

challenging or arbitrary, as shown in Figure 5.19. Features of the correlation function can lead to direct calculation of various useful quantities related to the two-phase structure, the values of which can be used to determine the relevant structure–processing–property relationships of polyurethanes. An example of such data is provided in Table 5.6, which gives available phase information from the scattering pattern and resulting correlation function of Figure 5.18.

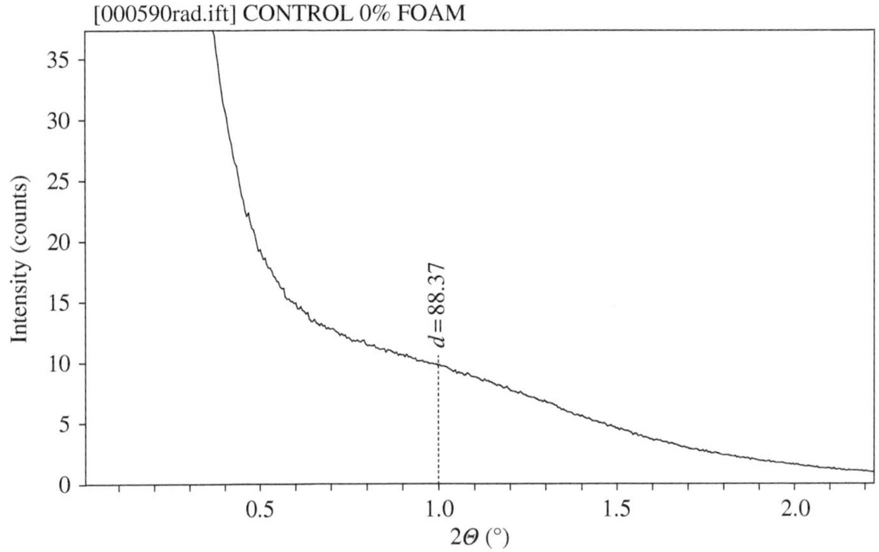


**FIGURE 5.19** SAXS data for a TDI flexible slab foam. The d-spacing determination requires more judgment than that required for data such as in Figure 5.17.

**TABLE 5.6 Data derived from a Porod analysis of Figure 5.17**

| | |
|---|---|
| d-spacing (Å) | 118 |
| $L_H$ hard block thickness (Å) | 26.3 |
| $L_S$ soft block thickness (Å) | 91 |
| $\phi_v$ bulk volume crystallinity | 0.23 |
| $\phi_H$ local crystallinity | 0.22 |
| σ interface thickness (Å) | 3.0 |
| Polydispersity | 0.23 |

The local crystallinity is the ratio $L_s$/d. The bulk volume crystallinity is determined from WAXS calculated (or approximated) by the area under the crystalline scattering divided by the total scattering intensity.
*Source:* Modified from [44].

## 5.3 MECHANICAL ANALYSIS

The mechanical analysis of polyurethanes is a very broad topic and it is difficult to cover in depth without a companion discussion associated with two-phase composite theory and the theory of block copolymers, which are covered in Chapter 4. However, most measurements associated with the performance of a specific polyurethane in a particular application begin with some kind of mechanical analysis. This is obviously true when one considers that a polyurethane foam must function under compression, be an elastomer under compression or tension, be an adhesive under tension or shear, be a coating under complex wear processes and chemical insult, and so on. Certainly, techniques such as x-ray scattering or spectroscopy may provide in-depth understanding of how or why a polyurethane structure exists. They do not, however, provide a predictive measure of how that particular structure will perform in a lap shear test, how long the structure will elongate under a particular load, or how much permanent set the structure will exhibit when sat upon in July in a humid or dry environment. For this information there are few substitutes for well-thought-out tests of mechanical performance. The fact is that, whether for foams, adhesives, elastomers, coatings, sealants, or any other imaginable application, the specific mechanical tests are often unique to the industry or even customer. The ASTM has numerous tests with specific methods for testing. Many tests such as those covered in Section 5.1 are well used and truly applied across industry and academia. Mechanical testing is handled differently because of regional differences in application requirements (i.e. Northern Europe versus South America) or because of historical testing protocols that pre-date standardized methods. Some tests simply reflect the biases of materials scientists represented within a certain company. In this section there will not be an attempt to cover all the tests that exist. This could probably fill a book in itself, and would represent a moving target as test requirements can change rapidly. Rather, this section will cover mechanical testing of materials in large-strain (displacement) modes, such as occur in tensile, tear, and elongation tests, and in small-strain modes, such as occur in DMAs. These tests each provide very useful and different information on material properties. In addition, DMA can be employed to provide information on reaction kinetics, rheology, frequency, and temperature-dependent properties of materials – but only at relatively low strains.

### 5.3.1 Tensile, Tear, and Elongation Testing

As indicated above, many application tests are specific to the application and even the customer. Many of these unique standard tests are modifications of ASTM tests. Some relevant ASTM tests are listed in Table 5.7.

For tensile tear and elongation information [84] the workhorse piece of equipment is an instrumented tensile testing machine. Such machines are made by several purveyors, including Instron Corporation and MTS Systems Corporation. These machines are essentially rectangular frames. Samples are positioned within the frame using specially designed fixtures. The fixtures are made to ensure that meaningful and reproducible results can be obtained according to precepts of good mechanical engineering and physics. Attached to one end of the fixture is a device called a load cell, which is essentially a device that measures force. Stress is obtained from the force measurement by normalizing to a geometrical factor such as sample cross-sectional area. At the beginning of a test the top or bottom end of the frame will move relative to the other three sides of the rectangle, applying stress to the sample. The stress is recorded by the load cell much as a scale will measure weight when standing upon it

**TABLE 5.7 Conventional polyurethane mechanical tests and ASTM procedures for measurement**

| Property | ASTM test and comments |
|---|---|
| Tensile properties of plastics | D638 for reinforced and unreinforced plastics (not carbon fiber composites) includes test for measuring Poisson's ratio at STP |
| Test methods for vulcanized rubbers and thermoplastic elastomer – tension | D412 overlaps D638 substantially but also covers a nondumbbell/rectangular geometry and unrecoverable set measurement |
| Standard test method for slab, bonded, and molded cellular polyurethane foams | D3574 includes method of sampling, indentation and compressive force, tear, airflow, dynamic fatigue, humid/wet aging |
| Standard specification for flexible polyurethane foams for furniture seating and bedding | D3453 guidance for acceptability and applicability for polyurethane furniture uses |
| Standard methods for microcellular foams | D3489 overlaps in many ways with other specifications for foams |
| Standard test method for abrasion resistance – static sample with moving abrasive | D1630 uses a standard testing machine executing a controlled abrasion of a stationary test sample against an abrasive paper |
| Abrasion resistance – moving sample and a moving abrasive | D5963 standard rotary drum abrader corresponds to the DIN abrader standard test 53516 |
| Rubber resilience by ball rebound | D2632 vertical rebound resilience test ratio of drop height to rebound height of a standard plunger. Correlates to a tan delta value at temperature |
| Elastomer deterioration by repeated flexure at application-specified temperature | D1052 cut growth by the Ross flex test for footwear. The test specifies STP conditions but many footwear applications demand conditions of 300,000 flexures at −15 °C |
| Hardness testing | D2240 durometer measurement using standard indenters spanning specific ranges of hardness. Correlates to polymer modulus |
| Compression set | D395 compression of a standard disk by a specified amount and amount of recovery. Although the test specifies STP conditions it is normally performed at elevated temperature or humidity conditions to gauge polyurethane network stability |
| Hydrolytic resistance | D3137 and D573 specific test for the effects of distilled water for a given time (i.e. 96 hours) at elevated temperature (i.e. 85 °C). Test can be run at buffered pH levels as well. Applies particularly to systems with polyester soft segments owing to susceptibility to hydrolysis |

DIN, Deutsches Institut für Normung; STP, standard temperature and pressure.

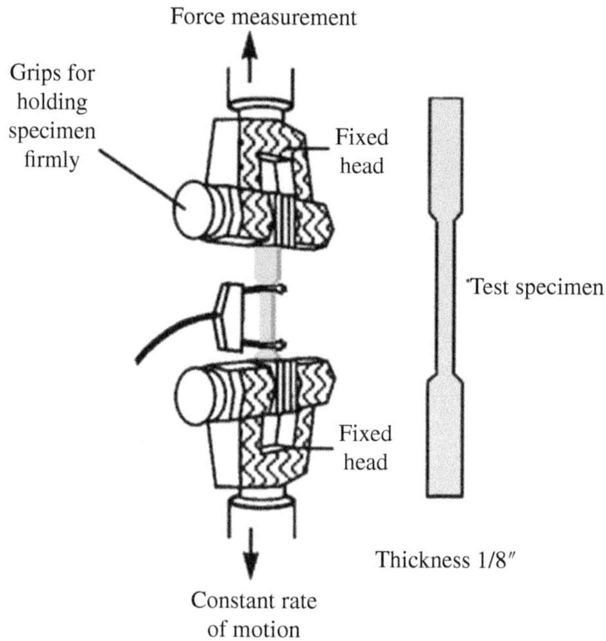


**FIGURE 5.20** Illustration of a conventional tensile measurement configuration. Stress on the sample can be applied from either the top or the bottom and the load cell measuring force can also be above or below the sample depending on the instrument manufacturer.

or when hanging a weight on the end of a spring. Figure 5.20 shows a diagram for a standard fixture for tensile measurements.

Representative stress–strain tensile data for a polyurethane elastomer and calculated output are provided in Figure 5.21. The instrument is capable of providing frequent sampling of the load cell reading with time, such that the data presentation is smooth and does not show significant loss of resolution as a result of time averaging. The straight line in the image is an instrument-drawn measure of the tangent to the initial part of the stress–strain curve termed the modulus. The slope at different strain intervals can also be captured and compared with the initial readings. In addition to the slope of the curve at any given interval, the data analysis can determine ultimate elongation, tensile strength (the maximum stress achieved by the sample), ultimate strength (the stress achieved at the sample break), the toughness of the sample (the integral under the stress strain curve), and numerous other tensile parameters.

Polyurethanes are unique in their versatility. They can be used effectively in almost any potential application, and that application will usually have a corresponding tensile fixture with explicit mechanical engineering so as to provide the most useful and accurate information. Figure 5.22 shows example fixtures designed specifically for application testing. Each fixture takes into account maintenance of

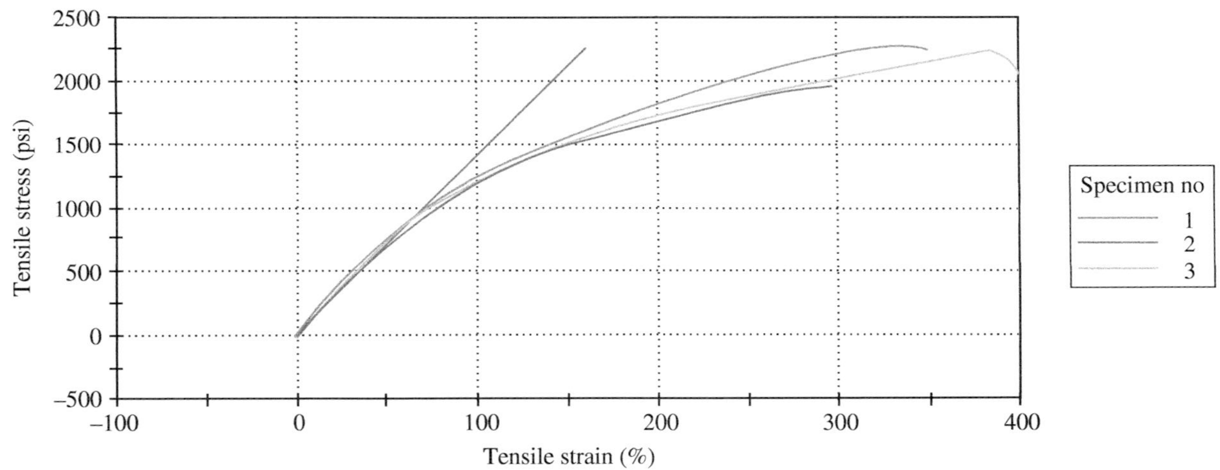


| | Specimen label | Modulus (automatic) (psi) | Modulus(Young's 0.63 mm - 1.64 mm) (psi) | Maximum tensile stress (psi) | Tensile strain at Maximum tensile extension (psi) | Maximum tensile strain (%) | Modulus (Segment 5% –25%) (psi) | Maximum tensile stress absolute peak (psi) |
|---|---|---|---|---|---|---|---|---|
| 1 | 40% HS 28-1–49 NOP TPU–A | 1352.46 | 1530.96 | 2236.470 | 351.1169 | 351.11687 | 1443.00 | 2268.822 |
| 2 | B | 1555.93 | 1862.76 | 1924.997 | 297.7810 | 297.78100 | 1713.48 | 1957.163 |
| 3 | C | 1412.35 | 1660.77 | 2029.442 | 399.9983 | 399.99831 | 1588.47 | 2245.898 |
| Mean | | 1440.25 | 1684.83 | 2063.636 | 349.6321 | 349.63206 | 1581.65 | 2157.294 |
| Median | | 1412.35 | 1660.77 | 2029.442 | 351.1169 | 351.11687 | 1583.47 | 2245.898 |
| Standard deviation | | 104.56634 | 167.20399 | 158.52715 | 51.12483 | 51.12482 | 135.36937 | 173.69741 |

**FIGURE 5.21** Representative stress–strain data for a polyurethane elastomer sample and an example of the data output.

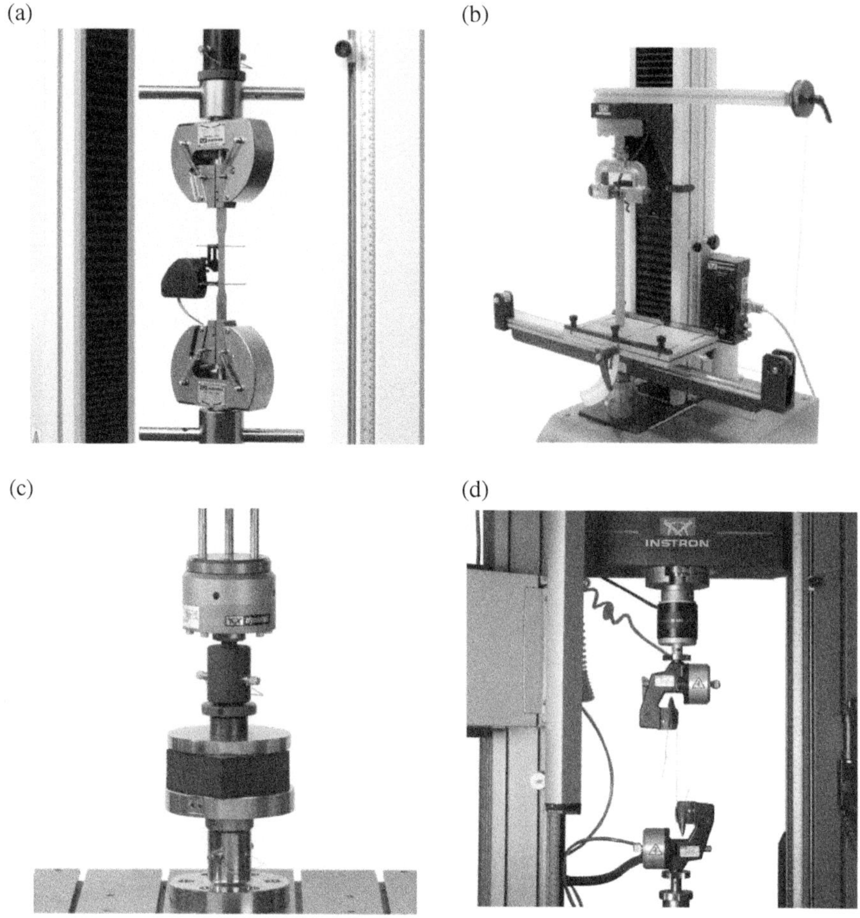


**FIGURE 5.22** Useful fixtures for mechanical testing of polyurethanes. (a) Standard tensile fixture for testing plastics; (b) 90° peel test fixture; (c) compression fixture; (d) cord-and-yarn fixture. Images courtesy of Instron®.

correct angles, mechanical compliance, and mass so that the loads to be measured are correctly matched to the load cell capability [85]. In this regard one would not want to test the compressive properties of a low-density foam specimen with a load cell used for compression of dense elastomers where the modulus can differ by orders of magnitude.

Software for obtaining information from stress–strain experiments is generally provided by the instrument vendors. The software available is highly advanced and becoming ever more intuitive in its user interface. It is in the correct use and setup of the individual test from which information is obtained that differentiates the quality of results. To do this one must use the equipment such that the load cell measuring force is in the correct range of the measurement (e.g. not measuring 50 g of force with a 5000 kg load cell); the fixtures that are used must be correctly sized, aligned to the sample, and must not damage the sample such that the grips predispose the sample to failure; extension must be correctly measured; and the experiment must be performed at an appropriate extension rate. There are clearly a large number of things to keep in mind and there is no substitute for training and experience. Figure 5.23 shows a typical tensile testing device in use. The fixtures in this case are those used for measuring the tensile properties of a film. The fixture is optimized to minimize damage to the film while at the same time providing a firm grip on the sample ends.

A common fixture for measurement of elastomer properties is that shown previously in Figures 5.20 and 5.22. In this case, the sample is formed such that the ends are wider, providing additional gripping surface by the hydraulically actuated flat or serrated grips. While the sample geometry is well known and widely used, a common artifact with this geometry occurs when the sample yields and draws material for extension from the wide sample ends. A device called an extensometer (depicted in Figure 5.22a) is sometimes employed to provide a true measure of sample extension within the narrow sample zone meant to be studied.

An example of the design concepts in the fixtures is given in Figure 5.22, which shows a fixture designed for testing a pressure-sensitive adhesive-type material. One can see that the fixture is designed to maintain a draw angle of 90° with respect to the adhered surface, and this is maintained by a pulley system that draws the adhered

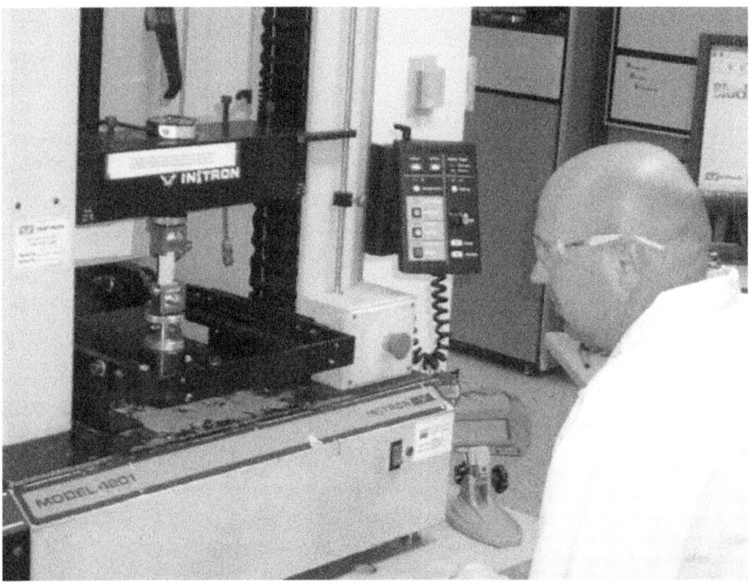

**FIGURE 5.23** A technician measuring a film sample using a tensile testing machine.

surface an amount equal to the length of adhesive pulled. The adhered surface and the drawing grip have sufficient stiffness that they do not vibrate or distort the signal representing the force of pulling the adhered surfaces. At the same time, the fixture, particularly the grip, is low mass and does not require the use of a large load cell in order to allow a tare or zeroing of the measured load prior to beginning the draw experiment. Similarly, the foam compression fixture is a low-mass and high-stiffness design, allowing for easy centering of the foam sample between the compressing platens.

An additional experiment often applied to polyurethane elastomers and flexible foams is the hysteresis measurement [86]. This is a cyclic loading–unloading experiment in which the sample is extended to a set fraction of gauge length (usually 100% or 300%) and the crosshead lowered back to the original gauge length. It is clear from Figure 5.24 that the sample has reached zero stress prior to the crosshead reaching its original position. This is an indication that the sample does not relax as fast as the crosshead was lowered and may have been permanently distorted. The sample is subsequently extended again. From its distorted state, with a longer length and polymer changes having occurred on the molecular level, the material does not achieve the same stress state on subsequent pulls. In general, it is deemed desirable that the area encompassed within the loop be as small as possible (having low hysteresis), which would suggest a high degree of network connectivity and fast chain-relaxation processes. A typical hysteresis experiment will often comprise three or five loops trying to achieve unchanging hysteresis during the stress–strain loop.

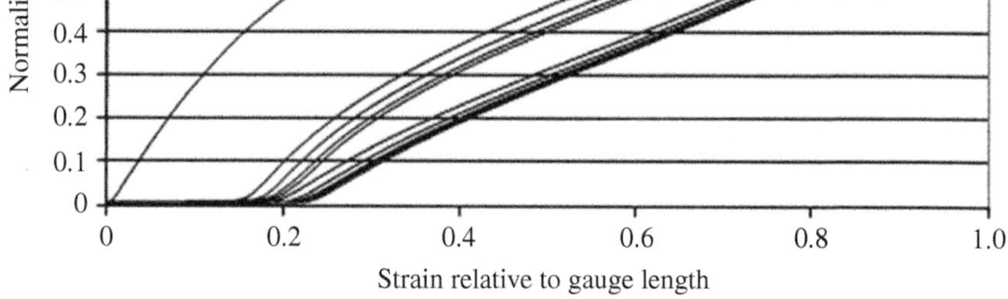


**FIGURE 5.24** Hysteresis loop of a polyurethane elastomer with 45% hard segment.

### 5.3.2 DMA

DMA, or dynamic mechanical spectroscopy (DMS), is a means of measuring the viscoelastic properties of a material, the rheokinetics of a reacting material, or the rheology of a simple or complex fluid [87]. The versatility of DMA has made dynamic mechanical analyzer use in polyurethane science very widespread, and it is certainly one of the most useful pieces of equipment in characterizing polyurethanes. Unlike a tensile test, the materials tested by DMA are rarely, if ever, tested intentionally to a failure condition. Instead, the sample is subjected to an oscillating or sinusoidal strain around a zero strain or a zero reference point. A so-called "strain sweep" is usually performed prior to a full analysis to ensure that the applied strain does not exceed the linear elastic limit of the material or the force limits of the load cell. This is similar to maintaining the strain in the very-low-strain conditions found at the beginning of a tensile measurement. If the experimentalist were to exceed these linear viscoelastic conditions the material would permanently deform and the measurement would not have a clear reference condition. Strains are often of the order of 1% of the gauge length (distance between grips) or a few degrees relative to a full 360° twist of the material, depending in each case on the thickness and hardness of the material being tested. An experimentalist will likewise often perform a frequency sweep of the material prior to beginning a full material analysis. This is to assure that the material properties are not significantly sensitive to the frequency of the applied sinusoidal strain. A frequency-sensitive material, or a material near a rheological event such as a glass transition, may show a frequency-dependent modulus, for instance. This frequency–temperature conflation is a process called time–temperature superposition and is well understood [88]. Problems associated with this condition in a frequency experiment such as DMA are easily avoided through good laboratory practice. A typical frequency is 1 Hz unless the point of the test is to explore the frequency dependence of the material (which is a common enough experiment). The conditions of the constant strain experiment apply to a rotational stress applied to a disk-shaped object in compression, or a torsional stress applied to a rectangular object in tension. In both cases a set minimal force in compression or tension is applied to assure fixture contact or sample rigidity, respectively, and the sinusoidal strain and measured force measured against this initial "zero force" reference condition.

Alternatively a constant sinusoidal stress can be applied to the sample and the resulting induced sample strain measured. This is a less common experiment than the constant strain experiment. In principle, the stress is defined by the strain and the strain defined by the stress; however, particularly in the case of polyurethanes, this expectation is rarely, if ever, realized.

The mechanical analysis of viscoelastic materials is derived from the definition of mechanical properties in these experiments. As oscillation proceeds, the material stress (i) increases with strain, (ii) decreases as the material is returned to its zero strain state, and then (iii) increases again as the material is strained in the opposite direction from the first oscillation. For a perfectly elastic response there is no energy lost within the material during the cycle and the modulus of the material is given by

$$E' = \sigma / \varepsilon \tag{5.11}$$

where $E'$ is the so-called storage modulus of the material or the energy put into the material that is immediately available to be returned, $\sigma$ is the applied stress, and $\varepsilon$ is the applied strain (the use of Greek letters is conventional in this field). In a viscoelastic material $E'$ represents the elastic component. Most nonideal materials such as polyurethanes will not be well represented by this equation as there are numerous molecular processes that are capable of dissipating some of the energy put into the sample during the stress phase of the experiment. In these oscillating experiments consisting of elastic energy storage $E'$ and energy losses $E''$ due to viscous processes, their moduli are defined by

$$E' = \frac{\sigma_0}{\varepsilon_0} \cos \delta \tag{5.12}$$

$$E'' = \frac{\sigma_0}{\varepsilon_0} \sin \delta$$

The ratio of the loss modulus ($E''$) to the storage modulus is termed the tan ($\delta$) (called the "tan delta"):

$$E'' / E' = \tan(\delta) \tag{5.13}$$

The tan $\delta$ is the phase lag between applied strain and measured stress, a very sensitive function of phase transitions. Representative data obtained from a DMA experiment on a polyurethane elastomer are shown in Figure 5.25. Figure 5.25b shows the tan $\delta$ spectrum for this elastomer clearly showing the $T_g$ (the peak with maximum at –25 °C in this case) and the onset of urethane reversion (see Chapter 3), beginning at approximately 120 °C.

More than 70% of all polyurethane chemicals are processed into foams. The DMA spectra of these materials have been shown to be very similar, if not identical, to the underlying elastomer, except they exhibit a much lower modulus, reflecting the well-known squared dependence on density [89]:

$$\frac{E_{\text{foam}}}{E_{\text{solid}}} \sim \left( \frac{\rho_{\text{foam}}}{\rho_{\text{solid}}} \right)^2 \tag{5.14}$$

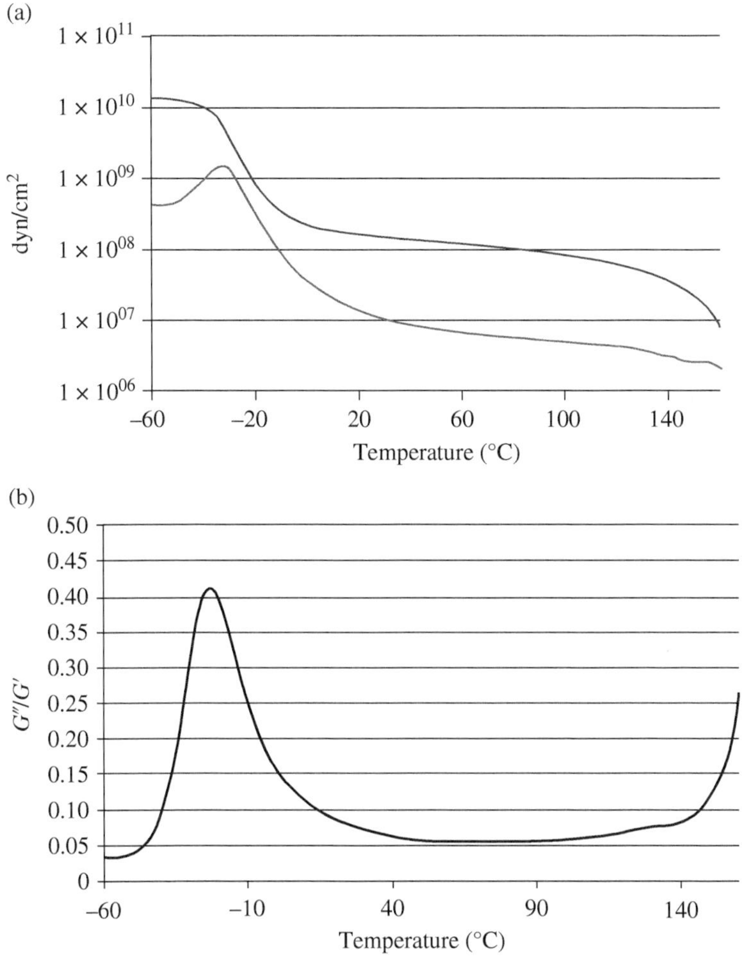


**FIGURE 5.25** Representative DMA for a polyurethane elastomer. (a) Storage (solid line) and loss modulus (dashed line). (b) The tan $\delta$ spectrum is the ratio of the loss to the storage modulus.

Experiments on foam samples are ordinarily done on disk-shaped samples. Examples of DMA fixtures for shear tension experiments are shown in Figure 5.26.

DMA experiments can also be employed to provide time- or temperature-dependent rheology for reacting systems. In this case, the components of the reaction are introduced into a cup-like fixture and a plate attached to the load cell submerged into the reacting components. The plate will oscillate through the reacting mixture, measuring the growth of molecular weight as an increasing resistance to plate rotation. Since the

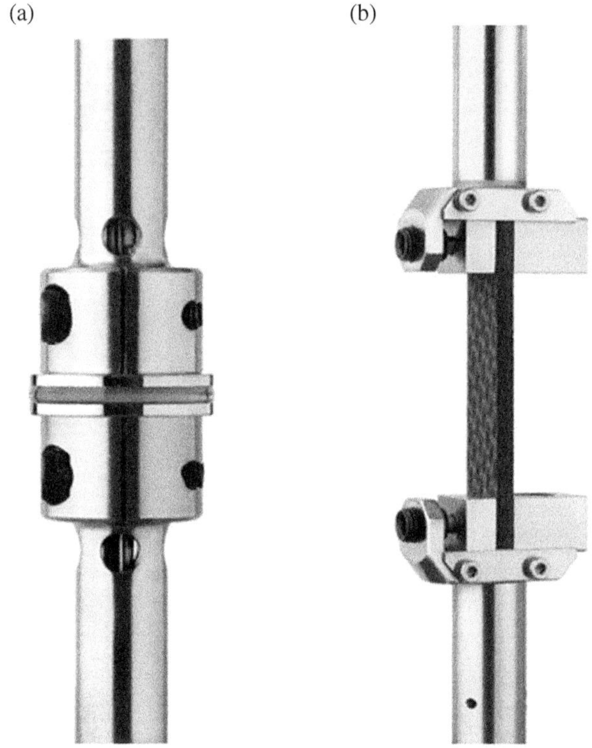


**FIGURE 5.26** DMA fixtures. (a) Parallel plate fixture for measurement of thin solid samples in shear or reacting liquids. (b) Torsion fixture for measurement of solid sample viscoelastic properties. Fixtures are carefully engineered and machined to provide minimal compliance and minimum mass. Images provided courtesy of TA Instruments, © 2013.

forces are often low in the initial state, the plate usually must oscillate with relatively higher amplitude through the mixture. A typical DMA is shown in Figure 5.27 and the data obtainable from a rheokinetic experiment are shown in Figure 5.28.

## 5.4 NUCLEAR MAGNETIC RESONANCE SPECTROSCOPY

Nuclear magnetic resonance (NMR) spectroscopy is an analytical technique that has been around for a relatively long time. However, apart from conventional chemical analyses, its application to polyurethanes and polymer structure has lagged other methods because of the technical difficulties associated with NMR spectroscopy of solids [90]. The basic concepts of NMR spectroscopy of liquids apply to the NMR spectroscopy of solids generally and polyurethanes specifically. They are well covered in numerous other publications [91, 92]. In the solid state the number of

**FIGURE 5.27** A technician preparing a DMA for analysis using a cup-and-plate fixture.

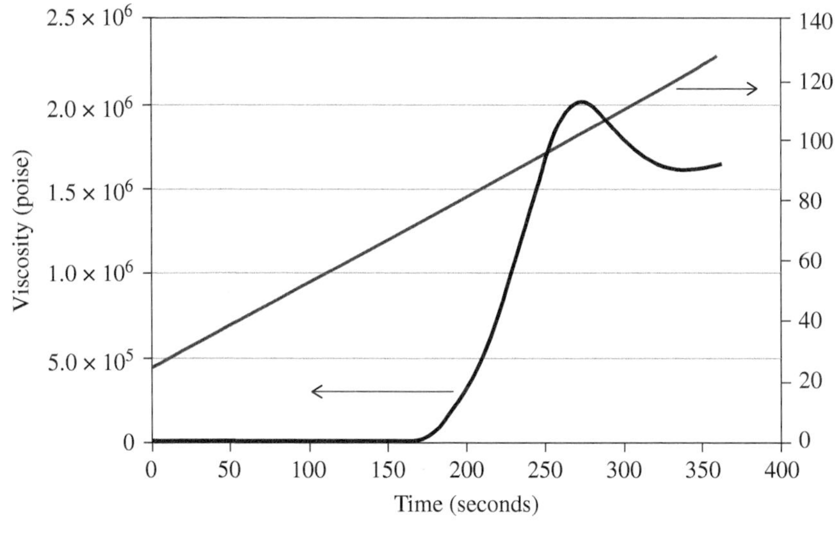


**FIGURE 5.28** Example of rheokinetic data obtainable with DMA for a two-part reacting system such as a polyurethane.

interactions of any nucleus with its environment (i.e. dipolar, quadripolar, j coupling, etc.) is very inhomogeneous and has a time dependence resulting in highly diffuse resonance absorption. The advent of magic-angle spinning (MAS) techniques has allowed for narrow resonant peak measurements to be made on solid polymers that would normally be too broad [93–95]. MAS spin rates are of the order of 5–10 kHz, and the magic angle of the sample placement in the magnet is 54.74° – the "magic angle". Typical nuclei investigated are $^{13}C$ and $^{1}H$, with $^{13}C$ providing superior resolution but $^{1}H$ requiring much shorter data acquisition times.

In addition to analytical improvements allowing for observations of nuclei and their chemical environments in the solid state by MAS, the physics of nuclear spin and diffusion within an NMR magnetic field has allowed accurate measurement of specific motions and characteristic times of those motions in the solid state. Thus, examples of critical information that can be provided by solid-state NMR include:

i. spatial order and connectivity
ii. small-molecule interactions
iii. surface interactions
iv. polymer dynamics and their temperature dependence
v. reaction monitoring
vi. site activity
vii. blend miscibility.

The limitations of solid-state NMR are that it:

i. is time intensive
ii. is expensive
iii. it has low sensitivity for some nuclei.

Since polyurethanes are often multiphasic materials consisting of low $T_g$ soft segments cross-linked by high glass transition hard segments, the opportunity for developing quantitative understanding of these motions and the resulting polymer properties is apparent. The use of specified pulse sequences such as Hahn spin echo techniques allows correlation of NMR-determined spin-oriented relaxation times with morphology and mobility [96–98]. Specific to polyurethanes, the mobile soft segments and the rigid hard domains should represent clearly distinguishable environments with substantially different relaxation times. In fact, it is observed that the spin–spin relaxation time (the product of a spin echo experiment, distinct from a spin–lattice relaxation) for hard segments is 7–20 μs while for soft segments it is 200–1000 μs [99]. As a matter of practicality, proton relaxation spectra are usually obtained for soft-segment dynamics because of their higher mobility and accompanying longer dipolar dephasing time constants. Hard-segment dynamics will often be probed by $^{13}C$ NMR owing to greater difficulty in resolving individual hard-segment $^{1}H$ peaks that often appear as a broad peak with spinning side bands extending over a wide frequency range [100].

The different scale of hard- and soft-segment relaxation times in the spin echo experiment allows for isolation of the two phases and analysis of each phase for its particular characteristics [101]. In the event of each phase existing in a homogeneous environment the relaxation decay will appear as a single exponential decay. In the event that a single phase comprises more than one environment, each environment results in a distinctive spin–spin relaxation time, and the relaxation may be well fit to a multiexponential decay described by Equation 5.15:

$$M(t) = M_0 \sum_i P_i e^{t/T_{di}} \quad (5.15)$$

where $P_i$ and $T_{di}$ are the percentage and dipolar dephasing time constant for the $i$th component, respectively [102].

An example of experimental data and analysis is shown in Figure 5.28. A time-resolved stack-plot of $^1$H MAS NMR spectra of selected resonances of a polyurethane soft segment is shown with the dipolar dephasing time indicated using the spin echo technique. The least squares analysis of intensity for a specific resonance versus dipolar dephasing time provides the best fit for the dipolar time constants, which can then be interpreted to infer the mobility of a percentage of soft-segment components and their subsequent role in measured polyurethane properties [16]. In the particular case of Figure 5.29, the multiexponential fit is well characterized by two time constants: one fast and one much slower. An interpretation of this observation might be that a percentage of the soft segment is mobile, reflecting the dynamics typical of an unencumbered soft segment, while another component is highly reflective of hard-segment dynamics (faster relaxation times) owing to dipolar and hydrogen bonding at the hard segment–soft segment boundary. The temperature dependence of the component percentage and time constants can provide additional insight into the molecular-level dynamics responsible for observed polymer properties.

## 5.5 FOAM SCREENING: FOAMAT®

As will be discussed in great detail in Chapter 6, optimized foam processes follow a highly reproducible timeline during which all the physics and chemistry that result in a particular foam occur. While it is necessary to understand polyurethane foam formation in the abstract, it is also necessary to quantify and identify the kinetics of these processes for any new formulation that may be experimented with. FOAMAT combines numerous sensors in a packaged system that can simultaneously measure the foam rise profile, the gas pressure in the expanding foam, the foam temperature increase owing to the adiabatic reaction enthalpy, and the mobility of the reacting system through a measurement of dielectric polarization [103]. The data in Figure 5.30b show the intuitively correct relationship between the halt in foam rise and the attainment of a rheological gel point coupled with the sudden decrease in

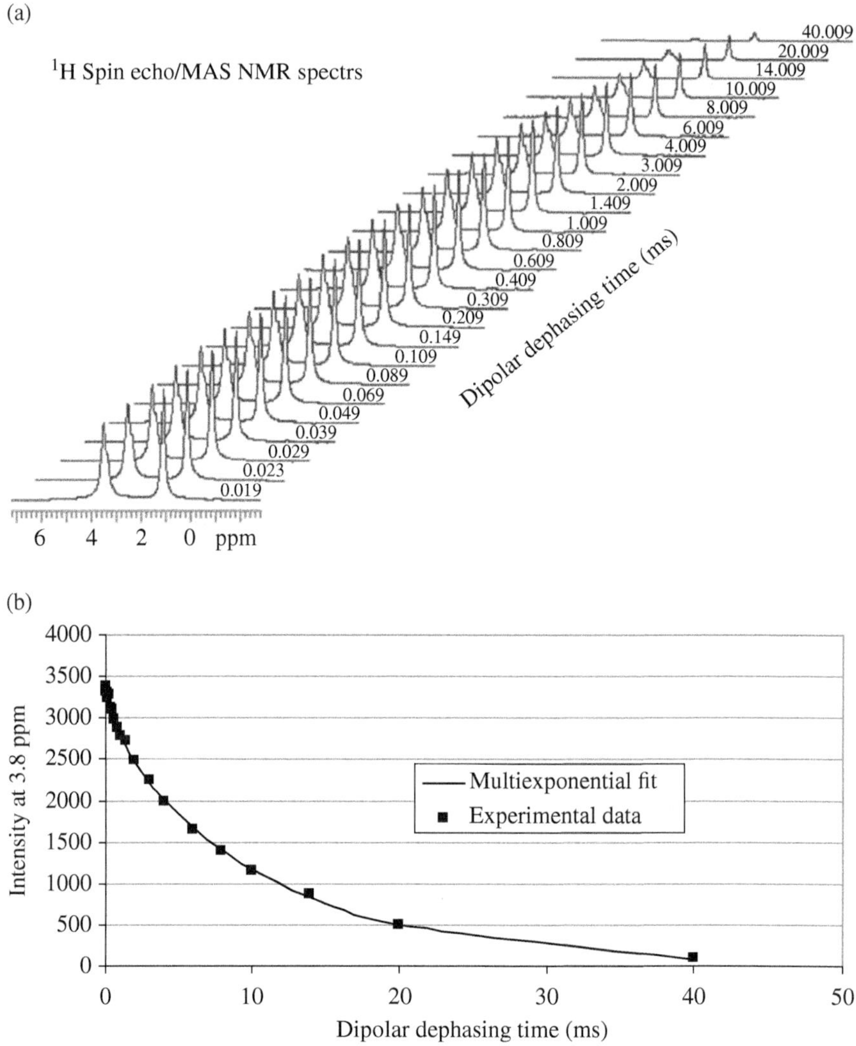


**FIGURE 5.29** (a) Representative stack-plot of $^1$H MAS NMR spectra of a polyurethane foam acquired at 38 °C with values of the $^1$H dipolar dephasing time indicated using the Hahn spin echo technique. The resonance at 3.7 ppm reflects components of propylene oxide and ethylene oxide from the soft-segment backbone, while the resonance at ~1 ppm is from the pendant methyl unit of propylene oxide. (b) Representative biexponential fit to data obtained from stack-plot data. Reprinted with permission from Ref. [16]. © Elsevier Pub.

(a)

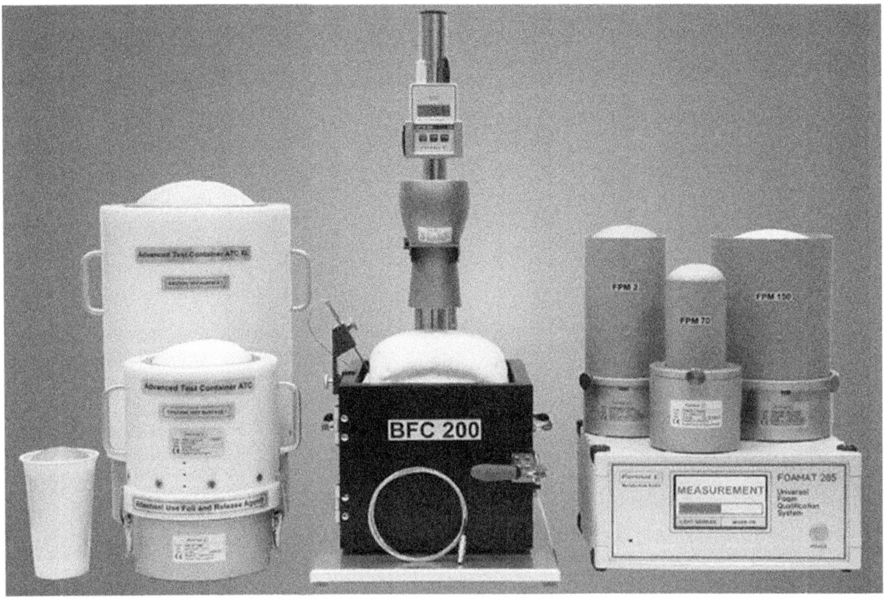


(b)

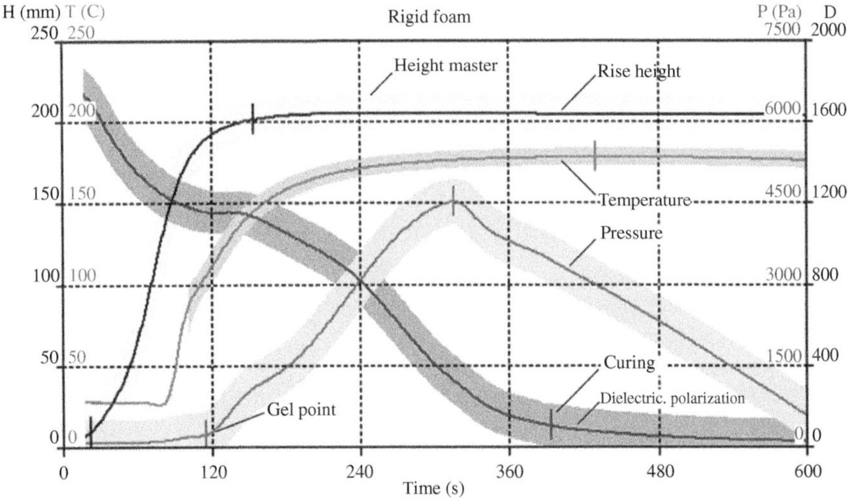


**FIGURE 5.30** The FOAMAT® foam qualification system for quantifying the physical processes occurring in a foaming system after mixing. (a) The appearance of a full system capable of preparing foams at different volumes and different thermal controls. (b) Sample data obtained simultaneously from a rigid insulation foam. The measurement of pressure and dielectric polarization occurs in the base, the temperature rise is measured by a thermocouple within the rising foam body, and the rise and gel conditions are determined by the sensor above the foam mass. Images courtesy of Format Messtechnik GmbH. (*See insert for color representation of the figure.*)

chain mobility by the inflection and decrease in dielectric polarization. When the foam ceases to rise the foam temperature continues to increase and is accompanied by the increasing internal gas pressure within the foam. The author has found this equipment to be invaluable in controlled comparisons between formulations, and its use is becoming industrially much more conventional and accepted as a point of reference for foaming performance.

## REFERENCES

[1] C Hynes and R Koile, *American Laboratory*, **40**, (2008), 20–21.
[2] J Bicerano, *Prediction of Polymer Properties*, 3 Ed., CRC Press, Boca Raton, FL, 2002.
[3] K Yamamoto, R Kakita, C Suzuki, M Meno, and Y Nozawa, *Applied Spectroscopy*, **5**, (2003)**7**, 353–355.
[4] R Herrington and K Hock, *Flexible Polyurethane Foams*, 2nd Ed., The Dow Chemical Co., Midland, MI, 1997.
[5] B Ping, N Hanzah, and H Soi, *Journal of Palm Oil Research*, **23**, (2011), 1172–1177.
[6] S Cho, H Chung, and Y Lee, *Microchemical Journal*, **80**, (2005), 189–193.
[7] S Godoy, M Ferraro, and A Gerbase, *Journal of the American Oil Chemists Society*, **84**, (2007), 503–508.
[8] E Zagar and J Grdadolnik, *Journal of Molecular Structure*, **658**, (2003), 143–152.
[9] K Lee, A Bunding, R Chylla, and T Janota, *Applied Spectroscopy*, **47**, (1993), 94–97.
[10] *Standard Practice for Polyurethane Raw Materials: Determining Hydroxyl Number of Polyols by Near-Infrared (NIR) Spectroscopy,* ASTM D6342, ASTM, West Conshohocken, PA, 2012.
[11] *Standard Test Method for Polyurethane Raw Materials: Determination of Monomer and Isomers in Isocyanates,* ASTM D7252-06, ASTM, West Conshohocken, PA; reapproved 2011.
[12] R Beasley and J Warner, *Analytical Chemistry*, **56**, (1984), 1604–1608.
[13] M Sonnenschein and B Wendt, *Polymer*, **53**, (2013), 2511–2520.
[14] T Choi, D Fragiadaikis, C Roland, and J Runt, *Macromolecules*, **45**, (2012), 3581–3589.
[15] M Ferrer, D Babb, and A Ryan, *Polymer*, **49**, (2008), 3279–3287.
[16] M Sonnenschein, *R Prange, and A Schrock*, **48**, (2007), 616–623.
[17] J Sheth, G Wilkes, A Fornof, T Long, and I Yilgor, *Macromolecules*, **38**, (2005), 5681–5685.
[18] E Christenson, J Anderson, A Hiltner, and E Baer, *Polymer*, **46**, (2005), 11744–11754.
[19] J Bicerano, R Daussin, M Elwell, H van der Wal, P Berthevas, M Brown, F Casati, W Farrissey, J Fosnaugh, R de Genova, et al., "Flexible polyurethane foams: a review of the state-of-the-art," in *Polymeric Foams*, S Lee and N Ramesh, Eds., CRC Press, Boca Raton, FL, 2004, 173–251.
[20] MF Sonnenschein, N Rondan, B Wendt, and JM Cox, *Journal of Polymer Science Part A: Polymer Chemistry*, **42**, (2004), 271–278.
[21] H Gruenbauer, J Bicerano, P Clavel, R Daussin, H de Vos, M Elwell, H Kawabata, H Kramer, D Latham, C Martin, et al., "Rigid polyurethane foams: a review of the state-of-the-art," in *Polymeric Foams*, S Lee and N Ramesh, Eds., CRC Press, Boca Raton, FL, 2004, 253–310.

[22] A Clarke and C Eberhardt, *Microscopy Techniques for Materials Science*, CRC Press, Boca Raton, FL, 2002.
[23] B Hayes and L Gammon, *Optical Microscopy of Fiber Reinforced Composites*, ASM International, Materials Park, OH, 2010.
[24] N Shaked, Z Zalevsky, and L Satterwhite, Eds., *Biomedical Optical Phase Microscopy and Nanoscopy*, Elsevier, Amsterdam, the Netherlands, 2012.
[25] R Mitchell, A Muller, A Boschetti-de-Fierro, D Fierro, V Lison, J Raquez, and P Dubois, *Polymer*, **53**, (2012), 5657–5665.
[26] M Tu, W Han, R Zeng, S Best, and R Cameron, *Colloids and Surfaces A: Physiochemical and Engineering Aspects*, **407**, (2012), 126–132.
[27] A Mishra and P Maiti, *Journal of Applied Polymer Science*, **120**, (2011), 3546–3555.
[28] J Wang, L Zheng, C Li, W Zhu, D Zhang, G Guan, and Y Xiao, *Industrial and Engineering Chemistry Research*, **51**, (2012), 10785–10792.
[29] J Culin, M Andreis, I Smit, Z Veksli, A Anzlovar, and M Zignon, *European Polymer Journal*, **40**, (2004), 1857–1866.
[30] C Fernandez, M Bermudez, R Versteegen, E Meijer, A Muller, and S Munoz-Guerra, *Journal of Polymer Science Part B: Polymer Physics*, **47**, (2009), 1368–1380.
[31] D Palima and J Gluckstad, *Contemporary Physics*, **51**, (2010), 249–265.
[32] K Shimizu and T Mitani, Eds., *New Horizons of Applied Scanning Microscopy, Springer Series in Surface Science*, Volume **45**, Berlin, Germany, 2010.
[33] R Brydsen, Ed., *Aberration Corrected Analytical Transmission Electron Microscopy*, Wiley, Chichester, UK, 2011.
[34] M Sonnenschein and W Koonce, "Polyurethanes," in *Encyclopedia of Polymer Science and Technology*, 4th Ed., Wiley, Hoboken, NJ, 2012, 1–63.
[35] J Armistead, G Wilkes, and R Turner, *Journal of Applied Polymer Science*, **35**, (1988), 601–629.
[36] A Chang, R Briber, E Thomas, R Zdrahata, and F Critchfield, *Polymer*, **23**, (1982), 1060–1068.
[37] H Xie, Y Gu, and H Ploehn, *Nanotechnology*, **16**, (2005), 5492–5501.
[38] P Eaton and P West, *Atomic Force Microscopy*, Oxford University Press, New York, NY, 2010.
[39] G Haugstad, *Atomic Force Microscopy: Understanding Basic Modes and Advanced Applications*, Wiley, Hoboken, NJ, 2012.
[40] A Opdahl, T Koffas, E Amitay-Sadovsky, J Kim, and G Samorjai, *Journal of Physics: Condensed Matter*, **16**, (2004), 659–677.
[41] T Koffas, E Amitay-Sadovsky, J Kim, and G Samorjai, *Journal of Biomaterials Science, Polymer Edition*, **15**, (2004), 475–509.
[42] S Morita, F Giessibl, and R Wisendanger, Eds., *Noncontact Atomic Force Microscopy*, Springer, Berlin, Germany, 2009.
[43] J Li, W Liang, G Meyers, and W Heeschen, *Polymer News*, **29**, (2004), 335–345.
[44] M Sonnenschein, B Wendt, A Schrock, J-M Sonney, and A Ryan, *Polymer*, **49**, (2008), 934–942.
[45] K Strawhecker, A Hsieh, T Chantawansri, and Z Kalcioglu, *Polymer*, **54**, (2013), 901–908.
[46] S Das, I Yilgor, E Yilgor, B Inci, O Tezgel, F Beyer, and G Wilkes, *Polymer*, **48**, (2007), 290–301.
[47] J Garrett, C Siedlecki, and J Runt, *Macromolecules*, **34**, (2001), 7066–7070.
[48] M Sonnenschein, V Ginzburg, K Schiller, and B Wendt, *Polymer*, **54**, (2013), 1350–1360.

[49] R Camargo, C Macosko, M Tirrell, and S Wellinghof, *Polymer Communications*, **24**, (1983), 314–315.
[50] M Elwell, A Ryan, H Gruenbauer, and H Lieshout, *Macromolecules*, **29**, (1996), 2960–2968.
[51] W Bras, G Derbyshire, D Bogg, J Cooke, M Elwell, B Komanschek, S Naylor, and A Ryan, *Science*, **267**, (1995), 996–999.
[52] D Dounis and G Wilkes, *Journal of Applied Polymer Science*, **66**, (1997), 2395–2408.
[53] V Furer, *Journal of Molecular Structure*, **476**, (1999), 215–222.
[54] A Samuilov, L Zenitova, Y Levin, A Kurdyukov, and Y Samuilov, *Russian Journal of Physical Chemistry A*, **82**, (2008), 1999–2004.
[55] G Raspoet, M Nguyen, M McGarraghy, and A Hegarty, *Journal of Organic Chemistry*, **6**, (1998)3, 6367–6877.
[56] M Spirkova, M Kubin, P Spacek, I Krakovsky, and K Dusek, *Journal of Applied Polymer Science*, **52**, (1994), 895–904.
[57] T Matsui, H Kamatani, Y Arimatsu, A Kaji, K Hattori, and H Suzuki, *Journal of Applied Polymer Science*, **42**, (1991), 2443–2452.
[58] D Bogg, G Derbyshire, W Bras, J Cooke, M Elwell, S Naylor, and A Ryan, *Nuclear Instruments and Methods in Physics Research B*, **97**, (1995), 536–538.
[59] B Chu, T Gao, Y Li, J Wang, C Desper, and C Byrne, *Macromolecules*, **25**, (1992), 5724–5729.
[60] L Zhang, H Jeon, J Malsam, R Herrington, and C Macosko, *Polymer*, **48**, (2007), 6656–6667.
[61] R Roe, *Methods of X-Ray and Neutron Scattering in Polymer Science*, Oxford University Press, Oxford, UK, 2000.
[62] D Dounis and G Wilkes, *Polymer*, **38**, (1997), 2819–2838.
[63] A Ryan, J Stanford, W Bras, and T Nye, *Polymer*, **38**, (1997), 759–768.
[64] M Sonnenschein, S Guillaudeu, B Landes, and B Wendt, *Polymer*, **51**, (2010), 3685–3692.
[65] R Rinaldi, M Boyce, S Weigand, D Londono, and M Guise, *Journal of Polymer Science Part B Polymer Physics*, **49**, (2011), 1660–1671.
[66] P Krol, *Progress in Materials Science*, **52**, (2006), 915–1015.
[67] W Li, A Ryan, and I Meier, *Macromolecules*, **35**, (2002), 5034–5042.
[68] A Aneja and G Wilkes, *Journal of Applied Polymer Science*, **85**, (2002), 2956–2967.
[69] Y Li, Z Ren, M Zhao, H Yang, and B Chu, *Macromolecules*, **26**, (1993), 612–622.
[70] R McKiernan, S Gido, and J Penelle, *Polymer*, **4**, (2002)3, 3007–3017.
[71] N Mahmood, J Kressler, and K Busse, *Journal of Applied Polymer Science*, **98**, (2005), 1280–1289.
[72] N Kasai and M Kakudo, *X-Ray Diffraction by Macromolecules*, Springer-Verlag, Berlin, Germany, 2010.
[73] A Kikheny and D Svergun, *FEBS Letters*, **589**, (2015), 2570–2577.
[74] G Strobl and M Scheider, *Journal of Polymer Science Polymer Physics Edition*, **18**, (1980), 1343–1359.
[75] J Koberstein, B Morra, and R Stein, *Journal of Applied Crystallography*, **13**, (1980), 34–45.
[76] G Porod, *Kolloid-Zeitschrift*, **124**, (1951), 83–114.
[77] G Porod, *Kolloid-Zeitschrift*, **125**, (1952), 51–57.
[78] G Porod, *Journal of Polymer Science*, **10**, (1953), 157–166.

[79] J Koberstein and R Stein, *Journal of Polymer Science Polymer Physics Edition*, **21**, (1983), 1439–1472.
[80] J Koberstein and R Stein, *Journal of Polymer Science Polymer Physics Edition*, **21**, (1983), 2181–2200.
[81] J Koberstein and R Stein, *Polymer*, **25**, (1984), 171–177.
[82] C Vonk, *Journal of Applied Crystallography*, **4**, (1971), 340–342.
[83] R Perret and W Ruland, *Journal of Applied Crystallography*, **4**, (1971), 444–451.
[84] V Shah, *Handbook of Plastic Testing Technology*, 2nd Ed., Wiley Interscience, New York, NY, 1998.
[85] W Grellmann and S Seidler, *Polymer Testing*, Carl Hanser Verlag, Munich, Germany, 2007.
[86] J Mark, "Physical states of polymers," in *Physical Properties of Polymers*, 3rd Ed., J Mark, K Ngai, W Graessley, L Mandelker, E Samulsi, J Koenig, and G Wignall, Eds., Cambridge University Press, Cambridge, UK, 2003.
[87] K Menard, *Dynamic Mechanical Analysis: A Practical Introduction*, CRC Press, Boca Raton, FL, 2008.
[88] I Ward and J Sweeney, *Mechanical Properties of Solid Polymers*, 3rd Ed., Wiley, Chichester, UK, 2013, 145–155.
[89] L Gibson and M Ashby, "The mechanics of foams: basic results," *Cellular Solids: Structure and Properties*, 2nd Ed., Cambridge University Press, Cambridge, UK, 1997, Chapter 5.
[90] A Brandolini and D Hills, *NMR Spectra of Polymers and Polymer Additives*, Marcel Dekker Inc., New York, NY, 2000.
[91] H Speiss, *Macromolecular Engineering*, **3**, (2007), 1937–1965.
[92] P Mirau, *Solid State NMR of Polymers in Rapra Review Reports*, **11**, (2001), 1–143.
[93] A Maryiez-Richa and R Silvestri, *Materials Science Forum*, **714**, (2012), 51–56.
[94] S Brown, *Solid State Nuclear Magnetic Resonance*, **41**, (2012), 1–27.
[95] A Aliev and R Law, *Nuclear Magnetic Resonance*, **39**, (2012), 227–267.
[96] H Walderhaug, O Soederman, and D Topgaard, *Progress in Nuclear Magnetic Resonance*, **56**, (2010), 406–425.
[97] R Huo, R Wehrens, J Van Duynhoven, and L Buydens, *Analytica Chimica Acta*, **490**, (2003), 231–251.
[98] G Cardenas, G Cabrera, E Taboada, and S Mirand, *Journal of Applied Polymer Science*, **93**, (2004), 1876–1885.
[99] D Idiyatullin, E Khozina, and V Smirnov, *Solid State Nuclear Magnetic Resonance*, **7**, (1996), 17–26.
[100] R Assink and G Wilkes, *Journal of Applied Polymer Science*, **26**, (1981), 3689–3698.
[101] J Moreland, G Wilkes, C Moreland, S Sankar, E Stejskal, and R Turner, *Journal of Applied Polymer Science*, **52**, (1994), 1175–1180.
[102] S Ghose, A Isayev, and E von Meerwall, *Polymer*, **45**, (2004), 3709–3720.
[103] L Vandenbossche, L Dupre, and J Melkebeek, *International Journal of Applied Electromagnetics and Mechanics*, **25**, (2007), 589–593.

# 6

# POLYURETHANE FLEXIBLE FOAMS: CHEMISTRY AND FABRICATION

Flexible polyurethane foam is the largest volume application for polyurethanes and the largest category of cellular polymeric materials [1]. The discussions in prior chapters relating to foam properties and chemistry apply completely to flexible foams from a fundamental standpoint. Polyurethane chemistry is flexible enough, and can incorporate enough components, that the potential variability in chemistry and process makes drawing hard rules about foam formulation a fruitless task. In fact, the properties of a particular foam can often be achieved by numerous combinations of foam components. Despite this challenge, it is still possible to generally discuss concepts and guidelines for foam production, and methods of making foams at various scales. Furthermore, even though there are many ways to make foam, there are many fewer ways to describe the product, its performance, and the physical paths that the reactants take to the final product. This chapter will broadly define design concepts for making polyurethane flexible foams, physical concepts relevant to the formation of foam from polyurethane components, and physical descriptors of the product. Discussion of the building blocks can be found in Chapter 2; the chemical reactions that occur during foaming in Chapter 3; theoretical concepts in polymer formation and foam structure in Chapter 4; and analysis of polyurethane foams in Chapter 5.

*Polyurethanes: Science, Technology, Markets, and Trends*, Second Edition. Mark F. Sonnenschein.

## 6.1 MAKING POLYURETHANE FOAMS

Industrial production of foams is a highly optimized and reproducible operation. The volumes of chemicals consumed in a commercial-scale plant can be startlingly large, and a large plant operating at capacity can be a supply chain challenge. The production of foam after receipt of raw materials subsequently becomes an exercise in accurate mixing and pouring. The necessary requirement is to perform these mixing operations in the right amounts, in the right order, and at the right time. When done correctly the physical processes that result in foam can be relied upon and reproduced such that variability between repetitions may not be detectable. Thus, the first step to making a foam is to assemble all the necessary reagents in sufficient amounts, and accurately dispense them into a reaction volume.

When making flexible foam, calculation of the correct mass to mix of a particular ingredient is normally based on the *equivalent* weight of the component rather than the molecular weight of the component [2]. The equivalent weight is simply the molecular weight of the component divided by its functionality. In setting formulation, the stoichiometry is set to match the equivalence of the isocyanate side with the total equivalence of the isocyanate reactive side because the materials to be mixed may have quite different functionalities; however, controlling stoichiometry is a necessary condition of getting a good result. The use of equivalence minimizes the likelihood of stoichiometry errors, although they do still occur.

As mentioned above, there are many desired flexible foam applications and a defined formulation for making each one. Flexible foams can be produced for applications such as mattresses, which may be cut from large blocks into defined shapes for the application. These foams are often called *slabstock foams* since the foams are made as enormous slabs from which the desired product is cut. Alternatively, the foam can be poured into a predefined mold to form a custom shape. Such foams are often found in furniture cushions and automotive seating, and for obvious reasons are commonly called *molded foams*. The distinctions between slabstock and molded foams form a large enough class boundary that their formulations can be distinguished by basic rules.

### 6.1.1 Slabstock Foams

Slabstock foams, like most foams, are prepared from polyols, isocyanates, and water. In addition to these basic ingredients, catalysts, surfactants, co-blowing agents, flame retardants, antioxidants, and sometimes colorants are used. All of the nonisocyanate components are premixed in combinations that are convenient for the foam manufacturer, then vigorously mixed with the isocyanates. The mixed and reacting system is applied to a moving belt (Figure 6.1). The large foam slab (sometimes also referred to as a "bun") is subsequently cut into manageable sizes and then set aside in a well-ventilated location to cool and complete its cure. The actual geometry of a particular slab foam production line may vary from that shown in Figure 6.1, but will be similar in essential details. The ventilation requirements are the result of the substantial exotherm from heat

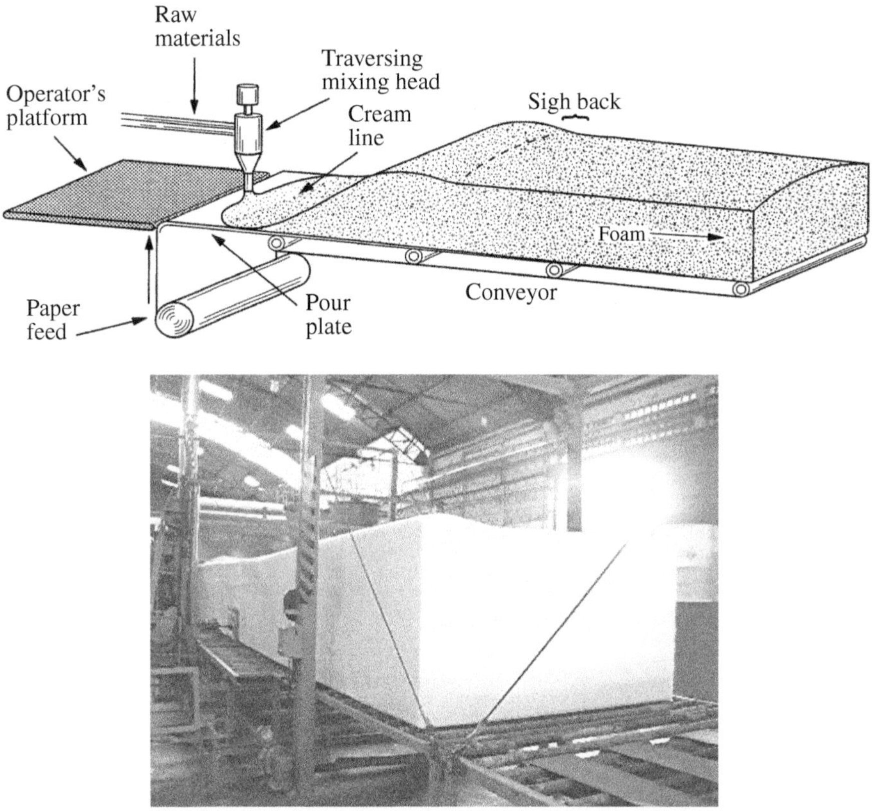


**FIGURE 6.1** Slab foam process. The image does not show the feed tanks, ventilation, or the substantial amount of electronic controls involved. Courtesy of the Dow Chemical Company. Reprinted with permission from Ref. [3]. © John Wiley & Sons, Inc.

of reaction and the volatility of the polyurethane components, particularly isocyanates, catalytic amines, and small molecule impurities from the polyol production – mostly aldehydes. Control of the exotherm is critical to prevent oxidative discoloration of the foam center, which is referred to as "scorch." In the extreme a foam may even catch fire if the foam interior temperatures are not controlled.

Typical foam formulations that could be attempted by an experimentalist with a high probability of success would be exemplified by those found in Table 6.1. These two formulations are distinguished by the amount of water and the amount of isocyanate, in this case toluene diisocyanate (TDI). The effect of increased water and concomitant TDI is to (i) reduce the foam density by increasing the amount of bubble formation due to the isocyanate–water reaction that produces $CO_2$ (Chapter 3), (ii) increase the foam stiffness as a result of additional urea formation, which further reinforces the foam with a high-modulus block (Chapter 4), and (iii) increase the exotherm produced by the foam as it rises.

**TABLE 6.1 Representative slab foam formulations**

| Component | Functionality | Molecular weight (g/mol) | Equivalent weight (g/eq) | Formulation 1 (g) | Formulation 2 (g) |
|---|---|---|---|---|---|
| Polyol | 3 | 3000 | 1000 | 100 | 100 |
| Water | 2 | 18 | 9 | 2 | 4 |
| TDI (110 index) | 2 | 174 | 87 | 30.6 | 52.1 |
| Surfactant DC5125 | | | | 1 | 1 |
| Triethylene diamine | | | | 1 | 1 |
| Tin dioctanoate | | | | 0.15 | 0.15 |

To make these foams in a laboratory place all components in a well-ventilated fume hood with an accurate scale. Mix all but the isocyanates in a beaker. Add isocyanate to the beaker, mix vigorously for 10 seconds, and pour into a box adequate for the volume expansion. Clean removal of the foam from the box will probably require the use of a release paper liner. Perform in a well-operating chemical hood with appropriate personal protection.
TDI, toluene diisocyanate.

The stoichiometry of the reactive components is matched so that the total equivalents of isocyanate and isocyanate reactive components are matched. There is a 10% excess equivalence of TDI added in this case (110% of equivalence, which is referred to as the "index"). This is done to assure that all of the active hydrogen components are utilized to make a complete network. Some of the excess isocyanate is volatilized, as previously mentioned. Some of the excess is employed to cover any loss of isocyanate functionality as a result of unintentional urea formation due to reaction with adventitious water, and cross-linking isocyanate–isocyanate reactions due to excess concentration fluctuations (Chapter 3). The formulations in Table 6.1 can be scaled up by proportionality, but the experimentalist should always be cautious because of the increasing adiabatic heating of the foam and its high insulation properties.

The quantification of hard segment in the foams in Table 6.1 can be described by either the percent hard segment or the hard-segment length. The percent hard segment is simply the sum of grams of water and isocyanate divided by the total mass of the foam. The length of the hard segment can be calculated from the sum of molecular lengths making up the hard segment multiplied by the average hard-segment oligomer number. Alternatively, it can be easier to simply refer to the hard-segment length based on the average number of urea repeat units between polyol soft segments, which is easily approximated by the equivalents of water divided by the equivalents of polyol in the formulation [4]. Table 6.2 provides an illustration of this variation for the foam formulations in Table 6.1. It is clear that there are no fractional hard-segment lengths such as 2.2 or 4.4. This number merely indicates that most hard-segment lengths will be about 2 or 4. As discussed in Chapter 3, one can expect that there will be a significant amount of isocyanate that is reacted with only polyol and there will be hard-segment blocks with a significantly larger number of repeat units greater than 2 or 4, but these will average out to approximately 2.2 or 4.4.

**TABLE 6.2 Alternative methods of describing the hard-segment content of a polyurethane foam**

| Formulation | $H_2O$ (g) | Polyol (g) | Hard segment (%) | $\frac{[\text{equivalents } H_2O]}{[\text{equivalents polyol}]}$ |
|---|---|---|---|---|
| 1 | 2 | 100 | 24.6 | 2.2 |
| 2 | 4 | 100 | 35.9 | 4.4 |

**TABLE 6.3 Property range of a conventional slabstock foam made from the formulations in Table 6.1**

| Property | Formulation 1 | Formulation 2 |
|---|---|---|
| Density ($lb/ft^3$) | 2.4–2.8 | 1.6–1.9 |
| Tensile strength ($lb/in^2$) | 20–24 | 18–22 |
| Tensile elongation (%) | 180–230 | 170–210 |
| Air flow ($ft^3/min$) | 2–4 | 3–5 |

The properties of the two foams defined in Tables 6.1 and 6.2 will be primarily determined in these particular cases by their density and the percent hard segment. On reflection it is apparent that these two quantities are not independent of each other since more water increases percent hard segment and reduces foam density as more $CO_2$ and heat are generated. The increasing temperature and gas pressure create expansion pressure on the growing foam, and so reduce density. However, while additional water creates additional hard segment and stiffening of the foam, a higher foam density obtained by reducing water will also make for harder foam by virtue of more massive struts making up the foam structure [5]. A range of selected properties that might be obtained from foam formulations 1 and 2 are given in Table 6.3.

High-resilience (HR) slabstock foams are made by a similar process to the conventional slabstock foams described above, but there are substantial formulation and performance differences. The HR designation reflects the increased elasticity of these foams, as measured by a ball rebound test (Section 5.31; ASTM test D2632 [6] replacing D3770). A standardized steel ball is dropped from a prescribed height and the rebound height measured. A typical conventional slabstock foam may have a rebound resilience of 40–50% while HR foams will usually be greater than 55%. HR foams are also referred to as "comfort" foams because of this enhanced energy return upon compression. The uses of HR foams are typically higher quality mattresses and furniture cushions. From the formulation standpoint the biggest difference is the polyol structure. As shown in Table 6.1, conventional slabstock foams typically have a hydroxyl equivalent weight of the order of 1000 g/eq and are triols made entirely from propylene oxide (PO). All PO polyols have only secondary hydroxyls to react with isocyanates. In contrast, HR slabstock foam employs polyols with functionality of 3 or greater and hydroxyl

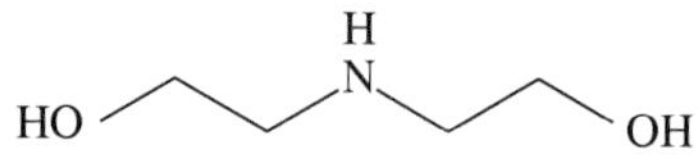

**FIGURE 6.2** The structure of the common polyurethane foam cross-linking agent diethanolamine (DEOA).

equivalent weight of between 1500 and 2000 g/eq. Since higher equivalent weight polyols have less functionality to react with isocyanate per unit mass, they are intrinsically less reactive than lower equivalent weight polyols of the same structure. To achieve similar reaction kinetics to conventional slabstock polyols, PO polyols are tipped with ethylene oxide (EO) to make triblock EO–PO–EO polyols, which have a significant amount of primary hydroxyl functionality. Primary hydroxyls are about a factor of 3 more reactive than secondary hydroxyls, and are part of the design options for optimizing foam rise and cure to meet a manufacturer's specifications. Despite the presence of primary hydroxyls, HR foam may still collapse if there is insufficient molecular weight build to prevent foam cell coalescence. In this case a cross-linker such as diethanolamine (DEOA) may be incorporated into the formulation [7]. DEOA has a secondary amine and two primary hydroxyls, creating a low-molecular-weight cross-linking agent for isocyanates and a means of rapidly building molecular weight (Figure 6.2). DEOA not only builds foam stability as a cross-linking agent; it can also have a significant negative effect on the ability of foam cells to rupture and reticulate throughout the foam body. The effect will be to reduce the foam's airflow and increase foam stiffness owing to hydraulic resistance to compression. This source of foam load bearing is not usually desirable since it is an artifact of processing and will change with use and age. For this reason, while DEOA is a very common constituent of foam formulations, it is used in small amounts, usually constituting less than 1% of the total isocyanate reactive composition. Similarly a silicone surfactant that does less to stabilize the cell wall will usually be added (medium to low efficiency) to facilitate cell window drainage to the cell interstitial boundaries (see Section 6.2.1.2).

While hydraulic resistance to foam compression is usually avoided, the formulator will often utilize copolymer polyols to increase foam stiffness and load bearing in HR foam (Section 2.1.4). A representative HR foam formulation and its properties are provided in Table 6.4.

### 6.1.2 Molded Foams

Preparation of molded foams is a widely practiced technology and the varieties of molded foam formulations are at least as varied as those for slabstock. The concept of molded foam production is simple: the components of the foam are mixed and injected or poured into a premade mold that the foam fills as it forms. The foam has a set residence time and is removed after a designed-for residence time in the mold. After removal from the mold the foam is set aside to finish curing. As practiced industrially, molded foam technology

**TABLE 6.4 HR foam formulation and its representative properties**

| | Weight |
|---|---|
| *Component* | |
| 15% SAN 1870 equivalent weight 1870 | 100 |
| Water | 2.2 |
| DEOA | 0.7 |
| Triethylene diamine (catalyst) | 0.36 |
| Dibutyl tin(IV) dilaurate | 0.25 |
| Medium-efficiency silicone surfactant | 1 |
| TDI (115 index) | 31.8 |
| *Property* | |
| Density (lb/ft$^3$) | 2.6 |
| Tensile strength (lb/in$^2$) | 17 |
| Elongation at break (%) | 150 |
| Resilience (%) | 60 |
| Tear strength (lb/in$^2$) | 1.4 |
| 90% compression set (%) | 5 |

DEOA, diethanolamine; SAN, styrene–acrylonitrile; TDI, toluene diisocyanate.

is highly optimized from the point of view of reproducibility of processing and properties. In contrast to slab foam production, in which parts must be cut from a block, molded foams produce the final part as part of the foaming operation. While this aspect reduces scrap rates and removes the cutting operation of making a finished part, the mass throughput of a molded foam plant is lower and requires the presence of molds, which can be quite expensive. In typical molded foam operations, molds are placed on carousel or oval racetrack fixtures that rotate, such that the timing of foam residence is matched to the time it takes for the carousel to make a cycle and the mold be refilled.

Besides the operational details, manufacturers choose to make molded foams because of the intricate shapes they are trying to fabricate, and the ability to overmold specific objects within the shape such as functional parts, decorative skins, or reinforcements. The structure is engineered to provide the correct appearance, function, and shape as a composite part.

Figure 6.3 shows the components of a molded foam production in terms of handling of raw materials and then mold filling. The isocyanate reactive parts (polyols, chain extenders, surfactants, catalysts, water, etc.) are put into their joint tank or individual tanks and then mixed with the isocyanate in a high-energy mixing head. The mixed components are injected into the mold, the mold closed, and the components allowed to cure.

Because of the need to optimize the efficiency of molded foam manufacturing and at the same time produce foams of exceptional durability, comfort, and variety, the formulation of molded foams is complex and could be described to some extent as an art. The speed of the molding, pouring/injecting, and curing process is all controlled

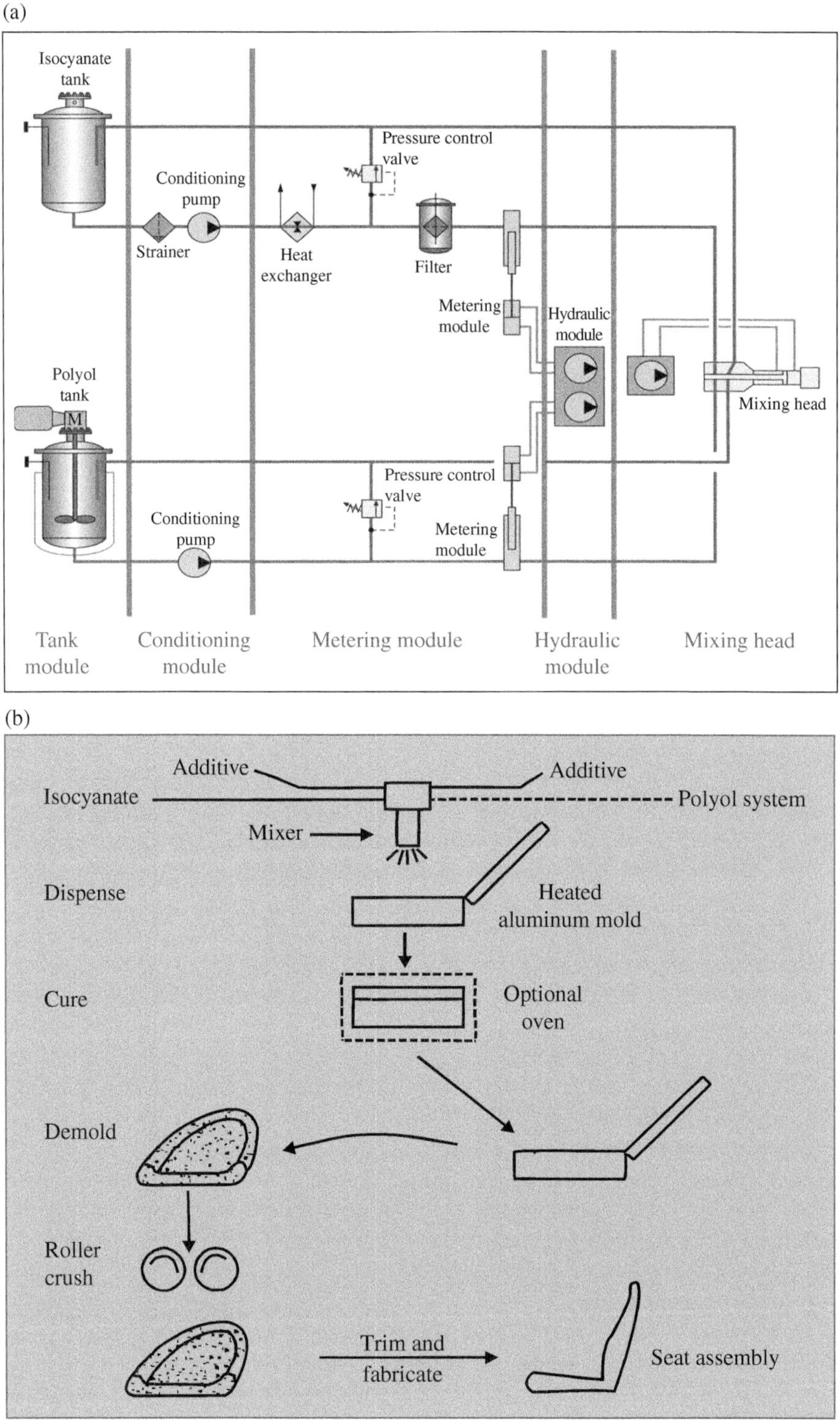


**FIGURE 6.3** Molded foam operations. (a) Raw material handling. (b) Foam fabrication. Reprinted with permission from Ref. [3]. © John Wiley & Sons, Inc.

by the chosen foam formulation. Similarly, the details of the manufacturing operation can have a significant influence on the formulation design. For instance, if the foam molds are heated, then a less reactive system can be employed; if the molds are unheated or heated to a lower temperature, then another formulation will be chosen. In some cases, the mold is partially filled with a polyurethane formulation to produce a specific characteristic and then filled subsequently with another formulation to produce the desired characteristics for the rest of the part. This flexibility of molded foam operations makes them very desirable for specialty part production, but also contributes greatly to the complexity of the formulation science.

As previously mentioned, examples of molded foam formulations are bound to be inadequate in reflecting the range of available options. However, the basic categories of components used in molded foams are similar to those used for HR slabstock foams. If the manufacturer is actively heating the mold, for instance by passing it through a hot oven, this will be termed a "hot cure" molded foam. In this case, the formulation will employ polyols that may be 3000 molecular weight triols (1000 g/eq OH), which will consist of a triblock of PO capped by EO. The polyol will be approximately 20% EO by weight (more by molarity) to assure that the great majority of end groups will be faster reacting primary hydroxyls. A manufacturer may use TDI, blends of TDI with polymeric 4,4′-methylene diphenyl diisocyanate (pMDI), or pMDI. The choice is dependent on price, preference of the manufacturer, and, in some cases, geography, as a particular isocyanate may be preferred based on convention. Table 6.5 offers examples of hot molded foam formulations to make foams of differing hardness. The formulation shows the role of copolymer polyol in hardening foam even in the case of lower overall hard segment. Of course, the reduced overall mass of hard segment is crowded into a smaller volume as well since the hard segment can only occupy the polymer space excluded by the SAN particles.

HR molded foams are also sometimes referred to as "cold cure" HR foam in distinction to the "hot cure" foams described above. They are a more recent industrial innovation relative to hot cure foams and are differentiated by mold and cure oven temperatures. These foams usually offer resiliency greater than 60%, and as high as 70% has been reported [8].

As with HR slabstock foams, HR molded foams utilize higher-molecular-weight polyols, typically 4000–6000 g/mol triols with relatively high amounts of EO capping (up to 85% primary hydroxyl functionality at 25% EO incorporation) to increase reactivity [9]. The higher reactivity of EO-capped polyols allows the formulator to use less gelation catalyst, but can dictate the intensified use of blowing catalyst to make sure that the urea hard segment forms in kinetic equilibrium with the urethane reaction that builds molecular weight. The lower temperatures and retarded blow conditions sometimes result in HR molded foams having a larger than desired population of foam cells with intact windows. It is then protocol in manufacturing to mechanically crush the foam after retrieval from the mold in order to break open cells, and so increase airflow and foam breathability. While it is, in principle, possible to formulate so that the crushing operation is superfluous, insuring completion of the gelation reaction so the foam does not collapse supersedes the nuisance of the

**TABLE 6.5 Representative formulations for making a "hot mold" molded foam of two hardnesses)**

| Component | Soft foam weight | Harder foam weight |
|---|---|---|
| 3000 molecular weight trio 25% EO | 80 | |
| 1300 equivalent weight 26% solids SAN copolymer polyol | 20 | 100 |
| High-efficiency silicone surfactant | 1 | |
| Medium-efficiency silicone surfactant | | 0.8 |
| Bis(dimethylaminoethyl)ether blowing catalyst | 0.1 | 0.05 |
| Triethylene diamine (strong gel catalyst) | | 0.1 |
| *N*,-dimethylaminoethyl morpholine (less strong gel catalyst) | 0.4 | |
| Tin dioctanoate | 0.05 | 0.1 |
| Water | 2.4 | 2.1 |
| TDI (100 index) | 30.9 | 27 |

EO, ethylene oxide; SAN, styrene–acrylonitrile; TDI, toluene diisocyanate.

mechanical manipulation of the foam to maximize openness. These are the kinds of considerations of practical foam formulation. An example of a TDI HR molded foam formulation is given in Table 6.6.

A common variation of the formulation in Table 6.6 is to substitute a blend of TDI and pMDI to provide foam properties intermediate between the foams produced by either isocyanate on its own. An example of such a formulation is provided in Table 6.7.

Some molded foam producers manufacture only with pMDI as the isocyanate. The choice is largely a matter of preference, including the better health and safety profile of pMDI owing to its lower volatility; however, there are several differences in foam properties between foams made with TDI and pMDI. One difference is that pMDI foams generally cure faster because of the uniform reactivity of MDI isocyanate functionality (the second isocyanate reaction on TDI is usually significantly

**TABLE 6.6 Representative formulation for a HR "cold mold" molded foam.**

| Component | Weight |
|---|---|
| 40% SAN EO-capped PO triol of ~1800 g/eq OH polyol | 70 |
| 40% solids EO-capped PO triol ~2350 g/eq OH polyol | 30 |
| DEOA (cross-linker to speed up gelation) | 1.25 |
| Triethylene diamine (gel catalyst) | 0.25 |
| Bis(dimethylaminoethyl)ether (blow catalyst) | 0.1 |
| Silicone surfactant for HR foam | 1.0 |
| Water | 4.0 |
| TDI (100 index) | 43 |

DEOA, diethanolamine; EO, ethylene oxide; PO, propylene oxide; SAN, styrene–acrylonitrile; TDI, toluene diisocyanate.

**TABLE 6.7 Representative formulation for a HR "cold mold" molded foam using blended isocyanates**

| Component | Weight |
|---|---|
| 40% SAN EO-capped PO triol of ~1800 g/eq OH polyol | 70 |
| 40% solids EO-capped PO triol ~2350 g/eq OH polyol | 30 |
| Silicone surfactant for HR foam | 0.7 |
| Triethylene diamine | 0.3 |
| Bis(dimethylaminoethyl)ether (blow catalyst) | 0.1 |
| Water | 3.5 |
| 80% TDI 20% pMDI with 136 g/eq isocyanate (100 index) | 42.7 |

EO, ethylene oxide; pMDI, polymeric methylene diphenyl diisocyanate; PO, propylene oxide; SAN, styrene–acrylonitrile.

slower than the first). While speed is a preferred attribute in molded foam operations, rapid viscosity increase can interfere with complete mold filling. However, when formulated effectively, pMDI foams can be made with shorter cycle times. This attribute can also allow manufacturers to lower their mold and oven temperatures to further balance chemical and process kinetics. The larger molecular volume of pMDI relative to TDI also means that the volume of hard segment produced for an MDI foam is higher than that for a TDI foam. This different molecular volume results in MDI-based foams having potentially higher load bearing at a given water and isocyanate index than TDI. It has been noted, however, that higher index molded foams may exhibit degradation in some dynamic properties while exhibiting higher hardness. A representative formulation is provided in Table 6.8.

A complicating aspect of foam formulation alluded to above is the conventional aspects of molded foam usage. This is greatly influenced by regional perceptions of comfort, which are then related to the foam stiffness, resilience, and breathability (related to airflow). Another foam feature dictated by regional preference is the rate

**TABLE 6.8 Representative formulation for a HR "cold mold" molded foam using pMDI**

| Component | Weight |
|---|---|
| 14% EO-capped 1600 equivalent weight PO triol | 98 |
| 74% EO-capped 1700 equivalent weight PO triol for enhanced cell opening | 2 |
| Silicone surfactant | 1 |
| DEOA | 0.5 |
| Triethylenediamine | 0.5 |
| Bis(dimethylaminoethyl)ether (blow catalyst) | 0.1 |
| Water | 3.4 |
| pMDI prepolymer 18.7% isocyanate (100 index) | 99.5 |

DEOA, diethanolamine; EO, ethylene oxide; pMDI, polymeric methylene diphenyl diisocyanate; PO, propylene oxide.

of change of resistance with increasing compression (sometimes called the "support" or "comfort" factor and defined quantitatively by the ratio of foam stiffness at two different compressions) [10]. Foam formulation factors that the foam designer may vary to meet these customer preferences include:

- foam density (influenced by water level and percentage hard segment)
- cell structure and foam openness (dictated in part by surfactancy and catalysis)
- isocyanate type and index
- polyol functionality and equivalent weight
- copolymer polyol and percentage solids
- additives, inserts, and seat design.

## 6.2 FOAM PROCESSES

While it is a nearly hopeless task to provide specific guidance for making foams because of the huge number of possible components and desired properties, it is possible to understand the basis for the foam formulator or designer to choose components. Regardless of the foam production format (i.e. slabstock or molded), the same physical processes occur, and the timing and intensity of the chemical reactions while the foam is forming must be controlled. Figure 6.4 shows a timeline for the various individual processes that the choice of components, catalysis, surfactancy, temperature, and myriad process details can affect.

Along the timeline, individual processes can be codependent (e.g. bubble entrainment, $CO_2$ generation, and bubble expansion), competitive (e.g. the water reaction versus the polyol reaction with isocyanate), and consistently coincidental, but with no clear connection (e.g. urea precipitation with cell opening) [11].

As indicated in Figure 6.4, the first process is mixing, which involves bubble entrainment. Surprisingly, this is a critical aspect of the foaming process since it has been shown that no new bubbles are formed after this step, and all subsequent processes are bubble growth and coalescence [12]. Thus, the work put into this step by mixing is a critical aspect, and mixer technology is a proprietary and protected intellectual asset. While $CO_2$ produced by the isocyanate–water reaction can, in principle, nucleate and form a bubble that can expand, this is in fact a highly unlikely event. The reason for this is that the bubble nucleus will of necessity be very small and the increase of free energy ($\Delta F$) to form the bubble has the following relationship to the radius of the bubble:

$$\Delta F > \frac{3\gamma}{r\rho} \tag{6.1}$$

where $\gamma$ is the interfacial surface tension, $\rho$ is the gas density, and $r$ is the bubble radius. The very small size of a nucleating bubble insures that the energy input to form those bubbles spontaneously will be immense [13, 14]. Video recording experiments have been performed that seem to confirm that no new bubbles are

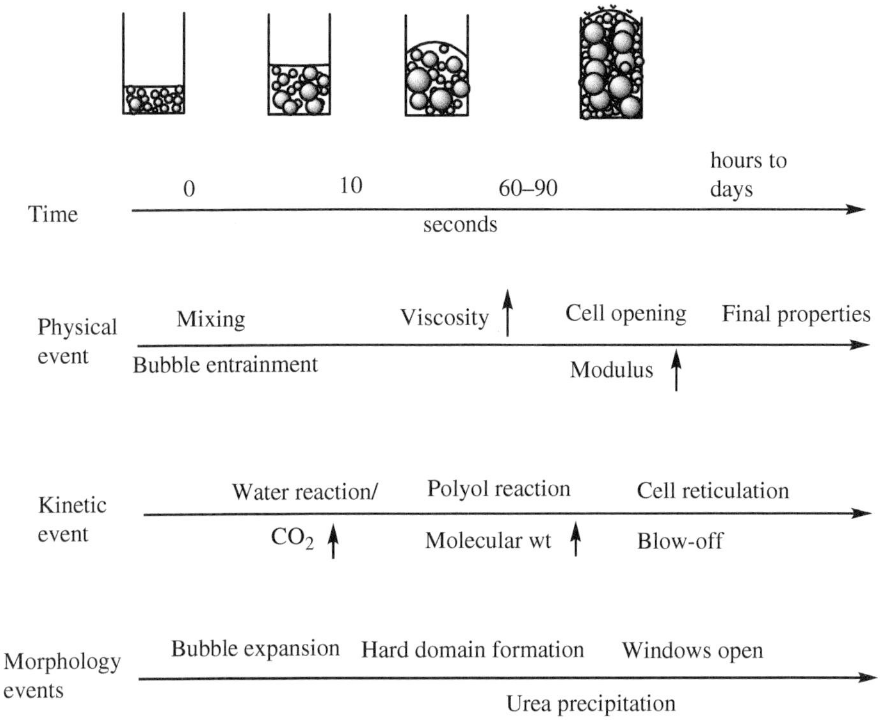


**FIGURE 6.4** Timeline for the physical and chemical processes occurring in polyurethane foaming. Reprinted with permission from Ref. [3]. © John Wiley & Sons, Inc.

formed in the foaming process after mixing. $CO_2$ produced from the isocyanate–water reaction diffuses to the already existing bubbles and solely affects the growth of those bubbles. Reacting polyurethanes that do not have bubbles produced by the mixing process are not observed to foam in any substantial way [15–17].

Following the timeline of Figure 6.4, with the initial rush of bubble expansion (and the increasing system temperature associated with the water–isocyanate reaction), the polyol–isocyanate reaction accelerates and the system viscosity increases as a result of increasing molecular weight. With increasing viscosity the system develops a viscoelastic response to the upward pressure of the expanding gas bubbles. As the gas bubbles contact each other their initial contact points preferentially increase in size, transforming the spherical bubbles into polyhedra. If unimpeded, the gas bubbles grow, coalesce, and burst prior to the system developing sufficient elasticity to stabilize the foam structure. If viscosity increases too quickly such that the cell windows cannot thin and retract to the plateau borders (Figure 6.5a) the cell windows will remain in place, resulting in closed cell or partially closed cell foam (Figure 6.5b). Flexible foams with a large number of unruptured cell walls will present a hydraulic resistance to foam compression (see Section 4.2.2.2 for a quantitative treatment), resulting in the foam being termed "tight."

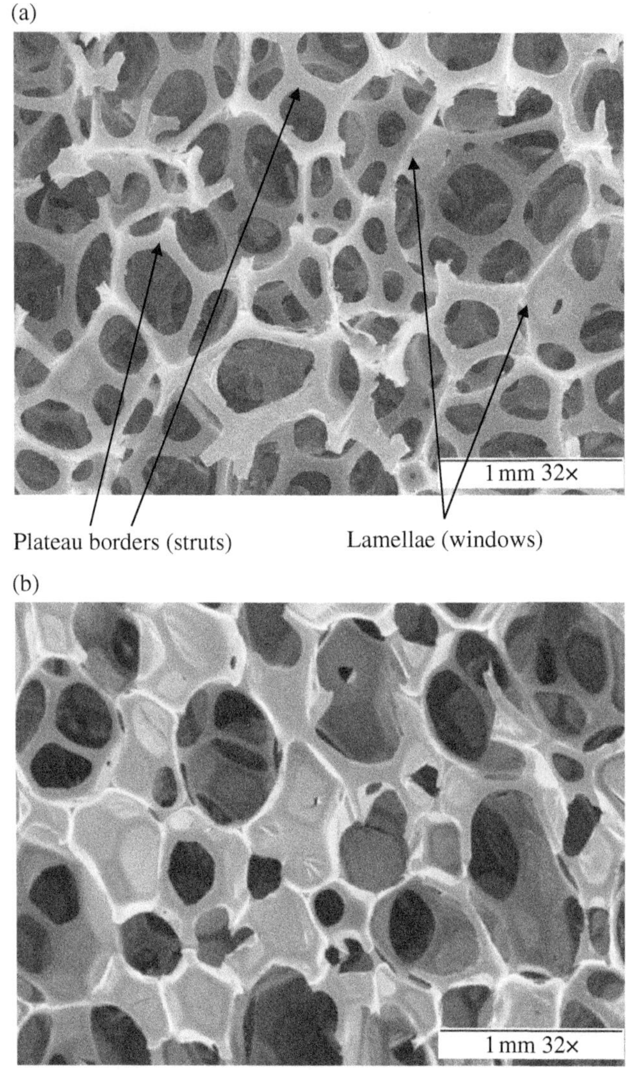


**FIGURE 6.5** Scanning electron microscope images of flexible foams. (a) An open cell foam that has had nearly complete drainage of the window materials into the bubble walls (termed the plateau borders (struts)). The remaining bubble windows are indicated. (b) A flexible foam showing the effect of kinetic imbalance such that the chemical processes were truncated before cell windows (lamellae) were drained into the plateau borders.

As discussed in Chapter 4, at some time the polyurea hard segment will phase separate from the chain-extending soft segment to assemble into a microphase structure dependent on the phase incompatibility (quantified by the Flory $\chi$ value) and the molecular weight of the phase. This precipitation event is often witnessed to occur just prior to the simultaneous cell wall breakages and foam reticulation into an open-celled foam. The reticulation event can in some cases be observed as an effervescence of gas off the top of the foam bun (termed "blow-off"). The coincidence of these events has led some researchers to speculate that the precipitation is itself the determining event for the timing of foam reticulation. However, arguments have been made that the timing of the events is coincidental and not physically dependent [11].

A method of visualizing and timing the events of polyurethane foaming is to perform an experiment, as shown in Figure 6.6 [3]. In this experiment all of the isocyanate reactive components and the catalysts are placed in one beaker and mixed with a high-shear mixer. For this experiment no surfactant is included so that the bubbles grow but collapse rapidly because of rapid drainage of the polymer bubble structure. The isocyanate is then poured into the first beaker and the polyurethane formulation again mixed with high shear for a short time (c. 10 seconds); the whole formulation is then poured into a disposable plastic bag, as shown in Figure 6.6. Following the

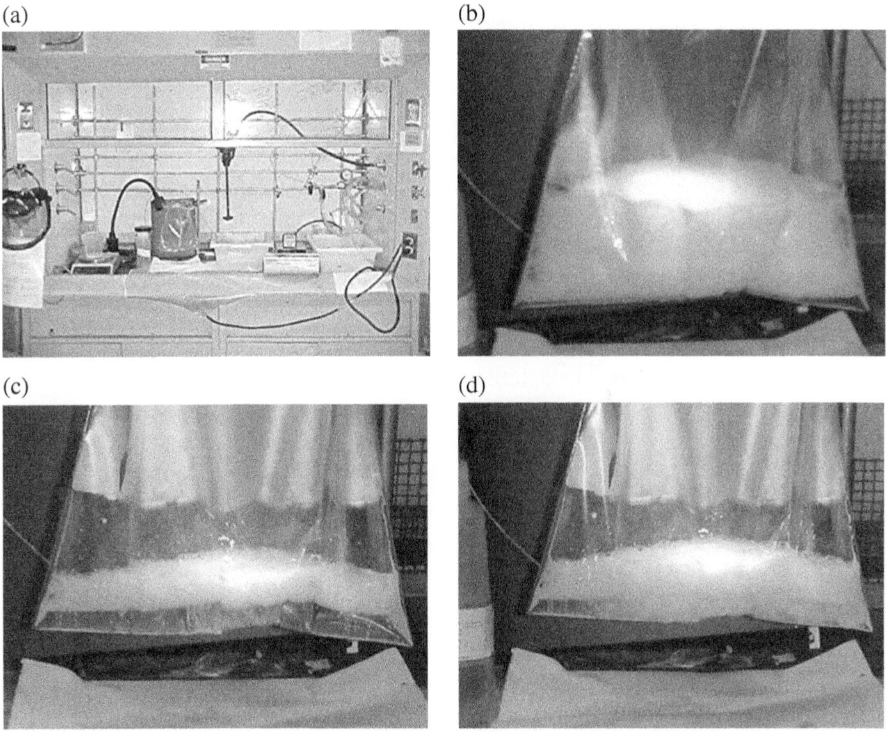

**FIGURE 6.6** A useful experimental technique for evaluating the timing of urea precipitation in a foam formulation. (a) The experiment is performed in a well-operating fume hood. (b) After mixing of the components and dumping into the disposable plastic bag (see text) the water reaction to urea and $CO_2$ formation is readily visible. (c) The water reaction begins to slow and the urethane reaction begins to accelerate. There has not been any microphase separation, as observable by the lack of precipitate in the bottom clear liquid. (d) Urea precipitation is clearly visible, marking the beginning of microphase separation. Reprinted with permission from Ref. [3]. © John Wiley & Sons, Inc.

timeline of Figure 6.4, all of the processes are visible except the growth of the foam into a bun. The bubbles grow and escape through the top of the bubbling mixture. At some point the polyurea hard segment exceeds its compatibility with the soft segment and precipitates as a white mass at the bottom of the bag. When this experiment is performed carefully, the reproducibility of the events is within 1%. This experiment is especially useful when comparing the foaming kinetics of a new foam component with a well-known benchmark foam composition (Table 6.9).

### 6.2.1 Surfactancy and Catalysis

***6.2.1.1 Catalysis*** The art of polyurethane foam formulation is balancing the competing processes by mixing complementary components such that all the processes in Figure 6.4 occur in amount and timing to result in foam of the correct structure and properties. Thus, along with the choice of polyol, isocyanate, amount of water, and any chain extender, the foam designer must also choose the catalyst, or often several catalysts, and a surfactant. Once the foam designer has chosen a surfactant and a suite of catalysts, the foam must then be screened by foaming at a small scale (typically hundreds of grams of raw materials poured into a cardboard box with a polyethylene or Teflon film interlayer). If the choice of surfactant and catalysis is not correct the screening foams may evidence poor foam openness (Figure 6.5b), shrinkage upon cooling, collapse during foaming, poor thermal properties, or even poor aging properties. If the screening foams are acceptable there is almost invariably additional iterations when the foam is scaled to larger volumes because of the different chemical energy release and foam geometry.

The structures and mechanisms of action of amine, organometallic, and trimerization catalysts are presented in detail in Chapter 3. Their incorporation into foam formulations is usually a matter of trial and error, in which certain process and foam property specifications are iteratively approached. Along with the representative but far from exhaustive catalogue presented in Chapter 3, there is a growing industry move away from utilizing organometallic catalysts towards using reactive catalysts or autocatalytic polyols. The purpose of this industry shift is to minimize contact

**TABLE 6.9 Effects of foam raw material structure variation on foaming and foam properties**

| Component | Structural variable | Effect on foaming | Effect on foam properties |
|---|---|---|---|
| Polyol | Equivalent weight | 1. Lower—increases gelation rate<br>2. Higher—decreases gelation rate | 1. Lower—increases $T_g$<br>2. Higher—increases resilience<br>3. Higher—increases tensile strength<br>4. Higher—increases d-spacing |

**TABLE 6.9 (Continued)**

| Component | Structural variable | Effect on foaming | Effect on foam properties |
|---|---|---|---|
| Polyol | EO capping | 1. More—increases gelation rate<br>2. More—increases compatibility with water | 1. More—increases $T_g$<br>2. More—increases degradation of humid age properties |
| Polyol | Functionality | 1. More—increases gelation rate<br>2. More—can increase formulation viscosity<br>3. More—can increase foam shrinkage | 1. Higher—can increase foam hardness<br>2. Less functionality can lead to less fatigue resistance<br>3. Lower—monol better durability<br>4. Lower—monol better dynamic properties<br>5. Lower—functionality better extension properties |
| Water | | 1. More—increases blow reaction<br>2. More—increases exotherm<br>3. More—decreases shrinkage<br>4. More—requires increased gel catalyst and surfactant | 1. More—increases hardness<br>2. More—increases aging degradation |
| Isocyanate | TDI | 1. More—lowers viscosity<br>2. Slower reacting/slower demold<br>3. High NCO content | 1. Lower density foams<br>2. Better tensile properties<br>3. Not a drop-in for pMDI<br>4. More stable in some molds |
| Isocyanate | pMDI | 1. Faster reaction<br>2. Skin formation in molds | 1. Higher durability<br>2. Harder foams<br>3. Improved compression set |
| Isocyanate | Index | 1. Higher—faster blow reaction<br>2. Higher index should decrease gelation catalyst<br>3. Higher—slows cure | 1. Greater than 100—decreases durability<br>2. Greater than 100—decreases tensile and elongation properties |
| Chain extender | Diol | 1. Faster cure<br>2. Faster demold | 1. Softer foam<br>2. More flexible skin (molded) |
| Chain extender | Diamine | 1. Faster cure<br>2. Faster demold | 1. Harder foam<br>2. Tougher skin |
| Cross-linker | | 1. More rapid cure<br>2. Reduced flowability<br>3. Faster demold | 1. Harder foam (at density)<br>2. Lower compression set<br>3. Lower airflow |

EO, ethylene oxide; pMDI, polymeric methylene diphenyl diisocyanate; TDI, toluene diisocyanate; $T_g$, glass transition temperature.

with fugitive heavy metals and to reduce the volatile organic components emanating from foam in day-to-day use.

Reactive catalysts are typically tertiary amines connected by a spacer group to an isocyanate reactive group, such as a primary hydroxyl group [18, 19]. They are added to the formulation as one would add a conventional catalyst, but typically in larger amounts. The increased usage rate is because of the lower tertiary amine equivalence per gram of catalyst, and because the catalyst becomes less effective once the reactive group has been incorporated into the polymer network and is no longer able to diffuse freely. Examples of reactive catalysts are provided in Figure 6.7. Each example has isocyanate reactive groups. The reactive group chosen for a catalyst has a significant effect on the catalyst function. For instance, a catalyst with secondary hydroxyls will react more slowly into the network than one with primary hydroxyls, so may maintain catalytic activity for a longer period of the foaming reaction, but risks not being incorporated at all and can result in a volatile amine. A reactive catalyst with amine groups is almost certain to react into the polymer network but may rapidly lose effectiveness owing to its inability to diffuse to new reactive sites. Some catalyst designs will ensure reaction into the polymer network by using a highly reactive group, but put a large spacer between the reactive group and the catalytic function that can diffuse through a relatively large spherical or conical volume even after integration into the network. These large elastically ineffective pendant groups have the potential to affect foam properties, especially aging properties.

In addition to reactive catalysts, manufacturers, especially polyol manufacturers, have attempted to incorporate catalytic activity into the polyol backbone [20–22]. A relatively straightforward way is to take a reactive catalyst and grow an alkoxylate chain off of the reactive group using PO or block polymerizing propylene and EO. An example of such a procedure is shown in Figure 6.8.

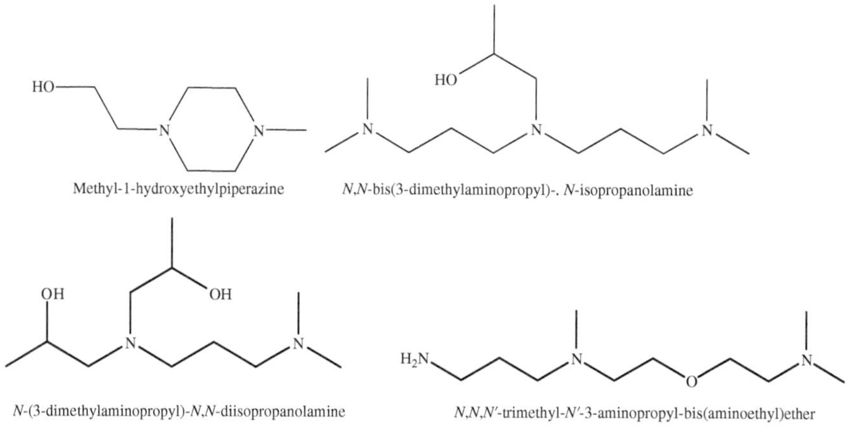


**FIGURE 6.7** Examples of reactive catalyst structures. The tertiary amines are catalytic for the polyurethane reaction.

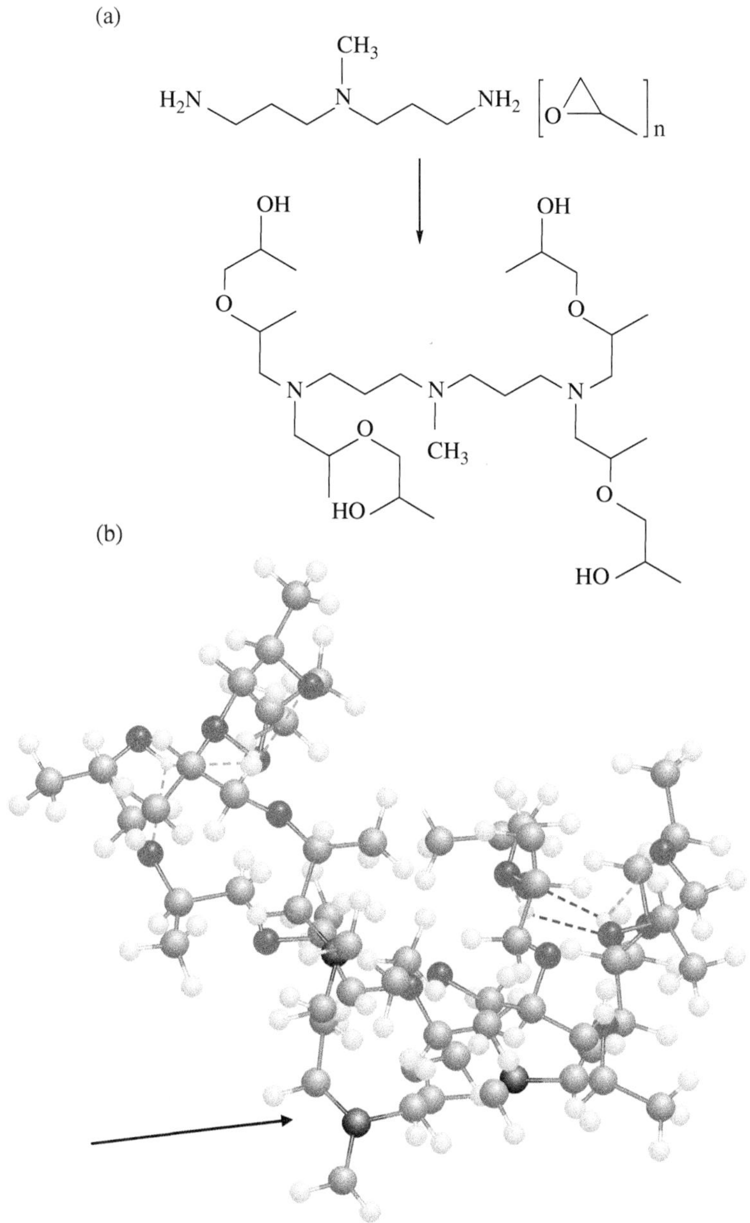


**FIGURE 6.8** (a) Example reaction for making a catalytic polyol. The reaction is usually base catalyzed. (b) Energy-minimized space-filling structure for a reactive catalytic polyol. The arrow indicates the position of the central catalytic tertiary amines. The alkoxylated amines are usually much less catalytic owing primarily to steric hindrance as well potentially electrostatic factors associated with local ether associations.

**TABLE 6.10 The effects of catalyst strength on the polyurethane foam process and foam**

| Activity | Effect on process and foam | | |
|---|---|---|---|
| Gelation | Faster gelation | = | Fast cure |
| | | = | Shorter demold times |
| | | = | Smaller and more closed cells |
| | | = | Higher density |
| | | = | Less flowability (faster viscosity build) |
| | | = | Lower air flow |
| | | = | More foam shrinkage during cooling |
| Blow reaction | Faster blow reaction | = | Faster creaming |
| | | = | Faster rise |
| | | = | Larger more open cells |
| | | = | Lower density |
| | | = | Lower stability (tendency to collapse) |
| | | = | Less shrinkage upon cooling |

The effectiveness of these reactive polyols in a foaming reaction is dictated by the same fundamental issues reflected in the behavior of any polyurethane catalyst, with the additional complication of steric hindrance of alkoxylate chains and their temperature-dependent interaction with the catalytic center.

In choosing a catalyst the foam designer must consider the characteristics of the catalyst in light of all of the processing and property attributes required by the customer. This does not always dictate the catalyst with the fastest or most balanced (blow/gelation) attributes. Table 6.10 indicates the effect of catalyst speed on the foaming process. One can see that if the preference is for a soft comfortable flexible foam one may choose a microcellular foam with good airflow. This would dictate a certain glass transition and a certain hardness dictated by the polyol, amount of isocyanate, and the amount of water (to control density). It might also dictate a catalyst that gels quickly to promote rapid gelation of the polymer prior to cell expansion, as well as a blow catalyst that could develop sufficient pressure to burst the cells at early reaction times. If molded foam producers need to increase throughput, they may want to increase overall catalysis, and especially the blow catalyst, to increase the early build-up of heat, but this may dictate that the mold closes much sooner than normally practiced since the foam rise may be much faster than normally accommodated.

***6.2.1.2 Surfactancy*** Virtually all polyurethane foams are made with the essential addition of surfactants. The role of the surfactant is still not well understood in specific detail, but a sufficient amount of controlled structure–property experimentation has been done to provide good understanding of desirable surfactant features for a specific foam [23]. The surfactant requirements for closed cell rigid foam are different in detail from those of an open cell flexible foam, but the general surfactant actions are the same. The specific surfactant requirements for rigid foams are covered in Chapter 8.

Conventionally, it is understood that the surfactant plays no intermediary role in the reactions of isocyanates with the other foam components [24]. This is inferred by the lack of a kinetic effect in the specific reaction rate as a function of surfactant composition. However, the surfactant plays a crucial role in the foam process kinetics and eventual structure and properties. The surfactant is believed to (i) lower surface tension at the polyurethane–air interface, promoting the generation of bubbles that evolve into foam cells, (ii) emulsify the incompatible polyurethane components (which does affect the polymerization rate by influencing available concentration), and, most importantly, (iii) control the stability of the cell polymer surfaces. The surfaces must specialize into cell struts and windows upon packing. At a specific point of the foam rise the cell windows must drain into the cell struts and open such that airflow through the foam is at the desired level. The details of this process are strongly dependent on the details of the surfactant structure.

The vast majority of surfactants used for making flexible polyurethane foams are of the class termed broadly as silicone surfactants. Silicone surfactants consist of polydimethylsiloxane (PDMS) backbones grafted with polyethylene oxide-co-polypropylene oxide side chains (Figure 6.9) [25, 26].

In the early stages of foaming, the isocyanate, polyols, and water may all be mutually immiscible. The surfactant can promote the initiation of the foaming reaction by specifically lowering the surface tension between the isocyanate and water phases. The structure of the silicone certainly affects this as the hydrophobic PDMS and the hydrophilic EO balance can lower the surface tension and increase the probability of reaction. As the reaction proceeds the system advances to a single phase and the surfactant's role as an emulsifying agent ends.

The physical mixing that occurs introduces bubbles into the reacting polyurethane components. In the absence of a surfactant or in the presence of a poorly chosen

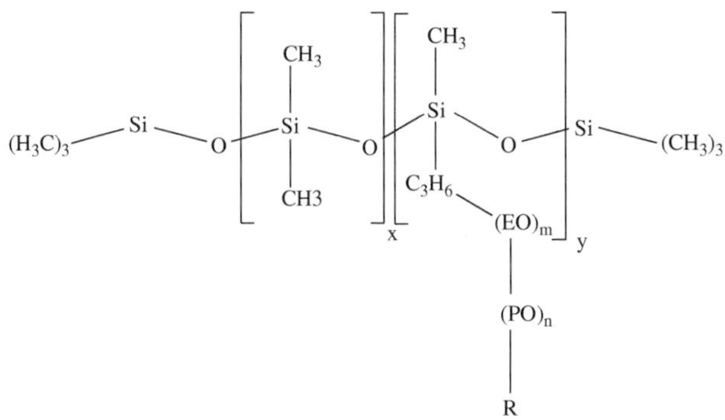


**FIGURE 6.9** General structure of a silicone surfactant, as might be used for polyurethane foaming.

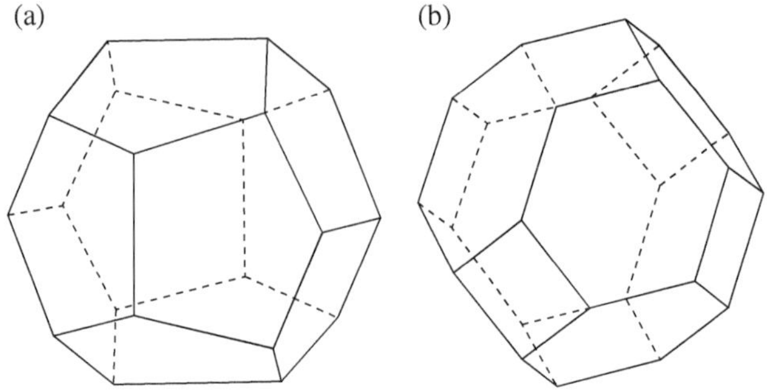


**FIGURE 6.10** Suggested polyhedral structures found in polyurethane foams based on space-filling considerations. (a) Pentagonal dodecahedron. (b) Tetrakaidecahedron. Compare with actual foam structures (Figure 6.5).

surfactant, these bubbles will coalesce, rapidly exceed the bubble wall's tensile strength, and escape from the foam. If the molecular weight of the foaming mixture has advanced, the reacting mixture may initially rise but will collapse. An efficient surfactant will serve to stabilize the foam structure throughout the foam process, and will also control film dynamics of the growing bubbles within the foam that lead to open and reticulated cells.

In the presence of adequate surfactancy, catalysis, and a suitable mixture of polyurethane components the foam will begin to rise. As the bubbles grow, the molecular weight of the polymer will increase, providing sufficient polymer tensile properties to contain the growing internal gas pressure. When the volume fraction of bubbles exceeds 74%, the bubble growth will become limited by bubble–bubble physical interaction. The bubble will cease to be spherical and will assume a polyhedral shape, as can be made out from the foam shapes in Figure 6.5. There is some controversy about the nature of the polyhedra formed (pentagonal dodecahedra or tetrakaidecahedra based on space-filling surface minimization; Figure 6.10); however, it is clear upon inspection that real foams are not regular shapes.

The successful fabrication of polyurethane foam requires the *artful* balance of polymerization kinetics, competing reactions, and physical processes racing to a finished product. Small maladjustments in achieving this balance are the difference between a beautiful finished foam product and a solidified puddle of raw material. The surfactant structure–property relationships for making successful foams have been broadly defined in the literature based primarily on model compound studies. This has been necessarily the case since the vast majority of silicone surfactants are produced by manufacturers who protect their structures as intellectual assets. Manufacturers instead define the surfactant in terms of their efficiency in certain foam classes (i.e. slabstock, hot molded, etc.) and in terms of their effectiveness at emulsification, cell size regulation, usefulness in polyester foams, copolymer polyols, specific isocyanates, etc. Table 6.11 provides general structure–property

**TABLE 6.11 Effects of silicone surfactant variables on foam processes and properties**

| Structural feature | Effect on foaming |
|---|---|
| Polyether content | Increase in surface tension |
| Siloxane content | 1. Increase—decreases surface tension<br>2. Smaller bubble size |
| Ratio of siloxane to polyether | 1. Increase—lowers surface tension<br>2. Increase—increases bubble count<br>3. Too high can be antifoaming, causing foam collapse<br>4. Too low can result in too low film elasticity, causing foam collapse |
| Increase of molecular weight at constant siloxane–polyether ratio | 1. No effect on surface tension<br>2. No effect on bubble count<br>3. No effect on bubble size<br>4. Higher film elasticity/slower drainage/more closed cells |

relationships for silicone surfactants as applied to flexible foam fabrication. As will be discussed in Chapter 8, along with silicone surfactants, block copolymers of alkoxylates have found commercial use in rigid foams.

## 6.3 FLEXIBLE FOAM FORMULATION AND STRUCTURE–PROPERTY RELATIONSHIPS

### 6.3.1 Screening Tests

The preparation of flexible foams will usually begin with the screening of formulations. It is not uncommon for initial attempts to fail spectacularly with a small foam (hundreds of grams of material) made in a box rising a few inches and then collapsing. As the iterative process continues in which catalysts, surfactants, components, and their amounts are adjusted the results will usually improve. At that point it is necessary to obtain quantitative data on the screening studies by which to compare relatively similar results. Among the techniques are measurement of the time dependence of the foam rise and the temperature rise during the experiment. While the measurement of foam rise is relatively straightforward, the measurement of temperature requires the use of a low-mass thermistor or thermocouple placed strategically in the foaming mass. The data obtained are typified by those shown in Figure 6.11. From the height data one can see that for some of the foams there is a distinct shrinkage that occurs shortly after reaching full height. This effect is termed "sigh back" in the industry. It is a result of the loss of foam pressure holding the inflated foam coupled with insufficient polymer strength to hold the weight of the polymer. While this is not always the point of highest chemical exotherm, as measured by the foam's

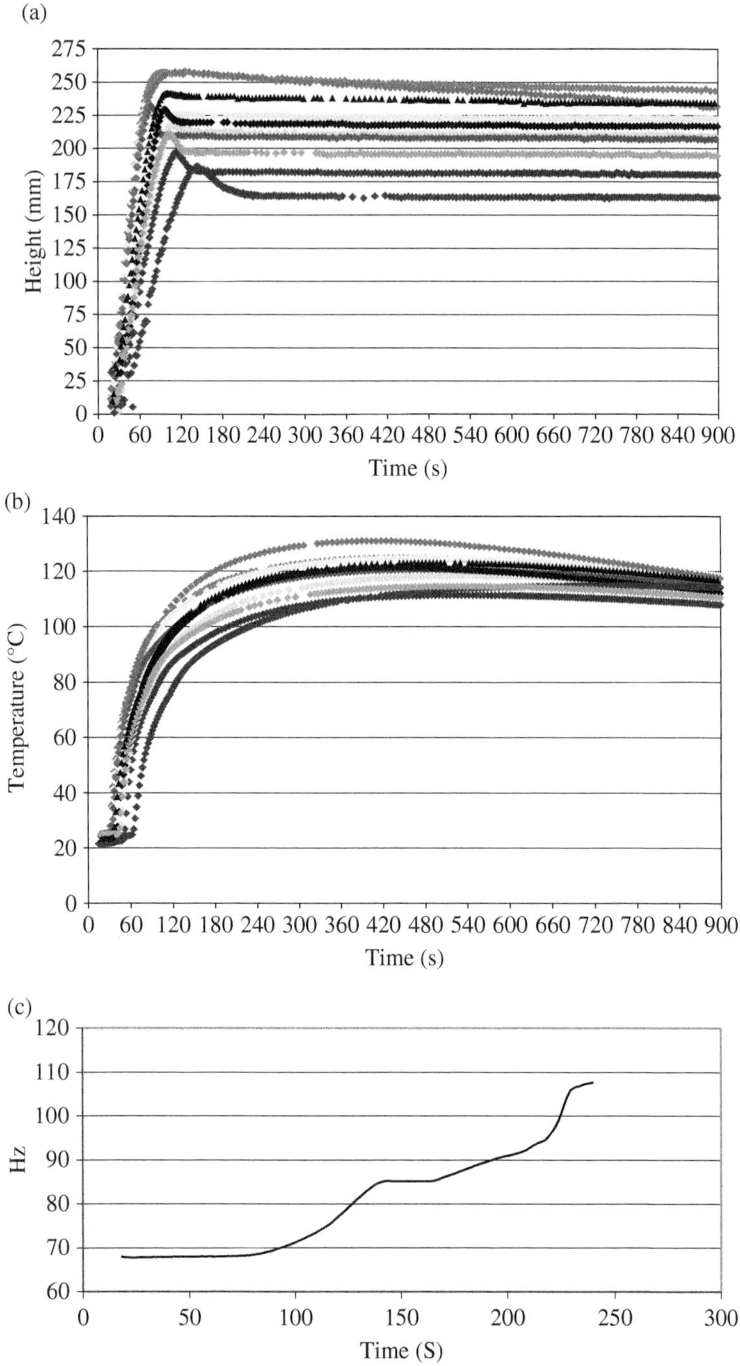


**FIGURE 6.11** Measured properties from small-scale polyurethane foam screening experiments. (a) Foam height change with time. (b) Foam temperature change with time. (c) Rheology change with time; the inflection at c. 150 seconds may reflect the precipitation of the hard segment. Reprinted with permission from Ref. [28]. © Elsevier Pub.

internal temperature, it is at elevated temperature and sets a relatively difficult to achieve criterion for the required polymer properties to maintain foam height. The temperature data are a direct measure of the rate of reaction, which is of course influenced by the formulation. It allows one to determine catalyst levels to meet the required timing of events, as previously described. It is not by coincidence that, in the illustrative dataset, the foam exhibiting the lowest temperature, increasing at the slowest rate, also had the lowest foam rise and severe sigh back afterwards. Additionally, one can obtain a measure of the changing foam rheology with time using one of several rheometers engineered for the purpose. These instruments typically use a vibrating probe that is submerged into the foaming mass. The change in polymer rheology affects the amount of energy needed to maintain the resonant energy or affects the frequency itself [27]. This is a notoriously tricky measurement and the experimentalist must obtain numerous measurements on a single formulation to be satisfied that there are interpretable data.

### 6.3.2 Foam Formulation and Structure–Property Relationships

As discussed in Chapter 5, the properties of a foam are largely defined by the tests that are performed [29–33]. Two foams may feel identical to the touch, but perform very differently in tests developed to measure the same property (e.g. compression set). In addition, subtle changes in test conditions can reveal significant differences in foam performance (Figure 6.12). This unavoidable fact regarding foam properties accentuates the need for very strict control of polymerization conditions to obtain optimized results. Additionally, it points out the need to make very careful measurements, especially with respect to temperature calibration since foam properties can have very nonlinear dependency.

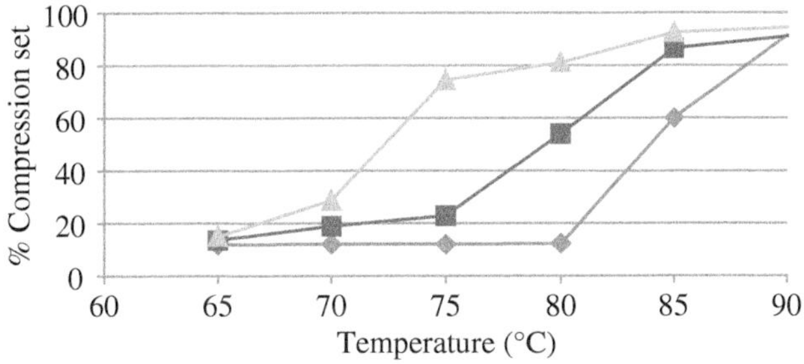


**FIGURE 6.12** Compression set for three molded foams measured following 75% compression for 22 hours (see ASTM 3574). All foams were four parts water and 105 index. The data show the sensitivity of foam performance to structural details, as well as the importance of careful control of temperature when making these measurements since errors or large fluctuations can result in artifacts in the data. Reprinted with permission from Ref. [34]. © Elsevier Pub.

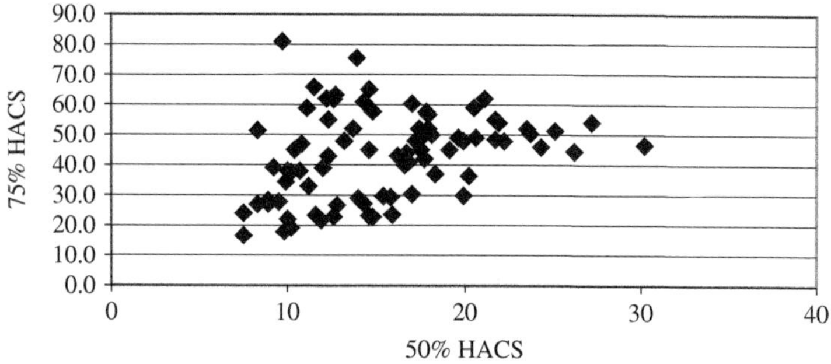


**FIGURE 6.13** Lack of correlation of 75% humid aged compression set (HACS) with 50% HACS for MDI molded foams. The two compression tests sample mostly different physical properties of the foam. Reprinted with permission from Ref. [28]. © Elsevier Pub.

It is generally conceded that the result of formulation changes is the effect of those changes on the underlying structure, as defined by the phase diagrams and phase separation mechanisms that determine that structure [28, 35]. Thus, if a test to measure humid aged compression set is measured at 50% compression the result may not correlate at all with the same measurement at 75% compression. In fact, it has been demonstrated that the 50% and 75% compression tests do not correlate at all in molded foams (Figure 6.13). While it is possible to pick points from the dataset that appear to show correlation, it is clear from the analysis of large datasets that these measurements probe different aspects of polymer structure. A 50% humid aged compression set may probe the ratio of loss modulus to elastic modulus or the stickiness of the humid aged foam walls, while the 75% compression set measurement may probe the cocontinuity of the hard-segment structure and its ability to regain its initial topology within the foam following severe deformation stresses. By the same token, different tests that do probe the same physical characteristics of the foam will correlate with each other, but not show correlation with a measurement that probes a different physical quantity. This is well illustrated by Figure 6.14, in which foam humid aged load loss (compressive softening to 75% compression after exposure to humid heat) correlates very well with 75% compression set but not at all with 50% compression set. As part of this analysis it has been shown that the foam structural property that correlates with these measurements at high deformation stresses is in fact the hard-segment thickness, as measured by the d-spacing, which is determined by small-angle X-ray scattering (Chapter 5) (Figure 6.15).

Finally, it is necessary to reiterate that the effect of formulation variables on foam properties is primarily a reflection of their effect on the foam's polymer structure. This is easily demonstrated by the lack of relationship between the content of DEOA, a polymer cross-linker, and a foam aging property, such as humid aged load loss, in the absence of control of other structural variables (Figure 6.16). While it may be possible to perform an experiment in which the progressive addition of DEOA can

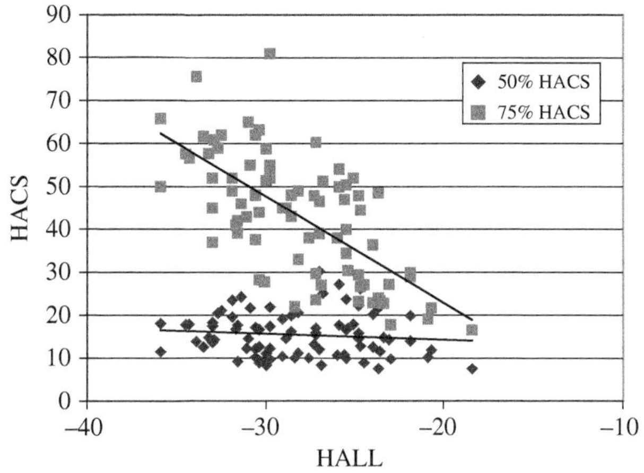


**FIGURE 6.14** Correlation of humid aged load loss (HALL), a test that ages the foam in humid conditions and then tests compression resistance to 70% compression, with humid aged compression set (HACS), where the foam is maintained in a compressed state for an extended period of time and then removed from compressive strain and allowed to relax to an equilibrium thickness. The two tests that probe at similar strains are highly correlated with each other. Reprinted with permission from Ref. [28]. © Elsevier Pub.

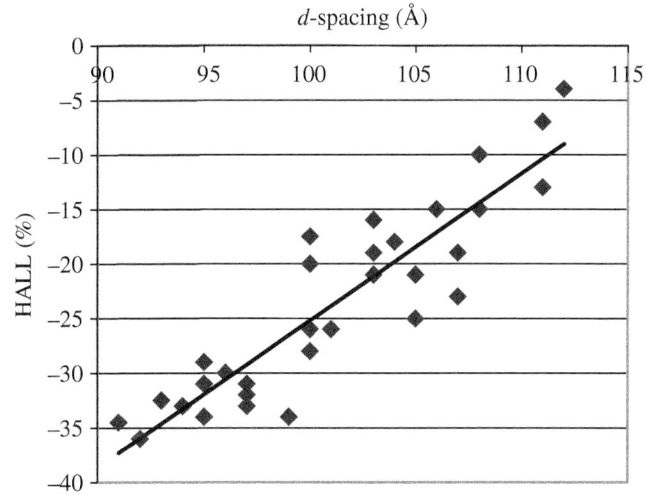


**FIGURE 6.15** Correlation of humid aged load loss (HALL) with the d-spacing, which is determined by small-angle X-ray scattering (see Section 5.2.3). All foams have the same hard-segment volume and demonstrate that the better the microphase separation, the better the aging properties when all other factors are controlled. Reprinted with permission from Ref. [28]. © Elsevier Pub.

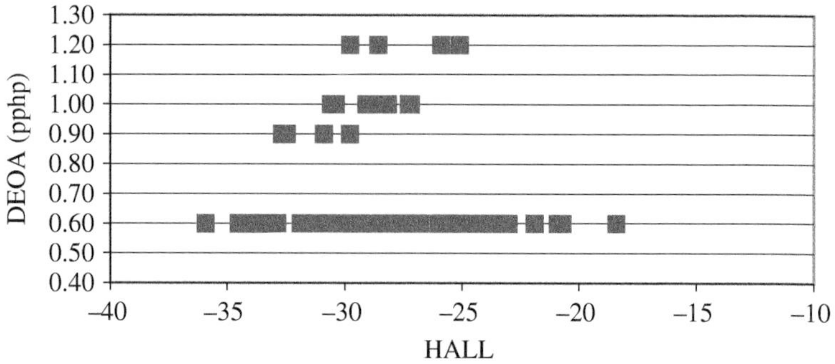


**FIGURE 6.16** Lack of correlation of diethanolamine (DEOA; a cross-linking agent that speeds gelation) content in foam formulation and humid aged load loss. DEOA is a casual, not a causative, factor of aging properties. Reprinted with permission from Ref. [28]. © Elsevier Pub.

affect foam properties, it must be remembered that the effect of DEOA is actually a measure of its effect on the foam kinetics and resulting polymer structure, which in turn are *directly* responsible and not casually responsible for foam performance.

## REFERENCES

[1] U Lochner, H Chinn, and Y Yamaguchi, *Polyurethane Foams Chemical Economics Handbook*, Market Research Report, 580.1600A, 2012.

[2] R Herrington and K Hock, "Foam calculations," in *Flexible Foams*, The Dow Chemical Co., Midland, MI, 1997, 6.1–6.10.

[3] M Sonnenschein and W Koonce, "Polyurethanes," in *Encyclopedia of Polymer Science and Technology*, 4th Ed., Wiley, Hoboken, NJ, 2012, 1–63.

[4] M Sonnenschein, V Ginzburg, K Schiller, and B Wendt, *Polymer*, 54, (2013), 1350–1360.

[5] J Moreland, G Wilkes, and R Turner, *Journal of Applied Polymer Science*, 52, (1994), 549–568.

[6] *Standard Test Method for Rubber Property – Resilience by Vertical Rebound,* ASTM D2632, ASTM International, West Conshocken, PA, 2008.

[7] B Kaushiva and G Wilkes, *Polymer*, 41, (2000), 6981–6986.

[8] J McEvoy, K Thakurta, S Pique, and J Bladon, *Proceedings of the 1996 SPI Polyurethane Conference*, Technomic Pub., Lancaster, PA, 1996, 504–511.

[9] A Aneja and G Wilkes, *Journal of Applied Polymer Science*, 85, (2002), 2956–2967.

[10] R Bros, F Casati, and R Herrington, "Endurance of polyurethane automotive seating under varying temperature and humidity conditions," in *Polyurethanes EXPO 1999: Innovation for the Next Millennium*, U Szewyk and T Griebe, Eds., Technomic Pub., Lancaster, PA, 1999.

[11] X Zhang, R Neff, and C Macosko, "Foam stability in flexible polyurethane foam systems," in *Polymeric Foams: Mechanisms and Materials*, S Lee and N Ramesh, Eds., CRC Press, Boca Raton, FL, 2004, 139–171.

[12] B Kanner and T Decker, *Journal of Cellular Plastics*, 5, (1968), 32–39.
[13] V La Mer, *Industrial and Engineering Chemistry*, 44, (1952), 1270–1277.
[14] M Owen and T Kendrick, *Journal of Colloid and Interface Science*, 27, (1968), 46–52.
[15] J Buist, G Ball, and G Woods, "Flexible foam: manufacture and properties," in *Advances in Polyurethane Technology*, J Buist and H Gudgeon, Eds., MacLaren and Sons, London, UK, 1968, 152–154.
[16] R Baumhakel, *Journal of Cellular Plastics*, 8/6, (1972), 304–310.
[17] S Lubetkin, *Chemical Society Reviews*, 24/4, (1995), 243–250.
[18] J Rothe, H Cordelair, and C Wehman, *Journal of Cellular Plastics*, 37, (2001), 207–220.
[19] H Yoshimura, Y Tamano, S Okozuno, and D Lowe, *Proceedings of the SPI Conference Polyurethanes*, 92, (1992), 304–372.
[20] F Aguirre, T Woods, L Cooper, and H LaRuelle "Catalytically active polyols for VOC emission reduction," in *Conference Proceedings Polyurethanes 2002*, Salt Lake City, UT, 2002, 387–392.
[21] J-M Sonney, F Casati, R Dawe, K Khameneh, T Jones, J Olari, and P Fielding, "Recent advances in the development of catalytically active polyols for automotive manufacture," in *Conference Proceedings Polyurethanes 2002*, Salt Lake City, UT, 2002, 206–214.
[22] H Larbig, W Hinz, and P Horn, "Catalytically active polyether polyols," in *Proceedings from Polyurethane 1995*, September 26–29, Chicago, IL, 1995, 212–215.
[23] X Zhang, C Macosko, H Davis, A Nikolov, and D Wasan, *Journal of Colloid and Interface Science*, 215, (1999), 270–279.
[24] K Yasanuga, R Neff, X Zhang, and C Macosko, *Journal of Cellular Plastics*, 32, (1996), 427–448.
[25] S Snow, U Pernisz, and R Braun, *Silicone Chemistry*, 3, (2006), 1–10.
[26] Y Shi, Y Wu, and Z Zhu, *Journal of Dispersion Science and Technology*, 24, (2003), 173–178.
[27] J McClusky, R O'Neil, R Priester, and W Ramsey, *Journal of Cellular Plastics*, 30, (1994), 224–241.
[28] M Sonnenschein, B Wendt, A Schrock, J-M Sonney, and A Ryan, *Polymer*, 49, (2008), 934–942.
[29] M van der Schuur, E van der Heide, J Feijen, and R Gaymans, *Polymer*, 45, (2004), 2721–2727.
[30] D Dounis and G Wilkes, *Polymer*, 38, (1997), 2819–2828.
[31] D Dounis and G Wilkes, *Journal of Applied Polymer Science*, 66, (1997), 2395–2408.
[32] J Armistead, G Wilkes, and R Turner, *Journal of Applied Polymer Science*, 35, (1988), 601–629.
[33] W Lee, *Journal of Cellular Plastics*, 21, (1985), 417–423.
[34] M Sonnenschein, R Prange, and A Schrock, *Polymer*, 48, (2007), 616–623.
[35] L Zhang, H Jeon, J Malsam, R Herrington, and C Macosko, *Polymer*, 48, (2007), 6656–6667.

# 7

# POLYURETHANE FLEXIBLE FOAMS: MANUFACTURE, APPLICATIONS, MARKETS, AND TRENDS

Inspection of available data suggests that aggregate demand for polyurethanes is easier to determine than volumes based on application (i.e. flexible foam, rigid foam, elastomer, etc.) and by region. Chapter 2 provides regional volume production estimates for polyurethane building blocks. These data are presumably easier to obtain since significant producers tend to be large, multinational, publicly held companies that are required to report their manufacturing data. The building block outputs can then be cross-referenced against the feedstock inputs, which are data that are also available in the public domain. This same information is not readily available for foam producers. Part of the reason for this is that there are thousands of such producers [1]. Producers tend to locate their foam production facilities relatively close to the next-in-line manufacturer within the value chain (i.e. furniture, automotive seating, mattresses, etc.). Furthermore, foam producers are a mix of smaller publicly held and privately held companies and can be highly economically sensitive. The economic downturn of 2008–2009 saw even relatively large foam producers enter into bankruptcy proceedings [2]. An aggregate of volume estimates for the flexible foam market and projected growth rate broken down by geography for 1991–2015 is shown in Figure 7.1. Within each geography the growth rate may be highly variable, with China, for instance, having twice as much growth in consumption of flexible foams as established economies, while Japan's growth experienced near zero

*Polyurethanes: Science, Technology, Markets, and Trends*, Second Edition. Mark F. Sonnenschein.

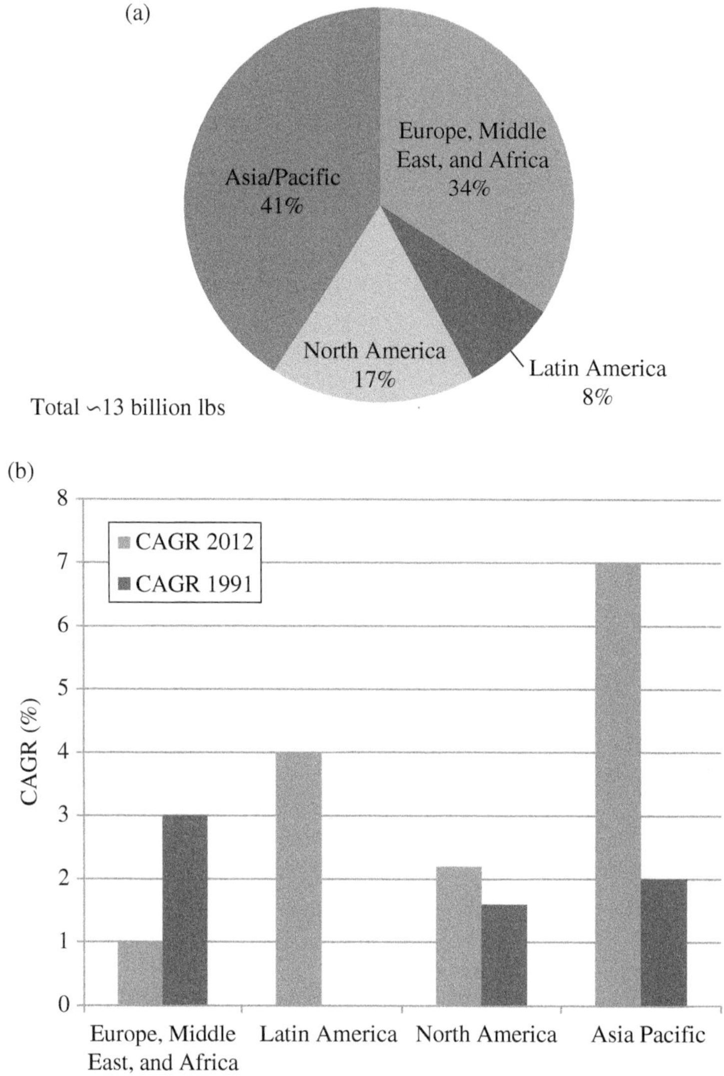


**FIGURE 7.1** Approximate volume breakdown for the production (a) of polyurethane flexible foams by geographical region for the year 2012 and (b) the compound annual growth rate (CAGR) of flexible polyurethane foam production for the years 1991 and 2012. Data for the year 2015 (c,d) are provided with greater geographical granularity.

to slightly negative growth in the flexible foam segment [3]. This historical discontinuity is dramatically shown in the data on the Asian growth rate, which are largely reflective of Japan's consumer demand in 1991 and China's in 2012. Parsed data for 2015 show the effects of Japan's aging population on economic growth.

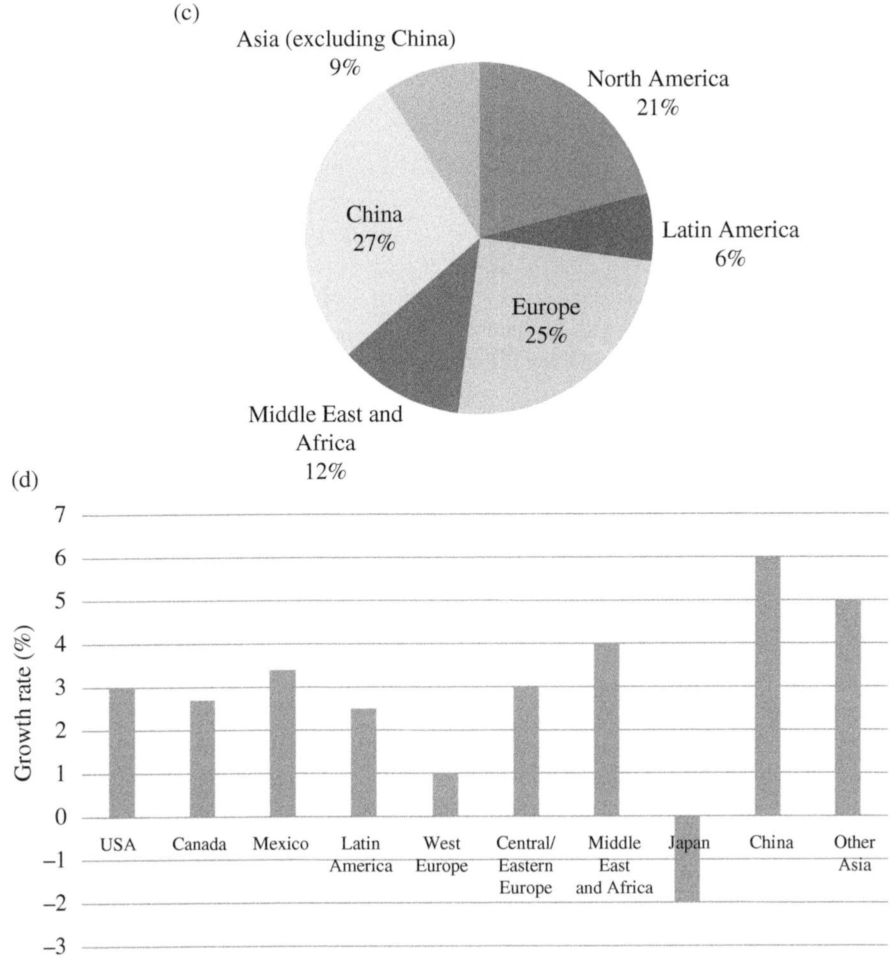


**FIGURE 7.1** (*Continued*)

Furthermore, there are only paltry data on Latin American demand for 1991, while in 2012 and 2015 this region reflects a substantial proportion of the total. This chapter will cover the flexible foam markets, commercial considerations, and future trends based on data collected from the past 16 years.

The maturity and complications associated with the flexible foam market belie the relatively prosaic nature of flexible foams and their range of applications. In general, flexible foams are associated with their ability to provide cushioning and comfort. Thus, they find large-volume uses in furniture cushions, mattresses, automotive seating, carpet underlayment, clothing, and packaging [4].

## 7.1 APPLICATIONS

Although applications for flexible polyurethane foams are well enumerated, the relative volumes broken down by geography and foam characteristics are difficult to obtain with accuracy. Problems developing a coherent picture are in part a reflection of the shifting use of descriptors and their uses. For instance, it is conventional wisdom that foam producers are under pressure to provide consumers with so-called "green" options for purchase. This is commonly understood to mean that the foam is based on feedstocks from agricultural rather than petrochemical feedstocks. While several agriculturally based polyurethane feedstocks have come on the market and there is anecdotal evidence for the growth in consumer demand, it is doubtful that the consumer is yet driving buying decisions based on feedstock origins. Rather, "green" technology market thrusts are largely based on governmental mandates surrounding air quality associated with minimization of volatile organic emanations, disuse of heavy metal catalysts, and supplementary blowing agents that may be employed to promote foam rise [5]. As will be shown later, there has been declining interest in acquiring intellectual assets (i.e. patents) in use of agricultural feedstocks over the past 5 years. Additionally, and specific to flexible polyurethane foams, there is a technology drive to implement flame-retardant solutions that do not involve the use of halogens [6]. Recently, the use of phosphorus-containing flame retardants has also come under scrutiny [7]. In 2011, there was also action by the US Environmental Protection Agency to explore restrictions of isocyanates in some applications [8]. European regulators have developed regulations that make automotive manufacturers liable for disposal of scrapped vehicles [9]. This burden has instigated activity for recovery and recycle of automotive polyurethane components, which is covered in Chapter 14. From the manufacturer's perspective, the drive is to provide equivalent cushioning and durability at lower densities translating to lower costs.

The breakdown of volumes applied to the various applications is also difficult to obtain, presumably because of poor tracking and a lack of motivation for data collation. A survey of data available for the USA is provided in Figure 7.2. Because of macroeconomic factors, as more chemical and end-use manufacturing is shifted to low-cost environments, the overall numbers can be expected to be volatile. While manufacturing may be shifted, it is likely that the snapshot provided by Figure 7.2 represents an equilibrium comparison between relative volumes required by mature economies for the various segments. A survey of similar data for 2005 [10] provided different absolute values but substantially similar relative numbers.

### 7.1.1 Furniture

Application of polyurethane flexible foams for furniture is a highly economically sensitive segment that follows the overall housing market [11]. The furniture segment uses flexible foams obtained from both slabstock and molded foam operations. However, with furniture manufacturing shifting to low-cost environments, and the

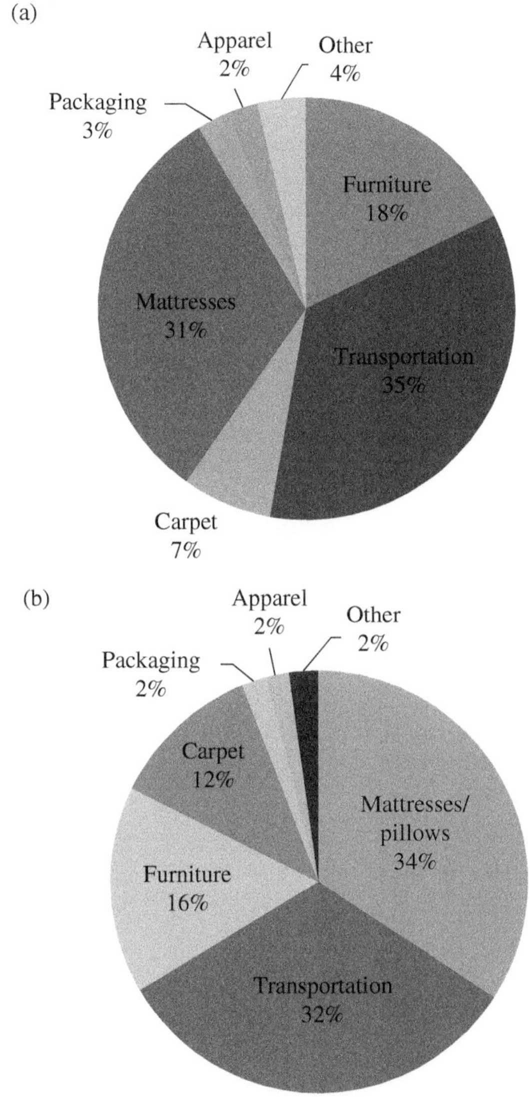


**FIGURE 7.2** Approximate breakdown for the volume usage of flexible polyurethane foams in the USA for (a) 2010 and (b) 2015.

cost-competitive nature of the furniture market, there is a significant concentration of furniture foams made in lower cost slabstock operations (Figure 7.3). The results show that the balance between molded and slabstock US consumption is stable, though absolute volumes have increased since the economically depressed period in 2010. Volume growth from 2006 to 2015 is nearly flat (approximately 2 billion

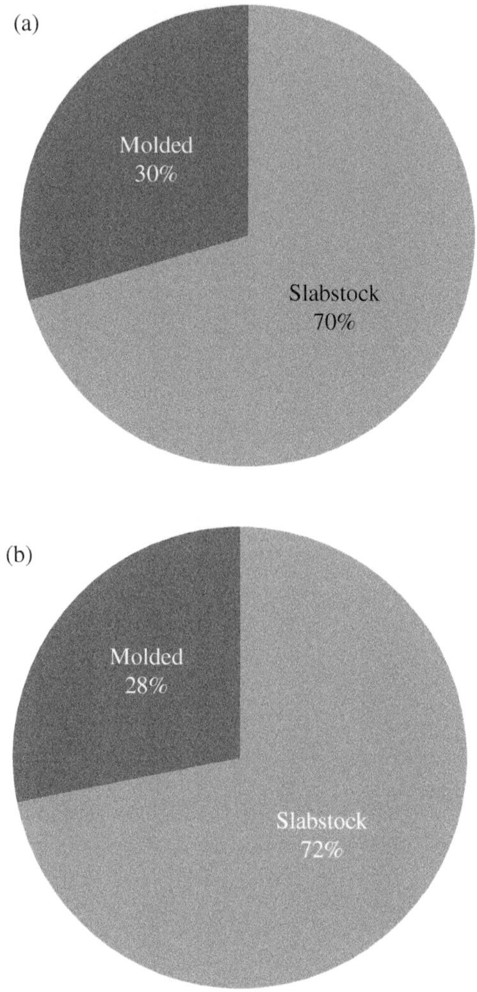


**FIGURE 7.3** Approximate split for the polyurethane flexible slabstock and flexible molded furniture foams in USA for (a) 2010 and (b) 2015. Total consumption was approximately 2 billion pounds in 2015.

pounds of flexible foam consumed). Examples of simple high-resilience slabstock formulations and properties are shown in Table 7.1.

The indentation force deflection (IFD, Table 7.1) measurement is a test that is unique to the comfort segment. This is not a test with a clear relationship to fundamental aspects of polyurethane structure or chemistry. It is an application test with a direct correlation with how the foam is used (usually sat upon). The test is specified in ASTM 3453 and D5672 [12]. A foam sample that is approximately the size of a

**TABLE 7.1 Simplified formulation and representative properties for high-resilience foam for furniture applications at three different foam densities**

| | | pphp | pphp | pphp |
|---|---|---|---|---|
| Sorbitol-initiated polyol EO-capped OH 29 | | 100 | 100 | 100 |
| Water | | 4.5 | 4 | 3 |
| Diethanolamine | | 2 | 1.5 | 1 |
| TDI index | | 100 | 105 | 105 |

| | | | | |
|---|---|---|---|---|
| Density | $lb/ft^3$ | 1.26 | 1.56 | 1.87 |
| IFD 40% | lb | 13.2 | 15.4 | 19.3 |
| Resilience | % | 60 | 60 | 62 |
| Compression set | % | 10 | 8 | 5 |
| Elongation at break | % | 150 | 135 | 120 |

Generally, the higher the foam density, the firmer and more durable the cushion.
EO, ethylene oxide; IFD, indentation force deflection; pphp, parts per hundred parts; TDI, toluene diisocyanate.

furniture cushion is compressed by a standard size circular press, as shown in Figure 7.4. Manufacturers of end-use products specify the performance of the foam in this test and the formulator and foamer will then design the foam to meet the test requirements. The specifications are usually the force measured by the load cell at a certain compression or several compressions (i.e. 25%, 40%, and 65% compression). The force value is normalized to the foam density. Careful experiment will also normalize the compression response to airflow, or at least assure that all compared foams have nearly the same airflow, to prevent the measurement being influenced by hydraulic resistance (see Section 4.2.2.2). The foam designer will adjust airflow properties by careful balancing of diethanolamine cross-linker, gelation, and blow catalysts within the formulation. An additional industry specification gauged from this test is the so-called "sag" factor, or comfort factor, which is calculated from the ratio of IDF at 65% and 25% compression. Some manufacturers will use alternative ratios or limitations when using these terms.

All polyurethane markets are competitive and the furniture market may be especially so [13]. Producer prices are always critical to market and manufacturer success, but intense competition assures that competent players rapidly respond to price changes in the market. Success in the furniture market usually requires an ability to meet all the customer requirements. Successful building block and foam manufacturers are able to produce the range of materials required by the next level in the value chain. The occasionally small-volume demands of particular applications have spawned the development of "systems houses." These businesses occupy

**FIGURE 7.4** Test fixture and sample geometry for testing flexible foam IFD, a common test for furniture applications. Image courtesy of Instron Corporation.

a middle-man position between the large building block producers and foam manufacturers. The systems houses rapidly develop formulations that meet the small-volume applications required by the foam manufacturer. From the market perspective the systems application must demand and justify the increased price of specific formulation development. The success of systems houses has inspired some large building block manufacturers to invest in or purchase their own systems companies.

Success in the furniture market also requires that suppliers within the value chain have credibility and historical position within the market. Intuitively it is clear that customers prefer suppliers with whom they have prior relationships and have a track record for delivering for the industry. This subjective requirement can act as a barrier to entry for new potential providers of products, as well as causing weakly positioned suppliers to fail when economic factors serve to differentiate among suppliers. In addition, history and position within the market provide the necessary up-to-the-minute intelligence of customers, competitors, and price, for a rapid response to market demands and fluctuations. Lack of such information creates so-called "market followers," who traditionally exhibit poorer profitability even under the best of circumstances [14].

The furniture industry is not considered one that makes significant demands for new polyurethane technology. Price and regulatory forces are the main drivers of new products for this segment. Development of polyols from seed oil feedstocks has created some adjustment of formulations as furniture producers try to develop a market demand for "green" products. Some producers will substitute 5–10% seed oil-derived polyol for petrochemically derived polyol into their formulation. Larger additions are shown to deteriorate properties using low-cost substitutions [15, 16], or unacceptably raise prices for more capable seed oil-designed products [17].

Another area of technology development is the need for improved environmental health and safety of polyurethane foams. These issues relate to improved fire resistance and inhibited smoke generation from foams without the use of halogenated flame retardants [18, 19]. Second choice flame retardants based on phosphorus are also potentially under scrutiny by regulators [20]. Reduction of volatile organic compounds from foams when in use is also considered a desirable feature by furniture manufacturers.

### 7.1.2 Mattresses and Bedding

As shown in Figure 7.1, mattresses and bedding (pillows, bolsters, mattress toppers) are a large and important industry segment for flexible foams. The market for mattresses is sensitive to the macroeconomic environment, and in some years mattresses and bedding will in fact be the largest industry segment for slabstock foams [21]. The number of mattresses (and so volume of flexible foam) and the type of mattresses (better or poorer quality/more or less foam) will vary significantly from year to year. Mattresses and bedding are largely based on slabstock polyurethane foam production. The types of slab stock foams used, however, cover a distribution of specifications and thus formulations.

It can be observed from Figure 7.5 that, despite the growth of the viscoelastic market, the ratio of production is relatively constant. This also reflects a constant balance in foam density market requirements since high-resilience foams tend to possess higher density. In Figure 7.5 the category of viscoelastic foam (also called "memory foam") is introduced, and this class will be fully developed below. The quality of the foam is in part a function of foam density. Many consumers will make an informed decision to purchase mattresses with higher quality foam, having experienced the permanent deformations of mattresses based on low-density/low-quality foam (Figure 7.6).

Mattresses are engineered structures [22, 23]. They are typically an ordered array of metal springs (the inner spring) covered by several inches of polyurethane foam. Depending on the size and density of foam covering the inner spring, a mattress may contain 10 pounds of polyurethane foam. Some mattresses (both high and low quality) will employ a polyurethane core in replacement of the inner spring. A high-quality foam core may utilize foam density as high as 4 $lb/ft^3$. The inner spring format still makes up almost 70% of the mattresses sold despite the substantial growth of solid foam core mattresses entering the market, as will be discussed below. Solid foam mattresses may weigh between 50 and 120 pounds, depending on the size and density – almost all of it high-density (3–4 $lb/ft^3$) polyurethane.

A significant segment for flexible foam growth has been viscoelastic foams [24, 25]. Unlike conventional or high-resilience foams, which are valued for their high efficiency

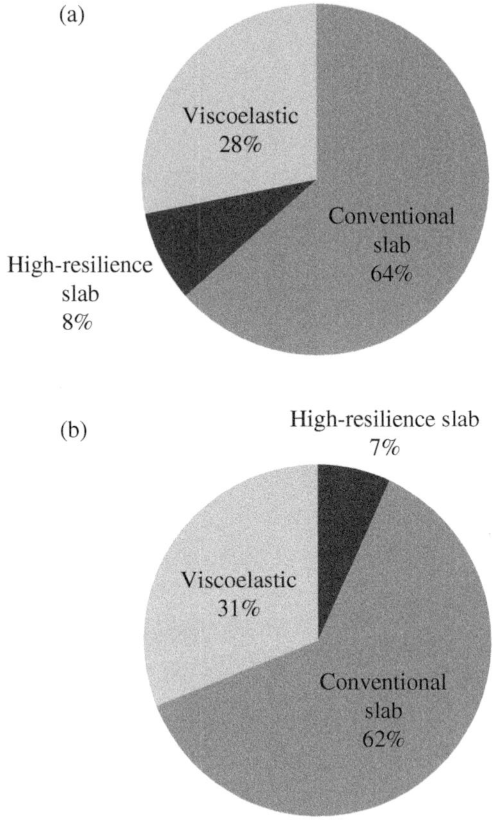


**FIGURE 7.5** Approximate breakdown for the types of foams used in the manufacture of mattresses for (a) average of years 2008–2011 (a period of lower economic activity) and (b) 2015. Despite the economic circumstances the differences are within the estimation error bar.

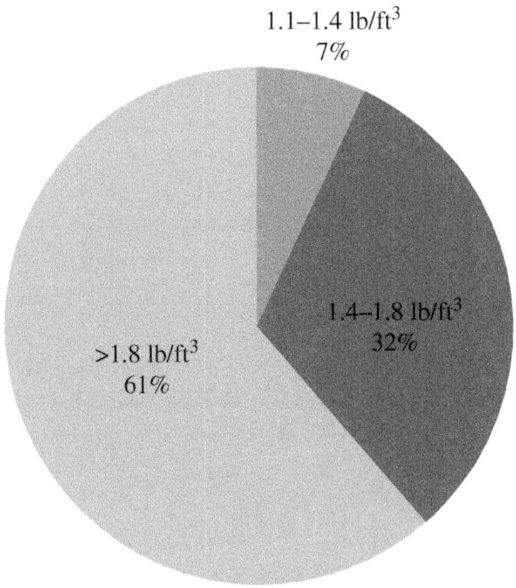


**FIGURE 7.6** Approximate breakdown of the flexible foam market based on foam density. These data can fluctuate based on economic conditions with cheaper/lower density foams increasing in poorer economic conditions and more expensive/higher density foams increasing in better economic conditions.

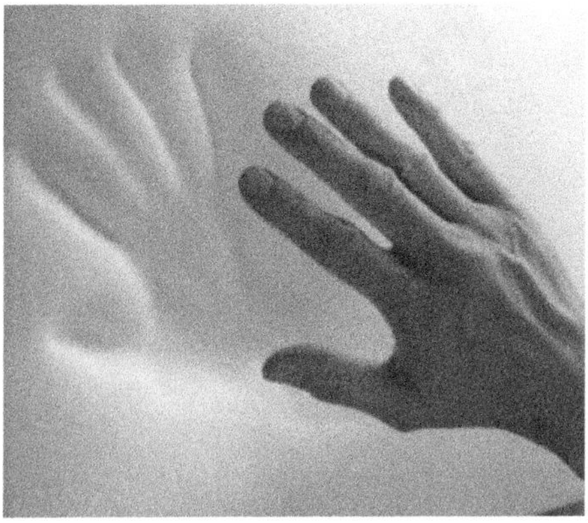

**FIGURE 7.7** Illustration of the delayed recovery response of a polyurethane viscoelastic foam. Image courtesy of The Dow Chemical Company.

in immediately returning energy, viscoelastic foams have a delayed recovery to deformation and are also called low-resilience foams. They are additionally known as "memory" foams because of their ability to temporarily retain the shape of what was just compressing them (Figure 7.7). Given time, the viscoelastic foam will recover its original shape. The behavior of viscoelastic foam is a result of the underlying polymer structure and is controlled by the formulation that the foam designer utilizes. Specifically, when designing a viscoelastic polyurethane foam the designer will formulate to result in a broad range of molecular weights between cross-links [26–29]. The effect of this molecular weight distribution is to broaden the glass transition temperature ($T_g$) such that the molecular motions of the polymer chains have a broad range of relaxation times. In Chapter 4 the effect of block copolymer structure on the $T_g$ was discussed, but the effect of cross-link density on the $T_g$ of a rubber is approximately given by Equation 7.1, where $M_c$ is the molecular weight between cross-links and k can be treated as a constant [30]:

$$T_g(n) \sim T_g(\infty) \times \left[ 1 + \frac{\mathrm{k}}{f(M_c)} \right] \tag{7.1}$$

Thus, as the molecular weight between cross-links evolves to a broad distribution, the $T_g$ will likewise reflect the breadth of that distribution. This will be reflected phenomenologically as a slow recovery of shape after deformation, particularly when the $T_g$ peaks between 0°C and 15 °C (Figure 7.8). Rheologically, the $T_g$ of a viscoelastic foam reflects the physically encumbered chain motions being activated over a broad range of temperatures. From a foaming point of view the broad and elevated tan ($\delta$) (tan delta) slows the foam rise and cell reticulation ("blow off") since the low molecular weight between cross-links will enhance gelation over blow reactions. The result can be reduced airflow of viscoelastic foams owing to a greater number of intact cell

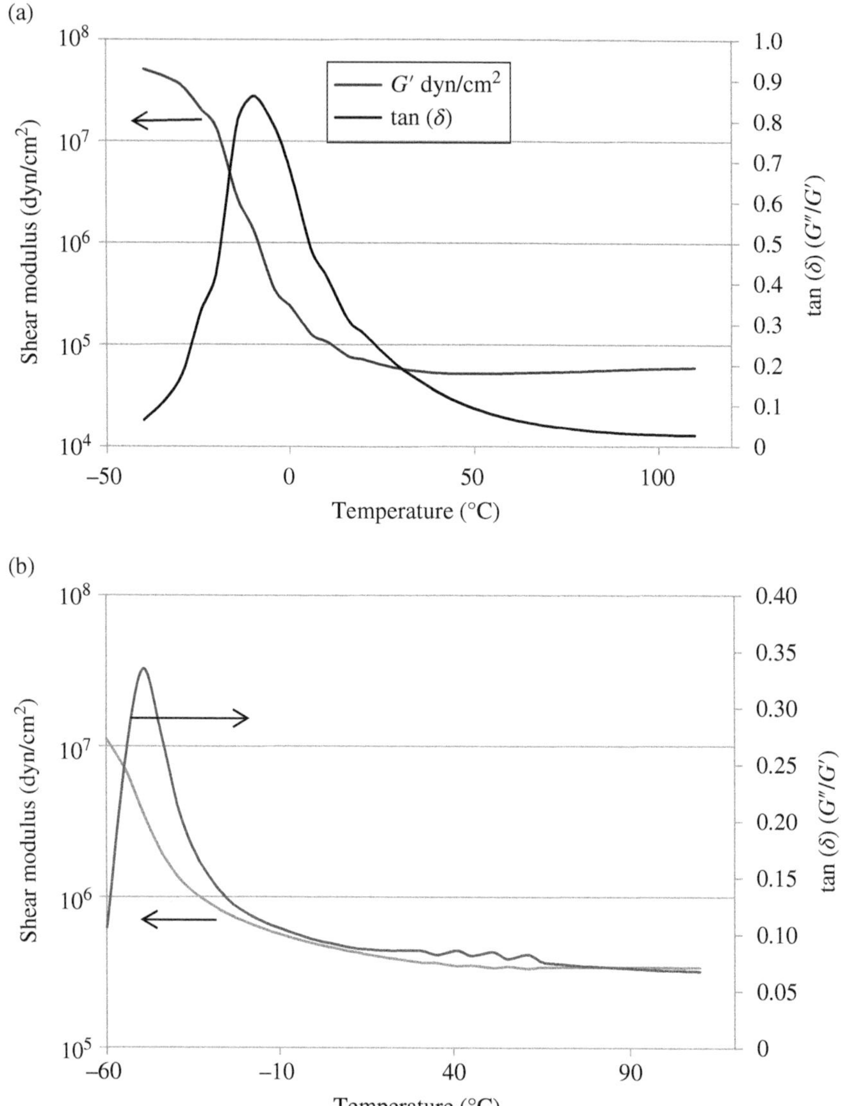


**FIGURE 7.8** Shear modulus and tan ($\delta$) (tan delta) spectrum of (a) a viscoelastic and (b) a comparative high-resilience flexible foam. The position, width, and amplitude of the tan ($\delta$) peak are typical of those whose applications demand viscoelastic response.

windows (Figure 7.9). The reduced airflow can contribute to the slow recovery response of the viscoelastic foam, but its contribution usually evolves as the foam ages. From the manufacturing point of view, the production of viscoelastic foam is not different from the production of other slab foams except that the designer will produce the desired effect by blending two polyols with substantially different equivalent

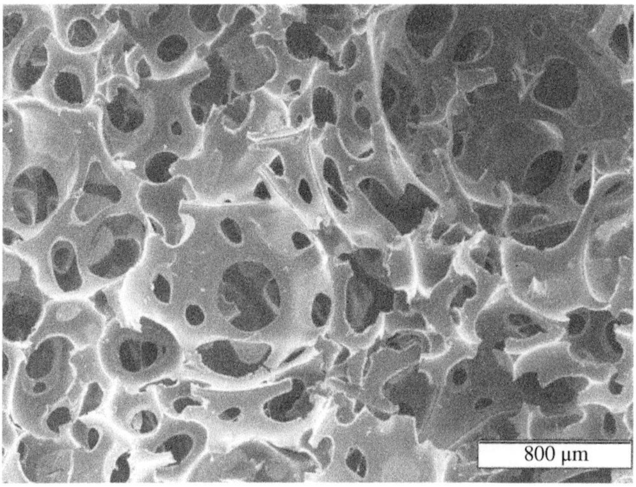


**FIGURE 7.9** Scanning electron microscope image of a typical viscoelastic polyurethane foam.

**TABLE 7.2 Representative formulations and property ranges for viscoelastic foams prepared at three different densities**

| Component | 2.5 lbs/ft$^3$ | 4 lbs/ft$^3$ | 6 lbs/ft$^3$ |
|---|---|---|---|
| Glycerine-initiated PO triol, equivalent weight 140 g/eq | 75 | 70 | 70 |
| Glycerine-initiated PO triol, equivalent weight 1000 g/eq | 25 | 30 | 30 |
| Water | 2.6 | 1.4 | 0.8 |
| Silicone surfactant for viscoelastic foam | 1.3 | 1.1 | 0.7 |
| Tin(II) catalyst | 0.09 | 0.05 | 0.03 |
| Balanced amine catalyst for slabstock | 0.7 | 1.5 | 1.9 |
| TDI index | 87.5 | 87.5 | 87.5 |

| Foam property | Value range |
|---|---|
| Density (lbs/ft$^3$) | 2.5–6 |
| IFD at 25% (lbf) | 13–45 |
| Recovery time (s) | 2–15 |
| Resilience (%) | 2–10 |
| Compression set at 75% (%) | <10 |

weights. For instance, in the representative formulation of Table 7.2, one polyol has a hydroxyl equivalent weight of about 140 g/eq, while the other is approximately 1000 g/eq. Miscibility of the polyols assures that there will be a broad distribution of molecular weights between cross-links. While viscoelastic foams have by far made their largest impact in mattresses and bedding, they are beginning to appear in some automotive seating systems as foam designers have learned to broaden out the tan ($\delta$) envelope to the required temperature range specified by manufacturers.

More recent trends have seen the rise of the "bed-in-a-box mattress," which is essentially a foam block in the shape of a mattress. Many of these mattresses are viscoelastic foams that have been packaged conveniently for shipping by post. They are usually purchased online and delivered, as the name implies, in a box. The viscoelastic response of the foam results in the foam needing one or more days to reacquire its intended shape. While the polyurethane industry as a whole grows slightly faster than gross domestic product (GDP), the mattress market grows at a rate that is more than twice GDP and is forecast to be 6–7% between 2019 and 2025. This reflects the faster purchase cycle of mattresses by the growing hospitality market, and the growing expectation of emerging middle-class consumers in developing economies. The global market for mattresses was US$30 billion in 2016, and is projected to eclipse US$40 billion per year by 2024 [31]. This rapid growth and significant market has resulted in heated competition by mattress manufacturers for market share, as well polyurethane producers on which much of this growth will depend. This has resulted in rapid innovation in polyurethane mattress foams. For instance, to improve airflow through the foam and thereby maximize coolness during sleep there has been innovation making viscoelastic foams possess airflows and appearance identical to flex foams, such as that shown in Figure 6.5, rather than that typically seen in Figure 7.9 [32]. An additional trend in mattress technology is the incorporation of phase change materials within the mattress itself which prevents the build-up of heat [33].

### 7.1.3 Transportation

Over 75% of flexible foam products targeted at the transportation market are designed for seating within automobiles, buses, trucks, airplanes, etc. [34]. Additional areas employing flexible polyurethane foams are instrument panels, headliners, door panels, arm and head rests, and also within structural members for the purpose of noise reduction [35]. Each application has specific requirements that are met by specific and highly optimized formulations.

As mentioned in Chapter 6, the comfort aspects of seating are subjective and different world regions have consistently developed formulations that reflect these preferences. Thus, the North American market typically seeks greater softness in seating, translating to lower density toluene diisocyanate (TDI)-based foams. The European market has generally delivered firmer seats using higher density foams based on 4,4′-methylene diphenyl diisocyanate (MDI). The Asian market has generally sought a middle ground between these consumer expectations [36]. In this regard, the aging demographics of the travelling public have evolved over time to seats that offer firmer support in seat backs [37]. Because of the importance of seat comfort to the purchasing decision, and the perception of delivered comfort, automobile manufacturers have specific standards that foams must deliver in every aspect of their static, dynamic, and aging properties. Furthermore, the specifications for driver, front passenger, and rear seats are usually different and in order of declining quality [37].

**TABLE 7.3 Representative formulation, molding conditions, and properties of a high-resilience toluene diisocyanate (TDI) foam that might be utilized for a seat cushion in the US transportation market**

| Component | pphp |
|---|---|
| 3.8 functional sucrose/glycerine mixed initiator 1730 hydroxyl equivalent weight 16% EO-capped PO polyol | 75 |
| 40% SAN 2500 equivalent weight copolymer polyol with similar composition to polyol | 25 |
| Water | 2.8 |
| Diethanolamine | 1.75 |
| Triethylenediamine (gelation catalyst) | 0.3 |
| (Dimethylaminoethyl)ether blow catalyst | 0.7 |
| Silicone surfactant | 0.8 |
| TDI | 100 index |

Molding: Hi-Tech Sure-shot 50
Block size: 38 × 38 × 11.4 cm
Pad weight: 695 g
Mold temperature: 75 °C
Demold time: 3 minutes
A/B steam pressures: 2300/2400 psig

| Test | Value |
|---|---|
| Density (kg/m$^3$) | 42 |
| Airflow (ft$^3$/min) | 1.4 |
| IFD 25% (kg) | 10 |
| IFD 50% (kg) | 19 |
| Resilience (%) | 71 |
| Humid aged compression set (%) | 18 |

Note the exceptionally high resilience measured for this formulation.
EO, ethylene oxide; IFD, indentation force deflection; PO, propylene oxide; SAN, styrene acrylonitrile.

Given the proliferation of specialty polyurethane foams sought by manufacturers and consumers for transportation applications, the best that can be offered are representative recipes for foams that have a good chance of working in typical equipment. While some transportation applications such as carpet underlay and headliners can be made from slabstock foams, almost all seats are prepared by molded foam operations. The specificity of molding equipment is an additional overlay of complexity in creating useful foam since different molding equipment may produce somewhat different foams with the same polyurethane formulation [38–40]. Table 7.3 provides a representative molded foam formulation based on TDI, molding conditions, and properties for an automotive seat cushion that might be found in a car manufactured in the USA.

**TABLE 7.4 Representative formulation, molding conditions, and properties of a high-resilience methylene diphenyl diisocyanate (MDI) foam that might be utilized for a seat cushion in the European transportation market**

| Component | pphp |
|---|---|
| 2000 equivalent weight glycerine-initiated PO polyol 15% EO cap 80% primary hydroxyl | 75 |
| 1675 equivalent weight glycerine-initiated EO/PO mixed feed Polyol (primarily used to enhance cell openness) | 4 |
| Water | 3.5 |
| Tris(dimethylaminopropyl)amine balanced blow/gel tertiary amine | 1 |
| Low-efficiency silicone surfactant | 1 |
| Polymeric MDI: 2.7 functionality, 137 isocyanate equivalent weight, 31% isocyanate content | 60 |
| MDI 30/70 ortho–para′/para–para (index 105)(see Chapter 2) | 40 |

Cannon Model A 60 w/FPL 18 mixing head
400 g/s output
2000 psi/2000 psi polyol/isocyanate pressure at 25 °C
Mold temperature: 50 °C
Opening time: 3 minutes
Demold time: 4 minutes

| Test | Value |
|---|---|
| Density (kg/m$^3$) | 51 |
| Resilience (%) | 57 |
| IFD at 25% (kg/m$^3$) | |
| Dry compression set (%) | 7 |
| Humid aged compression set (%) | 13 |
| Ultimate elongation (%) | 100 |

Pertinent differences relative to the toluene diisocyanate (TDI) foam of Table 7.3 are the density, resilience, and IFD.
EO, ethylene oxide; IFD, indentation force deflection; PO, propylene oxide.

An MDI-based molded foam formulation is presented in Table 7.4. While the conditions and results are somewhat different than those of the TDI foam in Table 7.3, these distinctions should not be viewed as a general representation of the differences. The opening and demold times can be as much about arbitrary process differences as about demands resulting from the specific formulation. Molded foam operations also commonly use prepolymers (Section 2.1.6) rather than pure polymer MDI (pMDI)/isocyanate monomer mixtures. The prepolymer route may ease certain process requirements related to mixing, pumping, and mold filling, justifying the added complexity and expense of employing a prepolymer. While the formulation choices and

**TABLE 7.5 General formulation guidelines for high resilience (HR) transportation seating molded polyurethane foams according to the isocyanate (iso) classification**

| Foam type | Clear polyol equivalent weight (g/eq) | % solids in polyol/ copolymer blend | Isocyanate |
|---|---|---|---|
| Hot cure | 1000 | 0–25 | TDI |
| HR iso blend | 1600–2300 | 5–35 | TDI/pMDI |
| TDI HR | 1600–2300 | 5–35 | TDI |
| MDI HR | 1600–2300 | 0–20 | pMDI/MDI or prepolymer |

MDI, methylene diphenyl diisocyanate; pMDI, polymeric MDI; TDI, toluene diisocyanate.

**TABLE 7.6 General structure–property relationships for flexible molded polyurethane foams related to choice of polyols**

| Formulation variable | Effect |
|---|---|
| Increased hydroxyl equivalent weight | Increased viscosity, increased elongation at break, increased resilience, lower load bearing |
| Increased polyol functionality | Increased viscosity, increased load bearing, lower elongation at break, lower compression set, lower airflow, increased durability |
| Lower monol content | Lower hysteresis, lower compression set, less vibration conduction, increased durability |

foam property variability can be mind-boggling, there are a few general rules for molded foam types relative to compositional choices (Table 7.5) and formulation variables (Table 7.6).

### 7.1.4 The Molded Foam Market

The dominance of the molded foam business by the transportation sector results in a unique commercial structure. The *slabstock* flexible foam market is highly driven by price, supply reliability, and the mutual interest of a varied and highly fractured foam market to absorb the high-volume output of an asset-driven, concentrated polyurethane building block industry. While the *molded* foam market is also highly price sensitive, it is equally driven by the competition between commercial relationships. Additionally there is the commercial tension brought on by low-margin large businesses (such as commodity producers) and the high cost of sales needed to supply specialized formulations for specific and niche applications. This conflicting need for price and specialization has resulted in the development of distributors who purchase building blocks directly from the building block manufacturer to service small accounts. Inevitable overlaps form with the range of direct sales from the major manufacturer and the distributors. In addition, specialization has spawned

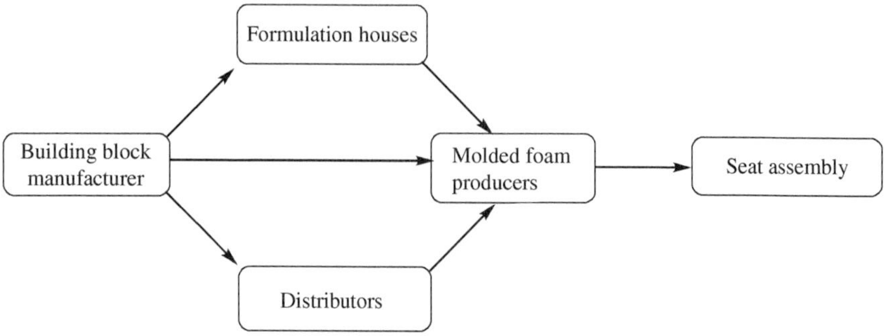


**FIGURE 7.10** Unique commercial relationships among suppliers to the transportation seating market. This situation is bound to evolve, but which channel prevails will depend on unforeseeable market and macroeconomic factors.

the creation of formulation houses (or "systems houses"), which simultaneously purchase polyols and isocyanates from building block manufacturers while competing with these same feedstock manufacturers as *customers* using fully formulated systems. In response to these pressures the large building block manufacturers have evolved by buying or developing their own distributor and formulation (or "systems") houses. Figure 7.10 provides a map of the relationship between the building block producers, distributors, and the formulation houses.

## 7.2 TRENDS IN MOLDED FOAM TECHNOLOGY AND MARKETS

While the patent literature is notably poor at providing insight into fundamental science, it is the clearest way to understand the future of a technology and market [41, 42]. This is because the patent literature represents a sizeable investment by a company or individual on what is believed will be desirable by customers and market competitors in the next-generation product. Furthermore, defense of a patent asset requires that the technology on which a patent is based be protected by patent prior to its commercial introduction.

While a single patent may not be instructive as to the direction of future innovation, a clear trend in industrial patent activity can be unmistakable. Figure 7.11 shows the patent activity of several large multinational polyurethane feedstock providers over the years 2000–2013. In this analysis the patents filed in the Chinese language represent the activity within the sizeable Chinese market by Chinese companies. For perspective, the total number of issued patents found within the search parameters was 1033, so it is clear that there are a large number of patents not accounted for by this analysis. Despite this, it is likely that the preponderance of unaccounted patents will reflect the activity of these highly experienced and influential participants (i.e. Dow, BASF, and Bayer/Covestro) within the polyurethanes market. Figure 7.11

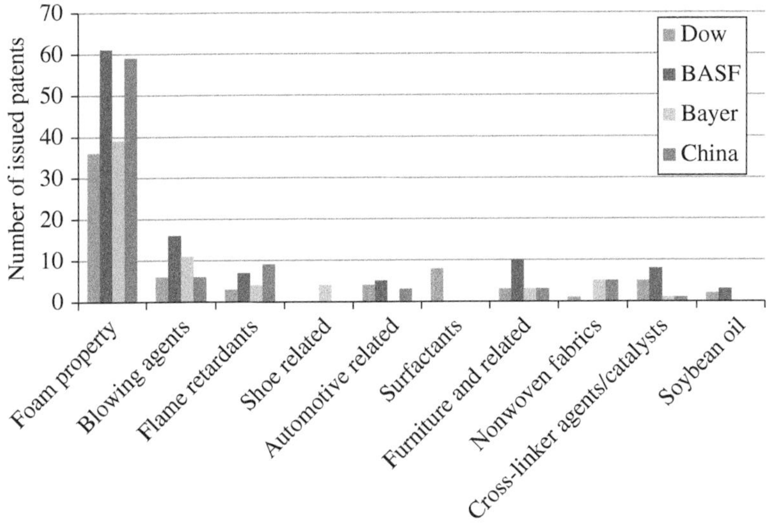


**FIGURE 7.11** Patent activities within the flexible polyurethane foam category of several large polyurethane market servicers and the patents filed in the Chinese language for the years 2000–2013. (*See insert for color representation of the figure.*)

shows that the majority of issued patents reflect the attempt to protect novel flexible foam properties and processes for a specific formulation or range of formulations. Arguably, this is suggestive of the richness of polyurethane chemistry and the fact that creative scientists are still finding new ways to achieve continuous improvements. However, it can also be argued that these patents cannot be considered as part of a more general future-oriented strategy since they reflect the properties found in pursuit of current applications. However, within this category there are also industrial trends.

Relative to foam properties, innovation is being pursued to provide thinner/lighter seat foams with identical support properties achieved by thicker/heavier foams. Some of this activity is being motivated by the desire of automobile manufacturers to provide lower profile automobiles for improved wind resistance without sacrifice of occupant head space. The advent of hybrid and electric vehicles has also created a premium for space that can otherwise be utilized to hold batteries.

Another foam property trend is the desire to provide foam that maintains a high level of support and comfort and at the same time dissipates vibrational energy that may otherwise be transmitted through the seat and degrade the perception of comfort. The balance is that many support and durability factors are highly correlated with low hysteresis. Low hysteresis, a result of network connectivity and phase separation that results in low energy loss upon compression, works against vibration damping. Foam innovation has resulted from the identification of those vibrational frequencies that are perceived to result in passenger discomfort. It has been established that discomfort is perceived at vibrational frequencies less than 6 Hz [37, 43–45]. Thus, foam innovation has focused on development of foams that under compression dampen frequencies below 6 Hz.

We can understand the innovations of tailored foam vibrational damping by consideration of the equations relating mechanical properties to vibration transmissivity (Equation 7.2) [46–48]. In Equation 7.2 K is the dynamic spring constant or stiffness (units of N/m), $g$ is the gravitational constant (m/s$^2$), $W$ is the applied load (N), and $C$ is the damping factor (N-s/m). From the perspective of designing a polyurethane foam to achieve a desired result, it is simplified at least to the extent that the only variable that can be impacted by formulation is K, the foam stiffness. The foam stiffness is a function of many possible design variables, and would have to be optimized within the boundaries of all of the other critical foam performance criteria. From the processing perspective, the damping factor can be influenced by foam structural properties such as pneumatic damping, suggesting that optimization of airflow resistivity can help to increase vibrational damping. Alternatively, the tan ($\delta$) spectrum can be optimized such that certain relaxation modes are inhibited at typical driving temperatures to retard energy return of certain frequencies (i.e. under 6 Hz) to the driver. This technique would have to be balanced with the comfort perception that highly resilient foams are closely associated with passenger comfort in the absence of vibrations.

$$\text{Resonance frequency} = \frac{1}{2}\pi\sqrt{\frac{\mathrm{K}g}{W}}$$
$$\text{Amplitude factor} = \frac{1}{C}\sqrt{\frac{\mathrm{K}W}{g}} \tag{7.2}$$

Figure 7.12 provides the same information as Figure 7.11 but without the patents protecting a specific foam property; instead, Figure 7.12 focuses on a general field of application or composition of matter. This analysis shows that the most active fields of patent activity are those guided by regulatory pressures associated with supplemental blowing agents and flame retardance. A somewhat hidden patent field is the catalyst category. The import of this activity is apparent by the fact that all companies are engaged, in spite of the fact that these companies are not specifically catalyst manufacturers. Their presence in this field may point to anticipated regulation by occupational, safety, and health agencies of certain catalysts such as those based on tin and mercury [49–50].[1,2]

[1] Di-octyl tin catalysts are currently classified as category 3 substances – a substance of concern for humans because of possible CMR (carcinogenic, mutagenic, or reprotoxic) effects but for which there is not enough information available to classify these substances and preparation into category 2 (a substance of very high concern) status. European Commission Decision 2009/425/EC claims organotin compounds pose a risk to human health and will expand the scope of restrictions beyond those currently in place on maritime applications.

[2] European regulation under REACH (Regulation, Evaluation, Authorization, and Restriction of Chemicals) number 1907/2006 severely limits the use of organomercury compounds such as polyurethane catalysts.

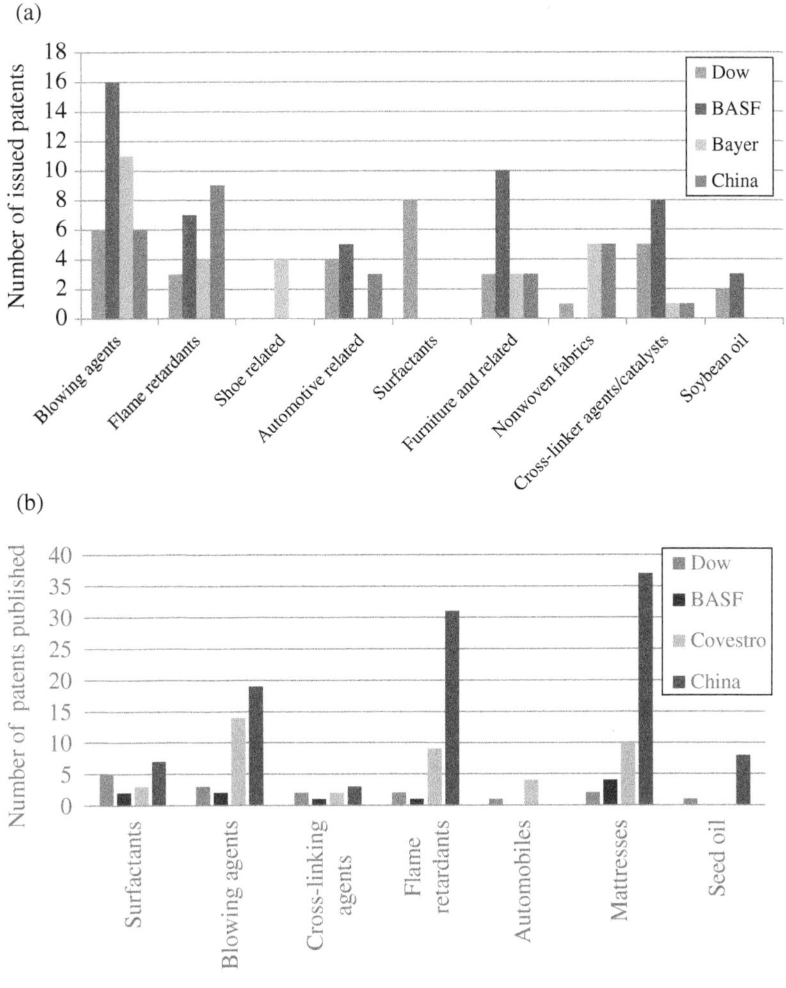


**FIGURE 7.12** (a) Data from Figure 7.11 with the "foam property" category removed to improve visualization of more specific subjects. (b) Data for the period 2015–2019 showing the changes in industrial investment focus. (*See insert for color representation of the figure.*)

Across the rest of the spectrum of activity, it is apparent that an individual company may pursue a specific intellectual asset strategy independent of the activity of its peers. Activity by a few companies in the use of soybean oil is a reflection of a strategy of pursuit of that agricultural oil as an alternative feedstock. However, the fact that not all companies are active in that area may suggest that not all companies share a soy-based-alternative-feedstock-for-polyurethane vision, or may be taking a strategy of waiting to see if alternative feedstocks come to commercial fruition before developing a market follower technology response. Several companies, including Bayer (now Covestro) and BASF, have also pursued an alternative feedstock vision employing ricinoleic acid (from castor beans; see Chapter 2), which has intrinsic secondary hydroxyl groups along the triglyceride chains. While ricinoleic acid has a process advantage by not requiring industrial introduction of hydroxyl groups onto the backbone, there is at least one serious disadvantage associated with a ricinoleic acid strategy for feedstocks (except for very small applications). Specifically, it relates to the limited volume of castor oil (the source of ricinoleic acid) in the world. Difficulty in growing and harvesting castor oil has narrowed the number of producing countries to primarily India. The annual volume of produced castor oil is less than 2 billion pounds [51]. While this sounds like a lot, available volume is in relatively short supply compared with alternative agricultural oils such as soya or canola. Acreage devoted to production is controlled by Indian governmental agencies and there is an approximate balance between current supply and demand at relatively elevated prices. Accessing ricinoleic acid volumes required to satisfy an alternative feedstock strategy would distort the market for castor oil, and raise prices – probably sufficiently to make the product economically untenable even with the lower capital cost allowed by an "isocyanate-ready" triglyceride or source of fatty acid. The extension of patent trends through 2103 to the years 2014–2019 is shown in Figure 7.12b. While much of the industrial focus has been consistently on the same topics there are a few interesting differences. One difference is that industrial interest in agricultural feedstocks has declined for a number of reasons, including based on cost, performance, and the perception that industrial use was a competitor of societal food requirements. It is also clear that patent activity is accelerating overall, and that industrial patent activity is increasing in the areas of greatest market interest overall.

Lastly, over 2000–2013 it would appear that patent activity is relatively muted for such a large industrial segment, and that BASF has been the most active company.

## REFERENCES


[1] H Chinn, Y Yoneyama, U Loechner, and X Xu, *Polyurethane Foams*, Chemical Economics Marketing Report, IHS, Englewood, CO, 2016, 8.

[2] American Chemical Council (ACC), End-Use Market Survey on the Polyurethane Industry-2010, Center for the Polyurethanes Industry, Washington, DC, 2011. Section 3.

[3] H Chinn, Y Yoneyama, U Loechner, and X Xu, *Polyurethane Foams*, Chemical Economics Marketing Report, IHS, Englewood, CO, 2016, 78.

[4] M Sonnenschein, W Koonce, "Polyurethanes," in *Encyclopedia of Polymer Science and Technology*, 4th Ed., Wiley, Hoboken, NJ, 2012, 1–63.

[5] D Miller, "Laws and regulations: what every polyurethane manufacturer needs to know," in *Proceedings of the Polyurethanes Conference*, October 13–16, Salt Lake City, UT, 99–105.

[6] S Kemmlein, D Herzke, and R Law, *Environment International*, **29**, (2003), 781–792.

[7] K Betts, *Environmental Health Perspectives*, **116**, (2008), A210–A213.

[8] US Environmental Protection Agency, *Action Plan for Methylene Diphenyl Diisocyanate (MDI) and Related Compounds*, Document RIN 2070–ZA15, US Environmental Protection Agency, Washington, DC, April 2011.

[9] European Union Council, *Directive of End-of-Life Handling of Vehicles #2000/53/EC*, Environment Directorate-General of the European Commission, Brussels, Belgium, September 18, 2000.

[10] J Miron, A Brief, S Epstein, and C Robinson, *Polyurethane IV Skeist Incorporated, Market Research Report*, Skeist Inc., Whippany, NJ, 2006, 359–431.

[11] American Chemical Council (ACC), *End-Use Market Survey on the Polyurethane Industry-2010*, Center for the Polyurethanes Industry, Washington, DC, 2011, 11.

[12] *Standard Test Method for Testing Flexible Cellular Materials. Measurement of Indentation Force Deflection, ASTM D5672*, ASTM International, West Conshocken, PA, September 1, 2009.

[13] (a) V Bruninghaus, "Flexible slabstock production: state-of-the-art technology and new development to meet the global challenge," in *Conference Proceeding PU Latin America*, August 28–30, Sao Paulo, Brazil, 2001, 1–3. (b) IBIS World, *Furniture Stores Industry in the U.S. – Market Industry Report*, IBIS World, New York, NY, June, 2019.

[14] Y Baek, H Lee, and H Jeong, *Physics Review E: Statistical, Nonlinear and Soft Matter Physics*, **82**, (2010), 026109.

[15] M Sonnenschein and B Wendt, *Polymer*, **54**, (2013), 2511–2520.

[16] L Zhang, H Jeon, J Malsam, R Herrington, and C Macosko, *Polymer*, **48**, (2007), 6656–6667.

[17] H Benecke, B Vijayendran, D Garbark, and K Mitchell, *Clean: Soil, Air, Water*, **36**, (2008), 694–699.

[18] S Lazar, F Carosio, A Davesne, M Jimenez, S Bourbigot, and J Grunlan, *ACS Applied Materials and Interfaces*, **10**, (2018), 31686–31696.

[19] M Modesti and A Lorenzetti, "Recent trends in flame retardancy of polyurethane foams," in *Flame Retardants*, P Merlani, Ed., Nova Publishers, Hauppauge, NY, 2010, 185–218.

[20] S Levchik and E Weil, *Journal of Fire Sciences*, **24**, (2006), 345–364.

[21] H Chinn, Y Yoneyama, U Loechner, and X Xu, *Polyurethane Foams*, Chemical Economics Marketing Report, IHS, Englewood, CO, 2016, 22–100.

[22] V Verhaert, B Haex, T De Wilde, D Berckmans, J Verbraecken, E de Valke, and J Vander Sloten, *Ergonomics*, **54**, (2011), 169–178.

[23] L Shen, Y-x Chen, Y Guo, S Zhong, F Fang, J Zhao, and T-Y Hu, *Work*, **41**, (2012), 1268–1273.

[24] K Rajan, D Prasad, R Manikandan, and N Veena, *Cellular Polymers*, **30**, (2011), 13–21.

[25] S Hager, R Skorpenske, S Triouleyre, and F Joulak, *Journal of Cellular Plastics*, **37**, (2001), 377–391.

[26] K Aou, A Schrock, V Ginzburg, and P Price, *Polymer*, **54**, (2013), 5005–5015.

[27] C Briody, B Duignan, S Jerrams, and S Ronan, *Polymer Testing*, **31**, (2012), 1019–1025.

[28] B Vaughan, G Wilkes, D Dounis, and C McLaughlin, *Journal of Applied Polymer Science*, **119**, (2011), 2683–2697.

[29] A Greco and F Lionetta, *Polymer Engineering and Science*, **49**, (2009), 1142–1150.

[30] J Bicerano, *Prediction of Polymer Properties*, 3rd Ed., Marcel Dekker, New York, NY, 2002.

[31] Mattress Market Size, Share & Trends Analysis Report By Product (Innerspring, Latex, Memory Foam), By Application (Domestic & Commercial), By Region, And Segment Forecasts, 2019–2025, Grandview Market Research, San Francisco, CA, 2019.
[32] K Aou, R Gamboa, and H Ma, Process for making high airflow and low compression set viscoelastic polyurethane foam. Assignee Dow Global Technologies, US patent 9029432 B2, (2015).
[33] K Aou, Y Gu, R Duggal, Y Srivastava, J Jacobs, Q Meng, G Cardoen, R Even, and M Springs, Viscoelastic polyurethane foam with coating. Assignee Dow Global Technologies, US patent 2019/0211171 A1, (2019).
[34] American Chemical Council (ACC), *End-Use Market Survey on the Polyurethane Industry-2010*, Center for the Polyurethanes Industry, Washington, DC, 2011, 96.
[35] P Farkas, R Stanciu, and L Mendoza, *Journal of Cellular Plastics*, **38**, (2002), 341–354.
[36] R Bros, R Herrington, and F Casati, *Journal of Cellular Plastics*, **36**, (2000), 207–245.
[37] B Neal, *Journal of Cellular Plastics*, **37**, (2001), 160–178.
[38] D Jackovich, B O'Toole, M Hawkins, and L Sapochak, *Journal of Cellular Plastics*, **41**, (2005), 153–168.
[39] D Hatchett, J Kinyanjui, and L Sapochak, *Journal of Cellular Plastics*, **43**, (2007), 183–196.
[40] S Geier, C Winkler, and M Piesche, *Chemical Engineering and Technology*, **32**, (2009), 1438–1447.
[41] R Sitarz, M Heneczkowski, and A Kraslawsi, *Computer Aided Chemical Engineering*, **31**, (2012), 570–574.
[42] P Chang, C Wu, and H Leu, *Scientometrics*, **82**, (2010), 5–19.
[43] N Hillyard, *Mechanics of Cellular Plastics*, MacMillan, New York, NY, 1983.
[44] N Hillyard, W Lee, and A Cunningham, *Cellular Polymers*, **20-22**, (1991), 187–191.
[45] Japanese Automotive Standards Organization (JASO), Test Code of Seating Comfort for Automobile Seats, *JASO* B407–87, 1987.
[46] H Nahvi, M Fouladi, and M Nor, *International Journal of Acoustics and Vibration*, **14**, (2009), 143–149.
[47] J Van Niekerka, W Pielmeierb, and W Greenberg, *Journal of Sound and Vibration*, **260**, (2003), 867–888.
[48] E Falou, J Duchene, M Grabisch, Y David, and L Langeron, *Applied Ergonomics*, **34**, (2003), 249–255.
[49] RPA, "Impact assessment of potential restrictions on the marketing and use of certain organotin compounds," *European Commission Directorate-General Enterprise and Industry, Document J570/Organotins*, Risk & Policy Analysts Ltd, Norfolk, UK, October 2007.
[50] E Foulkes, "Mercury," in *Patty's Toxicology, Volume 2*, E Binham, C Cohrssen, and C Powell, Eds., Wiley, New York, NY, 2001, 327–352.
[51] R Wu, L Zeng, L Viciu, and T Masuda, "Major fats and oils industry overview," *Chemical and Economics Handbook*, Report 220.5000A, IHS Group, Englewood, CO, April 2012, 11.

# 8

# POLYURETHANE RIGID FOAMS: MANUFACTURE, APPLICATIONS, MARKETS, AND TRENDS

About 50% of all polyurethane foam production is consumed by market demand for rigid polyurethane foam [1]. In contrast to flexible foams that are primarily an agent for comfort and cushioning, rigid foams are sought for their use in insulation and energy efficiency. To fulfill that mission polyurethane rigid foam formulation, physics, and methods are very different from those for flexible foams, even while the fundamental aspects of chemistry, reaction kinetics, structure development, and morphology are the same. Polyurethane rigid foam is a premier material for delivery of insulation efficiency, as illustrated by Figure 8.1 [2]. In contrast to other conventional materials, polyurethane provides the most insulation for unit thickness. As a two-component application it provides flexibility to where and when the building blocks are transformed from liquid to solid foam form. Because of the large number of building blocks and the commodity nature of the feedstocks to make them, rigid foams can be relatively low cost to produce and able to provide a significant amount of differentiation to suit specific application requirements.

## 8.1 REGIONAL MARKET DYNAMICS

The estimated production of rigid polyurethane foams for the year 2015 is shown in Figure 8.2. Rigid foams are principally composed of (i) those produced for building insulation primarily produced geographically close to where they are used and

*Polyurethanes: Science, Technology, Markets, and Trends*, Second Edition. Mark F. Sonnenschein.

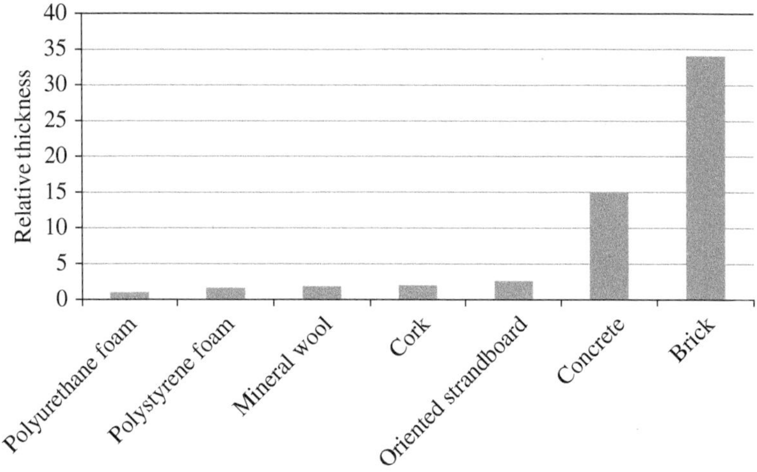


**FIGURE 8.1** Relative performance of common insulation materials and construction materials. It takes more than 2 inches of cork and 15 inches of concrete to provide the insulation of 1 inch of polyurethane rigid foam.

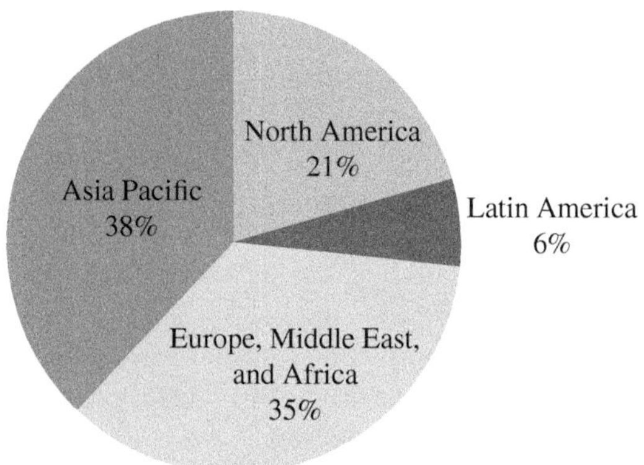


**FIGURE 8.2** Proportional production of rigid foams in 2015 by geographical region.

(ii) foams used for appliances that will represent the summation of regional consumption and those made for export. Many of the emerging economies are developing strong demand for appliances such as refrigerators, but certainly much of the production in the Asia Pacific region is for export to other geographies.

In mature economies such as those of North America and Western Europe the production of rigid foam for construction is about double that produced for appliance applications. In the emerging export economies the volume ratio of construction to appliance foams is very nearly 1, representing the overproduction of foams for appliances that are subsequently exported.

While useful, the data in Figure 8.2 are a simplification of a complex geographical puzzle. Year-to-year changes in production result not so much from shrinkage in markets but rather from greater or lesser growth. For instance, in the Asia Pacific region the polyurethane rigid foam market has grown, but notably slowed in the major economies outside of China. In the China market, the local Wanhua Group dominates local trade for isocyanate building blocks for making rigid foams. Chinese production of aromatic polyester polyols for rigid foams is much less mature than that for isocyanates but is growing at a rate even higher than that of the overall Chinese economy. Asian polyurethane producers monopolize more than 50% of the total Asia and Pacific market for chemicals. All the large multinational companies have a footprint in the Asia Pacific region with varying degrees of market penetration. As with many industries, of which polyurethanes is an example, the power of a historical business relationship between two companies (i.e. supplier and customer) is as important in many ways as price, and it is known that certain suppliers have entrenched relationships with customers that are resistant to simple price inducement to switch.

For illustrative purposes, another fascinating dynamic of the Asia Pacific market are the consequences of the phenomenal growth of these emerging and established economies [3]. In anticipation of the large Chinese population's demands for the conveniences and standards of modern life, there has been rapid expansion of industrial production. As Figure 8.3 shows, the supply–demand ratio for polyurethane chemicals has shown significant volatility, and predictions for industry supply and health have been notably inaccurate. There has been continued investment to maintain a healthy and growing industry. In any case, premised on long-term 80% asset utilization in China, and low trade barriers to export markets, it is assured that costs

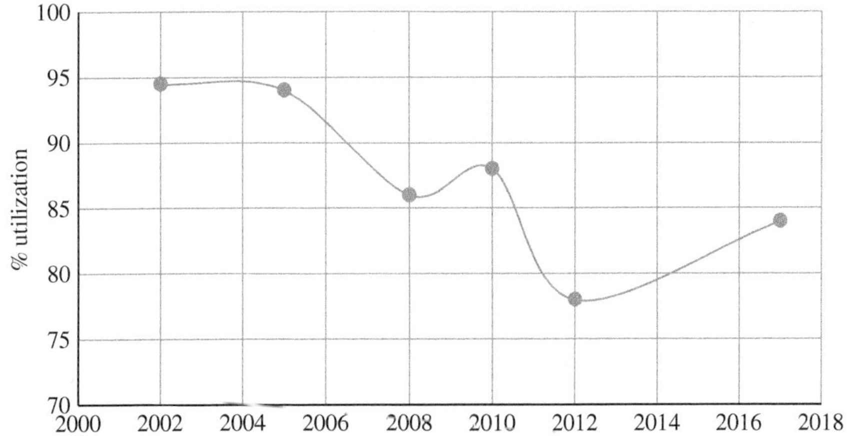


**FIGURE 8.3** Ratio of production and consumption volume in the Asia Pacific region showing the extent of regional investment in the polyurethane industrial base.

for polyurethane chemicals in global markets will remain low, and will contribute to low price inflation from finished products using polyurethane products.

A more complex market situation is presented by the European, Middle East, and Africa region. Beyond geographical proximity, this category is somewhat arbitrary, with the Middle East and Africa representing between 5% and 10% of the total addressable market for rigid polyurethane foams for this grouping [1, 4]. Within Europe there is a bifurcation in market activity between Eastern and Western European markets (Figure 8.4).

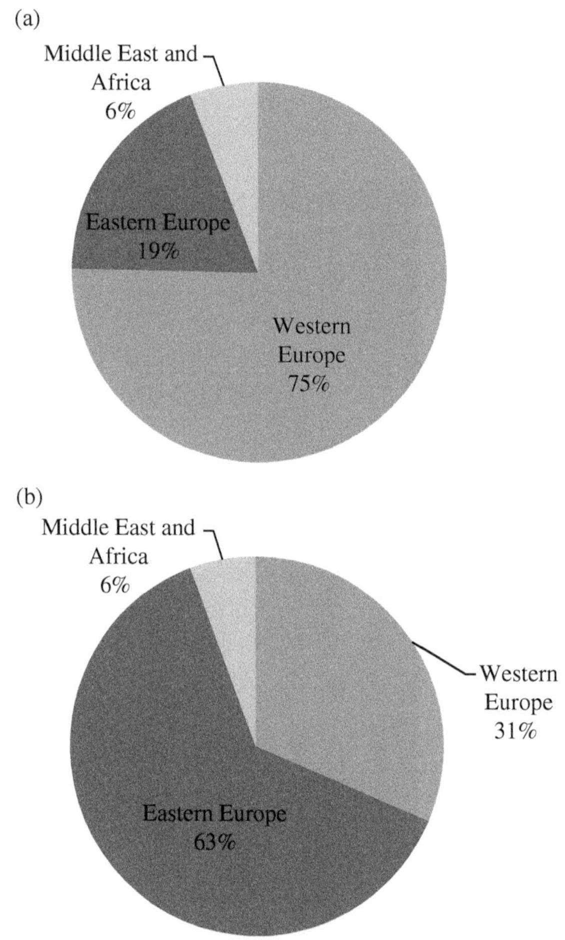


**FIGURE 8.4** Approximate European (a) construction and (b) appliance market sizes for 2012. Total volume of construction market is ca. 1 billion pounds and about 500 million pounds for the appliance market.

For rigid foam targeted towards the construction market, Western Europe represents about 65% of the total volume. This reflects greater emphasis on energy efficiency in the Western European construction market and willingness of consumers to pay for benefits accrued to resource conservation. Recent trends in Central and Eastern Europe have seen increased usage of polyurethane insulation foam in the construction market with growth rates about 15% higher than for Western Europe. In contrast, the market for polyurethane foams directed at appliances is relatively more important in Central and Eastern Europe. To some extent this reflects the faster growth of consumer culture purchasing appliances, but also reflects the lower labor cost environment of Eastern Europe attracting manufacturing assets for export to other geographies. To this point, Central and Eastern Europe export 2.5 times the volume (115 kMT) of foam as exported by Western Europe (45 kMT).

Like most Western economies, there is a high level of competition among multinational chemical producers. Unquestionably, the large German concerns, Covestro (formerly Bayer) and BASF, dominate the European commercial landscape, serving about 40% of the construction market and nearly 50% of the appliance market with planned expansion to increase total locally produced volume by over 10% by 2022. To be sure, other companies, including Dow and Huntsman, compete with Covestro and BASF in the European polyurethane chemical rigid foam markets. While the dominance of the German concerns may to a small extent reflect regional preference for a local producer, it also reflects the product mix that Covestro and BASF offer, which most closely suits local requirements. Polyurethane chemical feedstock producers from China are also developing a commercial presence in Europe, the main effect being to maintain a low-price environment for all competitors for the foreseeable future because of oversupply in Asia. Chinese firms currently supply about 10% of the locally produced polyurethane isocyanate building blocks.

The North American market for rigid foams (including the USA, Canada, and Mexico) totaled about 2.5 billion pounds in 2012 and about 3 billion pounds in 2015. This volume is dominated by the US market (Figure 8.5). The market dynamics are somewhat similar to those of Europe with the low labor cost Mexican economy producing a preponderance of appliance foams, which then get exported throughout North America. Large multinational concerns, such as Whirlpool, Samsung, LG, and Mabe (a company indigenous to Mexico), are significant customers for appliance foams targeted for local, but primarily export markets. The breakdown for market participation is provided is Figure 8.6. Canada and the USA are combined since their percentage breakdown by application is insignificantly different. The similarity between the participation of China and Mexico in their respective regions is striking and reflects their common participation roles in markets as purveyors of insulated appliances to mature and higher labor cost economies.

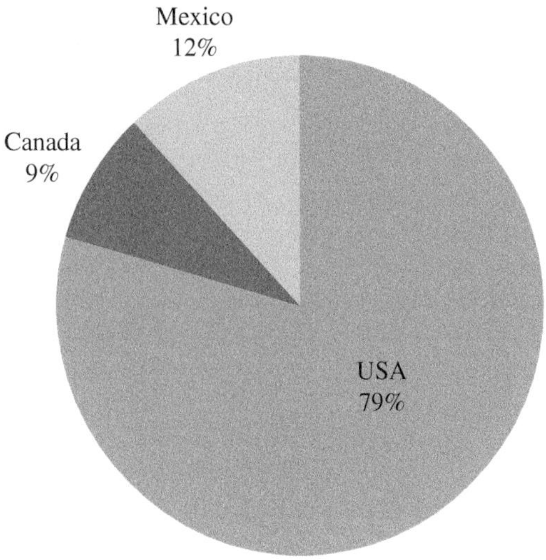


**FIGURE 8.5** Approximate breakdown for 2015 of the North American rigid foam market size.

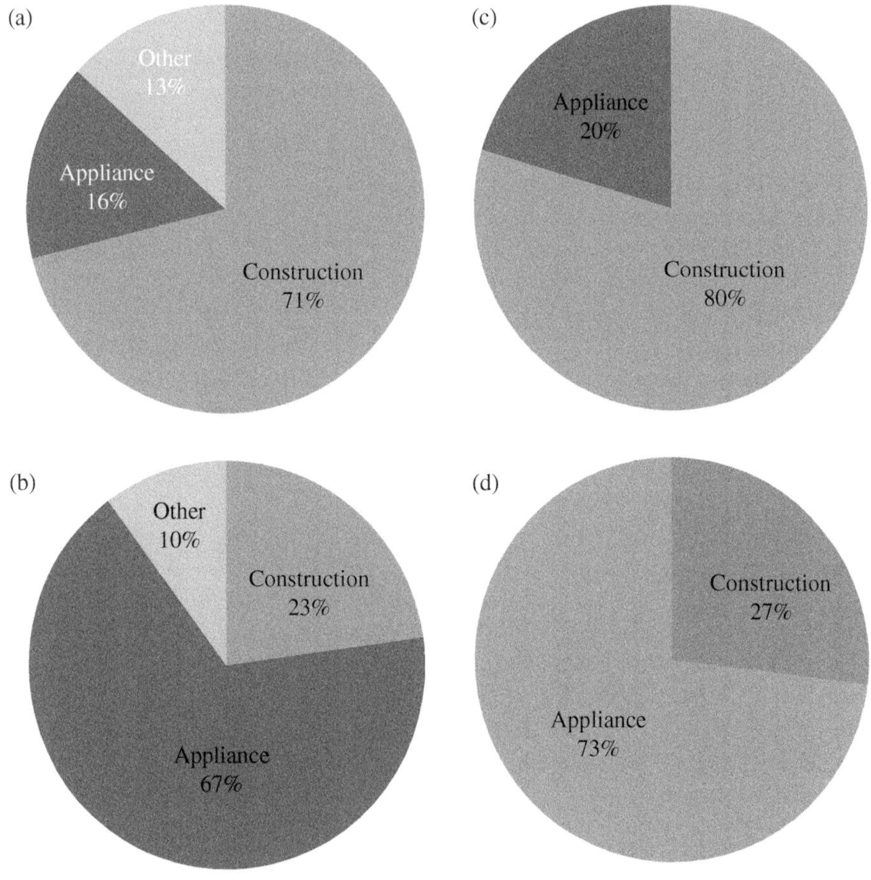


**FIGURE 8.6** Proportional consumption of polyurethane rigid foams in 2015 in (a) the mature consumer economies of the USA and Canada and (b) the consumer/export economies of Mexico; (c,d) a comparison with the similar economy relationships of Western Europe and China. "Other" represents automotive, pipe, packaging, and smaller applications.

## 8.2 APPLICATIONS

As mentioned in the previous section, the main use for rigid polyurethane foam is as a source of insulation. The primary industries utilizing rigid polyurethane foams are construction and appliance manufacturing [5]. Even applications under the heading "other" in Figure 8.6 serve the purpose of insulation for pipe and tank insulations or insulation from vibration and noise when used in automotive cavity filling. The foam differences in these different applications are minor compared with differences with, for instance, flexible foams. The differences that do exist are a result of optimization for the purpose of meeting performance or processing requirements unique to the particular application. For instance, spray foam insulations must provide a high degree of flame retardance and very rapid gelation so that they can be applied to vertical surfaces with minimal dripping or sag [6]. On the other hand, appliance foams must also form quickly, but not so quickly that the injected formulation cannot fill a mold cavity with any structural encumbrances to flow that may exist [7]. Furthermore, appliance rigid foams must contribute substantially to the overall appliance structural stability, but not deform the appliance exterior in any way. Finally, as a widely used consumer material with energy-saving functions, polyurethane rigid foams have been the target of governmental regulation, nongovernmental interest groups [8, 9], and performance requirements that have driven demand and innovation. The manner in which foam designers cope with these demands, some of which conflict with each other, is discussed in the following sections.

### 8.2.1 Construction Foams

Rigid polyurethane foams used for construction purposes must serve the insulation and energy conservation goals of the application and not act in any detrimental way to the overall structural integrity of the building [10]. Construction is an immensely complex market [11] and the development of polyurethane rigid foams has proliferated several methods of delivering insulation properties to the building site. Polyurethane rigid foams are produced in composite panels, in two-component spray foams, and more recently in aerosol spray cans.

***8.2.1.1 Polyisocyanurate Foams*** Virtually all urethane-based boardstock used in the construction industry, especially when used for roofing or walls, is based on polyisocyanurate chemistry [12]. The formation of isocyanurate structures is covered in Chapter 3. A high-functionality polymeric 4,4′-methylene diphenyl diisocyanate (pMDI) is employed. It is used in large stoichiometric excess in the overall foam formulation to facilitate reaction of one isocyanate with another. Isocyanate reaction to form ring structures is further promoted by employing catalysts known for specifically forming isocyanurate rings, as discussed in detail in Chapter 3. These catalysts are usually carboxylate salts or amine salts. Examples include potassium octoate and the 2-ethylhexoate salt of 2-hydroxypropyltrimethyl ammonium. The catalysts act by strongly complexing the electropositive isocyanate carbon, inducing a high enough negative charge at the isocyanate nitrogen to complex the carbon of another isocyanate propagating to an oligomer of 3 [13–15]. At that point the six-member ring closes to form the ring structure shown in Figure 8.7.

The isocyanurate structure is particularly well suited to construction applications because of its inherently greater flame retardancy and thermal stability than possessed by urethane or urea cross-links [16]. In addition, the high cross-link potential of isocyanurate structure (Figure 8.7) creates a more rigid board structure. The high isocyanate content used in the formulations furthermore creates the potential for very good adhesion to other components of a composite structure, such as might be employed to make a sandwich panel. Such composite materials could be aluminum or steel facers and fiber glass mats for additional dimensional stability and fire performance [5, 17].

Table 8.1 provides representative formulations for isocyanurate polyurethane rigid foams. Formulation 1 is what might be used in a commercial building and can be exposed to the building interior by virtue of the special design for flame retardance by using high levels of isocyanurate and added flame retardant. Formulation 2 is what might be utilized in residential insulation applications for external wall cavities. An aluminum-faced isocyanurate foam is provided in Figure 8.8; this shows the closed cells and the very thin aluminum facing employed for improved time dependence of the insulation properties and improved flame retardance.

The flexural modulus is also a measured property of the insulation foam composite structure. While the ultimate properties (i.e. strength, elongation at break, etc.) are useful for predicting conditions for catastrophic structural failure, the low strain properties are also useful for predicting the applicability of a structure for a specific application. As an isocyanurate sandwich board is a large composite structure the flexural modulus can provide this kind of practical information [18]. The equation for the modulus of a bending beam ($E_b$; as measured by a three-point bend test) is given by Equation 8.1, where $L$ is the beam length, $b$ is the beam width, $h$ is the beam height, and $m$ is the slope of the linear-elastic portion of the load deflection curve.

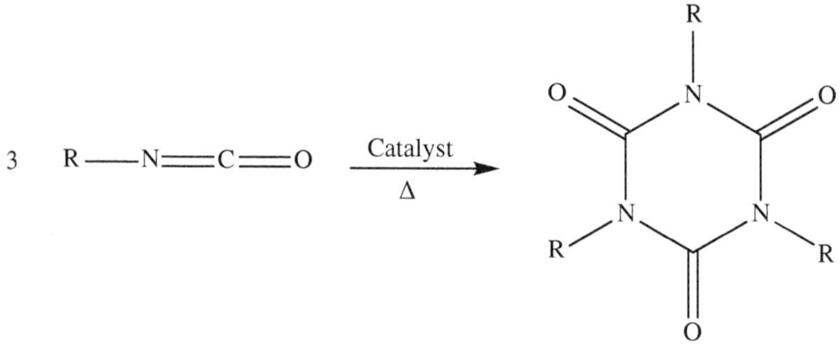


**FIGURE 8.7** Trimerization of isocyanate functionality to form an isocyanurate ring. For rigid foams, the R group would include additional unreacted isocyanate functionality capable of further reaction during polymerization.

**TABLE 8.1 Polyisocyanurate (PIR) foam formulations and properties for making insulation boardstock**

| Raw material | Formulation 1 | Formulation 2 |
|---|---|---|
| pMDI – average functionality = 3.2, iso equivalent weight = 138 | 173 | 346 |
| Cyclopentane/isopentane mixture (blowing agent) | 20–26 | 27–33 |
| Diol diester of phthalic anhydride with diethylene glycol (polyol) OH 227, functionality = 2 | 100 | 100 |
| Ester diol of tetrabromo phthalic anhydride/tris (1-chloro-2-propyl) phosphate flame retardant | 4 | 32 |
| Alkyl bromide (fire retardant) | 0 | 16 |
| Butylene oxide/ethylene oxide triblock surfactant | 2.31 | 4 |
| Water | 0.46 | 0.75 |
| Potassium acetate trimer catalyst | 3–6 | 5–7 |
| Isocyanate index | 2.75 | 4.5 |
| Density without facer ($lb/ft^3$) | 2.2 | |
| Compressive strength (with facer) (psi) | 20–40 depending on orientation of test relative to board manufacture and direction | |
| % closed cell (5) | >95 | |
| $R$-value/inch thickness depending on age of foam | 6–8 | |
| Flexural strength (psi) | 20–30 (depending on composite thickness) | |

Formulation 1 could be used in the slightly less demanding wall insulation application whereas formulation 2 could be used in a roofing application. In practice, many contractors use the most flame-resistant formula in all locations. The isocyanurate foam would normally be sandwiched between aluminum or polymer sheet and may have glass fiber matting within the composite structure.
pMDI, polymeric 4,4′-methylene diphenyl diisocyanate.

$$E_b = \frac{mL^3}{4bh^3} \tag{8.1}$$

In the case of an isocyanurate foam composite structure with the length and width of the structure held constant, the thickness of the insulation provides the main variable for controlling structural properties prior to failure. With the modulus varying as the inverse cubic power of thickness it is seen that the foam thickness dominates the properties of the structure (Figure 8.9) with small inclusions and flaws being minor perturbations.

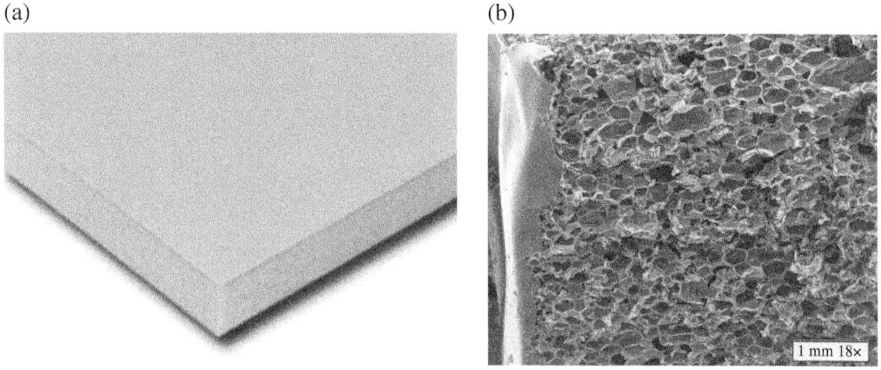


**FIGURE 8.8** (a) Photograph of an isocyanurate boardstock and (b) a scanning electron microscope image of the isocyanurate foam and the adhered aluminum film cover.

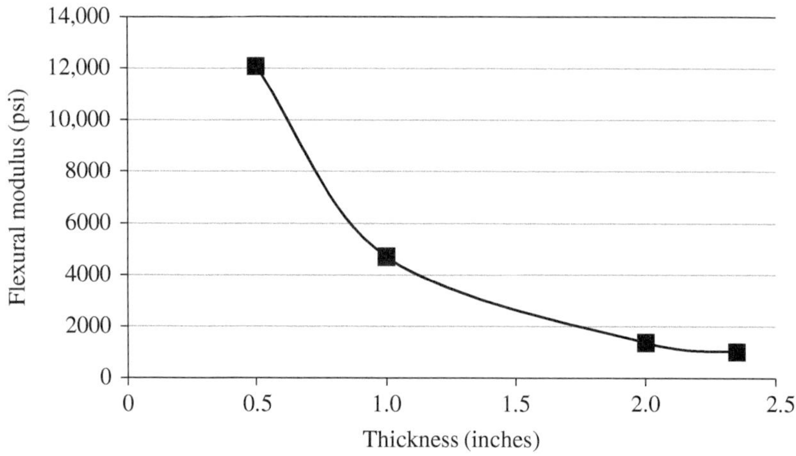


**FIGURE 8.9** Measured flexural moduli of isocyanurate board properties using 6 inch long by 2 inch wide samples with a deflection rate of 0.5 in/min.

Another property that a builder may try to influence is the overall insulation of a given structure. As Figure 8.10 shows, it is again the board thickness, which most affects the expected energy conservation, that can be designed in with the thickness–insulation relationship being linear. Isocyanurate foam (also called "iso" or "PIR" foam) is typically available in thicknesses from 0.5 inches to 2.375 inches. They are made in plant manufacturing operations such as illustrated in (albeit simplified) Figure 8.11. The operation is complicated by the numerous processes that are occurring simultaneously: the application of the reactants to the moving conveyor, the foam rise, the application of the laminating films, curing of the foam, development of adhesion to the film, and movement of the finished boards away from the process location. Economic efficiency requires the plant be used at high utilization and at high linear rates [19, 20]. Linear board rates of 60 m/min are possible; however, this feat of engineering optimization can still be economically disadvantaged if the plant capacity is not run at or near maximum utilization, since there are other insulations that, although less effective, may still be price advantaged since isocyanurate insulation is the most expensive insulation product commonly available. Isocyanurate foam may be delivered a commercial advantage if the California State Assembly Bill 127, signed into law in September 2013, results in further tightening of building flammability standards, as expected. In 2018, California enacted State Bill AB 2998 banning flame retardants from a list of consumer goods including mattresses.

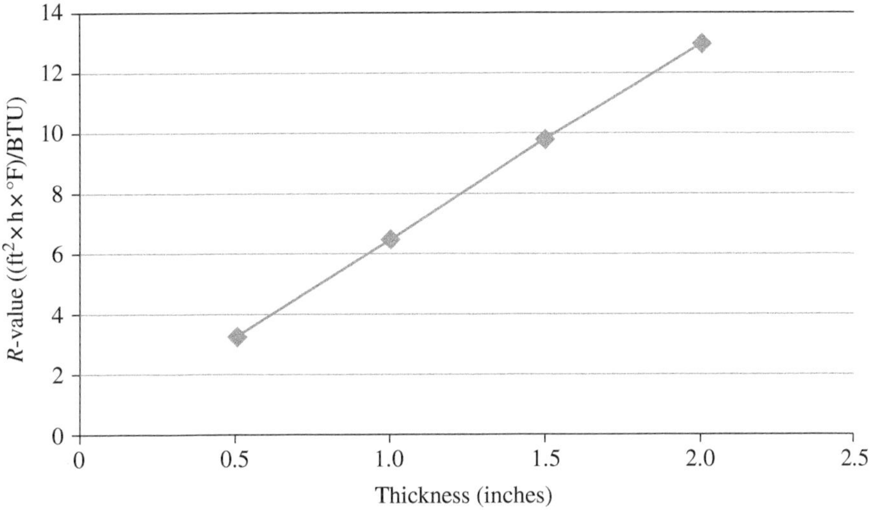


**FIGURE 8.10** Measured *R*-value (ASTM C518) as a function of board thickness for formulations 1 and 2 in Table 8.1.

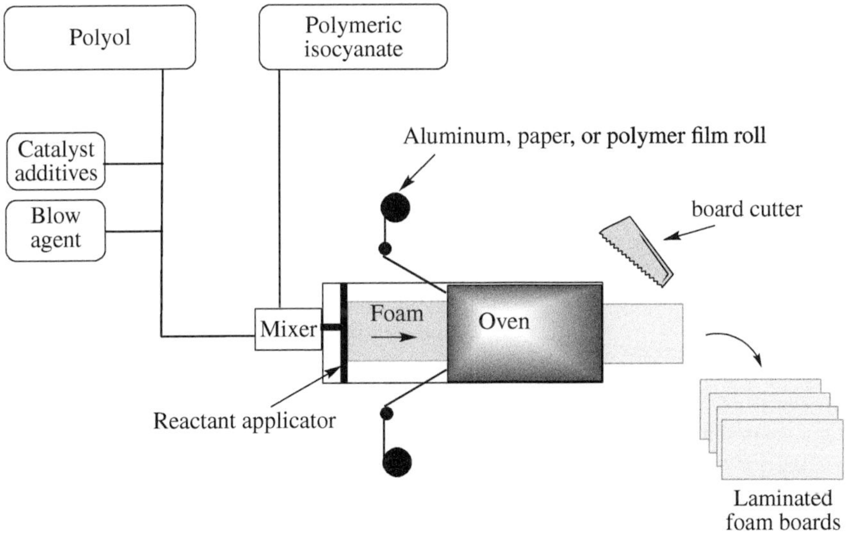


**FIGURE 8.11** Schematic of the process for making isocyanurate boardstock. A third roll is sometimes installed with glass matting that gets laid directly into the foaming mass.

*8.2.1.2* ***Spray, Poured, and Froth Foams*** While isocyanurate boardstock rigid foams dominate the construction market, other modes of application find significant uses. The most common alternative application methods are called "pour-in-place" (PIP) and spray or froth-spray. These methods both involve taking a formulated system of polyurethane building blocks, surfactants, catalysts, flame retardants, etc. and performing the foaming at the construction site. Since the applicator is transporting a relatively dense liquid to then make very-low-density foam, transportation costs are lower than with foam boardstock. Furthermore, these modes can be used to apply foam over wide irregular areas in the case of spray or to fill irregular cavities in the case of PIP.

*Spray Foams* Spray foams are employed for the manufacture of insulated structures where coverage of large areas is needed, where insulation of complex shapes is required, and where coverage without joints or seams is required. The foam is applied though a mixing head/nozzle connected by hoses to two or more tanks containing the foam formulation. Since the foam must be applied quickly over wide areas, the formulation exits the application nozzle at a high velocity after high-pressure impingement mixing. Typical operating pressures are from 700 to 2000 psi. Because of the need to flexibly deliver insulation from a central tank system of feedstocks to walls and roofs, hose lengths can be as long as 300 feet or more. Like many liquid delivery nozzles, the end of the nozzle has a fixture to deliver a specific pattern that can be adjusted and substituted. Foam delivered by a spray method is usually qualified by a battery of standardized tests among those listed in Tables 8.2 and 8.6. Specific qualifying spray foam tests are a reactivity profile, limiting oxygen index for flame ignition, flammability testing, smoke density testing, and trimer content (by Fourier transform infrared; see Chapter 5). The procedure for developing a spray foam formulation is not simple and can be laborious and highly iterative. Achieving a system that meets all qualifying tests simultaneously is a challenge. The task is usually broken into four parts, from which preliminary information and indications of performance can be gleaned prior to large-scale expensive flammability testing. This developmental procedure is summarized in Figure 8.12.

The rigid spray foam system is relatively simple in concept (Figure 8.13) but there are myriad details, which are difficult to convey in a picture. As with all polyurethane systems, the quality of what is obtained depends on strict control of stoichiometry, which in the case of spray systems requires careful calibration, proportioning, and control of liquid pumping systems. Furthermore, temperature and mixing conditions can likewise strongly influence foam properties. Lastly, it is the author's experience that skill in handling the spray gun and choice of application pattern can also have a strong influence on insulation appearance and quality.

The formulation of foam will to some extent be dictated by application, by building codes, by building insurance guidelines, and increasingly by nongovernmental interest organizations [21]. For many spray foams applied for roof insulation, the flammability inhibition must be very high (class 1; determined by the UL-94 test for flame spread and smoke generation), which can dictate the use of a relatively large amount of an optimized flame retardance/smoke-inhibiting additive package. Table 8.2 provides example formulations for spray foam roofing applications and the measured properties and performance of these formulations.

**TABLE 8.2 Formulation for producing spray foams**

| Composition | Formulation 1 | Formulation 2 |
|---|---|---|
| Mannich base-initiated, functionality = 3 | 16.1 | 46 |
| Equivalent weight = 124 equivalent weight = 155, functionality = EO/PO mixed feed | | |
| PO/EO alkoxylated ethylene diamine (EO capped) | 6.4 | 15.5 |
| Equivalent weight = 74 | | |
| Dimethyl terephthalate-initiated polyester polyol | 9.7 | 27.5 |
| Equivalent weight = 178 | | |
| Glycerine | 0.75 | 2.25 |
| Trichloroethyl phosphate | 4 | 11.5 |
| Dibromoneopentylglycol | 2 | 5.75 |
| 1,4-Triethylene diamine/dimethylethanolamine | 0.2 | 0.55 |
| Silicone surfactant (DC 193) | 0.35 | 1 |
| Water | 0.3 | 0.85 |
| Blowing agent | 10.5 | 30 |
| Polymeric isocyanate, functionality = 2.7 | 51.8 | 30 |
| Isocyanate index | 1.14 | 1.2 |
| Cream time (s) | 13 | 10 |
| Gel time (s) | 27 | 25 |
| Tack-free time | 36 | 35 |
| Density (lb/ft$^3$) | 1.5 | 1.93 |
| Compressive strength (psi) | | |
| Parallel to rise | 30.4 | 29.2 |
| Perpendicular to rise | 13.6 | 21.9 |
| *K*-factor (insulative value) | 0.155 | 0.129 |
| Dimensional stability 70 °C/95% humidity % volume change | 1.6 | 1.6 |
| ASTM E906 Burn test | | |
| Smoke | 2712 | 1927 |
| Heat | 303 | 406 |
| % Weight loss | 63 | 65 |

While the formulations are relatively different, the resulting foams only differ in some dimensions of performance, showing that the formulation will be driven by the particular needs of the structure.
EO, ethylene oxide; PO, propylene oxide.

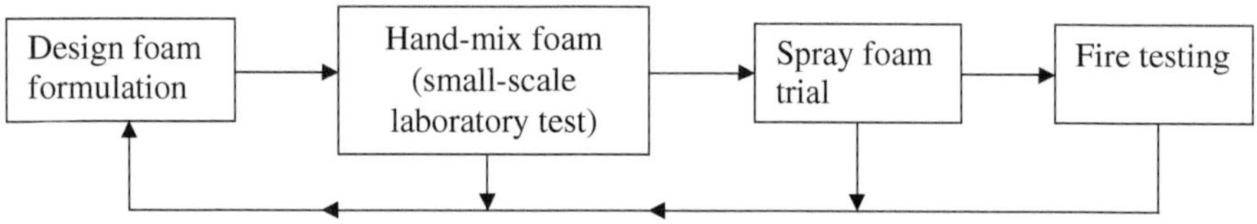


**FIGURE 8.12** Typical process for developing a polyurethane spray foam formulation.

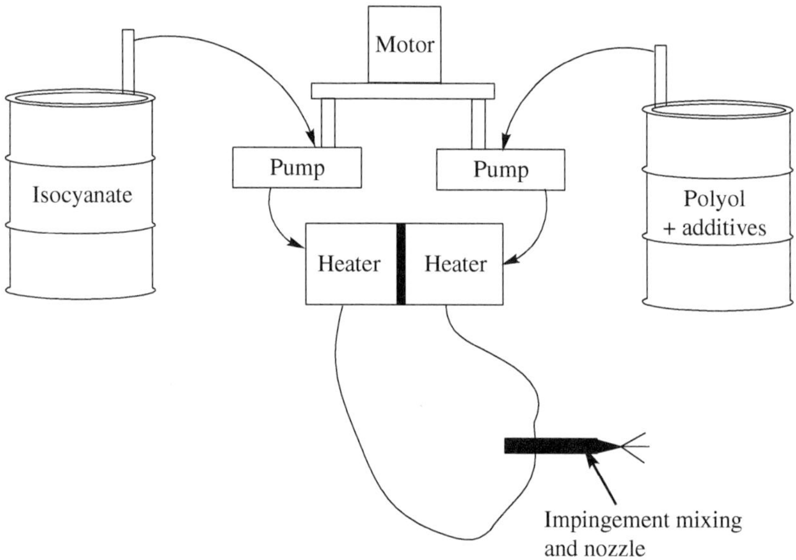


**FIGURE 8.13** Schematic of the components for application of a spray foam.

A specific polyol initiated from a Mannich base [22] is specified in the formulations of Table 8.2. This is a unique polyol used almost exclusively in spray foam applications. It is unique in that it is (i) autocatalytic (see Chapter 2), (ii) has high functionality and low equivalent weight, and (iii) is aromatic. These attributes contribute to the requirements of spray foam technology by (i) speeding the reaction to enable fast rise at low cure temperatures, (ii) increasing reaction rate and decreasing gel times to improve reaction speeds and allow application to vertical surfaces, and (iii) contributing to the required flame retardance of the overall system. Figure 8.14 details the synthesis and structure of the polyol. The Mannich polyol most established for the spray foam application is the reaction product of nonyl phenol, formaldehyde, and diethanolamine. Manufacturers of Mannich polyols provide formulation flexibility by varying the ratio of formaldehyde, ethanolamine, and alkoxylate to provide a specific equivalent weight and functionality.

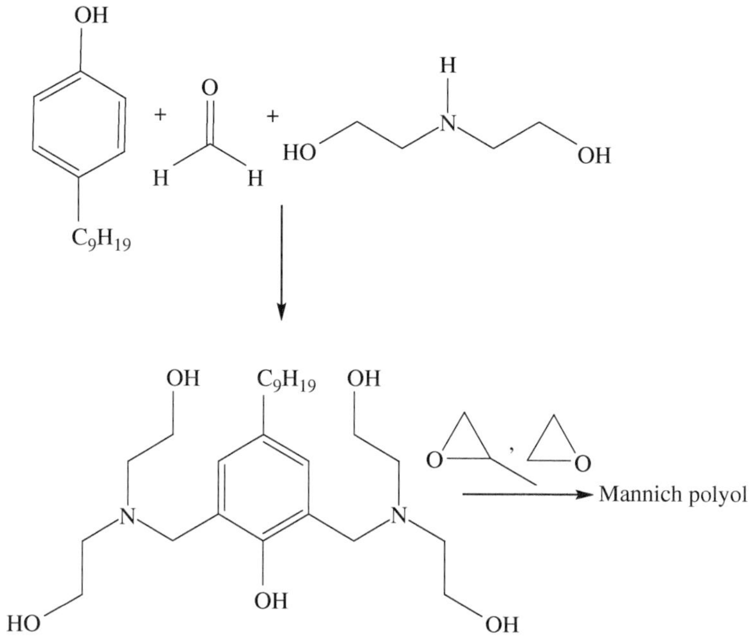


**FIGURE 8.14** Process for the production of Mannich polyols for use in spray foams.

*Froth Foams* Spray foams can be more efficient than board foams in some applications because of the inefficiencies associated with transportation of bulky low-density materials. However, in some circumstances spray foams can also be inefficient. For instance, spray foam applications tend to be large volume and large surface area. For applications where there are many relatively small areas to apply foam, a product has been established providing needed product convenience and cost effectiveness. Rather than the high-pressure charging of materials in conventional spray foam combined in impingement mixing, the low-pressure application (called froth foam) product is charged at a relatively low pressure to a nozzle that has a static mixer contained within the nozzle handle [23]. The mixed polyurethane streams exit the nozzle in a relatively lower volume targeted spray, which might allow insulation in a small crawl space ceiling, around a window frame, to a section of piping, to a small vessel, or other place for a spot application of insulation. The two-component mixture is called froth foam because of the formulation components being charged to their vessels with an insulating gas-blowing agent, and often a pad of inert gas such as nitrogen. The tank pressure is in the range of hundreds of pounds per square inch (psi) and this gas pressure is the motive for the movement of materials from the tanks to the spray gun. Flows are controlled by valves to make sure the relative proportions of the two components are at the levels specified by the manufacturer. As indicated, both components of the froth foam system are under pressure. As the gas/polyurethane system flows into the relatively lower pressure environment of the nozzle handle, gas bubbles will begin to evolve from the flowing streams, creating the froth. The frothing streams must mix in the static mixer within the handle. While the forming bubbles create some agitation in the liquid, their formation does not necessarily facilitate mixing. Foams are notoriously poor as a mixing medium [24–26] and the static mixer and handle design is a highly engineered device and an object of highly competitive and fiercely protected intellectual property. In fact, more than the formulations comprising the foam, the nozzle and the ergonomics of the froth foam packaging are the most intensely patented aspects of this technology [27–31].

While the nozzle design is highly protected as a marketing asset, there are certain basic features that make up a typical nozzle. Such a generic nozzle is pictured in Figure 8.15 and the basic form of the product is shown in Figure 8.16. The contents of the froth foam formulation are unique in that the foam does not depend on the chemical reactions that occur to make the polyurethane reduce density. While some froth foam formulations will have water, the levels are small, usually of the order of 1% or less. Instead, the froth foam depends on the expansion of the pressurizing gas and the blowing agent to create a porous structure. Some exemplary formulations are presented in Table 8.3 and a comparison of typical spray foam versus froth foam properties is given in Table 8.4.

(a) (b)

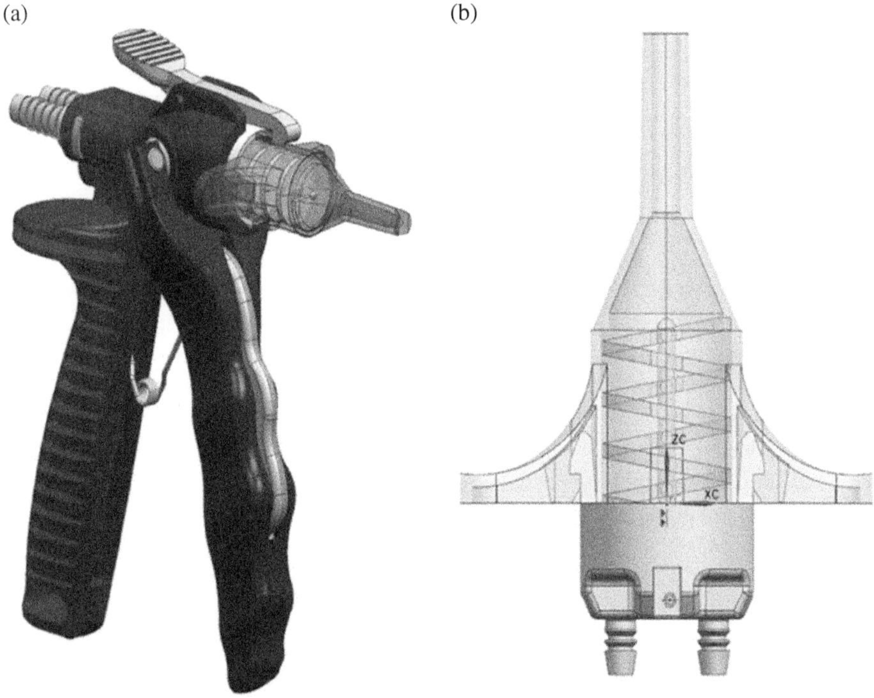


**FIGURE 8.15** Drawings of (a) the external view of a froth foam gun and nozzle and (b) cutaway showing the froth chamber and static mixer. Image courtesy of Peter Schultz, The Dow Chemical Co.

(a) (b)

**FIGURE 8.16** Form (a) of a purchased froth foam system and function (b) of a froth system in use. Image courtesy of The Dow Chemical Co.

**TABLE 8.3 Formulations for froth foam that might be use as a window sealant (Formulation 1) and as insulation foam for an attic space (Formulation 2).**

| Component | Formulation 1<br>Sealant applications | Formulation 2<br>Insulation applications |
|---|---|---|
| *Polyol side* | | |
| Polyester polyol terephthalate ester of diethylene glycol equivalent weight = 178 | | 17 |
| Sucrose/glycerine-initiated polyether polyol equivalent weight = 114, functionality = 4.3 | 12 | |
| Aromatic polyester polyol based on recycled PET 160 equivalent weight, functionality = 2 | 15 | |
| Sucrose/glycerine-initiated PO polyol equivalent weight = 151, functionality = 7 | | 22.6 |
| Sucrose/glycerine-initiated PO polyol equivalent weight = 155, functionality = 4.5 | 14 | |
| Trichloroethyl phosphate – flame retardant | 23 | 13.2 |
| Triethyl phosphate – flame retardant | 2 | 3.6 |
| Brominated diphenyl ether – flame retardant | | 7 |
| Dibutyl tin dimercaptide – gel catalyst | 0.7 | 1 |
| Potassium octoate in DEG – trimer catalyst | 2 | 3.4 |
| Penta methyl dipropylene triamine – blow catalyst | 0.7 | |
| Silicone – surfactant | 2 | 1.1 |
| Water | 0.8 | |
| 1,1,1,2-Tetrafluoroethane – blowing agent | 19 | 19 |
| 1,1,1,3,3-Pentafluoropropane – blowing agent | 8.4 | 12 |
| *Isocyanate side* | | |
| pMDI, functionality = 2.7 | 92 | 92 |
| 1,1,1,2-Tetrafluoroethane – blowing agent | 7 | 7 |
| Isocyanate stable silicone/polyether copolymer – surfactant | 1 | 1 |

DEG, diethylene glycol; PET, polyethylene terephthalate; pMDI, polymeric 4,4′-methylene diphenyl diisocyanate; PO, propylene oxide.

**TABLE 8.4 Comparison of typical properties of high-pressure spray foam and froth foams**

| Property | Spray foam (class I) | Froth |
|---|---|---|
| Gel time at 75 °F (s) | 8–10 | 8–10 |
| Rise time at 75 °F (s) | 12–15 | 40 |
| Aged *R*-value/in ($ft^2$ × h × °F/BTU) | 6.2 | 5.4 |
| Aged 90 days @ 140°F | | |
| Water vapor permeability ASTM E96 (perm-in) | 3 | <5 |
| Compressive strength ASTM D1621 ($lb/in^2$) | 25 | >15 |
| Closed cell content ASTM D 6226 (%) | 90 | 90 |
| Dimensional stability ASTM D2126 (% linear change) | 9 | <12 |
| Aged 7 days at 158 °F 97% relative humidity | | |
| Surface burning characteristic ASTM E84 | Class 1 | Class B (at 2 inches thick) |

*PIP Foams* PIP rigid construction foams fill a particular application space that other less expensive techniques cannot be modified to fill. PIP is a discontinuous foam production process that places foams in complex cavities less expensively than low-volume froth foam techniques, and much less expensively than highly capital-intensive continuous techniques [32, 33]. Like froth foams, the components are taken to the application site, mixed, and poured or sprayed into cavities. The polyurethane building block chemicals then foam to fill the cavities. The requirements of this process are unique but can overlap the other methods depending on the details of "the pour." One distinguishing feature of PIP is that the foaming and cure are somewhat slower than spray and froth applied foams since the foam must be able to flow into all the cavities and irregularities. Foam thickness in PIP applications can be significantly larger than other rigid foam applications. The applications that PIP methods find their largest uses in are shown in Figure 8.17.

The details of the cavity dictate the processing technique. The cavity-filling technique must effect uniform distribution of the mixed and reacting polyurethane, optimize flow to all parts of the cavity, allow for displaced air to escape, eliminate entrapped air, minimize void formation, and minimize cell orientation (anisotropy) [34]. Like high-pressure spray techniques, the two components must be mixed, and this is usually achieved using impingement techniques such as occur in spray polyurethane foam operations [35–37]. However, unlike spray foams, PIP exits the nozzle with a relatively slow forward velocity and with laminar flow. Also, unlike spray foam, PIP operates with a programmed time- or volume-controlled shot size. The foam is then injected into the cavity using procedures optimized for the particular shape. An alternative technique uses a lance

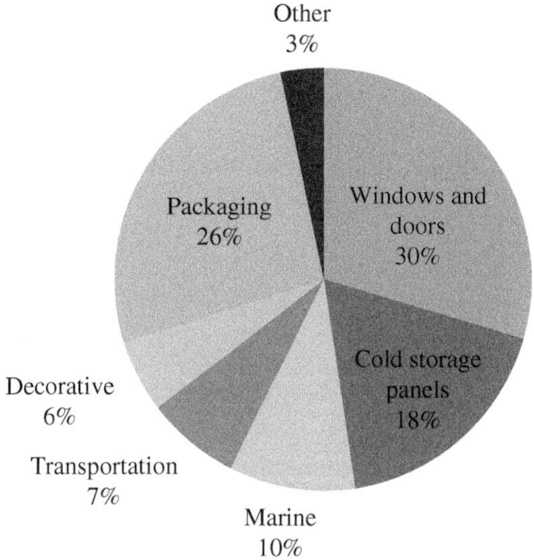


**FIGURE 8.17** Pour-in-place (PIP) application segments (nonappliance) in 2015. If included, PIP appliance applications would be about half of the total applied.

withdrawal approach. In this method, a hollow wand directs the polyurethane components into the cavity; the wand is moved about the volume and withdrawn as the components are deposited. This technique minimizes the flow requirements of the dispensed foaming mixture and is useful for narrow and long cavities of moderate to lower volumes.

As mentioned, relative to spray techniques, a discontinuous technique such as PIP requires slower reactivity. Relative to spray application, which must gel in less than 10 seconds, PIP foams gel in 90–120 seconds for larger cavities and 30–50 seconds for smaller cavities such as entry doors. The ideal foam insertion initiates volume expansion after completion of liquid injection to minimize pouring (and concomitant partial distortion and collapse) into a rising foam. The reacting foam should vitrify shortly after rise to minimize cell anisotropy (stretching in the rise direction) and resulting weakening of the foam. Lastly, the PIP foam must fill the space intended, but not deform the object containing it. For example, when poured into a window frame, a hollow entry door, a reefer container, or a boat hull cavity, it must not deform the object containing it as a result of overexpansion of the foam by overfilling or as a result of overpressure of the blowing gas. Such perfect performance is not usually realized, but is approached by the appropriate formulation choices. Table 8.5 provides example formulations of PIP foams for different uses. The flame-retardant requirements are often less for PIP applications than for wall and roof insulations, which have a larger surface, whereas PIP foams are encased in a containing envelope that also protects the foam from an initial ignition source [38].

**TABLE 8.5 Formulations and properties for three pour-in-place applications**

| Component | Water blown | Marine formulation | Door cavity formulation |
|---|---|---|---|
| *Isocyanate side* | | | |
| pMDI isocyanate equivalent weight = 134 g/eq, functionality = 2.7 | 163.5 | 136.5 | 136.3 |
| 1,1,1,2-Tetrafluoroethane blowing agent | | 0.83 | 0.83 |
| Isocyanate stable silicone/polyether copolymer surfactant | | 4.7 | 5.25 |
| *Polyol side* | | | |
| Aromatic phthalate polyester polyol OH equivalent weight = 211 g/eq, functionality = 2.3 | | | 54 |
| Mixture of sucrose/glycerine and glycerine-initiated polyols OH equivalent weight = 167, functionality = 3.6 | 100 | | |
| Sucrose/glycerine-initiated equivalent weight = 155, functionality = 4.5 | | 50 | 27 |
| *o*-Toluene diamine-initiated equivalent weight = 143, functionality = 4 | | 50 | |
| Glycerine-initiated PO polyol OH equivalent weight = 85, functionality = 3 | | | 5.5 |
| Mix of sucrose/glycerine and sucrose-initiated PO/EO-capped polyol equivalent weight = 200, functionality = 7 | | | 13.5 |
| Tris (2-chloropropyl)phosphate flame retardant | 5 | 18 | 17 |
| Bis aliphatic epoxy | | | 1.4 |
| Dimethylethanolamine reactive catalyst | 0.1 | | |
| Pentamethyl diethylene triamine catalyst | | 0.1 | |
| *N,N*-dimethyl cyclohexyl amine | | 0.06 | |
| 70/30 1,2-dimethylimidazole in DEG | | | 0.25 |
| Silicone surfactant, i.e. DABCO™ DC 193 | | 1.9 | 2.1 |
| $EO_{38}/BO_{46}/EO_{38}$ triblock polyol surfactant | 5 | | |
| Water | 4 | | |
| 1,1,1,2-Tetrafluoroethane blowing agent | | 22 | 18 |

| Property | Water blown | Marine formulation | Door cavity formulation |
|---|---|---|---|
| Gel time (s) | 365 | 150 | 220 |
| Tack-free time (s) | 731 | 237 | 418 |
| Density (lb/ft$^3$) | 2.3 | 1.6 | 1.8 |
| Initial *K*-factor | 0.21 | 0.16 | 0.16 |

The water blown formulation would be considered for applications where insulation is not a primary consideration as in automotive support members.

BO, butylene oxide; DEG, diethylene glycol; EO, ethylene oxide; pMDI, polymeric 4,4′-methylene diphenyl diisocyanate; PO, propylene oxide.

Water-blown PIP technologies exist in the marketplace. In some cases, they can have a price advantage over new-generation fluorocarbon blowing agent formulations owing to the high prices of blowing agents. Water-blown PIP can be disadvantaged by having significantly longer cure times, higher densities, and higher $K$-values (lower $R$-values – poorer insulation properties). In addition, they can often have significantly higher viscosity, which complicates the efficiency of mixing and nozzle delivery. Of course, blowing agent formulations along with being more expensive also require the use of pressurized tanks to maintain the system sides in a liquid state.

### 8.2.2 Rigid Construction Foam Market Segments

Reliable data on the segment volumes are difficult to obtain with a wide range of values available in market surveys. All surveys find that the majority of polyurethane rigid foams in the construction market are isocyanurate with volume percentage ranging from c. 35% to c. 65%. Available data are incomplete, but a rough average is presented in Figure 8.18, which shows historical context and consistency over time and the role of geography in application preferences.

The construction rigid foam market is primarily segmented on the structural part being insulated. This segmentation breaks into roofing, walls, tanks and pipes, refrigerated buildings, and metal or fiber-glass doors. The breakdown in utilization is provided in Figure 8.19. While the details of this distribution change slightly over time, the percentage breakdown given in the figure has been relatively stable for 20 years. Door insulation is well under 1% of the total rigid foam usage and is neglected in this figure to simplify the representation.

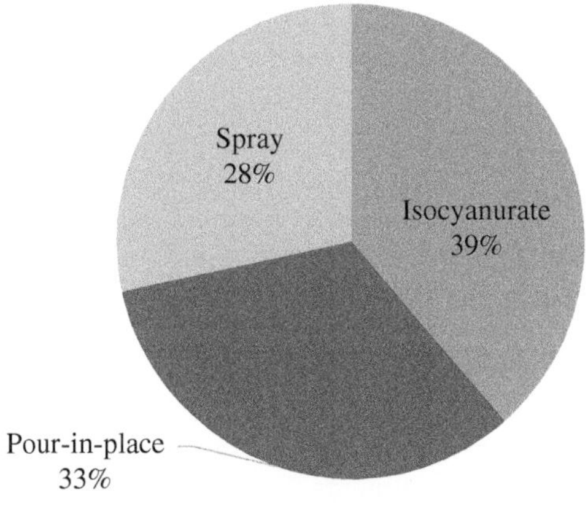


**FIGURE 8.18** Global segmentation of rigid construction foams globally in 2015. Geographical differences may reflect capital investment avoidance for spray systems in some regions.

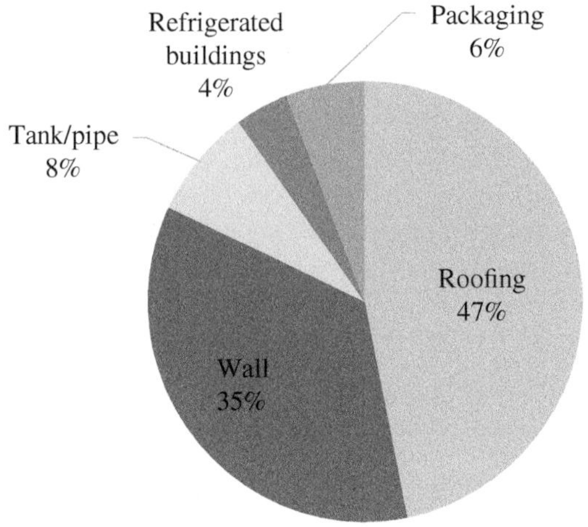


**FIGURE 8.19** Segmentation of the polyurethane rigid foams for construction market by volume consumed in 2015.

With roofing and wall insulation representing virtually all of the construction market for polyurethane rigid foams, the regulatory standards for performance are critical to successful commercial positioning. Table 8.6 is a list of relevant tests for determining the suitability, performance, and risk associated with insulation systems such as polyisocyanurate systems. Individual states and countries may require additional testing or performance to meet local requirements. Along with standard testing methods issued by national organizations such as ASTM (formerly American Society for Testing and Materials) and European and Asian testing organizations, there also exist trade organizations such as ASHRAE (American Society of Heating, Refrigerating and Air-Conditioning Engineers) promulgating standards. Aside from characterization of the foam structure, their standards apply to the overall performance of the structure which the foam insulation must be a part of meeting targets [39]. Several important tests commonly applied to construction foams are listed in Table 8.6.

### 8.2.3 Appliance Foams

Appliance foams have similarities to PIP foams in that they are used to fill a cavity and that they must not deform the cavity during foam rise as a result of expansion pressures or during contraction as the foam cools. The majority of polyurethane appliance foams are used for refrigeration, including refrigerators, freezers, coolers, refrigerated trucks, and railcars [40]. Another category is used for water heaters. Figure 8.20 breaks down the approximate volume ratios that occupy these two market segments. While there is a continuing demand for these appliances, the aggregate demand follows the construction market, and particularly new home construction. The household appliance import–export ratio in the USA is 6.2 – one of the most

**TABLE 8.6 List of standard tests and descriptions relevant to the polyurethane rigid foam construction applications**

| Test | Standard description |
|---|---|
| ASTM C1289–13 | Standard specification for faced rigid polyisocyanurate thermal insulation board including general industry requirements for materials and performance. Only for faced boards |
| ASTM C591–12 | Standard specification for unfaced preformed rigid cellular polyisocyanurate thermal insulation including general requirements, specifications, and dimensions for unfaced boards only |
| ASTM C1303–12 | Standard test method for predicting long-term thermal resistance of closed-cell foam insulation. Only applicable to unfaced or permeably faced insulation. Uses accelerated aging conditions |
| ASTM E1730–09 | Standard specification for rigid foam for use in structural sandwich panel cores. Structural testing of composite structural insulation panels, including impact, flame resistance, thermal conductivity, compressive, tensile, and shear strength |
| ASTM E84–13 | Standard test method for surface burning characteristics of building materials. Determines the relative flame spread and smoke density generation produced. Uses long (24 ft) and wide (20 in) specimens |
| ASTM E662–13 | Standard test method for specific optical density of smoke generated by solid materials. Quantification of the attenuation of a light beam by smoke in a closed chamber |
| ASTM E1465–08 | Standard practice for radon control option for the design and construction of new low-rise residential buildings. Primarily a piping system for detouring radon from penetrating the structure. Piping may need insulation, and radon buildup may be made more severe by improved insulation |
| ASTM E1678–10 | Standard test method for measuring smoke toxicity for use in fire hazard analysis. Collection and analysis of smoke followed by animal testing for $LC_{50}$ for exposure to smoke |
| ASTM E906 | Also called "the OSU burn test"; standard methods for heat and visible smoke release rates for materials and products |
| DIN 4102–1 B3 | German standard fire test to building materials [sic]; standard defines fire behavior classes and specifies requirements and test methods for each class. Urethane-based materials generally fall under B3 classification, suggesting intrinsic flammability |
| ISO 5660 | International Organization for Standardization; reaction-to-fire test, heat release, smoke production, and mass loss rate (cone calorimeter method). Measure time to ignition and heat release rate based on oxygen consumption in a calibrated gas flow |
| UL-94 | Underwriters Laboratory test for plastics flammability using small specimens (5 in × 0.5 in); measures tendency of a material to self-extinguish or spread a flame once ignition occurs |
| ASHRAE 90.1 | Energy standard for building except low-rise residential buildings. States apply the standards differently over the range of building purposes |
| BS EN 13823 | British Standard for materials of construction, relatively small scale; measures heat output and smoke generation in a corner geometry |

$LC_{50}$ is the concentration of a compound that is lethal for 50% of an exposed population.

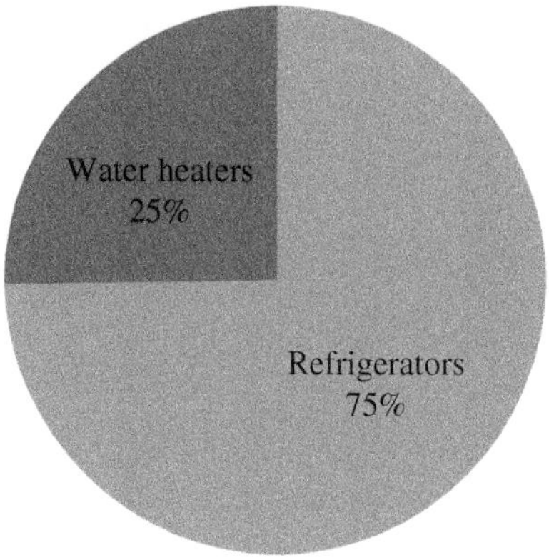


**FIGURE 8.20** Percentage consumption of polyurethane rigid foams for appliance insulation in 2015.

adverse trade ratios in manufacturing. Commercial trends in these markets relate to consumers eating at commercial establishments, obtaining second refrigerators and freezers, growing use of refrigerated vending machines and display cases, and worldwide rising disposable income [41].

The typical components of appliance foams are also similar to PIP foams with the notable exception that appliance foams are expected to provide little in the way of flame retardance relative to construction foams, or even flexible foams used in furniture. What they are expected to provide is (i) the ability to rise and cure rapidly with no deformation of the refrigerator or water heater cabinet and (ii) a high degree of insulation performance. Along with energy savings to the customer, highly efficient insulation allows the manufacturer to specify a smaller, less expensive compressor for refrigeration to maintain a larger capacity. Optimum design can also help make up for poorer insulation efficiency arising from the use of environmentally friendly, if less efficient, blowing agent/insulating gases. A discussion of insulating gases comes later in this chapter. Table 8.7 provides example appliance foam formulations.

The data in Table 8.7 are actual values for the foam formulations given, but the properties should be viewed as representative and illustrative of the fact that blowing agents are a major component of the formulation and can have a major influence on the formulation and the properties. Furthermore, different compartments of an insulating unit (i.e. the freezer cabinet versus the freezer door) may require more or less efficiency and the manufacturer may inject more than one foam formulation to optimize performance and price.

Table 8.8 provides a general guide to property requirements for this segment. Properties achieved with any particular formulation are variable and will depend greatly on numerous process specifics, including the details of the injection, the shape of the mold to be filled, the ambient temperature, and even the geographic elevation of the manufacturer making the insulation foam. While the properties may vary, the general trend in formulation optimization is indicated in Table 8.8 as designers continuously try to enhance foam efficiency along all performance dimensions.

**TABLE 8.7 Formulations and properties for use in appliance insulation applications**

| Formulation 1 components | Formulation 1 (wt%) | Formulation 2 components | Formulation 2 (wt%) |
|---|---|---|---|
| Sucrose/glycerine initiated equivalent weight = 108 g/eq all PO, functionality = 5.2 | 33.8 | Propoxylated sorbitol equivalent weight = 116 g/eq, functionality = 6 | 63 |
| Propoxylated ethylene diamine equivalent weight = 87, functionality = 4 | 10.35 | Propoxylated glycerine equivalent weight = 360 g/eq, functionality = 3 | 25 |
| Toluene diamine initiated equivalent weight =121, functionality = 4 | 7 | Propoxylated ethylene diamine equivalent weight = 112 g/eq, functionality = 4 | 5 |
| Aromatic polyester PO phthalic anhydride with diethylene glycol equivalent weight = 185, functionality = 2 | 17.2 | Water | 2.5 |
| Pentamethyldiethylene triamine (blow catalyst) | 0.55 | 2,4,6-Tris-dimethylamino phenol (delayed trimer catalyst) | 0.7 |
| *N,N*-dimethylcyclohexyl amine (gel-balanced catalyst) | 0.8 | Pentamethyldiethylene triamine (blow catalyst) | 1.3 |
| Potassium acetate in ethylene glycol (trimer catalyst) | .2 | Potassium octoate in diethylene glycol | 0.1 |
| 2,4,6-Tris-dimethylamino phenol (delayed trimer catalyst) | 0.8 | Polyether-modified polysiloxane copolymer (surfactant) | 2.5 |
| Polyether-modified polysiloxane copolymer (surfactant) | 1.8 | 85%/15% Cyclopentane/isopentane (blowing agent) | 14.5 |
| Water (blowing agent and chain extender) | 1 | | |
| 1,1-Dichloro-1-fluoroethane (blowing agent) | 26.5 | | |
| pMDI equivalent weight = 134 g/eq (iso index = 1.3), functionality = 2.7 | 118 | pMDI equivalent weight = 134 g/eq (iso index = 1.3), functionality = 2.7 | 146 |
| Molecular weight per cross-link | | | 348 |
| Cream time (s) | 6 | Cream time (s) | 4 |
| Gelation time (s) | 39 | Gelation time (s) | 47 |
| Tack-free time (s) | 45 | Tack-free time (s) | 60 |
| Density ($lb/ft^3$) | 1.9 | Density ($lb/ft^3$) | 2.0 |
| $K$ (25 °C (BTU in)/(h $ft^{-2}$ °F) | 0.12 | $K$ (25 °C (BTU in)/(h $ft^{-2}$ °F) | 0.16 |
| Compressive strength, $x$,$y$-axis (psi) | 22.5, 32 | Compressive strength, $x$-axis (psi) | 19, 34 |

Formulation 1 employs a chlorofluorocarbon blowing agent and has superior insulation properties, while formulation 2 uses a hydrocarbon blowing agent. pMDI, polymeric 4,4′-methylene diphenyl diisocyanate; PO, propylene oxide.

**TABLE 8.8 Typical range of rigid polyurethane foams for appliances and the direction foam designers are trying to improve their foams in that dimension**

| Property | Units | Value range | Direction of improvement |
|---|---|---|---|
| Thermal conductivity | (BTU in)/(h ft$^{-2}$ °F | 0.12–0.16 | ↓ |
| Density | lb/ft$^3$ | 1.8–2.2 | ↓ |
| Compressive strength | psi | >16 | ↑ |
| Demold time | s | 190–240 | ↓ |
| Post demold expansion | (%) | 0–3 | ↓ |

While there are ASTM tests for measurement of thermal insulation panel performance such as ASTM C518, in the case of polyurethane rigid foams for appliance applications, the manufacturer will often rely on internal tests of performance. Rather than a screening test on the foam component, the test will incorporate the entire molded and foamed cabinet. This makes sense since energy leakage from, or into, a container such as a refrigerator may depend on the envelope materials and the overall design of the cabinet. Many of these tests are performed to conform to regulatory agency mandates requiring explicit evaluation and communication with consumers about expected energy consumption. An example of such protocols is the US Department of Energy's "Uniform Test Method for Measuring the Energy Consumption of Electric Refrigerators and Electric Refrigerator-Freezers" 10 CFR Appendix A1 to Subpart B of Part 430. A screening test for this battery of standardized tests and conditions is the test outlined by the Association of Home Appliance Manufacturers (AHAM), which is referred to as the reverse heat leakage (RHL) test. The RHL test is performed on a complete cabinet without an installed compressor. A light bulb and energizing circuit are placed inside the insulated cabinet, which is then placed in a cold room (Figure 8.21). The light bulb is energized and the amount of energy required to maintain the interior at an elevated temperature is recorded. While still a bit involved and requiring a high level of control, this is a relatively low-cost method to screen the thermal efficiency of a foam-insulated appliance. The overall refrigerator conductance ($R_c$) is calculated from this experiment using Equation 8.2, where the temperature is usually an average of several measurements made in the corresponding chambers [42].

$$R_C = \frac{q_{\text{heater}}}{T_I - T_A} = \frac{q_{\text{walls}}}{T_I - T_A} \tag{8.2}$$

The reverse heat loss measurement correlates very strongly to the thermal conductivity of the blowing agent insulating gas employed for making the rigid foam, as shown for a controlled dataset in Figure 8.22 [43].

Figure 8.22 shows the importance of the blowing agent on the insulation performance and on the manufacturer's ability to meet regulator and customer expectation.

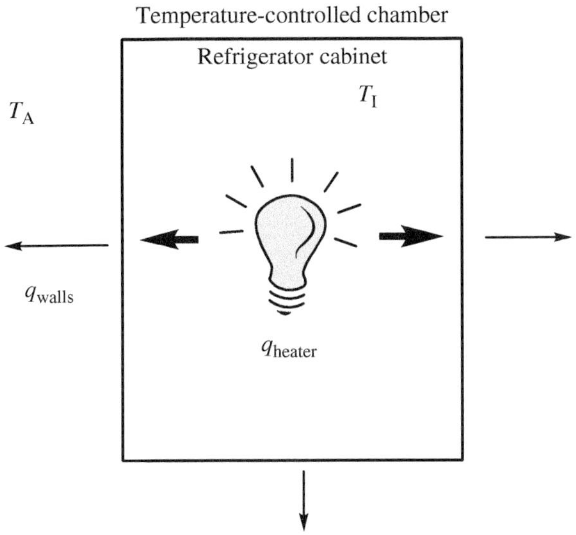


**FIGURE 8.21** Illustration of the concepts in a reverse heat loss test for refrigeration insulation efficiency.

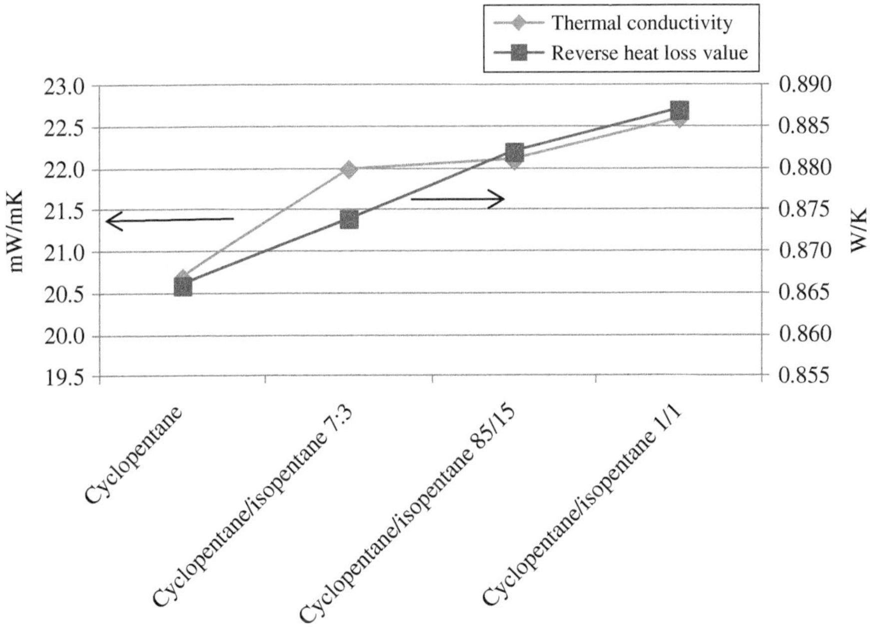


**FIGURE 8.22** Relationship between measured blowing agent thermal conductivity and the results for the reverse heat loss test using foams blown with the same blowing agents.

In fact, as will be shown later, blowing agent technology and associated formulation adjustments make up the field of most intense patent activity for the large polyurethane building block providers in spite of the fact that none of them is a blowing agent manufacturer [44]. In addition to the RHL test procedure, the US Department of Energy and the US Environmental Protection Agency operate the "Energy Star" program. Energy Star certification utilizes a different testing protocol from the RHL test, instead measuring energy demand response to a number of normal operations of a working refrigerator such as making ice.

## 8.3 BLOWING AGENTS AND INSULATION FUNDAMENTALS

### 8.3.1 Blowing Agents

As Figure 8.22 shows, the manufacturer and consumer have a significant stake in the choice of blowing agent that is made. The manufacturer must meet regulatory, cost, and quality requirements in production. At the same time, the economics of ownership are affected greatly by insulation efficiency [45–47]. In addition to these prosaic considerations, there is the recognized influence that blowing agents can have on the global environment. There is particularly recognition that certain blowing agents that were previously ubiquitous are capable of depleting ozone present in the upper atmosphere [48–50]. While ozone in the lower atmosphere is regarded as a pollutant, in the upper atmosphere (stratosphere) ozone serves as a necessary filter on high-energy ultraviolet (UV) B radiation. In 1973, chemists F Rowland and M Molina showed that the high vapor pressure and stability of chlorofluorocarbons, commonly used as blowing agents and propellants, could be dissociated by UV radiation, releasing a chlorine atom that proceeded to break down ozone [51]. The removal of ozone from the upper atmosphere, responsible for filtering of UV radiation, would result in a greater flux of UV radiation at the Earth's surface, increasing the potential for skin cancer and cell damage in living things.

In 1987 an international treaty was signed [52] to protect the ozone layer by phasing out production and the use of chemicals capable of depleting stratospheric ozone. Since then it has undergone numerous revisions and updates and has been ratified and upheld by almost all nations. Among the revisions, the protocols have recognized that, in addition to their potential for ozone depletion (ODP), chemicals useful for blowing agents can be scrutinized for their contributions to global warming (GWP) with several blowing agents showing orders of magnitude more global warming potential than $CO_2$. Recent attention has also focused on the atmospheric lifetime of blowing agents. Since the effect of this treaty and its revisions has a major effect on the industrial practice of making polyurethane foams for insulation, it is of major interest to practitioners of polyurethane science.

Table 8.9 is a partial compendium of blowing agents that have been used for preparation of rigid polyurethane insulation foams. Other blowing agents have been explored and have been mixed in attempts to obtain synergistic effects or dilute the different negative issues one agent has without harming the positive attributes that may come in the pairing.

**TABLE 8.9 Properties of rigid polyurethane foam blowing agents, several of which are banned or scheduled for phase-out**

| Blowing agent | Chemical formula | Molecular weight (g/mol) | Boiling point (°C) | *K*-factor (mW/mK) | ODP | GWP |
|---|---|---|---|---|---|---|
| CFC-11 | $Cl_3CF$ | 137.5 | 23.8 | 7.9 | 1.0 | 3400 |
| HCFC-123 | $C_2HCl_2F_3$ | 152.9 | 27.9 | 9.5 | 0.02 | 93 |
| HCFC-141b | $CH_3$-$CCl_2F$ | 116.9 | 32 | 8.7 | 0.11 | 610 |
| HCFC-142b | $CH_3$-$CF_2Cl$ | 100.5 | –9.8 | 11.6 | 0.065 | 1600 |
| HCFC-134a | $C_2H_2F_4$ | 102 | –26.5 | 14.5 | 0 | 1320 |
| HFC-245fa | $CHF_2CH_2CF_3$ | 134 | 15 | 12.4 | 0 | 1020 |
| HFC-365mfc | $C4H_5F_5$ (1,1,1,3,3-pentafluorobutane) | 148 | 40 | 10.5 | 0 | 782 |
| FEA™-1100 | 1,1,1,4,3,44-hexafluoro-2,3-butene | 164 | 33 | 10.7 | 0 | 9 |
| Carbon dioxide | $CO_2$ | 44 | –78 | 16.5 | 0 | 1.0 |
| Air | $N_2/O_2$ | 28.8 | –192 | 29 | 0 | 0 |
| *n*-pentane | $C_5H_{12}$ | 72 | 36 | 14.8 | 0 | 20 |
| *Iso*-pentane | $C_5H_{12}$ | 72 | 28 | 14 | 0 | 11 |
| Cyclopentane | $C_5H_{10}$ | 70 | 50 | 12.9 | 0 | 10 |
| HFO-1233zd | $CF_3$-CH=CClH | 130 | 19 | 10.2 | 0 | 0 |
| HFO-1234ze | *Trans*-1,3,3,3-tetrafluoropropene | 114 | –19 | 13 | 0 | 6 |

CFC, chlorofluorocarbon; GWP, relative global warming potential; HCFC, hydrochlorofluorocarbon; HFC, hydrofluorocarbon; HFO, hydrofluoroolefin; ODP, relative ozone depletion potential.

### 8.3.2 Blowing Agent Phase-Out Schedule

Government regulation and special interest group activism drives a significant amount of technology innovation for companies involved with rigid polyurethanes. As will be discussed in the next section, insulation efficiency requirements and blowing agent restrictions are inextricable. Approximately 75% of the insulative potential of a foam depends on the foam structure and the properties of the gas within the closed cells, both of which the blowing agent has strongly causative relationships with. Regulations governing industrial use of blowing agents are particularly relevant to polyurethane technology because they govern not only current use but also future use, since whole classes are designated for banned use in future years [53]. Table 8.10 is a compendium of relevant historical and future restrictions. Future restrictions will provoke a significant amount of reformulation of polyurethane systems along with development of new classes of blowing agent. Along with insulation potential, a blowing agent must also possess sufficient solubility in foam formulations to result in low-density foam, must be acceptable from an environmental health and safety

**TABLE 8.10 Promulgation of regulations for blowing agents previously used for polyurethane rigid foams**

| Date | US Regulation |
|---|---|
| 1/1/1994 | Banned production of bromochlorodifluoromethane (halon 1211), bromotrifluoromethane (halon 1301), and dibromotetrafluoroethane (halon 2402) (and isomers) |
| 1/1/96 | Production of CFCs banned except for very critical applications such as emergency inhalers and research |
| 1/1/2004 | Restricted use of 1,1-dichloro-1-fluoroethane (HCF-141b) by 35%; 75% reduction by 2010; and total phase out by 2040. With alternatives available, the US EPA *does not* consider it an acceptable substitute for CFCs in polyurethane foam blowing. No addition production or importation of HFC-141b |
| 2015 | No production or importation of any other HCFC except as refrigerants in equipment manufactured prior to 2020 |
| 2020 | No production or importation of HCFC-142b and HCFC-22 |
| 2030 | No production or importation of HCFC, resulting in 100% reduction in HCFC consumption and production from baseline at signing of Montreal Protocol |
| **Date** | **European Regulation** |
| 1/1/2005 | Phase out of 1,1,dichloro-1-fluoroethane and other hydrochlorofluorocarbons. Different nations within the EU adopted different schedules (often faster). Allowing for only very limited production and consumption after 1/1/2010, and for some banned after 2020 and others banned after 2030 |
| 2010 | No production or importation of HCFC-142b and HCFC@-22 except for use in equipment manufactured before 2010 |
| 1/1/2012 | Broad regulation affecting HFCs and perfluorocarbons proposed, including HFC-245, HFC-134a, and HFC-365mfc.This is a tightening of regulations previously proposed in 2006 |
| 2015 | Reclaimed and recycled HCFC can no longer be used for maintenance or servicing of refrigeration or air-conditioning equipment within the EU |
| 2017 | The EU achieved 2020 goals for phase-out of ozone-depleting substances by 2017; the European regulations are considered more stringent than those of the Montreal Protocol (Regulation (EC) no. 1005/2009) |
| **Date** | **Japanese Regulations** |
| 1/1/1994 | Phase-out of bromochlorodifluoromethane (halon 1211), bromotrifluoromethane (halon 1301) and dibromotetrafluoroethane (halon 2402) (and isomers) |
| 1/1/1996 | All CFCs phased out |
| 1996 | Began phase-out of HCFC consumption with gradual phase-out to 65% of 2003 consumption by 2009, 35% of 2003 consumption by 2014, 10% of 2003 consumption by 2019, and total phase-out by 2020 |
| 2016 | Signs on to amended Montreal Protocol to phase out HCFCs by 85% by 2036 |

*Source:* United States Environmental Protection Agency (EPA) 8/2019.
CFC, chlorofluorocarbon; EU, European Union; HCFC, hydrochlorofluorocarbon;
HFC, hydrofluorocarbon.

standpoint, must have acceptable flammability characteristics, and must be compatible with equipment and other formulation additives. The dates listed are not universally implemented on the same time schedule but, given the commercial importance of the US and European markets for polyurethane rigid foams, they certainly are milestones for industrial use of blowing agents. In fact, some countries have implemented restrictions on accelerated time schedules.

An alternative view is to look at which blowing agents are allowed and the typical regulated blowing agent it can replace (Table 8.11). A new generation of blowing agents comprising of hydrohaloolefins, such as those prefixed by HFO in Table 8.11, have produced a new burst of formulation innovation. These new blowing agents have good environmental and insulation performance but were found to produce poor quality polyurethane foams when conventional formulations were stored in a conventional manner. Such aged compositions produced collapsing foam. It was found that the cause of the failure stemmed from reactions of amine urethane catalysts with the unsaturation present in HFO blowing agent. In addition to the reaction product of

**TABLE 8.11 Status of acceptability/unacceptability for a variety of foam blowing agents for spray, refrigeration, and appliance foams**

| Blowing agent | Acceptable/unacceptable |
|---|---|
| $CO_2$ | Acceptable |
| HFO-1234ze | Acceptable |
| HFO-1336mzzZ (hexafluoro-2-butene) | Acceptable |
| HCFC-123 | Acceptable |
| HFC-152a | Acceptable |
| HCFC-124 | Unacceptable |
| HCFC-141b | Unacceptable |
| HCFC-365mfc | Unacceptable after 2020 |
| HCFC-245fa | Unacceptable |
| HFC-134a | Unacceptable after 2020 |
| Methylal (dimethoxymethane) | Acceptable |
| Formacel® TI | Unacceptable after 2020 |
| Light hydrocarbons (pentanes) | Acceptable |
| Formacel®-Z-6 | Unacceptable after 2020 |
| Trans-1-chloro-3,3,3-trifluoroprop-1-ene | Acceptable |
| Methyl formate | Acceptable |
| HCFC-141b | Unacceptable |
| Water | Acceptable |
| HCFC-22 | Unacceptable |
| HCFC-142b | Unacceptable |
| Electroset technology | Acceptable |
| Exxsol | Acceptable |
| Transcend Technologies | Acceptable |

*Source:* US EPA 6/2019.

HCFC, hydrochlorofluorocarbon; HFC, hydrofluorocarbon; HFO, hydrohaloolefin.

the amine and blowing agent that depleted the catalyst, there was also observed degradation of the silicone surfactant necessary [54] for acceptable foam structure. Suitability of a blowing agent will additionally require considerations of solubility in the foaming mixture and boiling point relative to process temperature.

## 8.4 INSULATION FUNDAMENTALS

The worldwide acceptance and use of polyurethane foams for insulation applications focuses competition and innovation towards making the best product at the best price. As research organizations mobilize to improve the use of insulation materials, it is natural to focus on those aspects that can exhibit the most improvement. Analyzing the performance of an insulation foam means quantifying the physical contributions of each structural and chemical feature for its influence on thermal conduction. Thermal conductivity, or the "*K*-factor," is the proportionality constant between the heat flow per unit area ($q$) and the thermal gradient through the substance in question (Equation 8.3) [55]

$$q = \frac{\partial Q / \partial T}{A} = K\left(T_1 - T_2\right) \tag{8.3}$$

The lower the thermal conductivity of a substance (in our case a polyurethane foam), the lower the heat flow rate per unit area across the sample for a given temperature gradient, and the better its performance as an insulation material. Commercial insulation materials are normally identified by their "*R*-value," which is product specific based on its thickness $L$, and is related to $q$ by Equation 8.4 [56]:

$$R = L / k \tag{8.4}$$

$$q = \frac{T_1 - T_2}{R}$$

The units of $K$ in the SI system are watts/(meter × Kelvin) and are also often expressed in terms of imperial units (BTU × in)/(h × ft$^2$ × °F), where a BTU (British Thermal Unit) is the energy to raise a pound of water by 1 °F or about 1055 joules. For simplicity, the numerical relationship between "*R*" and "*K*" is $K$ (w/mK) = 144.2/(*R*/in).

When a polyurethane foam is applied as an insulation it displaces air and substitutes the gas entrapped in the foam (95% of the substitution by volume) and the polyurethane material. The foam walls and windows serve to scatter and impede radiative conduction of heat. At the same time, the polyurethane acts as a conduction path for heat transport through the material. Not counting the influence of one path on another (an interaction term) the *K*-factor for a polyurethane foam is the sum of three conduction terms (Equation 8.5)

$$K_{\text{foam}} = K_{\text{solid}} + K_{gas} + K_{\text{radiative}} \tag{8.5}$$

where $K_{\text{solid}}$ is the solid conduction component of the foam *K*-factor, $K_{\text{gas}}$ is the gas conduction component of the foam *K*-factor, and $K_{\text{radiative}}$ is the radiative component of the foam *K*-factor [57–59].

Given the very low densities of rigid foams (typically >95% pore volume) the solid component is usually ignored in a first approximation but can account for about 10% of the total heat transfer across a foam [60]. The radiative component is that associated with temperature difference and not gradient. It is commonly associated with black-body emission and is thus a function of foam thickness and density, but not the material itself. This factor can account for about 20% of the heat transfer across a foam [61]. The remaining heat conduction is associated with that due to transfer through the gas phase. Consequently, both the solid and radiative components are relatively minor contributors to the overall thermal foam conduction, and even significant improvements in their *K*-factor contribution would be minor in overall effect. Accordingly, the gas conduction component has traditionally been the focus of almost all attempts to improve the insulation properties of polyurethane rigid foams [62–64]. However, with greater constraints on the properties of blowing agents, innovation in this area has been mitigated. Instead, there has been increasing work on ways to reduce cell size and so reduce the radiative component of heat conduction. Work has largely focused on the use of foam nucleating agents, which serve to increase the number of cells; at constant density, this can significantly reduce cell size [65–67]. Another method has been to utilize a soluble blowing agent at elevated pressure such as supercritical $CO_2$, which upon rapid depressurization can produce a profusion of cell nuclei and create conditions for thermal conductivity to occur under Knudsen [68] conditions, which will be discussed below.

A simple first-order expression for the thermal conductivity of gases is obtained from kinetic molecular theory (Equation 8.6) [69]:

$$K_{gas} = \frac{f}{2\pi b^2}\left[\frac{\mathrm{k}_{\mathrm{B}}^3 N_0 T}{\pi m_0}\right]^{1/2} \tag{8.6}$$

where $T$ is absolute temperature, $m_0$ is the molecular weight of the gas molecules, $b$ is the molecular diameter, $f$ is the number of degrees of freedom per molecule, $N_0$ is Avogadro's number ($6.023 \times 10^{23}\text{mol}^{-1}$), and $\mathrm{k}_{\mathrm{B}}$ is Boltzmann's constant ($1.38 \times 10^{-23}$ J/K). This equation predicts that the thermal conductivity of gases is independent of pressure, proportional to the square root of the system temperature, and has inverse proportionality to molecular mass. This equation is confirmed by experiment at ambient temperatures and pressures [70]. Refinements of Equation 8.6 have been developed by taking into account the potential for energy transfer between molecular vibration or rotational modes into translational energy [71]. Equation 8.6 can be expressed in terms of gas viscosity $v$, density $\rho$, pressure $P$, molecular speed $u$, mean free path $l_o$, and the gas constant R (8.314 J/molK), as shown in Equation 8.7:

$$k_{gas} = \tfrac{1}{2} f \rho u l_o \frac{R}{2m_o} \tag{8.7}$$

Terms in Equation 8.7 can be calculated using the relationships in Equation 8.8.

$$\rho = \frac{Pm_o}{RT} \tag{8.8}$$

$$u = \left[\frac{8RT}{\pi m_o}\right]^{1/2}$$

$$I_o = \frac{k_B T}{\sqrt{2}\,\pi b^2 P}$$

$$v = \frac{1}{2}\rho u I_o \text{ or by substitution}$$

$$= \frac{1}{\pi b^2}\left[\frac{k_B m_o T}{\pi N_o}\right]^{1/2}$$

For air, the mean free path (in micrometers) is $2.24 \times 10^{-4}$ $T(K)/P(\text{atm})$ or about 0.07 μm at room temperature.

Rigid foam designers have employed gas mixtures as a means to improve overall insulation cost and performance. The thermal conductivity ($K$-factor) of a mixture of gases at normal temperatures and pressures has been described by the Riblett equation [63] (Equation 8.9). Variations and refinements have been proposed to take into account observed deviations from Equation 8.9.

$$K_{gas} = \frac{\sum P_i m_i^{1/3} k_i}{\sum P_i m_i^{1/3}} \tag{8.9}$$

While mixing gases has been one route to reduce the gas conductivity of polyurethane insulation foam, another route has been to take advantage of the Knudsen effect [72–75]. The Knudsen effect takes into account the fact that thermal transport by the gas phase assumes that that the mean free path between molecular collisions is negligible compared with a foam cell size. The calculated mean free path for air calculated above is about 70 nm. Compared with the typical foam cell size (on the scale of a millimeter), it is safe to assume that the temperature of the insulation gas and the neighboring foam wall are the same. The Knudsen number (Equation 8.10) is a dimensionless number that characterizes the pressure dependence of the thermal conductivity of a gas, where $L$ is the length over which energy and momentum are transported – in this case the foam cell size.

$$K = \frac{I_o}{L} \tag{8.10}$$

The typical case as outlined above is the case where the Knudsen number is very small. However, in the case of very low pressure or very small $L$ (cell size) (Figure 8.23) the pressure and corresponding temperature of a cell gas at any point will be more nearly the same since direct (or specular) and diffuse reflection of gas molecules off the cell wall surface will be effectively isotropic (or more so) and no surface will be significantly hotter than any other. Thus, at very small foam cell size, less than 100 nm, the foam is effectively acting as if the gas has a lower ratio of heat

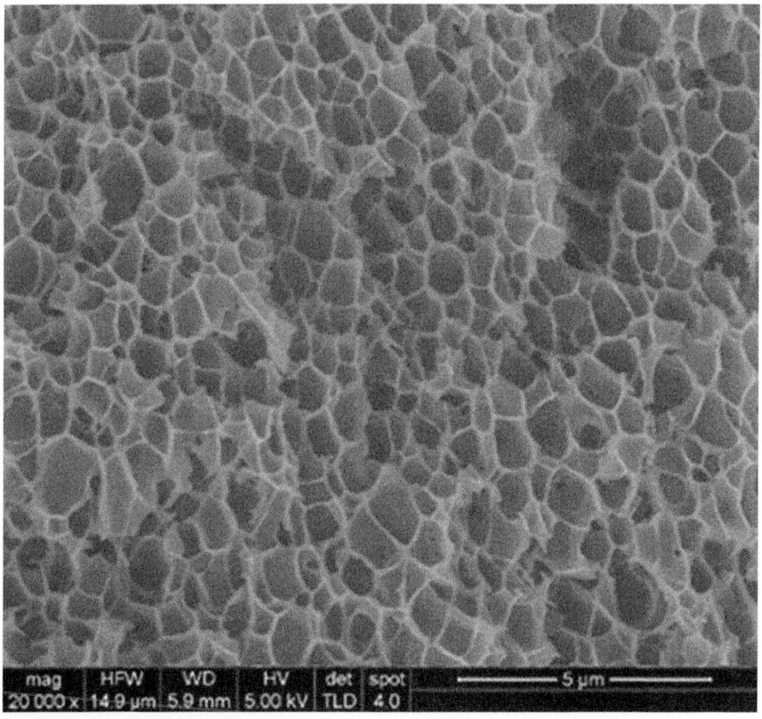


**FIGURE 8.23** Nanocellular polyurethane foam produced using supercritical $CO_2$. A 500 nm average cell size imputes to a thermal conductivity of 20 mW/m-K. Image courtesy of Matt Larive and Tirtha Chatterjee, The Dow Chemical Co.

transport to temperature gradient than would be expected. In the limit of this phenomenon the insulation foam with a very large Knudsen number would appear to be a foam with evacuated cells and the $K_{gas}$ of Equation 8.7 would tend towards zero.

As will be discussed in the later section on rigid foam technology trends, this phenomenon has been a research path numerous academic and industrial laboratories have pursued to most influence insulation technology. Some more exotic technologies such as aerogels and xerogels [76–78] have provided very small pore sizes, but generally have relatively poor structural integrity and high cost. Obtaining very small pore sizes in conventional insulation materials such as polyurethanes has proven difficult without sacrificing low foam density, which then causes an undesirable increase in the *K*-factor contribution from solid conduction.

## 8.5 TRENDS IN RIGID FOAMS TECHNOLOGY

While the patent literature is notably poor at providing insight of fundamental science, it is the clearest way to understand the future of a technology and market [79, 80]. This is because the patent literature represents a sizeable investment by a company or individual on what is believed will be desirable by customers and market competitors in the next-generation product. Furthermore, defense of a patent asset requires that the technology on which a patent is based must be protected by patent prior to its commercial introduction.

While a single patent may not be instructive as to the direction of future innovation, a clear trend in industrial patent activity can be unmistakable. Figure 8.24 shows the patent activity of several large multinational polyurethane feedstock providers over 2000–2013. In this analysis the patents filed in the Chinese language represent the activity within the sizeable Chinese market by Chinese companies. For perspective, the total number of issued patents found within the search parameters was 1451, so it is clear that there are a large number of patents not accounted for by this analysis. Despite this, it is likely that the preponderance of unaccounted for patents will reflect the activity of these highly experienced and influential participants (i.e. Dow, BASF, and Covestro) within the polyurethanes market. All of the illustrative market participants are actively patenting technologies to protect current and future market strategies and commercial opportunities, Leadership by BASF during the first decade of the twenty-first century has been recently ceded to Covestro, and to an even greater extent to the intense pursuit of intellectual assets in China. Also notable about Figure 8.24 is that companies patenting in the Chinese language are more concentrated in rigid foam technologies than in the same analysis for flexible foams discussed in Chapter 7.

Figure 8.25 shows that the majority of issued patents are concentrated around improving properties that define rigid foam applications and those that are required by regulation. Dow, BASF, Covestro, and Chinese-based companies have all shown a strong desire to invest in protecting future technologies associated with blowing agents and improved insulation properties. In this analysis it is likely that there is some overlap between categories, since blowing agent technology can strongly impact thermal insulation properties, as discussed in the previous section. Additionally, it is apparent that fire proofing of polyurethane foams is a continuing patent field of strategic importance to all commercial concerns related to rigid insulation foams for construction applications. While smaller but significant concentrations of rigid foam patenting are associated with advantaged catalysts and automotive applications, the coherence of this trend among market participants has dissipated in recent years. This activity and its intensity are common to those found in flexible foams.

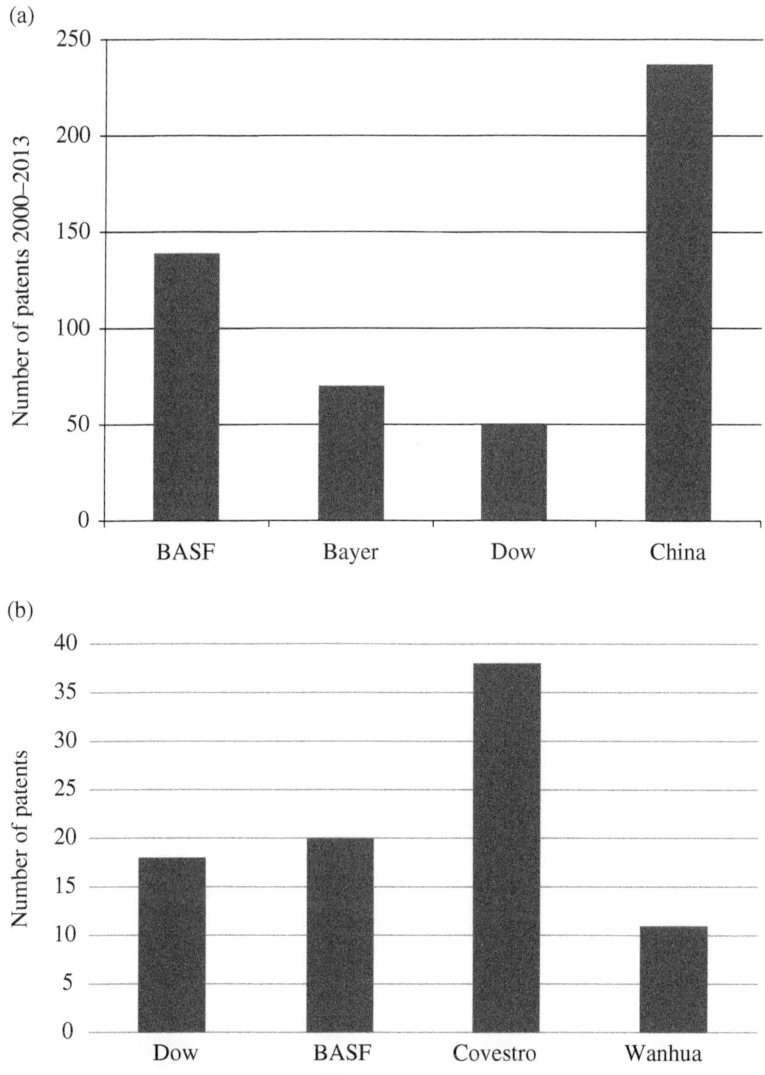


**FIGURE 8.24** Relative patent activity by three large polyurethane chemical producers and patent activity written in Chinese. (a) 2000–2014; (b) 2015–2019, substituting the Wanhua Group for all patent sources in China. Total Chinese patent activity for 2015–2019 was over 550 patents.

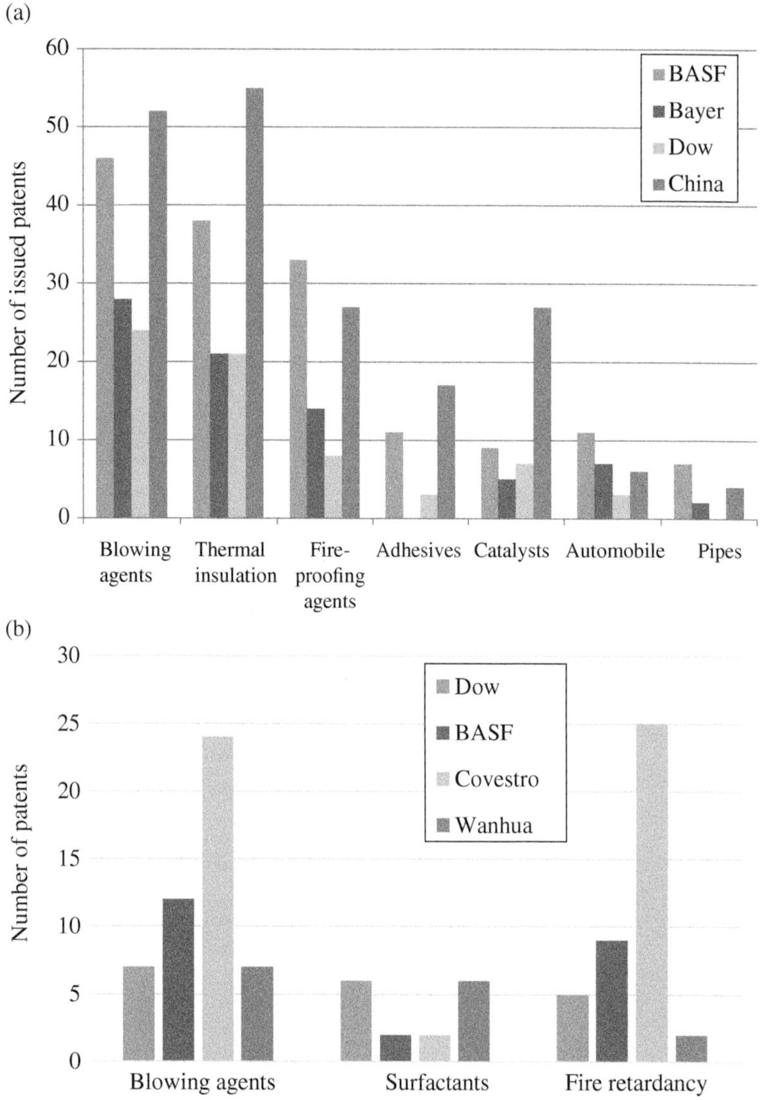


**FIGURE 8.25** Patent activity of three large polyurethane chemical producers and written in Chinese in the field of rigid foams by topic. (a) 2000–2014; (b) 2015–2019, substituting the Wanhua Group for integrated activity in China. (*See insert for color representation of the figure.*)

It is difficult to pick out coherent trends within the very recent patent literature – even within the bounds of keyword analysis shown in Figure 8.24. Dow has protected specific isocyanurate foam formulations that also enumerate effective blowing agents. BASF has protected formulations in which new-generation blowing agents possess superior shelf life stability. Covestro has patented specific formulations containing polyethercarbonate polyols, which presumably possess superior thermal and hydrolytic stability to polyether or polyester polyols. In addition, Covestro has patented formulations containing new-generation blowing agents. Wanhua has patented novel rigid foam formulations initiated from specific classes of polyols or having incremental improvements in performance.

Technology trends for future polyurethane rigid foam advancements appear to be highly incremental and existing line extension improvements. It is useful to understand this fact, and equally important to understand what is not being worked upon. Gaps in activity can represent opportunities that are being missed by "the herd" or traps to those uninitiated on the difficulty and risks of bringing forward new technologies for a very conservative market place. The current analysis does not predict what the future will hold but it is telling that taking advantage of the Knudsen effect for improved gas thermal conductivity for foam *K*-factor does not appear in this analysis up to the year 2019 of polyurethane technologies. Similarly, although there has been significant industrial and academic activity in the area of seed oil feedstocks for polyurethane synthesis, it barely registers as a field of pursuit in the patent literature of rigid polyurethane foams. In fact, for the years 2015–2019 there are no found patents with the keywords "seed oil polyol" or "biorenewable" connected to rigid polyurethane technology. There have been, however, seven patents related to vacuum panel insulation using open cell polyurethane foams during the period 2015–2019. It would appear that the primary driving forces for future rigid foam technology differentiation has been, and continues to be, in order (i) energy improvement, (ii) productivity increase, and (iii) regulation compliance.

## REFERENCES


[1] H Chinn, M Yoneyama, U Loechner, and X Xu, *Polyurethane Foams, Chemical Economics Handbook*, IHS Markit, London, UK, 2016.

[2] R Koschade, *Sandwich Panel Construction*, Ernst and Sohn Publishers, Weinheim, Germany, 2002.

[3] *The Market for Polyurethane Chemicals and Products – East Asia*, IAL Consultants, London, UK, 2013.

[4] *PU Chemicals and Products in Europe, Middle East and Africa (EMEA)*, IAL Consultants, London, UK, 2012.

[5] H Gruenbauer, J Bicerano, P Clavel, R Daussin, H de Vos, M Elwell, H Kawabata, H Kramer, D Latham, C Martin, et al., "Rigid polyurethane foams: a review of the state-of-the-art," in *Polymeric Foams*, S Lee and N Ramesh, Eds., CRC Press, Boca Raton, FL, 2004, 253–310.

[6] D Schiff, C Koehler, K Parsely, and D Bhattacharjee, *Journal of Cellular Plastics*, **32**, (1996), 274–297.

[7] S Schilling, *Journal of Cellular Plastics*, **36**, (2000), 190–206.

[8] T Zywicki, *Case Western Law Review*, **53** (2), (2002), 315–351.

[9] M Livermore and R Revesz, *Environmental Law*, **45** (1), (2015), 1–17.

[10] A Lyons, *Materials for Architects and Builders*, 3rd Ed., Elsevier, Oxford, UK, 2007.

[11] E Linak, A Kishi, M Guan, and U Buchholz, *Construction Chemicals*, IHS Chemicals Update Program, Englewood, CO, 2019.
[12] R Richter and T Bodnar, "Poly(urethane isocyanurate)," in *Reaction Polymers*, W Gum, W Riese, and H Ulrich, Eds., Hanser Publishers, Munich, Germany, 1992, 358–377.
[13] M Roman, B Andrioletti, L Bruno, MB Lemaire, J-M Schwartz, JP Barbeau, *Tetrahedron*, **67**, (2011), 1506–1510.
[14] M Spirkova, M Kubin, P Spacek, I Krakovsky, and K Dusek, *Journal of Applied Polymer Science*, **52**, (1994), 895–904.
[15] R Romero, R Grigsby, E Riser, J Pratt, and D Ridgway, *Journal of Cellular Plastics*, **41**, (2005), 339–359.
[16] H Singh and A Jain, *Journal of Applied Polymer Science*, **111**, (2009), 1115–1143.
[17] P Dournel and L Zipfel, *Journal of Cellular Plastics*, **39**, (2003), 211–227.
[18] W Michaeli and D Obeloer, *Journal of Polymer Engineering*, **31**, (2011), 353–359.
[19] A Poncet and M Morari, Value for money: making more with less-optimizing asset utilization, ABB Review 2/2006, 39–43.
[20] A Howell, K Hanson, V Dhole, and W Sim, *Chemical Engineering Progress*, **98**, (2002), 54–63.
[21] S Emmitt and D Yeomans, *Specifying Buildings: A Design Management Perspective*, CRC Press, Boca Raton, FL, 2012.
[22] S Moore and S Williams, *Journal of Cellular Plastics*, **36**, (2000), 57–69.
[23] J Backus, "Rigid polyurethane foams," in *Polymer Foams and Foam Technology*, 2nd Ed., D Klempner and V Sendijarevic, Eds., Hanser Publishers, Munich, Germany, 2004, 121–140.
[24] M Davis and B Fungtamasan, *International Journal of Multiphase Flow*, **16**, (1990), 799–817.
[25] K Cole, K Waters, X Fan, S Neethling, and J Cilliers, *Minerals Engineering*, **23**, (2010), 1036–1044.
[26] P Gutman, K Hildebrandt, V Alstaedt, and A Mueller, *Journal of Cellular Plastics*, **46**, (2010), 239–258.
[27] T Wishneski, T Ognonowski, M Jean-Rejouis, and M Hunter, Polyurethane spraying system used to minimize emissions of a polyisocyanate. Assignee BASF. US patent 20130119152 A1. 2013.
[28] J Schoenberger and S Doerr, Dispersed two-component polyurethane foams. Assignee Bayer. WO 2010083953 A1. 2010.
[29] T Monma, T Kurita, and M Kazan, Apparatus and method for manufacture of polymer foams with uniform fine cells, no voids, and good thermally insulating properties using pentanes as blowing agents. Assignee Achilles Corp. Japanese Kokai JP 2009079173 A. 2007.
[30] T Fishback, D Caffoe, A Taylor and J Shoemaker, Method for filling wall cavities with expanding foam insulation. Assignee Fomo Corp. US patent 20120121805. 2012.
[31] J Jennings, T Perry, R Jenkines, and A Atkinson, Low density attached polyurethane foams made by containment of blowing agents during foam processing. Assignee Dow Chemical. US patent 20070197674 A1. 2007.
[32] T Morley, F Casati, A Birch, H Kramer, and P Athey, Polyol mixtures containing a 1,3-and/or 1,4-bis(aminomethyl) cyclohexane-initiated polyol and at least one of a methylene bis(cyclohexylamine)-initiated polyol and a cyclohexanediamine-initiated polyol, and polyurethanes made therefrom. Assignee Dow Chemical. US patent 2011041163 A1. 2010.

[33] H Kramer, Polyol formulations for production of rigid polyurethane foams and use of rigid polyurethane foams. Assignee Dow Chemical. WO 2010111021 A1. 2010.

[34] M Bomberg and J Lstiburek, *Spray Polyurethane Foams in External Envelopes of Buildings*, CRC Press, Boca Raton, FL, 1998.

[35] P Trautmann and M Piesche, *Chemical Engineering and Technology*, **24**, (2001), 1193–1197.

[36] S Baser, D Shetty, and D Khakhar, *Polymer Engineering and Science*, **33**, (1993), 1611–1618.

[37] P Kolodziej, W Yang, C Macosko, and S Wellington, *Journal of Polymer Science Part B: Polymer Physics*, **24**, (1986), 2359–2377.

[38] D Baerman, "Thermal insulation," in *Time Saver Standards for Architectural Design*, D Watson and M Crosbie, Eds., McGraw Hill, New York, NY, 2005.

[39] M Owen, *ASHRAE Handbook – Fundamentals*, ASHRAE Publisher, Atlanta, GA, 2013.

[40] K Ashida, *Polyurethane and Related Foams*, CRC Press, Boca Raton, FL, 2007, 85–88.

[41] P Parker, *The 2016–2021 World Outlook for Electronic and Fabricated Plastic Products for Household and Commercial Appliances*, ICON Group International, Las Vegas, NV, 2015.

[42] C Melo, L da Silva, and R Pereira, "Experimental evaluation of the heat transfer through the walls of household refrigerators," *8th International Refrigeration and Air Conditioning Conference*, Paper 502, July 25–28, Purdue University, West Lafayette, IN, 2000.

[43] T Heinemann and K Lorenz, "Polyurethane system solutions for the global appliance market," in *PU Latin America Rigid Foam Session*, Day 3, Paper 3, Sao Paolo, Brazil, August 28–30, iSmithers Rapra Publishers, London, UK, 2001.

[44] ENERGY STAR® Program Requirements Product Specification for Residential Refrigerators, Refrigerator-Freezers, and Freezers Test Method to Validate Demand Response, May, 2013.

[45] S Singh, Blowing Agents for Polyurethane Foams, Rapra Review Reports, Volume 12, Number 10, Smithers Rapra Press, Shawbury, UK, 2002.

[46] M Modesti, V Adriani, and F Simoni, *Polymer Engineering and Science*, **40**, (2000), 2046–2057.

[47] S Katz and A Lindner, *Journal of the Air and Waste Management Association*, **53**, (2003), 469–477.

[48] M Thirumal, D Khastgir, N Singha, B Manjunath, and V Naik, *Cellular Polymers*, **28**, (2009), 145–158.

[49] M Han, S Choi, J Kim, Y Kim, W Kim, H Lee, and J Sung, *Macromolecular Research*, **17**, (2009), 44–50.

[50] S Kim, H Lim, J Song, and B Kim, *Journal of Macromolecular Science Part A: Pure and Applied Chemistry*, **45**, (2008), 323–327.

[51] M Molina and F Rowland, *Nature*, **249**,(1974), 810–812.

[52] The Signatories to the United Nation Treaty for the Protocol on Substances, Protocol on Substances that Deplete the Ozone Layer, signed September 16, 1987, Montreal.

[53] S Eury and J Wu, "Blowing agents for rigid polyurethane foam: current status and future perspectives," in *PU Latin America Conference Proceedings*, August 21–23, Sao Paulo, Brazil, Crain Communications, London, UK, 2001, P20/1–P20/9.

[54] Y Bin, K Yiu, and D Williams, Stabilization of foam polyol premixes containing halogenated olefin blowing agents. Assignee Honeywell International Inc. US patent 20170313806 A1. 2017.

[55] B Bird, W Steward, and E Lightfoot, *Transport Phenomena*, 2nd Ed., Wiley, New York, NY, 2007, Chapters 9–11.

[56] A Ostrogorsky and L Glicksman, *ASTM Special Technical Publication Insulation Material Testing*, **1030**, (1990), 109–120.
[57] L Glicksman, *Cellular Polymers*, **10**, (1991), 276–293.
[58] J Wang and C Hagentoft, *Journal of Thermal Envelope and Building Science*, **22**, (1998), 110–131.
[59] E Placido, M Arduini-Schuster, and J Kuhn, *Infrared Physics and Technology*, **46**, (2005), 219–231.
[60] A Biedermann, C Kudoke, A Merten, E Minogue, U Rotermund, H Seifert, H Ebert, U Heinemann, and J Fricke, *High Temp-High Pressures*, **33**, (2001), 699–706.
[61] O Almanza, M Rodriguez, and J Saja, *Journal of Polymer Science Part B: Polymer Physics*, **38**, (2000), 993–1004.
[62] T Stovall and M Bogdan, *Thermal Conductivity*, **29**, (2008), 71–87.
[63] R Dohrn, J Fonseca, R Albers, J Kusan-Bindels, and I Marrucho, *Fluid Phase Equilibria*, **261**, (2007), 41–49.
[64] R Dohrn, R Treckmann, and T Heinemann, *Fluid Phase Equilibria*, **158-160**, (1999), 1021–1028.
[65] J Owens, Z Zhang, and R Dams, Polymeric foams including fluorinated oxiranes, methods of preparation and use of same. Assignee 3M Innovative Properties Company. US patent 2014/0058003A1. 2014.
[66] M Schutte, M Elbing, J Klassen, F Fechner, W Wiegmann, M Klostermann, J Venzmer, C Eilbracht, M Glos, and C Schiller, Composition that is suitable for producing polyurethane foams and that contain at least one nucleating agent. Assignee BASF SE and Evonik Industries AG. WO 2015/091801 A1. 2014.
[67] Y Lee, M Jang, K Choi, and W Kim, *Journal of Applied Polymer Science*, **133**, (2016), 43557.
[68] S Costeaux, *Journal of Applied Polymer Science*, **131**, (2014), 41293.
[69] J Jeans, *An Introduction to the Kinetic Theory of Gases*, Cambridge University Press, New York, NY, 1967, 185–198.
[70] F Devienne, "Low density heat transfer," in *Advances in Heat Transfer*, Volume **2**, J Hartnett and T Irvine, Eds., Elsevier, Amsterdam, the Netherlands, 1965, 271–356.
[71] E McLaughlin, "Theory of thermal conductivity of fluids," in *Thermal Conductivity*, Volume **2**, R Tye, Ed., Academic Press, New York, NY, 1969, 1–64.
[72] J Wolf and W Streider, *AIChE Journal*, **40**, (1994), 1287–1296.
[73] K Raed and U Gross, *Thermal Conductivity*, **29**, (2008), 356–373.
[74] V Felix, Y Jannot, and A Degiovanni, *Review of Scientific Instruments*, **89**, (2012), 1–8.
[75] P Ferkl, R Pokorny, M Bobak, and J Kosek, *Chemical Engineering Science*, **97**, (2013), 50–58.
[76] A Rigacci, J Marechal, M Repoux, M Moreno, and P Achard, *Journal of Non-Crystalline Solids*, **350**, (2004), 372–378.
[77] G Reichenauer and H Ebert, *NanoS*, **1**, (2008), 17–23.
[78] L Li, B Yalcin, B Nguyen, M Meador, and M Cakmak, *Applied Materials and Interfaces (ACS)*, **1**, (2009), 2491–2501.
[79] R Sitarz, M Heneczkowski, and A Kraslawsi, *Computer Aided Chemical Engineering*, **31**, (2012), 570–574.
[80] P Chang, C Wu, and H Leu, *Scientometrics*, **82**, (2010), 5–19.

# 9

# POLYURETHANE ELASTOMERS: MANUFACTURE, APPLICATIONS, MARKETS, AND TRENDS

Of the nearly 800 billion pounds of plastic produced every year about 6% of that total is polyurethane, and about 10% of that 6% is used purely for elastomer applications (Figure 9.1). Polyurethane elastomers are very high performing within the general class of synthetic materials and their applications run the gamut from footwear to automotive body panels, truck bed liners, rollerblade wheels, polishing pads for integrated circuit production, gaskets, blood bags, cattle ear tags, and polymer films to name just a few. They are particularly specified when abrasion resistance, tensile strength, dynamic properties (periodic stress/strain), and low-temperature properties are paramount. They are usually not selected for applications where price is overriding, where routine temperature service greater than 120 °C is required, or where resistance to particular solvents is expected.

The polyurethane elastomer business is divided into thermosetting elastomers, which are cross-linked and made by casting techniques; thermoplastic, which are made by reactive extrusion techniques; and elastomeric fibers, which are made via numerous methods but generally spun from concentrated solutions. The fiber markets are handled separately since the applications are quite divergent from the other elastomers produced. Of thermosetting and thermoplastic polyurethane (TPU) elastomers, the approximate ratio for their production and consumption is 3:1 (Figure 9.2) [1]. Apart from fibers, the large producers of polyurethane building blocks are significant participants in the polyurethane elastomer business, most of them having a global presence in this market.

*Polyurethanes: Science, Technology, Markets, and Trends*, Second Edition. Mark F. Sonnenschein.

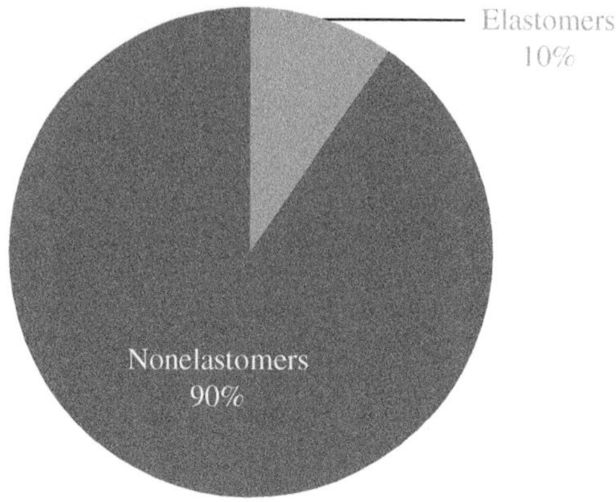


**FIGURE 9.1** Approximate volume relationships between polyurethane elastomers and all polyurethanes consumed in 2016.

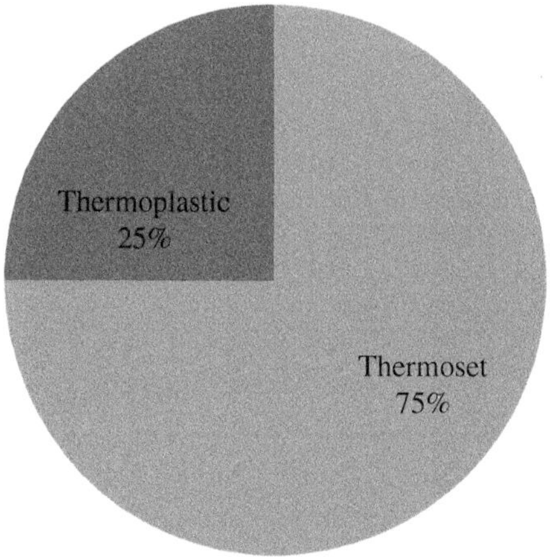


**FIGURE 9.2** Approximate volume relationships between polyurethane thermosetting elastomers and thermoplastic polyurethanes consumed in 2016.

However, there are numerous (thousands) of much smaller participants in the polyurethane market that have niche product or regional presence and can compete quite effectively within specific segments. Furthermore, what concentration does exist within the polyurethane elastomer business may in the future be prone to dilution since commoditization of products makes them less attractive to large publicly held manufacturers [2].

## 9.1 REGIONAL MARKET DYNAMICS

The polyurethane elastomer business, like the other polyurethane businesses, is in flux because of the growth of the Asian manufacturing economies and the movement of production assets to Asia (Figure 9.3) [3, 4]. For instance, the growth of exported chemicals (Figure 9.4) from Asia into developed manufacturing economies has a significant influence on pricing and the health of manufacturing in the regions of pricing competition. At the same time the growth of production capacity globally and regionally is not clearly in balance now, and may be more or less so in the future. This is most clearly illustrated by the polyurethane elastomer fiber market, which has seen prices go from US$8 per pound of fiber to US$3–4 per pound of fiber in a span of about 15 years. This deflation is occurring in spite of the inherent costliness of polyurethane fiber processes and a global growth rate of ca. 8% per annum. Polyurethane elastomeric fibers will be covered in detail later in this chapter, but the remarkable commoditization of chemicals and products is also apparent in the case of polyurethane elastomers.

The regional dynamics of polyurethane elastomer markets are strongly influenced by macroeconomic factors of overall economic activity, which, in turn, influence the more direct factors of asset utilization within the polyurethane industry. This is usually measured by comparing the ratio of production capacity and product consumption as a function of history, and projecting with more or less confidence into the future based on announced plant shut downs, expansions, and new capacity. The data in Figure 9.5 show the trend and projection in both Europe and Asia for oversupply of isocyanates. While this analysis does not cover polyols, isocyanates are of course required for every polyurethane elastomer, and the assortment of polyols for elastomers is significantly more varied, complicating what should be a simple point. That is, for as long as there is oversupply in the polyurethane industry, prices will be depressed for feedstocks as well as for final products [5, 6]. Furthermore, the relatively high-capacity utilization in China has to be viewed in light of the very high China export volumes to other geographies.

Should the global economy pick up substantially, a condition of short supply of feedstock could emerge, as was observed in the early years of the analysis in

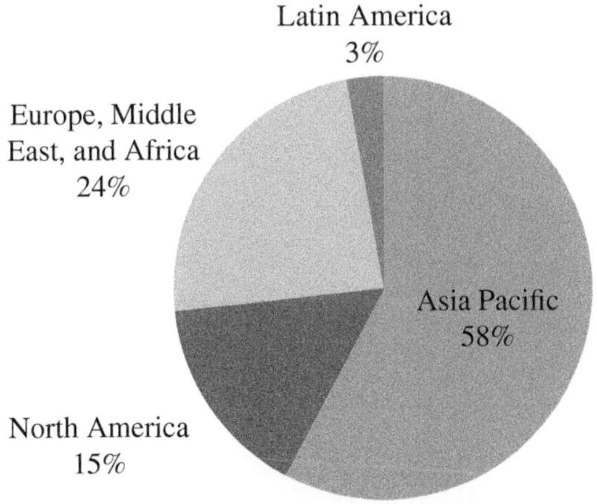


**FIGURE 9.3** Approximate geographical industrial consumption of polyurethane elastomers for 2016. The pie chart represents industrial consumption and not consumer consumption.

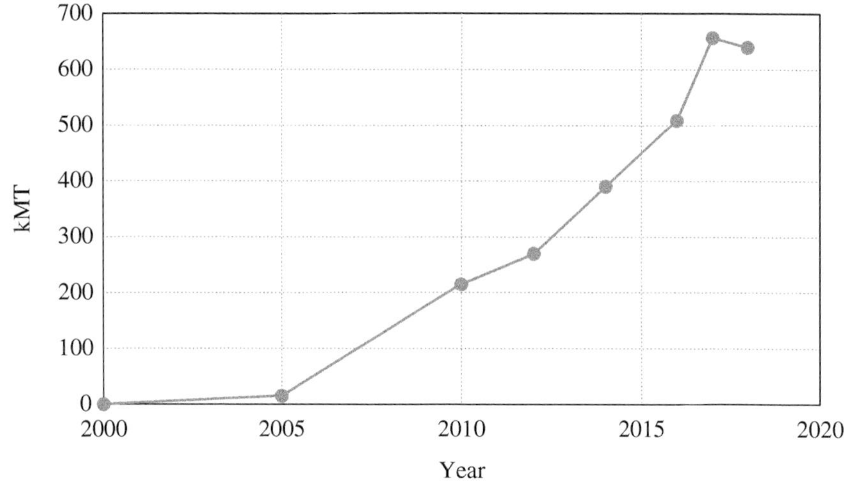


**FIGURE 9.4** Chinese global exports of monomeric and polymeric methylene diphenyl diisocyanate (pMDI). kMT, thousand metric tons.

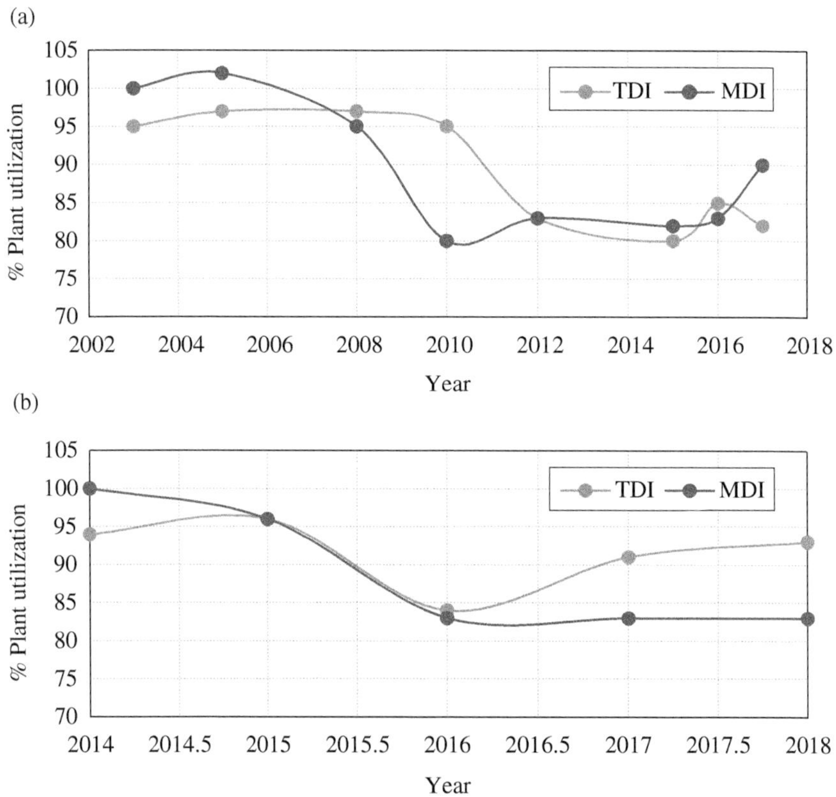


**FIGURE 9.5** Capacity utilization for isocyanate production in (a) Western Europe and (b) China. Low asset utilization results in low prices or production assets being shuttered. TDI, toluene diisocyanate.

Figure 9.5, and prices could rise, as will company profitability. However, if there is capital for expansion, manufacturers may gamble by adding still more capacity in an attempt to corner more revenues in a potential future boom, and frighten off capital-poor competitors from gaining economic health by raising prices [7, 8]. Thus, the relatively dramatic point-to-point fluctuations of Figure 9.5 reflect the convolutions of many factors, both intentional and unintentional, within the industry.

As alluded to earlier, while the number of applications that polyurethane elastomers can be applied to is quite large, the elastomers themselves can be conveniently divided into thermosetting or thermoplastic. The ratio of their consumption is about 75% thermoset and 25% thermoplastic. While it is impractical to break regional consumption into relatively small application segments, it is still informative to simplify

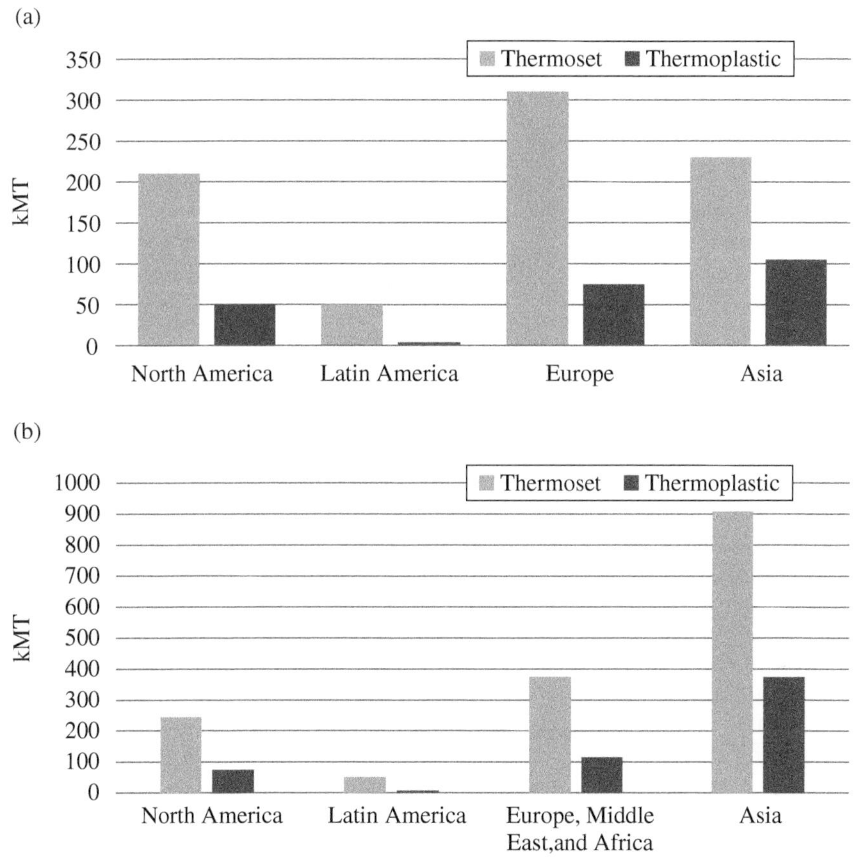


**FIGURE 9.6** Regional industrial consumption of polyurethane elastomers for (a) 2002 and (b) 2016, reflecting the migration of thermosetting elastomers, especially for shoe soles, to Asia. kMT, thousand metric tons. (*See insert for color representation of the figure.*)

the categories into just thermosetting and thermoplastic elastomers. Figure 9.6 shows the changes in regional polyurethane elastomer consumption as certain key segments, particularly footwear, has transferred manufacturing from Europe to Asia. The mature economies of Europe and North America (dominated by Western Europe and the USA, respectively) have nearly identical proportional consumption of thermosetting to thermoplastic elastomers. That consumption is somewhat skewed towards thermoplastics because of disproportionate use of thermoplastics within the automotive industry [9].

## 9.2 APPLICATIONS

Elastomer applications are varied and it is difficult to draw sweeping generalizations. It is especially difficult to provide meaningful geographical data regarding consumption since the applications are globally relevant to consumers, but manufacturing of specific products, such as footwear, is highly concentrated (in China). Another development has been the near disappearance of reaction injection molding (RIM) applications from the automotive segments and its stagnant market development in nonautomotive applications.

Segmentation of applications by regional production exemplifies the changing global concentration of low-cost manufacturing of smaller devices and parts in low-cost economies, versus the high-cost manufacturing but dominant commercial markets in developed economies (Figure 9.7). The dominance of transportation-related applications in North America reflects the compatibility of higher cost materials with more complex value-added assembly manufacturing. In contrast, the dominance of polyurethane elastomers for footwear in China demonstrates the importance of low-cost manufacturing and the reduced impact of transportation of small items to final consumer cost. This macrotrend is further illustrated by the decline in polyurethane footwear manufacturing in Western Europe, as the opposite trend emerged in China (Figure 9.8).

### 9.2.1 Footwear

The footwear segment is a large and varied market and probably worthy of a more thoroughgoing treatment than can be afforded in this book. Polyurethane materials find uses throughout the footwear industry from work boots to ski boots, to running shoes, to high heels. Each segment has many optimized formulations based on cost and performance requirements. The footwear industry has moved quickly to provide ever greater performance in terms of low-temperature properties, abrasion resistance, and appearance. Certainly one of the areas of greatest innovation has been the pursuit of excellent properties at lower densities, resulting in lighter weight. Lastly, there has been significant progress in making TPUs with the required low modulus to result in adequate traction on all surfaces.

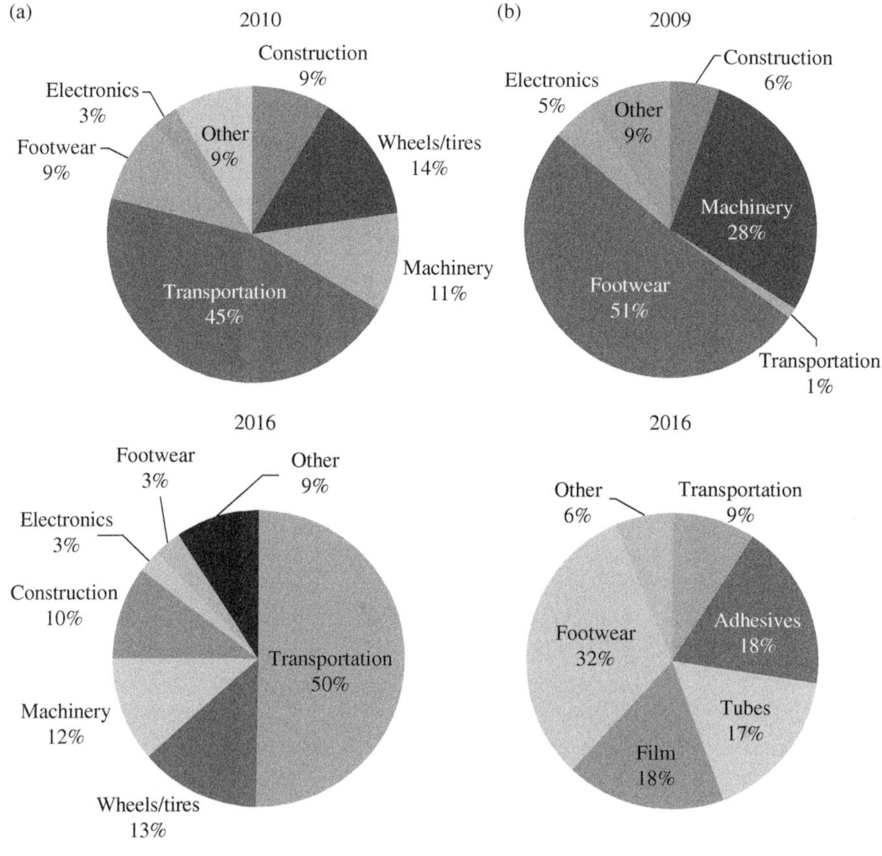


**FIGURE 9.7** Volume fraction industrial usage of polyurethane elastomer in (a) North America in 2010 and 2016 and (b) China in 2009 and 2016.

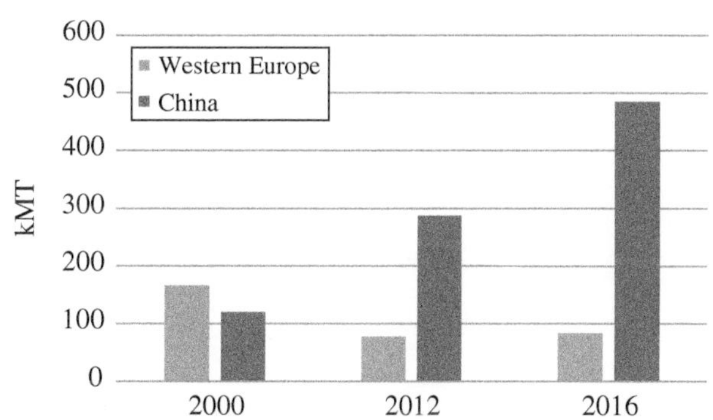


**FIGURE 9.8** Consumption of microcellular polyurethane elastomer primarily used in footwear applications, comparing data for Western Europe and China for 2000, 2012, and 2016. *Source:* [10].

Relative to all polymer materials, polyurethanes account for only about 6% of the total shoe-soling materials employed. This still accounted for over 1 billion pounds of polyurethane in 2016, but by comparison is dwarfed by materials such as polyvinylchloride (~5 billion pounds), natural vulcanized rubber (about 4 billion pounds), the triblock copolymer of styrene–butadiene–styrene, also referred to as thermoplastic rubber (TPR; about 2 billion pounds), and ethylene vinylacetate ("EVA"; about 1.5 billion pounds). For perspective, the same amount of polyurethane is used as is leather for shoe soling. Polyurethane is favored by shoe designers for its excellent low-temperature flexibility, its very good wear properties (low abrasion), and the ability to produce it at a wide variety of densities by introducing porosity.

The polyurethane component is broken up into thermosetting polyurethane produced by injection molding/microcellular methods and those using TPU. The ratio for these is approximately 5:1 (Figure 9.9).

Thermosetting polyurethanes for footwear applications are usually made from preformulated urethane systems developed to make a specific product with a designed hardness, density, cost, and wear performance. Many shoe sole manufacturers begin with a system called a quasi-prepolymer [11, 12]. A quasi-prepolymer is essentially a prepolymer (see Chapter 2 and Section 9.2.2.1 on cast elastomers) with excess isocyanate. To make a quasi-prepolymer, isocyanate (usually monomeric 4,4′-methylene diphenyl diisocyanate [MDI] or an added uretonimine-modified MDI to maintain fluidity; see Chapter 3) is reacted with a 1000–6000 g/mol polyester or polyether polyol, or a mixture of polyols with total NCO content ranging from 12% to 28% (see Chapter 2 for calculation of percentage NCO in a prepolymer). This makes one component of a two-component formulation. The other component will have a variable amount of polyol, chain extender, and additives such as blowing agent,

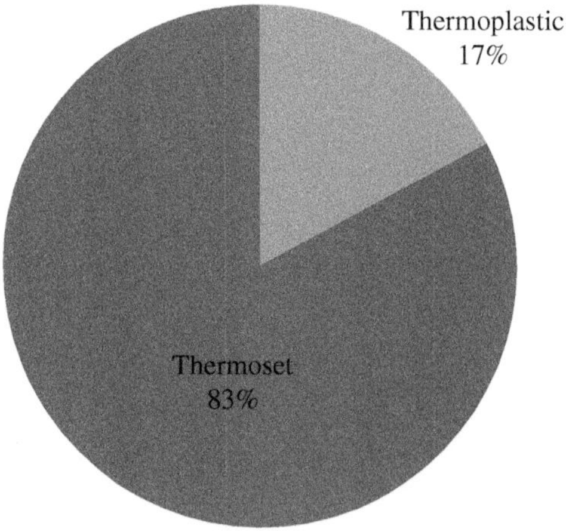


**FIGURE 9.9** Approximate volume fraction consumption of thermoset versus thermoplastic polyurethane elastomers for footwear in China in 2016.

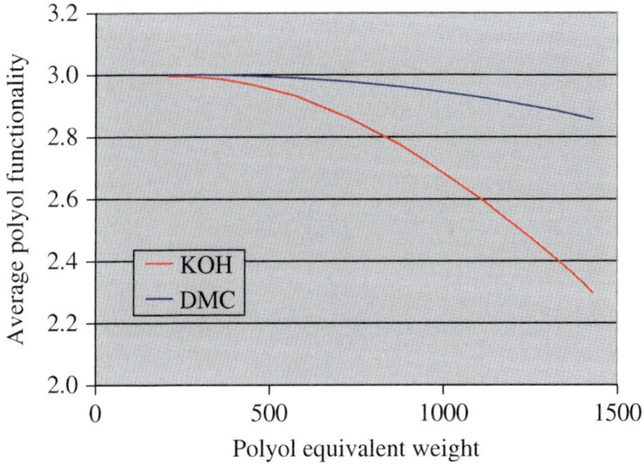


**FIGURE 2.14** Difference in measured polyol functionality attained by KOH and DMC catalysis for a triol of variable equivalent weights. The difference becomes appreciable at higher equivalent weights as the number of end groups decreases.

*Polyurethanes: Science, Technology, Markets, and Trends*, Second Edition. Mark F. Sonnenschein.

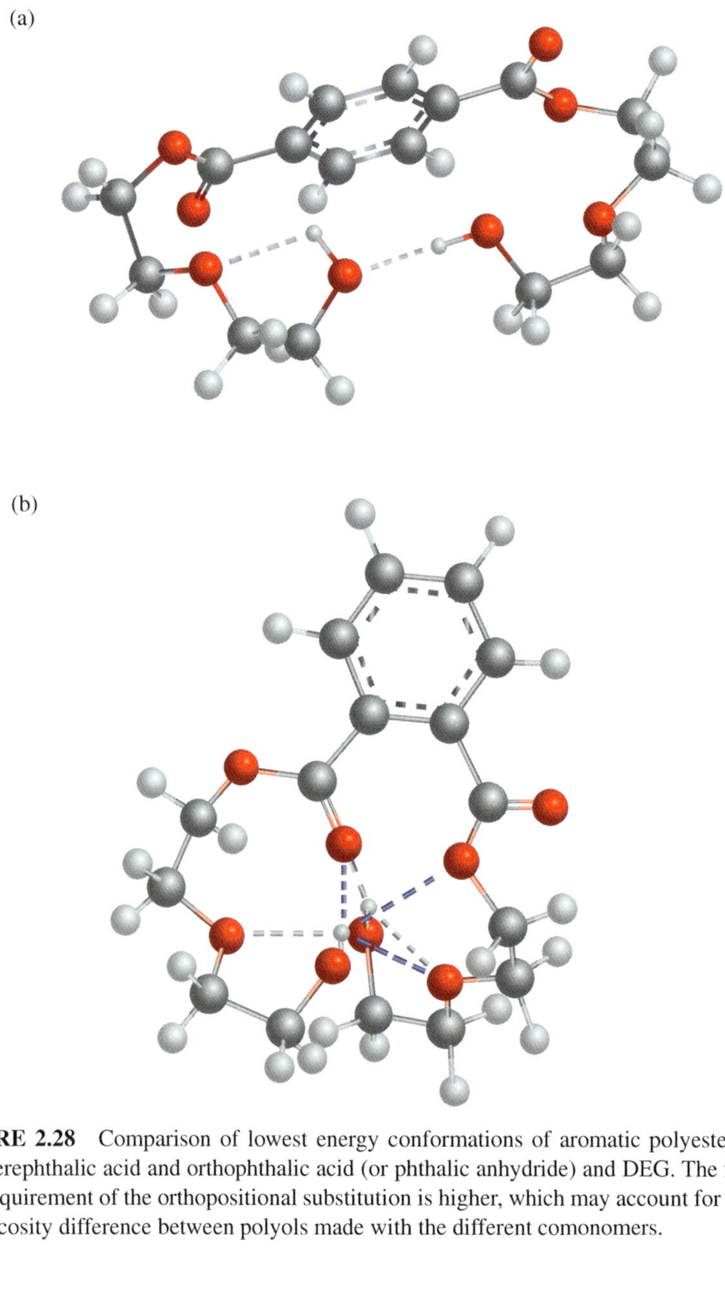


**FIGURE 2.28** Comparison of lowest energy conformations of aromatic polyester polyol from terephthalic acid and orthophthalic acid (or phthalic anhydride) and DEG. The free volume requirement of the orthopositional substitution is higher, which may account for much of the viscosity difference between polyols made with the different comonomers.

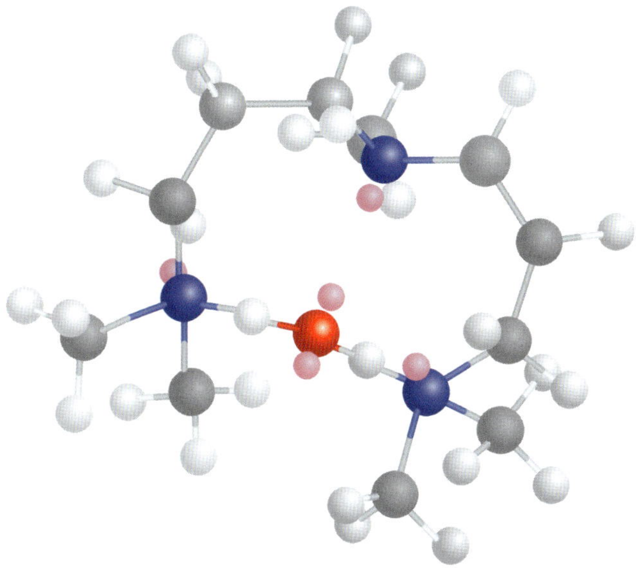

**FIGURE 3.8** Energy-minimized interaction between water and pentamethylene dipropylenetriamine demonstrating the favorable water interactions that result in good blow catalyst properties (light red surfaces indicate nonbonding electron pairs; red is oxygen and blue is nitrogen).

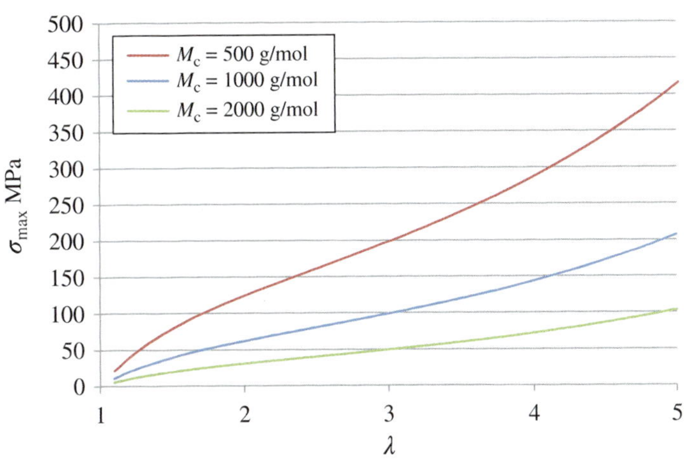


**FIGURE 4.18** Graphical representation of Equation 4.17 calculating the maximum stress as a function of maximum elongation and molecular weight between cross-links ($M_c$). The maximum elongation at break is arbitrarily, though reasonably, set at 400%.

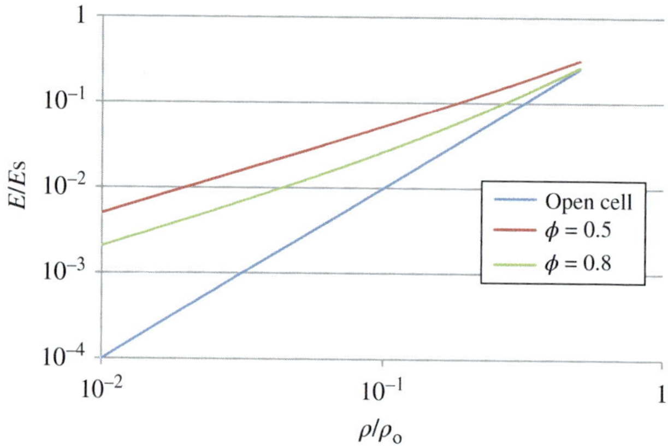


**FIGURE 4.19** Graphical representations of Equations 4.18 and 4.24 showing the modulus ratio of foam to its solid polymer as a function of density and the fraction of polymer in the cell edges ($\phi$).

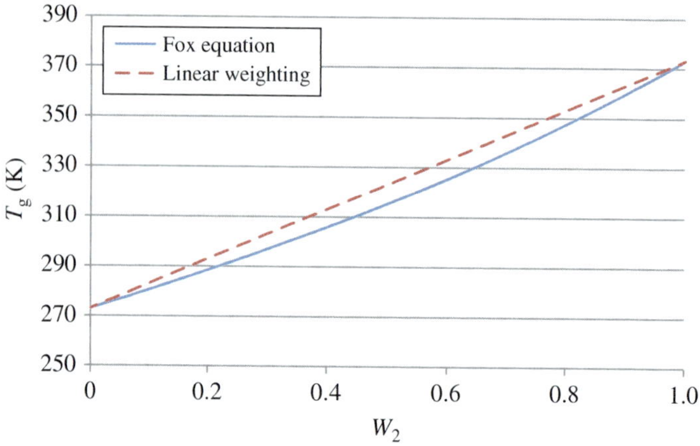


**FIGURE 4.20** Graphical representation of the Fox equation (Equation 4.25) relative to a simple mass-weighted linear combination of pure material $T_g$ values. For illustration $T_{g1}$ was chosen to be 273 K and $T_{g2}$ was chosen to be 373 K – relevant to many polyurethane systems.

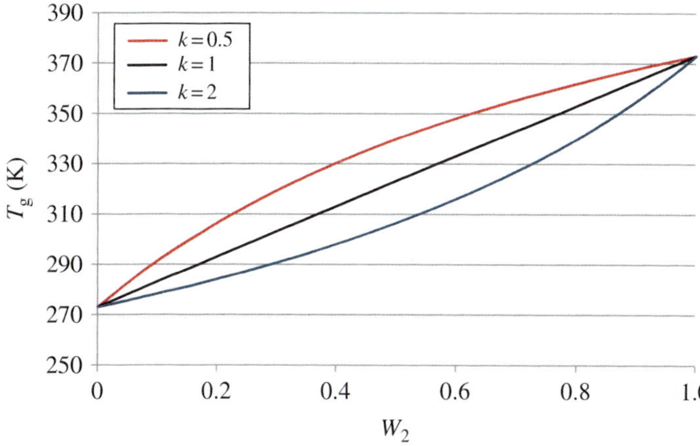


**FIGURE 4.21** Graphical representation of the Gordon–Taylor equation for varying values of $k$, yielding either positive or negative deviations from a mass-weighted linear combination of pure material $T_g$ values.

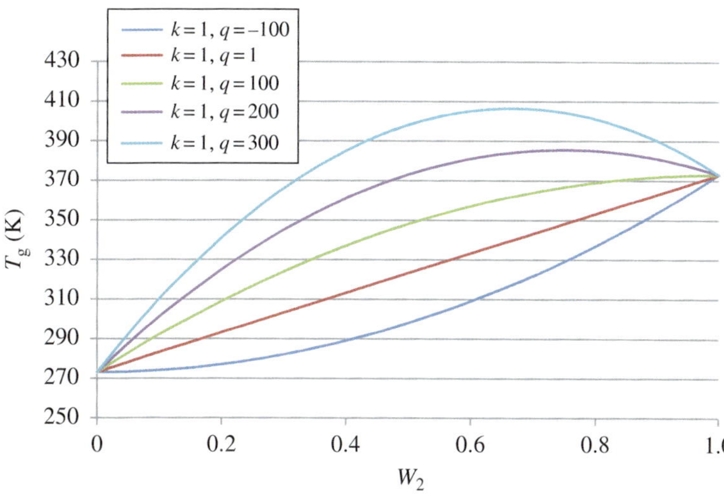


**FIGURE 4.22** Graphical representation of the Kwei equation, keeping $k$ at or near 1 (as in the Gordon–Taylor formalism) and varying $q$ to demonstrate its effect on the functional form.

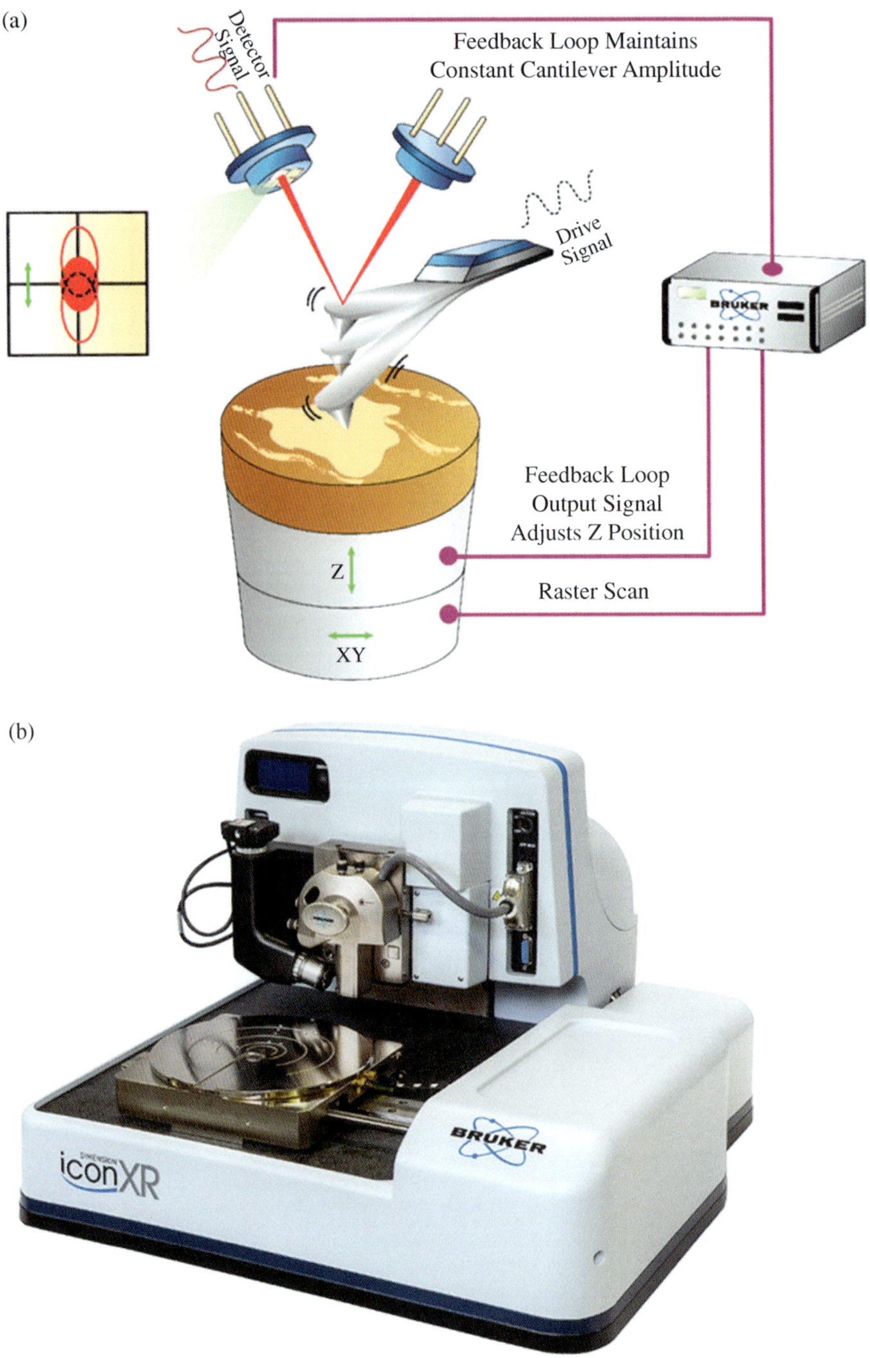


**FIGURE 5.7** Simplified illustration showing sample and signal detection controls for a tapping-mode atomic force microscope and the appearance of a working instrument. Images courtesy of Bruker Nano Surfaces.

(a)

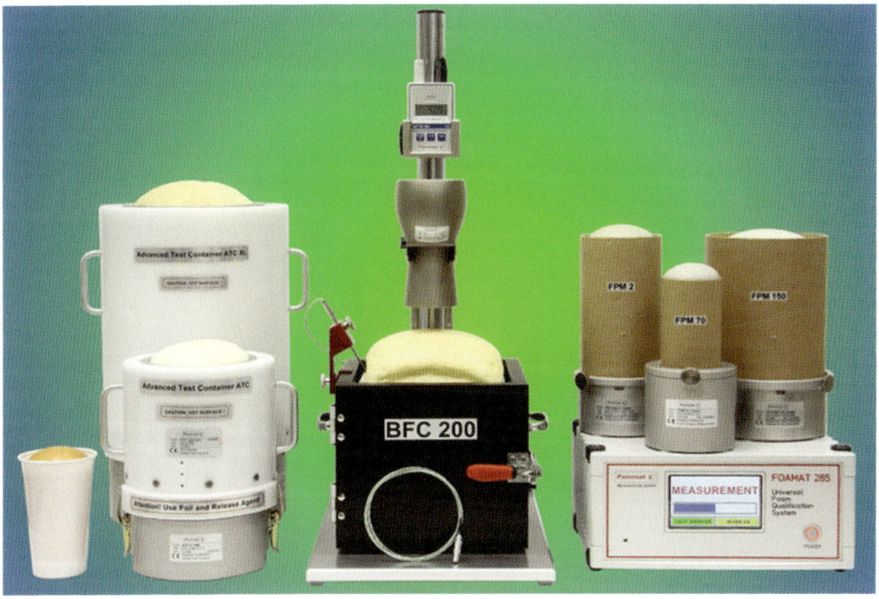


(b)

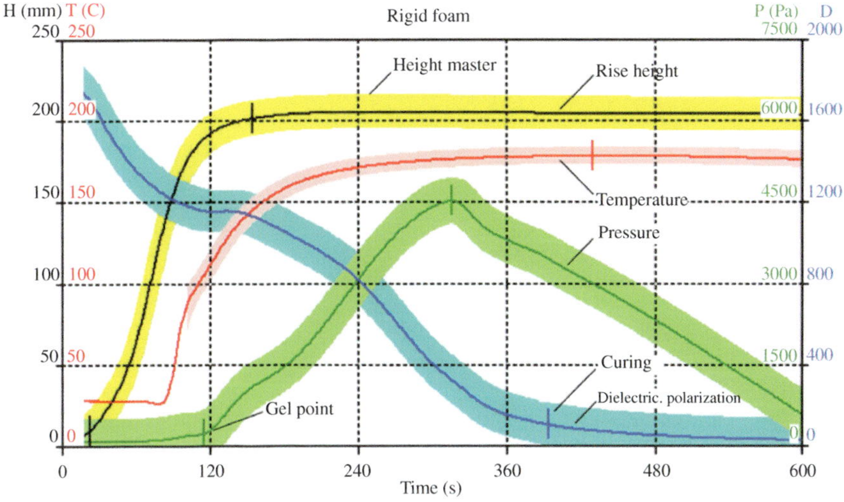


**FIGURE 5.30** The FOAMAT® foam qualification system for quantifying the physical processes occurring in a foaming system after mixing. (a) The appearance of a full system capable of preparing foams at different volumes and different thermal controls. (b) Sample data obtained simultaneously from a rigid insulation foam. The measurement of pressure and dielectric polarization occurs in the base, the temperature rise is measured by a thermocouple within the rising foam body, and the rise and gel conditions are determined by the sensor above the foam mass. Images courtesy of Format Messtechnik GmbH.

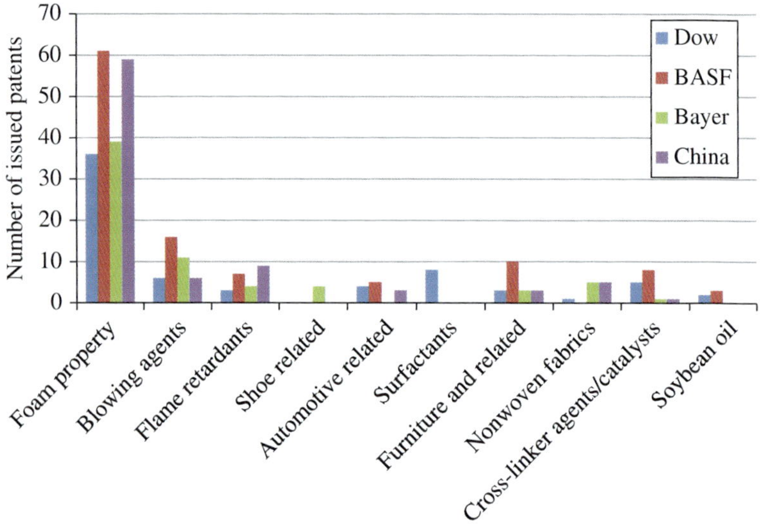


**FIGURE 7.11** Patent activities within the flexible polyurethane foam category of several large polyurethane market servicers and the patents filed in the Chinese language for the years 2000–2013.

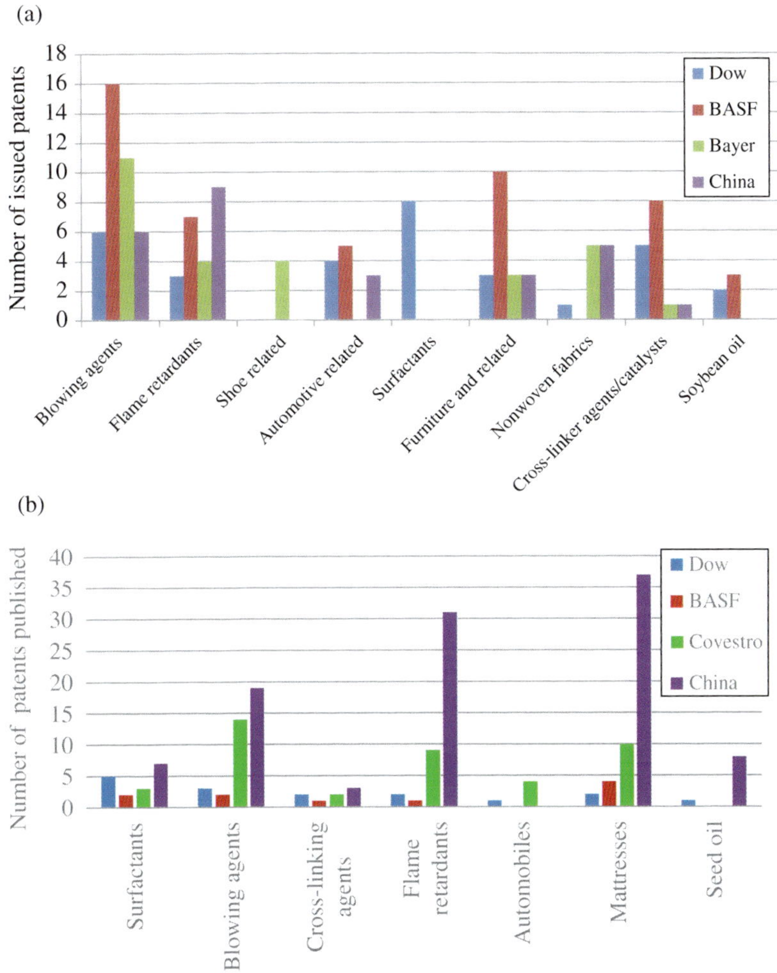


**FIGURE 7.12** (a) Data from Figure 7.11 with the "foam property" category removed to improve visualization of more specific subjects. (b) Data for the period 2015–2019 showing the changes in industrial investment focus.

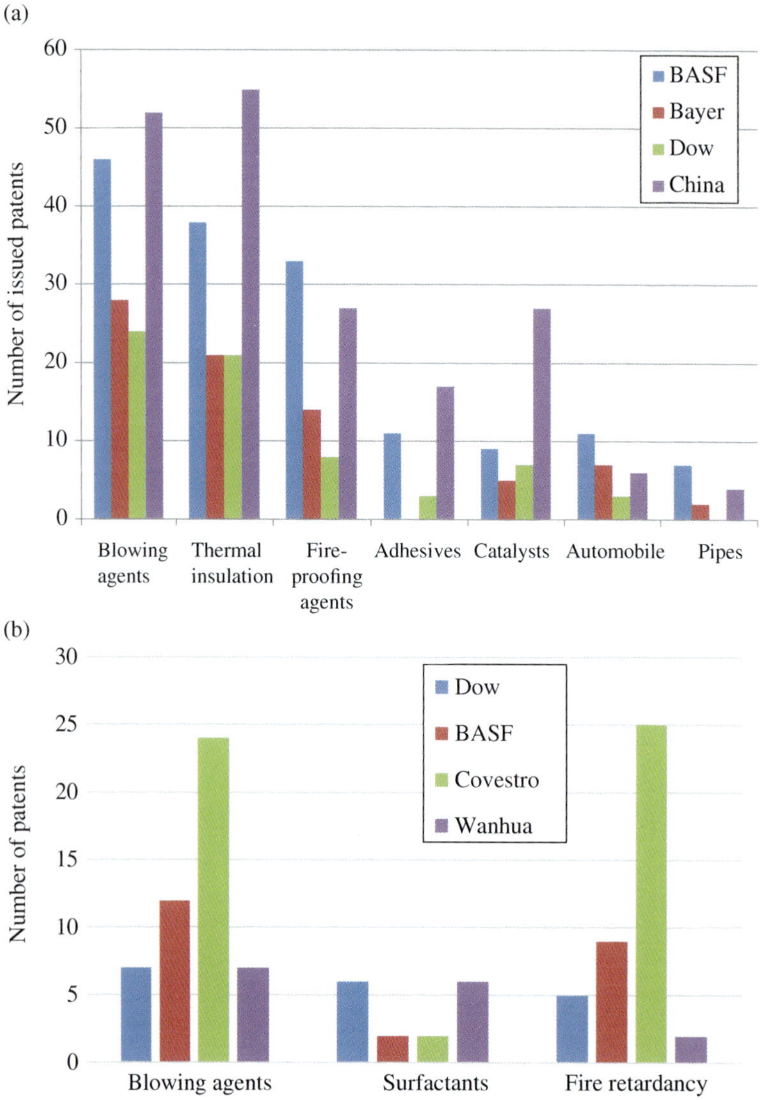


**FIGURE 8.25** Patent activity of three large polyurethane chemical producers and written in Chinese in the field of rigid foams by topic. (a) 2000–2014; (b) 2015–2019, substituting the Wanhua Group for integrated activity in China.

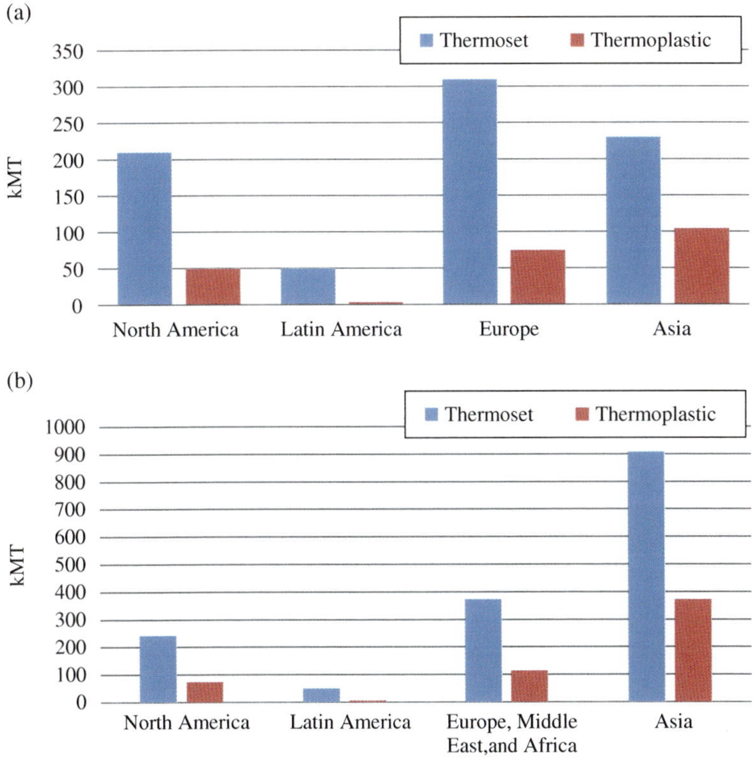


**FIGURE 9.6** Regional industrial consumption of polyurethane elastomers for (a) 2002 and (b) 2016, reflecting the migration of thermosetting elastomers, especially for shoe soles, to Asia. kMT, thousand metric tons

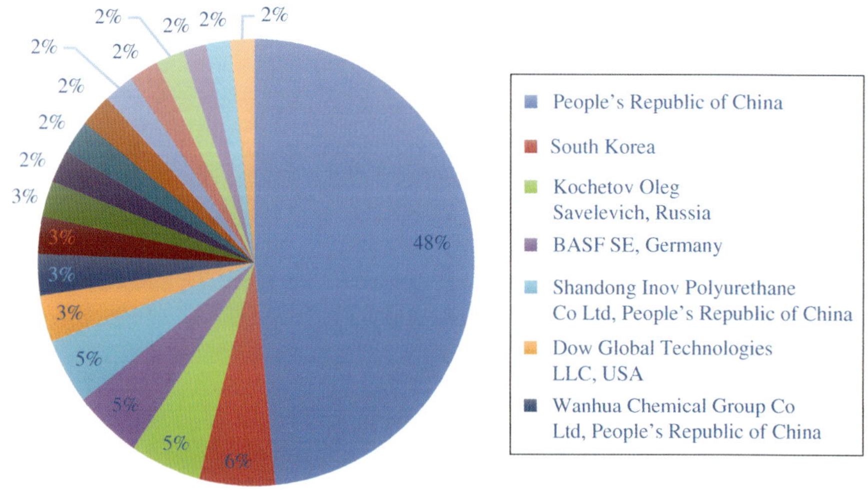


**FIGURE 9.29** Pie chart of patent activity filed from 2015 to 2019 of top 18 assignees relative to the activity of the top seven assignees. The top seven account for about 15% of all polyurethane elastomer patents filed. Many Chinese companies and universities will stipulate their national origin in their names, which then gets recorded as the assignee by the search engine.

stabilizers, dyes, etc. to result in a final soling elastomer. Aliphatic polyester polyols certainly are the predominant choice for shoe sole systems. This is because of the higher toughness of polyester polyols in these systems, the attractive pricing of polyester polyols, and the ability to easily achieve the soft materials footwear applications demand for improved traction. The deficits of polyester systems relate to suboptimal low-temperature flexibility, the requirement for using heated lines to maintain adequate fluidity, and relatively poor polyester hydrolytic stability.

Some of the problems associated with shoe soles made using polyester-based polyurethane formulations can be overcome by using polyether formulations. In the past, polyether formulations were encumbered by an inability to fabricate soft elastomers that could meet durability requirements (Table 9.1). Relatively long demolded times also hampered use of polyether systems from a manufacturing standpoint. Recent formulation and design advances have solved some of these problems, and the attractive low-temperature performance, production handling, material feel, and price attributes of polyethers have led to significant growth. Although reliable data for the relative proportion of polyester- versus polyether-based systems in shoe sole applications are lacking, it is unquestionable that polyester systems represent more than 80% of the polyurethane shoe sole volumes, and also quite likely that polyether systems will continue to gain ground as systems with better performance are created. The use of polyether systems will also increase as designers try to obtain the best of both polyester and polyether worlds by mixing the systems. There is evidence [13] that polyester/polyether mixtures actually result in systems handicapped by the weaknesses of both systems rather than compensated by the strengths, but this will not stop attempts at innovation by artful blending of components.

Polyurethane footwear made from prepolymer or quasi-prepolymer two-part formulations are pumped from their respective tanks and mixed by a high-speed (6000–20,000 rpm) mixing mechanism. The mixed and reacting polyurethane is then either injected or poured into a mold to form the desired shape. The mixing head may have numerous ports allowing for sequential injection of different formulations to get specific performance, or gas to introduce porosity and reduce density, or different dye packages to achieve specific cosmetics [14].

**TABLE 9.1 Comparison of a polybutylene adipate ester and an ethylene oxide-capped polypropylene oxide polyether-derived shoe sole material.**

| Property | Ester-based shoe sole | Ether-based shoe sole |
|---|---|---|
| Density (g/cm$^3$) | 1.21 | 1.08 |
| Shore A hardness | 65 | 65 |
| Tensile strength (N/mm$^2$) | 28 | 11 |
| Tear strength (N/mm) | 83 | 59 |
| Abrasion (mm$^3$) | 80 | 350 |
| Flex fatigue resistance (cycles at –20 °C) | >200K | >200K |

Until recently [15] footwear made from TPUs had been limited to injection molding of relatively hard footwear components such as ski boots and base plates for soccer cleats (and the cleats as well), air bags for shoe cushioning, and decorative appliqués, for instance. This was because preparation of sufficiently soft TPU had been unobtainable in pellet form without the pellets melding upon sitting in a bag or a hopper. Minimum hardness conventionally obtained for TPUs had previously been about Shore A 70. Upon superficial inspection, a Shore A 70 TPU pellet will seem soft, but it is not soft or flexible enough for shoe sole application. To obtain adequate comfort and traction on a smooth surface shoe soles typically require Shore A 45–65. Breakthroughs in TPU formulation [16] and production have in part eliminated these restrictions and TPUs are becoming more freely employed for various footwear constructions. In some cases this means use of conventional injection molding techniques. In other cases, to avoid degradation of the TPU due to overheating, manufacturers essentially pour the TPU melt into a standard casting mold rather than heating the plastic to a sufficiently high temperature that the viscosity is low enough to be injected.

Formulation and innovation of polyurethane for footwear application must always be performed with respect to specified tests. Specified tests provide a common framework for comparison of different materials and also precisely prescribe the methods for measurement. In the footwear industry certain tests and measurements are fundamental for understanding suitability. Such tests include density, hardness (Shore A for moderately soft to moderately hard or Asker C for very soft), abrasion resistance, tear resistance, the Ross flex test to measure resistance to low-temperature cut growth, and compression force deflection. Other tests may be required by specific manufacturers or for specific applications or may be more prevalent in one region than another. Other tests, such as "feel," are subjective and resistant to quantification. Table 9.2 gives a list of commonly employed tests.

For passing shoe manufacturer tests, the formulation of shoe sole systems can be as much art as science [17]. Shoe soles have become highly complex structures in the case of athletic shoes, and very-high-performance materials in the case of work and safety shoes. It would be futile to offer a shoe sole formulation or table of performance attributes that could be understood as anything other than illustrative. However, there is little doubt that certain properties are exemplary of many shoe soles, and that certain formulations possess components common to the application. From a fundamental standpoint one can begin by drawing common relationships between material properties and critical shoe sole requirements [18–21]. For instance, a shoe sole must possess a preferred coefficient of friction (COF) on dry and wet surfaces. While a high COF may be preferable, maximizing the COF for the material may entail an unacceptable tradeoff in other dimensions of performance. Table 9.3 offers the effect of designing material properties on this particular variable. While not all aspects of this comparison are intuitively consistent (i.e. varying effects of resilience and hysteresis, which are normally inversely coupled), they are

**TABLE 9.2 Representative standard tests for characterizing, comparing, and ranking materials, including polyurethanes for footwear applications**

| Test | Standard description |
|---|---|
| ASTM D3851–07 | Standard specification for urethane microcellular shoe-soling materials. Provides a list of standard tests for specific areas of performance and a table of typical values for three hypothetical grades distinguished by tensile strength and abrasion resistance |
| ASTM D624 | Test method for tear strength for vulcanized rubber and thermoplastic elastomers |
| ASTM D1938 | Test method for tear propagation resistance by trouser tear method |
| ASTM D792 | Test method for measuring density and specific gravity by displacement. Can substitute ISO 845 |
| ASTM D1052 | Ross flexural test – machine and sample preparation are specified. The test specifies a test temperature of 23 °C; however, most manufacturers test at –10 °C to –20 °C for 100,000–300,000 flex cycles to determine low-temperature flexural performance. Can substitute DIN 53543 |
| ASTM D2240 | Test method for measuring durometer hardness. Can substitute DIN 53505 |
| ASTM D5963 | Standard method for abrasion resistance using a rotary drum abrasion covering the German DIN 53516 abrasion test |
| ASTM D1630 | Test method for abrasion measured by a footwear abrader – test uses a rotary abrader like Standard D5963, but the specifics of the abrasion pattern differ |
| ASTM D395 | Test methods for determining compression set |
| ASTM D575 | Test method for measuring elastomer properties under compression. Can substitute DIN 53543 |
| ASTM D3137 | Test method for measuring rubber hydrolytic stability. Simply a high-humidity test measured at 85 °C after 96 hours using *distilled* water. Results at nonneutral pH could be substantially different |
| ASTM F2333 | Standard test method for traction characteristics of the athletic shoe–sports surface interface – application-specific test but does not require specific instrument |
| ASTM F2913 | Standard test method for measuring the coefficient of friction for evaluation of slip performance of footwear and test surfaces/flooring using a whole shoe tester. Test requires precise control using specified piece of equipment and procedures |
| ASTM F2370 | Standard test method for measuring the evaporative resistance of clothing using a sweating manikin. Can be used to quantify breathability |
| ASTM F1291 | Standard test method for measuring the thermal insulation of clothing using a heated manikin. Can be used to quantify the insulative qualities of footwear |

DIN, Deutsche Institut für Normung; ISO, International Organization for Standardization.

**TABLE 9.3 Design guidelines for the effect of a change in material property on wet and dry coefficients of friction (COFs)**

| Polymer property | | Effect on dry COF | Effect on wet COF |
|---|---|---|---|
| Modulus | ↑ | ↓ | ↑ |
| $T_g$ | ↑ | ↓ | ↓ |
| Hysteresis | ↑ | ↑ | ↓ |
| Resilience | ↑ | ↑ | ↑ |

$T_g$, glass transition temperature.

generally found to be true by experience. This suggests that the polymer structural characteristics, which are normally reflected in consistent polymer properties, are not being *tested* consistently and so provide relationships that are apparently at odds with each other [22].

An example of a polyurethane shoe sole formulation based on polyester polyols and associated properties is given in Table 9.4. The effects of ester hydrolysis are pronounced on tests that stress the degree of network continuity that hydrolytic chain scission most effects. An example of a polyurethane shoe sole formulation based on polyether polyols and associated properties is given in Table 9.5.

TPU for shoe soles represents only about 20% of the total polyurethane shoe sole market, but the total polyurethane market is only about half as large as the feedstock nonvulcanized styrene–butadiene–styrene (SBS) copolymer. This makes TPU only about one-seventh as large as the competitive TPR SBS [23–26]. While a small fraction of the shoe sole market, it still represents about 380 million pounds of polyurethane consumed – a fairly significant absolute volume of material. The general details of polyurethane manufacture will be covered later in this chapter. TPU is commonly considered a high-performance material of shoe construction because of its low-temperature flexibility, good abrasion and scratch resistance, and its ability to fill even very complex shoe molds with very high fidelity resulting in desirable aesthetic values. These attributes have made TPUs desirable components in the construction of high-quality work, safety, and hiking shoes [26–28]. Limiting factors have been cost and the relatively poor slip performance associated with difficulty making TPU at low enough hardness while still meeting customer requirements. To lower cost and reduce polymer hardness, plasticizers have been added.

Most TPU shoe soles are based on polybutylene adipate polyols with monomeric MDI and butane diol chain extender [29–31]. These materials have very good properties, good clarity allowing for easy dyeability, and relatively low cost.

TPR, which is compounded SBS copolymer (Figure 9.10), is the dominant thermoplastic rubber within the shoe sole market. Compounding components vary, but sulfur and zinc oxide are common additives employed to produce some cross-linking among the SBS chains through the butadiene moieties. Along with shoe soles, TPR finds

**TABLE 9.4 Example shoe sole formulation based on a polyester quasi-prepolymer and associated properties**

| Component | Weight | Comments |
|---|---|---|
| *Quasi pre-polymer side* | | 18% NCO |
| MDI monomer | 54.6 | 125 g/iso equivalence |
| Uretonimine-modified MDI | 6.7 | 169 g/iso equivalence |
| Polyester polyol of ethylene glycol/ diethylene glycol and adipic acid | 32.4 | 1000 g/OH equivalence |
| Polyester polyol of diethylene glycol/ trimethylolpropane and adipic acid | 2.7 | 935 g/OH equivalence lightly branched |
| Polyether polyol triblock of PO/EO end caps 20% EO by weight | 3.4 | 2000 g/eq diol |
| Benzoyl chloride | 0.2 | Acid stabilizer |
| *Polyol/chain extender side* | | |
| Ethylene glycol | 7.67 | Chain extender to define the hard segment 18 g/eq OH |
| Polyester polyol of ethylene glycol/ diethylene glycol and adipic acid | 90.4 | 1000 g/eq OH |
| Water | 0.35 | |
| 33% Triethylene diamine in ethylene glycol | 1.43 | catalyst |
| Silicone surfactant for shoe sole application | 0.13 | Helps stabilize density, decreasing bubbles created by isocyanate–water reaction |
| *Properties* | | |
| Cream time (s) | 8 | |
| Gel time (s) | 21 | |
| Minimum demold time (min) | 3.25 | |
| Density (molded, g/L) | 550 | |
| Hardness (Shore A) | 55 | |
| Tear strength (N/mm) | 7.2 | |
| Tear strength after hydrolysis test ISO 20345 (N/mm) | 6.4 | |
| Flex test (DIN 53543) (cycles) | 100,000 | |
| Flex test after hydrolysis test (ISO 20345) | 30,000 | |
| Abrasion ($mm^3$) | 80 | |

Shoe sole formulation based on polyester polyols. Quasi-prepolymer–polyol ratio = 94:100. DIN, Deutsche Institut für Normung; EO, ethylene oxide; ISO, International Organization for Standardization; MDI, 4,4′-methylene diphenyl diisocyanate; PO, propylene oxide.

common usage in tire construction. Several positive attributes of TPR include (i) price, (ii) durability, (iii) good aesthetics, and (iv) relatively easy processability [32]. This last point is a very clear advantage in the eyes of manufacturers since much less can be asked or expected of machine operators filling molds, especially in

**TABLE 9.5 Example polyether polyol shoe sole formulation based on a polyether soft segment**

| Component | Weight | Comments |
|---|---|---|
| *Quasi pre-polymer side* | | 18.4% NCO |
| MDI monomer | 65 | 125 g/iso equivalence |
| 20/80 EO/PO triblock polyether polyol OH = 27.5 | 15.5 | Diol 2040 g/eq OH |
| 14.7/85.3 EO/PO triblock polyether polyol OH =27.5 | 15.5 | Glycerine-initiated 2040 g/eq OH |
| Dipropylene glycol | 4.0 | 67 g/eq OH |
| Benzoyl chloride | 0.01 | Acid stabilizer |
| *Polyol/chain extender side* | | |
| 20/80 EO/PO triblock polyether polyol OH = 27.5 | 63 | Diol 2040 g/eq OH |
| 14.7/85.3 EO/PO triblock polyether polyol OH =27.5 | 15 | Glycerine-initiated 2040 g/eq OH |
| 17.6/83.4 EO/PO triblock polyether polyol OH =21.3 40% SAN-grafted SAN copolymer | 10 | |
| 1,4-Butanediol | 8.75 | Chain extender |
| 33% triethylene diamine in ethylene glycol | 1.5 | Catalyst |
| Silicone surfactant for shoe soles | 0.15 | |
| Polybutadiene liquid | 1 | Abrasion reducer |
| Water | 0.43 | |
| Dibutyl tin dilaurate | 0.02 | Catalyst (for formulations able to use tin catalyst) |
| *Properties* | | |
| Cream time (s) | 8 | |
| Gel time (s) | 16 | |
| Minimum de-old time (min) | 3.5 | |
| Density (molded, g/L) | 550 | |
| Hardness (Shore A) | 51 | |
| Tear strength (N/mm) | 5.2 | |
| Flex test (DIN 53543) (cycles) | >100,000 | |
| Abrasion (mm$^3$) | 130 | |

Quasi-prepolymer–polyol ratio = 67:100.
DIN, Deutsche Institut für Normung; EO, ethylene oxide; MDI, 4,4′-methylene diphenyl diisocyanate; PO, propylene oxide; SAN, styrene–acrylonitrile.

operations in emerging economies [33]. However, this simplicity attribute is becoming less consequential as automation becomes more prevalent and EVA and TPU options are more available and cost competitive. This is particularly evident when it is considered that along with beneficial properties TPRs suffer inferior tensile

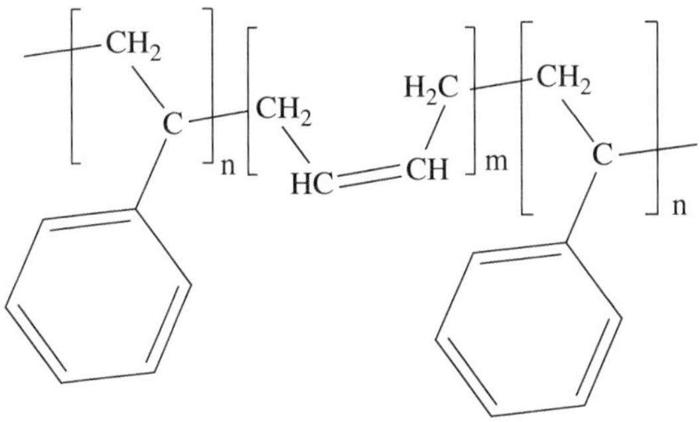

**FIGURE 9.10** Styrene–butadiene–styrene triblock copolymer, the major component of TPR used in shoe soles.

properties, abrasion resistance, resilience, and slip resistance at equivalent hardness in comparison with TPUs.

***9.2.1.1 Trends in Footwear Applications*** Although consumers rarely ask questions about their shoe sole materials, it would surprise many if informed about the level of innovation and technology that is invested into shoe soles. While patenting on shoe sole technology is a global pursuit, since 2000 the majority of patents have been filed from concerns located at the shoe manufacturing centers. Since the year 2000, 952 patents were filed specifically on polyurethane shoe soles; however, 362 of those were filed during 2015–2019. Since the year 2000, 412 were on behalf of companies entirely of Chinese origin. Among large global manufacturers BASF filed 91 patents on shoe sole technology, Bayer/Covestro 62, and Dow Chemical filed 21 patents. The technology covered in these patents relates to improved hydrolysis, improved abrasion, improved cosmetics and design, improved manufacturing, electrically conductive shoe soles, shoe soles of complex density gradients, shoe soles with very low density and excellent resilience, improved mold release agents, and odor mitigation. Certainly, the predominant technology trend is pursuit of high performance at reduced densities [34, 35]. The evidence for this trend is validated by casual inspection of shoes displayed at retail outlets.

### 9.2.2 Nonfootwear Elastomer Applications and Methods of Manufacture

Apart from shoe soles, elastomer categorization relative to applications is an exercise in rapidly diminishing returns. Analysis of large sectors such as "transportation" reveals that the applications are myriad, none of which by themselves would merit significant exposition, or are much like other unrelated applications. A partial list of application and application areas is provided in Table 9.6. For instance, the requirements of conveyor belts [36] and tire treads [37, 38] are quite similar but apply to vastly different consumers and compete with different companies and alternative materials. Rather, these applications are differentiated by their use of specific

**TABLE 9.6 A representative listing of polyurethane elastomer applications**

| | | |
|---|---|---|
| Tire fill | Instrument panel covers | Pipeline pigs |
| Print rollers | Shock absorbers | Pipeline insulation |
| Conveyor belts | Nonpneumatic tires | Roof membranes |
| Roller-skate wheels | Run-flat tires | Rubber tougheners |
| Cattle tags | Pump impellers | Mechanical polishing pads |
| Toothbrushes | Track and field surfaces | Suspension bushings |
| Blood bags | Potting/encapsulation | Railway pads |
| Medical devices | Water skis | Bicycle helmets |
| Seals | Bicycle seats | Electrical jacketing/tubing |
| Gaskets | Tool hand covers | Breathable films for apparel |
| Spray coatings | Belts and pulleys | Airline emergency slides |

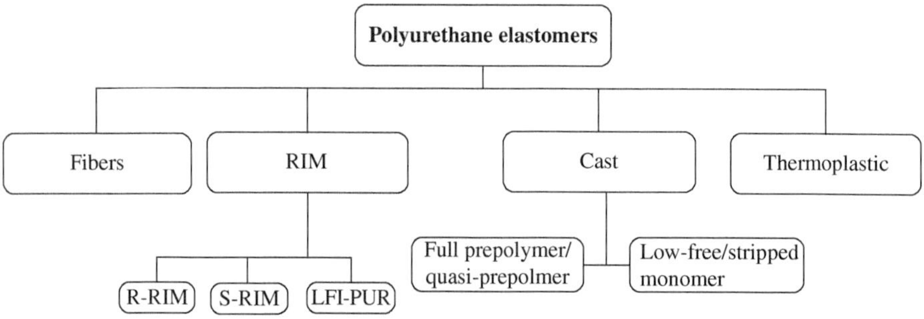


**FIGURE 9.11** Segmentation of polyurethane elastomers not including footwear. Elastomers applied as coatings are also not included here. RIM designation defined in Section 9.2.2.3

polyurethane components and the way these components are then used to make parts. A graphical way of depicting these components of construction is shown in Figure 9.11.

***9.2.2.1 Cast Elastomers*** within applications. The custom nature of the applications and resulting formulations makes for a complex commercial environment. Since the applications are relatively small and dispersed among numerous end users, cast elastomers seem like an ideal commercial space for systems houses to flourish. Indeed, polyurethane-formulating systems houses do occupy an important distribution channel. However, the availability of raw materials such as polyols, isocyanates, and curatives at low cost invites one-stop-shop distributors to serve as alternative lower cost middle men, as well for molders to try and fill the space of these middle men, and to purchase directly from the large chemical manufacturers (Figure 9.12). The ability of molders to choose their mode of purchase creates a very competitive market, and results in very rapid commoditization of formulation innovation or methods of production [1] and movement of manufacturing to the lowest cost environment.

The choice of commercial relationship will also depend on the technical sophistication of individual molders and their interest in workplace health and safety

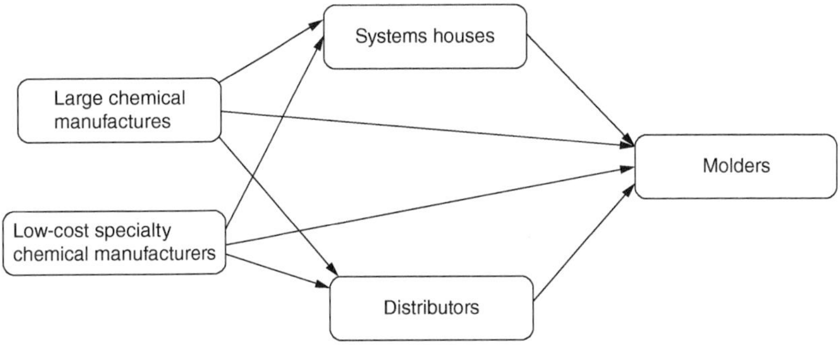


**FIGURE 9.12** A simplified representation of polyurethane elastomer market positions.

controls. For cast polyurethane elastomers the primary means of production that has created equilibrium relationships among all of the participants has been via use of prepolymer and stripped (or low-free isocyanate) prepolymer formulations.

For clarity, the prepolymer approach can be contrasted to the so-called "one-shot" approach. When employing a one-shot approach an elastomer producer will take a predetermined amount of polyisocyanates, polyols, curative (or chain extender), and additives, mix thoroughly, and pour them into a mold, where reaction proceeds to completion.

In a prepolymer process, polyol is reacted with a stoichiometric excess of isocyanate such that there is a minimal amount of polymerization that can occur since all the available hydroxyl groups are reacted with isocyanates and there is no available hydroxyl functionality to react with isocyanate (Figure 9.13) [39–41]. These fully reacted polyols possessing isocyanate end groups then make up one side of a two-component formulation. The other side will usually consist of curative (chain extender) or polyol and chain extender and other additives.

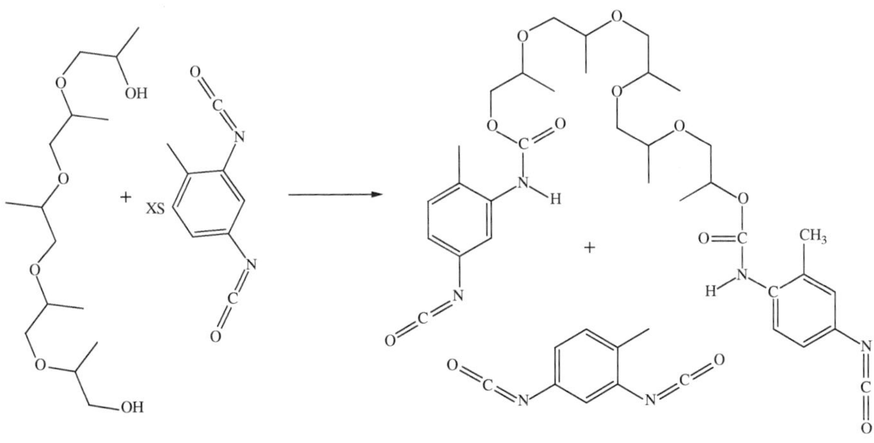

**FIGURE 9.13** The prepolymer production concept.

**TABLE 9.7 Characteristics of the types of commercially available polyurethane prepolymers**

| Prepolymer variety | Amount of free isocyanate (%) | Comments |
|---|---|---|
| Quasi-prepolymer | 12–28 | Requires formulating with polyol and curative – provides flexibility to molder to vary final elastomer properties |
| Full Prepolymer | <12 | Provides very high control to molder for a specific final polymer result, but each specific prepolymer can only make a single elastomer |
| Low-free isocyanate | ~1 | Less expensive than stripped prepolymer |
| Stripped prepolymer | <0.1 | Most expensive, lowest hazard, usually does not require isocyanate hazard labeling |

From this process, four different prepolymer varieties have established themselves: the full prepolymer, the quasi-prepolymer, the low-free isocyanate prepolymer, and the stripped prepolymer [42]. These prepolymer varieties differ from one another merely in the amount of free isocyanate left in the prepolymer side to then be reacted with curatives or curatives and polyols (Table 9.7).

In a quasi-prepolymer preparation, a relatively small amount of polyol is added to a large stoichiometric excess of polyisocyanate, such as described in Figure 9.14.

While Figure 9.14 depicts the preparation of a toluene diisocyanate (TDI) quasi-prepolymer, MDI quasi-prepolymers are commonly found and are predominant in manufacturing environments trying to limit worker chemical exposure to isocyanates. The full prepolymer is characterized by a much lower content of free isocyanate, which somewhat simplifies handling but also limits the number of different materials that can be prepared from a single full prepolymer. The quasi-prepolymer allows the molder to vary end products simply by varying the ratio of curative and polyol to the second component. The same can be accomplished to a more limited degree with the full prepolymer, but, because of the low-free isocyanate available, the molder will usually only add curative as a second component in order to build hard-segment volume. While the full prepolymer and quasi-prepolymer approaches have dedicated users in all geographies, it is more common for European molders to request quasi-prepolymer systems and North American formulators to request full prepolymer systems.

While full prepolymer and quasi-prepolymer formulated systems are predominant in most cast elastomer applications, there is growing awareness, in all geographies, that free isocyanate can present an undesirable exposure risk to workers in environments that might otherwise have acceptable environmental controls. For these customers, producers have created products that have the excess isocyanate removed from the prepolymer. The thoroughness of the distillation will create either the low-free (monomer) prepolymer or the stripped prepolymer. There is not much reliable information on the relative volumes of low-free versus stripped prepolymer, and they are often

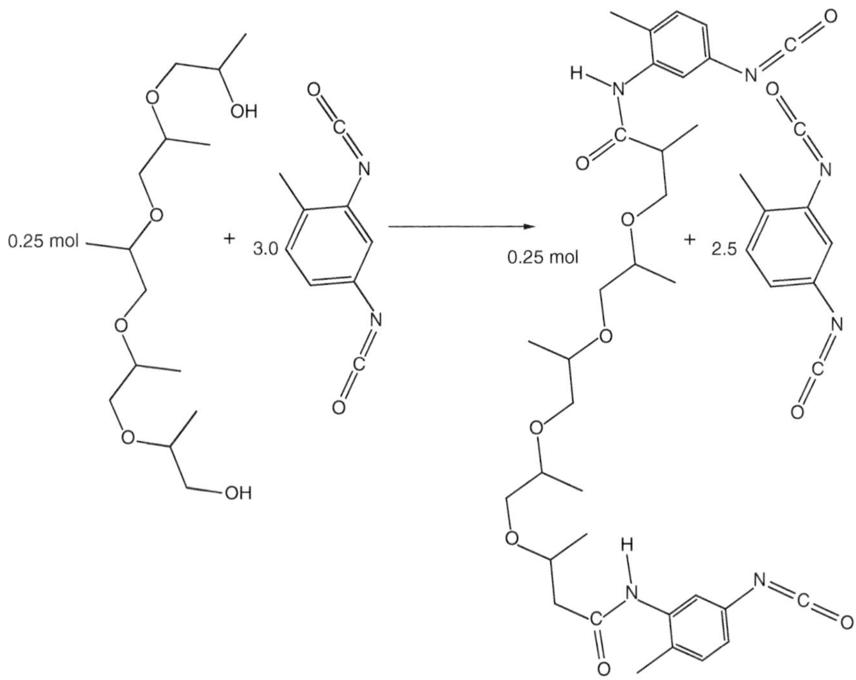


**FIGURE 9.14** Example of the quasi-prepolymer concept.

considered as a single class. Furthermore, owing to the relatively greater difficulty distilling low-vapor-pressure MDI from prepolymer systems [43] the prevalence of stripped and low-free TDI prepolymers is much greater than the same for MDI. Figure 9.15 shows that the North American market for stripped/low-free prepolymers (all TDI based) is approximately 30% of the total prepolymer market.

The overall size of the prepolymer market is difficult to estimate with a high degree of confidence and available published market estimates are often unwarranted extrapolations from unrelated data. A reasonable estimate based on analysis of numerous sources is that the overall prepolymer market, irrespective of the isocyanate, is about 400 million pounds split in half between TDI and MDI with a trend toward increasing MDI incorporation. The overall choice of customers to use a low-free or a stripped isocyanate prepolymer is to some extent arbitrary, though two things are certain: (i) there is a substantial price increase, as much as US$1.0 per pound, in reducing free isocyanate from 1% to less than 0.1% and (ii) that stripped prepolymers do not have to possess warning labeling on their packaging, while labeling of low-free prepolymers varies from country to country [45–47]. For customers who are insensitive to labeling issues, the price differential generally drives them toward low-free prepolymers since the absolute difference in resulting elastomer properties between low-free and stripped prepolymers is generally anecdotal, and likely reflects subjective preference.

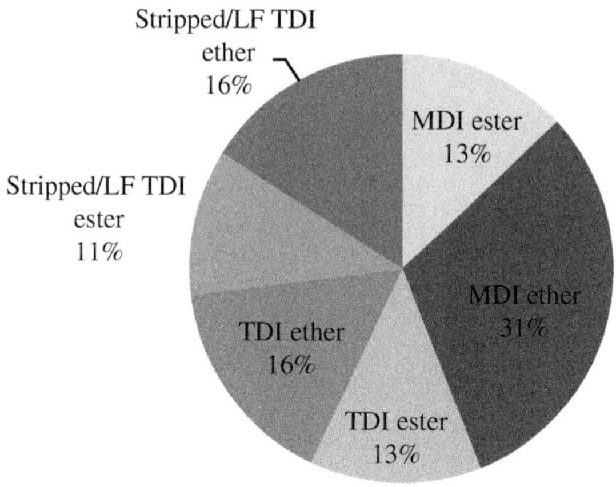


**FIGURE 9.15** North American consumption of prepolymers by type. LF, low free isocyanate. Reprinted with permission from Ref. [44]. © John Wiley & Sons, Inc.

The decision to use a quasi-prepolymer or a full prepolymer has consequences that the molder must consider before making a process commitment. The basic equipment is common for both quasi-prepolymers and full prepolymers except for the possible use of a third tank when working from a quasi-prepolymer in order to vary curative and polyol volumes independently (Figure 9.16) [48]. Other differences are provided in Table 9.8.

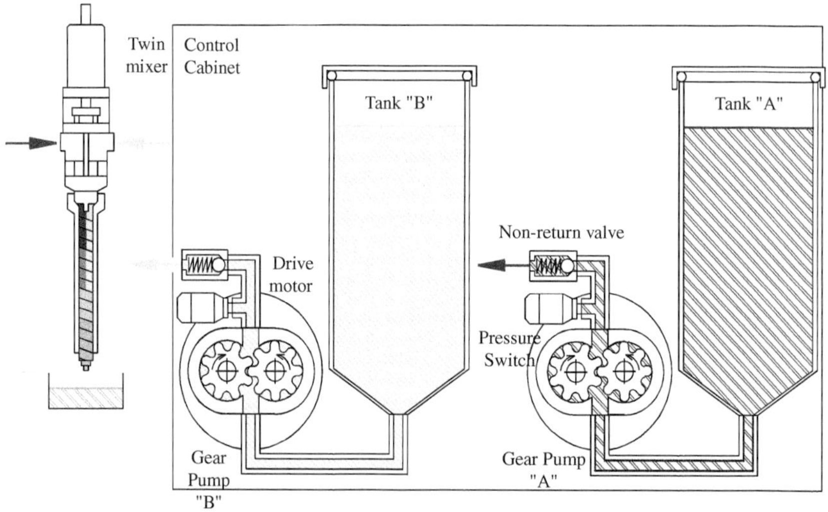


**FIGURE 9.16** Simplified diagram of a cast elastomer molding system with two tanks. A quasi-prepolymer system might often employ a third tank to allow for easy variation of curative and polyol.

**TABLE 9.8 Differences between quasi-prepolymer and full prepolymer systems and resulting elastomers**

| Property | Full prepolymer | Quasi-prepolymer |
|---|---|---|
| Viscosity | Higher (often solids) | Lower |
| Shelf life | Longer | Shorter owing to dimerization of MDI monomer |
| Exotherm during reaction | Lower (many fewer reactions to achieve cure) | Higher and may promote unwanted side reactions |
| Shrinkage during cure | Lower | Higher (owing to greater exotherm) |
| Flexibility in making materials | Very low – one full prepolymer can make a very limited range of materials, usually just variation by changing curative | Very high by simultaneous variation of polyol and curative/ chain extender/shorter turnaround times with equipment |
| Catalyst required | Often not when using aromatic amine curatives | Yes – requires high gelation activity |
| Mechanical properties | Usually better defined and more predictable | Also good but phase separation dynamics may be highly sensitive to process details |
| Low-temperature handling | Usually requires significant amount of heating of tank and heating of transfer lines | Much lower heating requirements |
| Worker exposure issues | Higher with TDI-based prepolymers, but otherwise lower owing to lower overall isocyanate | Lower with MDI quasi-prepolymer systems |
| Abrasion (at equal hardness) | Higher | Lower (owing to resistance of hard segment) |
| Chemical stability [49] | Reported to be lower | Reported to be relatively better |

MDI, 4,4′-methylene diphenyl diisocyanate; TDI, toluene diisocyanate.

In no way representative of all full or quasi-prepolymer systems, Table 9.9 provides a simple comparison of formulation and resulting elastomer properties. A representative experimental procedure for making prepolymers and making the cast mold plaque for testing is provided in the Table 9.9 footnote. The simplicity of the procedure is largely responsible for wide adoption of cast elastomer techniques; however, proper protection from exposure to isocyanates should always be employed. Elastomers made with process and formulation optimization can be employed in applications such as those listed in Table 9.10.

***9.2.2.2 Thermoplastic Polyurethanes*** TPUs are generally regarded as providing the desirable properties of polyurethane elastomers with the desirable processing of thermoplastics. This means that TPUs can be processed by extrusion, injection molding, casting, and casting from solution. TPUs achieve this process-

**TABLE 9.9 Prepolymer formed at between 70 °C and 80 °C under $N_2$ at desired NCO content for 3–6 hours**

| Component | Quasi-prepolymer | Full prepolymer |
|---|---|---|
| Polyether polyol | 16% NCO (MDI) | 12% NCO (TDI) |
| Curative/chain extender | BDO | MOCA |
| Hardness range (Shore A) | 60–90 | 40 (Shore A)–60 (Shore D) |
| Tensile strength (Shore A) (psi) | 4800 | 5400 |
| Elongation at break (Shore A 90) | 450 | 650 |
| Compression set (%) | 30 | 30 |
| Ball rebound (%) | 50 | 56 |

The extent of reaction is determined by amine titration (di-*n*-butylamine) followed by degas at 70 °C. Prepolymers mixed with the second part in a speed mixer followed by open-air cure at 100 ° in a heated mold followed by post-cure heating (annealing) at 100°C in a convection oven.
BDO, 1,4-butane diol; MDI, 4,4′-methylene diphenyl diisocyanate; MOCA, 4,4′-methylenebis(2-chloroaniline); TDI, toluene diisocyanate.

**TABLE 9.10 Applications defined by the typical Shore A hardness required**

| Application | Shore A | Shore D |
|---|---|---|
| Papermaking rolls | | 70 |
| Metal forming wiper dies | | 65 |
| Mallet heads | 95 | 55 |
| Run-flat tires | 95 | |
| Solid tires | 90 | 45 |
| Metal forming die pads | 85 | 40 |
| Idler rolls | 80 | 35 |
| Abrasive handling pads | 75 | |
| Auto tire treads | 65–80 | |
| Silk screen wiper blades | 60 | |
| Door seals | 55 | |
| Inner tubes | 50 | |
| Printing rolls | 25 | |
| Rubber bands | 20 | |

ing flexibility, and other polyurethane systems do not, by avoiding cross-linking of the polymer chains. By and large, all building blocks used for making TPUs have a functionality of 2.0. It has been shown that TPU processing is not noticeably harmed when employing soft-segment functionalities as high as 2.05 (by incorporation of small amounts of a three-functional polyol), but soft-segment functionalities as low as 2.1 cannot be processed as a melt [50]. As discussed in the section on cast elastomers, the market for TPUs is also fractured among applications (Figure 9.17) and customers. Each application segment has numerous subsegments

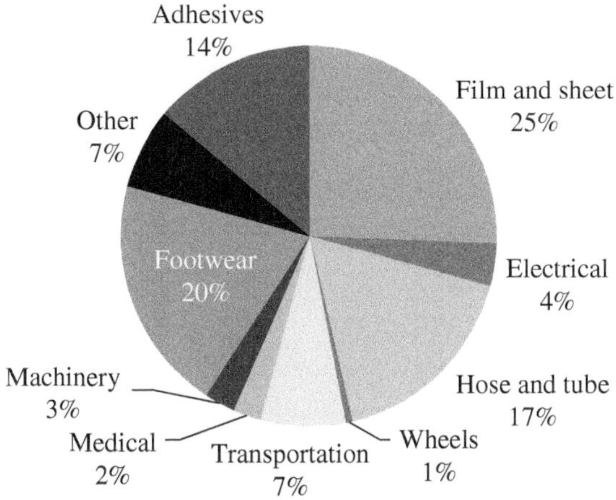


**FIGURE 9.17** Global segmentation of the 2016 TPU market by consumption volume.

demanding unique solutions. In keeping with China's dominance in parts manufacturing, and particularly in footwear, it is consistent that China should be the dominant consumer of TPUs, and the growth of its consumer culture coupled with increasing demand for automotive transportation assures that China should continue to see a high TPU growth rate (estimated at about 5% per year) for the near future.

While production of TPUs is not very technologically advanced, it is capital intensive, requiring large twin screw extruders for reactive processing and a significant amount of specialized processing equipment for before and after the reactive extrusion [51]. Furthermore, since the number of applications and requirements within each application segment are numerous, the requirement for rapid and frequent formulation change-over and inventory storage complicates manufacturing and customer service. North American production of TPUs is dominated by Lubrizol Corporation, which manufactures about 65% of North American production. The other producers are low-cost global producers of TPU feedstocks (particularly MDI required for its 2.0 functionality) such as BASF, Covestro, and Huntsman. The same is true for Europe, where they are joined by COIM, a major polyol producer. These same producers also maintain a footprint in China, often with local partners. Yantai Wanhua, China's largest producer of MDI, is also the largest producer of TPU in China, but for perspective, in a more fractured market, this reflects only about 15% of all Chinese production (Figure 9.18).

The capital requirements for TPU production are not conducive to service by formulation systems houses operating on a low capital asset business model. However, production of TPU grades for large accounts can be made more profitable by selling additional inventory to smaller, more difficult to service accounts through distributor channels (Figure 9.19). The success of this structure is complicated by the drive to

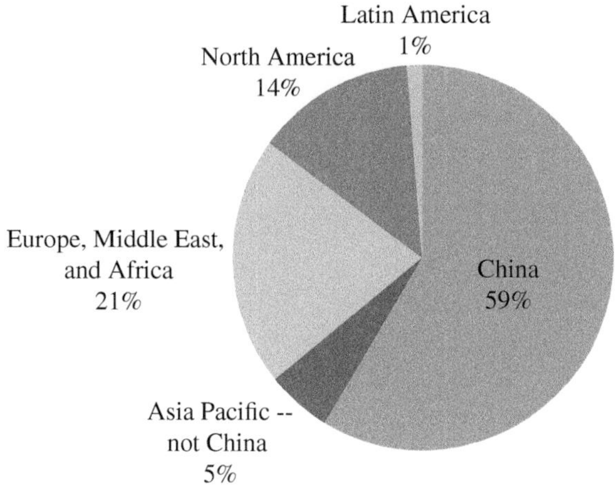


**FIGURE 9.18** Segmentation of the 2016 global TPU market by region.

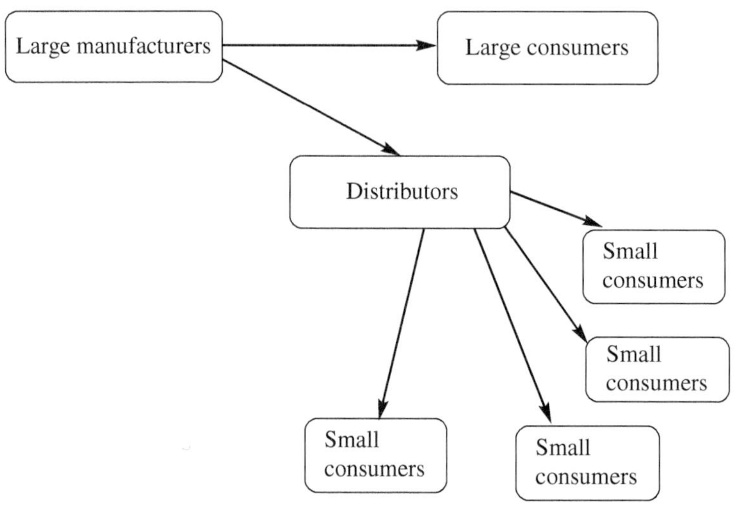


**FIGURE 9.19** Simplified diagram of the commercial relationships between manufacturers and molders of TPUs. Systems houses do not participate in this segment owing to the large capital costs associated with polymer production.

low-cost supply and the competing cost for distributor profit. It is additionally complicated by small consumers' need for technical assistance and the uneven ability of distributors to supply such service.

TPU technology is relatively straightforward and highly optimized. TPUs are produced by reactive extrusion, in which all of the ingredients are simultaneously introduced to a twin screw extruder; under temperature and mixing action within the

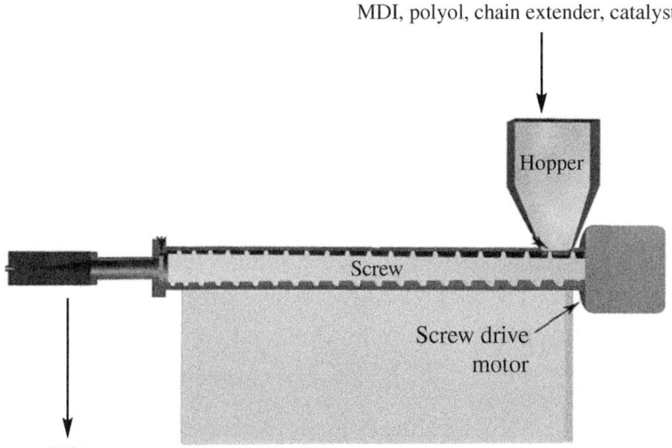


**FIGURE 9.20** Simplified illustration of reactive extrusion of TPUs. *Source:* [44].

screw elements, the extruder produces a fully formed polymer at the extruder exit (Figure 9.20). Almost all TPUs are made with MDI and butane diol for the hard segment. Soft-segment differences have been the major area of differentiation in performance and price. The reaction is catalyzed using hydrolyzable tin catalysts such as tin octanoate. Hydrolytically stable catalysts such as dibutyl tin dilaurate are not used industrially in order to avoid complications of a live catalyst in the polymer for long periods. In the past, tin catalysts were preferred to amine catalysts because of lower color formation, lower odor, and specificity for the urethane reaction; however, recent restrictions, particularly in Europe, have generated increased reliance on amines [52].

While it is a simple matter to substitute TPU reactive components, most commercially available TPUs are composed of MDI and 1,4-butane diol (BDO) hard segments. Manufacturers will use different soft segments depending on the application. For hydrolytically sensitive applications manufacturers will typically use polyether polyols. The vast majority of these use polytetramethylene ether glycol (PTMEG), which is prepared from the polymerization of tetrahydrofuran (see Chapter 2). PTMEG has many good attributes contributing to TPU properties. Among these is that, as a linear molecule, it is able to accommodate a balance of hard-phase separation with hard-phase dispersion. The phase separation is accommodated by PTMEG only interacting with hard segment by hydrogen bonding through the ether linkage [53, 54]. This limited soft segment–hard segment interaction minimizes the competition with hard segment–hard segment interaction. Also, as an unbranched chain, PTMEG does not preferentially compete for free volume in the polymer, optimizing hard-segment phase dispersion. Another benefit of PTMEG is that the terminal hydroxyls of PTMEG are all primary, maximizing the reaction rate with isocyanates. PTMEG is also able to strain crystallize at higher molecular weights and so increase the ultimate tensile strength of the polymer. Finally, because of the method of preparation (see Chapter 2), all the polymer chains in PTMEG are bifunctional, maximizing network connectivity and elastomer properties. TPUs based on PTMEG are generally regarded as possessing a very good balance of properties but are among the most expensive owing to the market price of PTMEG, which is tied to the price of tetrahydrofuran as opposed to the less expensive propylene oxide (see Chapter 2) and a more expensive process (see Chapter 2). PTMEG is a polyether, and therefore thermally limited by a relatively low ceiling temperature. At the ceiling temperature the polymer will depolymerize back to starting materials; for polyethers, this is usually between 230 °C and 250 °C [55].

Polyethers based on polypropylene oxide make poorer TPUs than does PTMEG; this is a result of macrophase separation of hard segment caused by the free volume requirement of polypropylene oxide pendant methyl groups [56]. Equally problematic for TPUs prepared from polypropylene oxide is the slow reactivity of the terminal secondary hydroxyl groups of these soft segments. The reactivity ratio of secondary to primary hydroxyls is at least a factor of 3 (see Chapter 3), which makes it difficult to complete polymerization during the short reaction time in the reactive extruder, and also drives the equilibrium of hydroxyl and isocyanate to urethane unfavorably back to reactants [57]. Lastly, because of reaction isomerization there is a net probability for some of the end groups of polypropylene oxide to not create terminal hydroxyls, but instead create terminal allylic groups (see Chapter 2 for a discussion of "monol" production). Since the formation of allylic groups is a statistical process, and they are polymerization chain stoppers, their proportional importance is greater as soft-segment molecular weight increases [58, 59].

TPUs prepared with polyethylene oxide soft segments are elastomeric, but phase separation of hard and soft segments is hindered by the large number of ether groups in the soft segment available to hydrogen bond with the hard segment [60, 61]. This hard–soft segment interaction competes with intrahard-segment hydrogen bonding, promoting *phase mixing* rather than phase segregation. Also, polyethylene oxide soft segments are relatively hygroscopic. TPUs with polyethylene oxide absorb up to 15% of their weight in water from the atmosphere [62], whereas TPUs with PTMEG soft segments will absorb about 1% water by mass at standard test conditions. Water plasticization of polyethylene glycol TPUs greatly degrades TPU tensile properties.

TPUs are also commonly prepared with polyester soft segments. Polyester soft segments are preferred for applications where potential hydrolysis of the material is not an issue for long-term performance and where cost is of paramount concern. Most polyester TPUs employ soft segments made from condensation polymerization of adipic acid and butane diol. Along with a price that is less than half that of PTMEG, polyester polyol creates acceptable and transparent thermoplastic elastomers. Furthermore, polyesters offer superior thermal stability to polyether polyols, but can lose molecular weight in the presence of moisture and heat, reverting to reactant acids and alcohols. This degradation pathway can be catalyzed by the presence of acid end groups, which are an inevitable artifact of the polyester polymerization.

These acid end groups also degrade polyurethane network connectivity by effectively acting as chain stoppers.

The carbonyl groups of polyester polyols are efficient competitors with urethane carbonyls for hydrogen bonding with the urethane N–H. The result of this is that a higher volume percentage of hard segment is required to effect hard segment–soft segment phase separation, and the equilibrium mixing of nonphase-separated urethane groups is relatively high [63, 64] (see Chapter 4). This can result in a higher soft-segment glass transition temperature ($T_g$) in the polymer; however, in many cases, this effect is not significant enough to greatly hinder low-temperature toughness of these materials. Additionally, the high-temperature transurethanification reaction produces alcohol groups from hard-segment diols as well soft-segment diols. At high temperature, scrambling of the alcohol units can contribute to additional phase mixing [65].

With alternative thermoplastic elastomers available, the parts designer must try to evaluate the suitability of a particular material for its role. In general, TPUs are considered very high-performing elastomers with a very good balance of properties, but not ideal for all applications. Table 9.11 shows comparative properties for TPUs from polyether and polyester soft segments compared with a generic block copolymer polyester elastomer and an olefin elastomer. While many properties are good for each

**TABLE 9.11 Comparison of properties for typical thermoplastic elastomer classes**

| Property | Copolyester elastomer | Olefin elastomer | Polyether polyurethane | Polyester polyurethane |
|---|---|---|---|---|
| Transparency | 0 | +/0 | + | + |
| Tensile strength | + | +/0 | + | + |
| High-temperature tensile properties | 0 | – | 0 | 0 |
| Modulus | + | 0 | + | + |
| Abrasion resistance | 0 | – | + | + |
| Compression set | 0 | – | + | + |
| Tensile set | 0 | – | + | + |
| Tear strength | + | + | + | + |
| Low-temperature resilience | 0 | + | + | + |
| Chemical resistance | 0 | + | 0 | 0 |
| High-temperature stability | 0 | – | + | + |
| Water resistance | – | + | – | + |
| Weatherability | + | + | +/– aliphatic/aromatic | +/– aliphatic/aromatic |
| Density | – | + | 0 | – |
| Cost | – | + | 0 | + |

The choice of which elastomer to use will depend on performance and price requirements. + indicates good performance; 0 indicates adequate performance in limited situations, – indicates relatively poorer performance.

**TABLE 9.12 Example formulations for TPUs of various hardnesses (modulus) and from various soft segments**

| Component | Shore A 70 | Shore A 80 | Shore A 91 | Shore A 92 |
|---|---|---|---|---|
| PTMEG 2000 | 72 | | | |
| PTMEG 1000 | | 60 | | |
| Polycaprolactone | | | 60 | |
| Polybutylene adipate | | | | 60 |
| BDO | 5 | 6.5 | 8.4 | 8.4 |
| 4,4′ MDI | 23 | 33.7 | 31.6 | 31.6 |
| Modulus (psi) | 450 | 770 | 1300 | 800 |
| Elongation at break (%) | 655 | 560 | 1700 | 2200 |
| Break stress | 4700 | 6800 | 5650 | 9000 |
| Abrasion (mm$^3$) | 37 | 33 | 27 | 35 |
| $T_g$ (°C) | −44 | −25 | −18 | −14 |
| Isocyanate index | 1.02 | 1.02 | 1.02 | 1.02 |
| % Hard segment | 28 | 40 | 40 | 40 |

MDI, 4,4′-methylene diphenyl diisocyanate.

elastomer class, there is no doubt that TPUs distinguish themselves with regard to tensile, dynamic, and abrasion dimensions of performance.

As mentioned previously, it is a simple matter to prepare new TPU formulations as long as the materials can form a thermoplastic (all components are two-functional), they are dry, they are matched stoichiometrically, and they are properly catalyzed for reaction during their residence within the reactive extruder. Table 9.12 provides illustrative formulations for making TPUs of a given hardness. They are conveniently prepared by their simultaneous introduction into the feed throat of a reactive extruder such as that shown in Figure 9.20. Along with the reactive components the manufacturer will often introduce catalysts, antioxidants, processing aids such as waxes for easy transit through the process and mitigating interpellet stickiness, and also small amounts of monofunctional alcohols to limit molecular weight if a higher melt index polymer melt rheology is desired.

***9.2.2.3 RIM Elastomers*** Reaction injection molding (commonly referred to as "RIM") is a processing technique for forming relatively large urethane elastomer structural panels. To provide adequate structural rigidity it is common to incorporate glass or mineral fiber fillers into the polymer structure [66, 67]. As with all composites, process innovation has created numerous hyphenated forms of RIM. The basic process is superficially similar to that of some cast elastomer processes and thermoplastic injection molding. The components of the urethane matrix are energetically mixed and injected or poured into a mold. In many cases the mold is prearranged with a glass fiber mat that provides composite reinforcement. Other embodiments may have the glass or other filler applied to the casting polymer solution and the mold subsequently closed. The details of the injection process and the composite phase employed is coded into the prefix used to identify the RIM species. For instance,

- R-RIM (reinforced-RIM): the process in which short mineral fibers are dispersed into the polyol phase and mixed with the isocyanate phase prior to injecting the reacting mixture into a closed mold.
- S-RIM [68] (structural-RIM): the process in which, prior to injection of the reacting mixture, the mold is prelaid with a woven glass mat or prepreg. Usually a closed mold but not required. This technique may also be termed LD-RIM for low-density-RIM.
- LFI-PUR [69]: the process in which chopped long glass fiber is randomly applied to a mold simultaneously with the reacting polyurethane resin and the mold left either open or closed.

A pultrusion operation is yet another way of making a polyurethane composite [70]. In a pultrusion operation a fiberglass roving, yarn, or mat, is drawn through a molten polymer or reactive system, and cured by continuous draw through an oven. The result is a material of constant cross-section. While often anisotropic in properties, unidirectionality can be offset by the use of off-axis fiber alignments.

Historically, the primary use of RIM polyurethane products has been for automotive structures such as bumpers, vertical structural panels, and under-the-hood applications that are not in direct contact with the engine. However, the increasing use of formulated thermoplastic olefin elastomers (referred to as TPOs) has seen the utilization of RIM decrease dramatically (Figure 9.21). The dramatic drop in

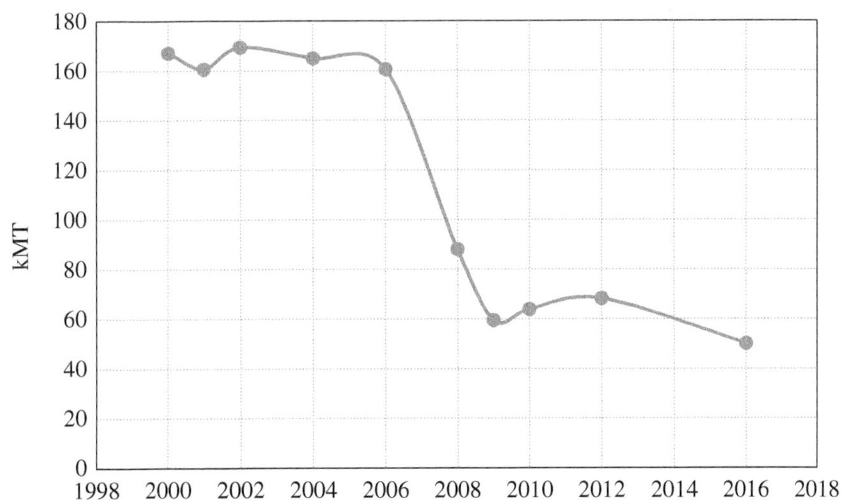


**FIGURE 9.21** Volume production of polyurethane building blocks for RIM applications. kMT, thousand metric tons.

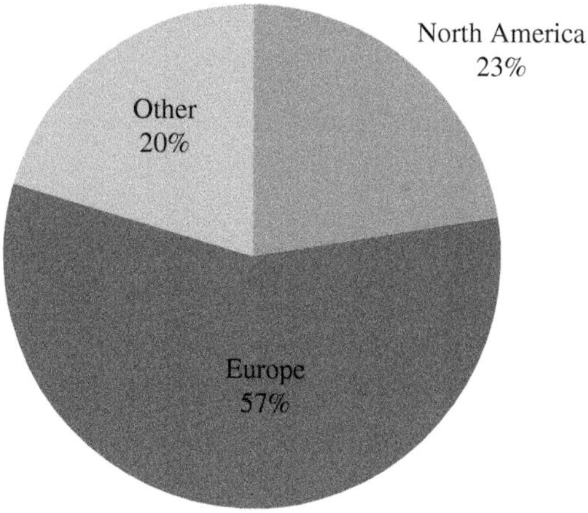


**FIGURE 9.22** Geographical segmentation of RIM production in 2016.

RIM consumption reflects a significant drop in automotive production at that time. In the 1990s there was a similar fall-off in RIM-based polyurethane consumption because of the rapid adoption of the less expensive, lighter, and less complex TPOs in the automotive industry. Other uses for RIM have remained approximately steady in terms of volumes, including continued use in some automotive platforms, tractor panels, snowmobile panels, and furniture [71] (Figure 9.22).

In terms of a competitive landscape for material solutions, polyurethane composites can be compared with technologies employing epoxy matrices in resin transfer operations or sheet molding compound (SMC) usually employing an unsaturated polyester resin. A qualitative comparison of properties obtained by these techniques is provided in Table 9.13 [72].

The processing conditions of a RIM part are as much a part of formulation conditions as final composite properties. The ability to rapidly and reliably produce high-quality parts at competitive costs will be implied in the choice of each component and how the overall system is employed (Table 9.14).

The trends in RIM polyurethanes are those mostly expressing the interests of manufacturers of RIM components rather than RIM consumers. From 2008 through 2013 approximately 35 patents were filed, and from 2015 to 2019 10 patents on RIM technology were filed. The subject matter of these patents varied widely but was primarily on the application of RIM to specific automotive parts. While an average of about four patent filings per year is indicative of the general lack of commercial excitement in RIM technology, a focus was on the application of RIM to foamed shaped automotive parts for weight reduction [73]. In nonautomotive applications there were several patents filed on the application of RIM methods to making windmill components [74]. In terms of RIM methods, there is activity protecting methods of fabrication using chopped long fiber to make class A finished body panels, and

**TABLE 9.13 Comparison of properties obtained for various plastic composites**

| | R-RIM | S-RIM | LFI-PU | RTM | SMC |
|---|---|---|---|---|---|
| Resin | PU | PU | PU | Epoxy | Polyester |
| Cycle time | Fast | Slow | Slow | Slow | Fast |
| Applicable to large parts? | No | No | Yes | Yes | Yes |
| Easily automated? | Yes | Partial | Yes | Partial | Yes |
| Surface quality[a] | Good | Good | Good | Good | Good |
| Modulus | Low | High | High | High | High |
| Toughness[b] | High | High | High | Low | Medium |
| Thermally stable? | Moderate | Moderate | Moderate | High | High |

PU, polyurethane; R-RIM, PU reinforced with mineral fiber; S-RIM, PU reinforced with glass mat; LFI-RIM, PU reinforced with long chopped glass fiber; RTM, epoxy resin reinforced with a glass prepreg in a resin transfer molding operation; SMC, a sheet molding compound of glass fiber reinforced with a polyester resin.

[a] Class A or near class A obtainable.

[b] Defined as the integral under a stress–strain curve.

**TABLE 9.14 A modified S-RIM formulation that could be scaled for making an automotive door panel**

| Component | Weight | Comments |
|---|---|---|
| Polyether triblock 85% PO/15% EO cap | 34 | 1600 equivalent weight, functionality = 3 |
| Propoxylated ethylene diamine | 50 | 88 equivalent weight, functionality = 4 |
| Ethylene glycol | 14 | Chain extender |
| Water | 1.5 | Chain extender/blowing agent |
| Silicone surfactant | 1 | |
| Pentamethyldiethylene triamine | 1 | |
| Dimethylethanolamine | 0.75 | Reactive catalyst |
| Triethylene diamine | 0.75 | Gelation catalyst |
| Milled glass fiber | 50 | Average 200 μm |
| pMDI (105 index) | 175 | 135 isocyanate equivalent weight, functionality = 2.7 |

| Condition | Value |
|---|---|
| Throughput | 750 g/s |
| Shot time | 1.5 s |
| Material temperature | 35 °C |
| Mold temperature | 65–80 °C |
| Air loading | 30% |
| Cure time | 30–60 s |
| Demold time | 45 s |
| Mold size | 400 mm × 600 mm × 45 mm |

(*Continued*)

TABLE 9.14 (Continued)

| Property | Value |
|---|---|
| Density | 0.38 g/cm$^3$ |
| Flex modulus[a] | 830 MPa |
| Bending strength[b] | 14.8 MPa |
| Izod impact[c] | 6 kJ/m$^2$ |

EO, ethylene oxide; pMDI, polymeric 4,4′-methylene diphenyl diisocyanate; PO, propylene oxide.
[a] See Chapter 8.
[b] $= \frac{3PL}{wt^2}$, where $P$ = break load, $L$ = distance between bending fulcrums, $w$ = sample width, and $t$ is sample thickness.
[c] According to ASTM D256.

techniques to improve polyurethane wetting of glass mat preforms in S-RIM operations [75]. Additionally, there was a patent on the use of biodegradable polyols in RIM formulations [76].

***9.2.2.4 Polyurethane Elastomer Fibers*** The global changes in the polyurethane elastomeric fiber (referred to generically as "spandex" or "elastane" fibers) industry are emblematic of significant changes in overall global manufacturing, and polyurethane manufacturing specifically. Figure 9.23 shows the rapid growth of spandex

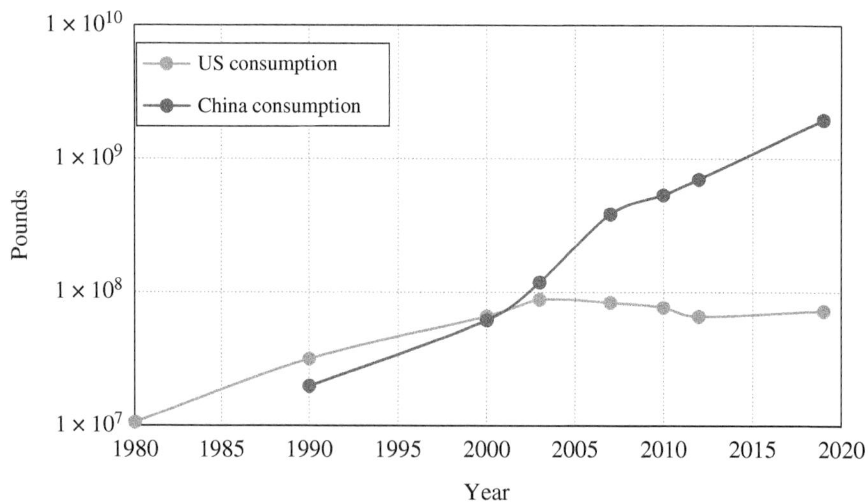


**FIGURE 9.23** Industrial consumption of polyurethane elastomeric fiber.

manufacturing in China coupled with the tailing off of spandex manufacturing in the USA [77]. The trend for Europe is nearly identical to that for the USA over this time period. The market consumption points to several trends. One is that the overall market for spandex fiber is very strong (growth variously estimated at between 7% and 9% per year) and usage is expanding into numerous areas of everyday life, including nearly all clothing segments, upholstery, disposable diapers, and medical garments. The growth of consumption was mirrored, and probably caused, by growth in excess manufacturing capacity and concomitant downward price pressures. The worldwide manufacturing asset utilization was only about 73% in 2010 and still only about 72% capacity in 2019, matching growth in Chinese production capacity and at the beginning of an increased rate of consolidation in US and European capacity [78]. During the period of record for Figure 9.23, the price of spandex fibers went from a high of about US$10/lb in 1986 to a low of about US$3.50/lb in 2018. The fluctuating price of building blocks comprising spandex fibers, particularly PTMEG (see Chapter 2), can force normalization of fiber price or limit supply over time. While it is assured that spandex prices will continue to fluctuate, the high prices of the 1980s are unlikely to be seen again. A floor for spandex pricing is created by the costs of component building blocks and manufacturing costs.

Spandex fibers are technically polyurethane–urea polymers employing diamine chain extenders, MDI, and either polyether or polyester soft segments. The need for polyurea hard segments is a result of the requirement for higher thermal stability than offered by polyurethane hard segments [79–81]. The need for elastomeric fabric stability in clothes dryers and under ironing conditions is critical in this regard. Furthermore, owing to the strength of *polyurea* hard segments, minimization of hard-segment volume and maximization of soft-segment volume in the polymer is achieved, creating a relatively soft polymer with very good elastomer retractive properties. Equivalent soft-segment retracting power would require a larger amount of polyurethane hard segment and result in an overall harder fiber, which might reduce the comfort goal these fibers have.

Production of spandex fibers is via a solution dry spinning process using a prepolymer route [82]. Producers manufacture prepolymer by reacting the polyols with an excess of MDI (virtually all polyurethane elastomeric fibers are made with 4,4′ MDI) (Figure 9.24). The amount of excess MDI is controlled to minimize chain extension whereby soft-segment chains are polymerized with MDI, but not so much that too much hard segment is produced in the next step.

The prepolymer is subsequently dissolved in an aprotic polar solvent, usually dimethylformamide, and reacted with a diamine such as ethylene diamine (Figure 9.25). The dilute (but still very viscous) solution is subsequently spun through a spinneret into a long, heated shroud able to remove the significant amount of solvent involved, leaving the elastomeric fiber. The resulting polymer has a high density of interactive N–H units and carbonyl units that are capable of interchain interactions. The isocyanate–diamine hard-segment phase separates from the soft segment by virtue of phase incompatibility (see Chapter 4) and hydrogen bond interaction

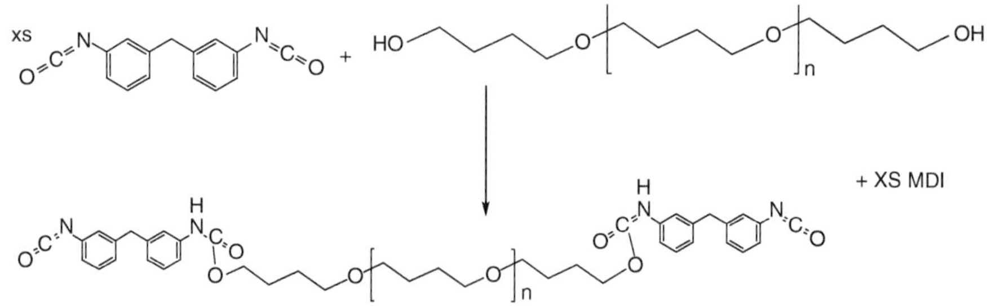


**FIGURE 9.24** Synthesis of the prepolymer for making spandex. The soft segment shown is polytetramethylene glycol. Manufacturers may use polyester soft segments, though these potentially suffer from hydrolytic instability.

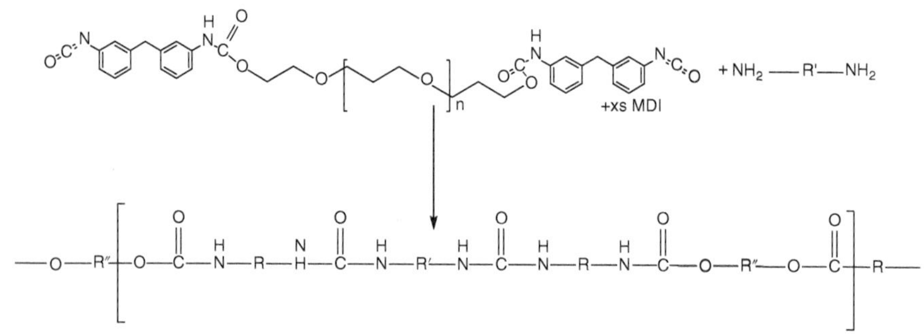


**FIGURE 9.25** Preparation of spandex polyurethane urea polymer from prepolymer and diamine. R = methylene bisphenyl from MDI; R′ is a linear moiety such as ethylene from ethylene diamine; and R″ is the PTMEG soft segment. Reprinted with permission from Ref. [44]. © John Wiley & Sons, Inc.

between the N–H and carbonyls. The short distance between amine functions on a chain extender such as ethylene diamine allows for the unpaired electrons of carbonyl groups to simultaneously entertain two hydrogen bonds from the urea moiety. Thus, the interchain cohesive energy between urea hard segments is as much as two times that of urethane hard segments, and allows for stronger hard-segment interactions with less hard-segment volume for the same cross-link strength [83]. However, the high cohesive energy density of these hydrogen bonds precludes melt spinning of these materials owing to decomposition of polyether soft segments at temperatures below the polyurea hard-segment melting point (Figure 9.26). The phase-separated urea segments act as cross-links tying the soft segments together. From this standpoint, polyurethane elastomer behavior is well understood by simple application of concepts of rubber elasticity.

A trend in spandex technology, besides lower cost, is to produce designed fibers that fill performance gaps in current spandex performance. Particularly, there is continuing effort to produce spandex fibers that show less susceptibility to

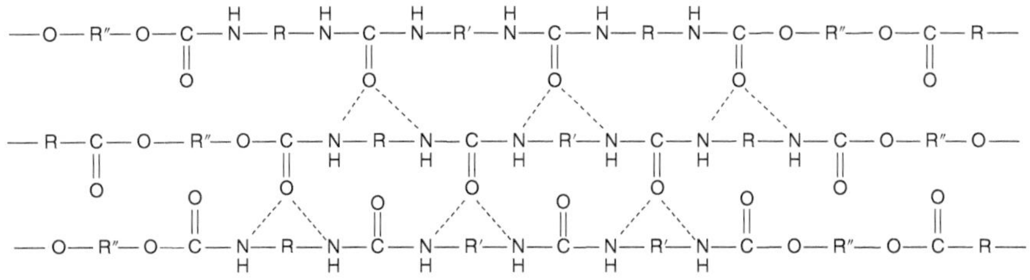

**FIGURE 9.26** Hard-segment structure stabilization through germinal hydrogen bond interaction between hard segments on different chains. Reprinted with permission from Ref. [44]. © John Wiley & Sons, Inc.

degradation upon exposure to chlorine. Chlorine exposure is known to greatly reduce the overall strength of polyurethane elastomer fiber via a substitution of Cl for the carbamate hydrogen followed by eventual scission of the urethane bond, or by substitution for hydrogen along the polyether soft segment. There has been polymer design work to make the spandex backbone more workable by mixing diamine extenders and diol chain extenders, attempts to use bioderived and recycled feedstocks, and preparation with cool-feel haptics [84–86]. There is also a trend for fiber manufacturers to produce fibers compatible with various wrinkle-free technologies. Another trend of the past several years has been to find new ways to incorporate elastomer fibers in more complex fabric structures and also to develop new melt-blown elastomeric film structures [87–89].

## 9.3 TRENDS IN POLYURETHANE ELASTOMERS

While the patent literature is notably poor at providing insight into fundamental science, it is the clearest way to understand the future of a technology and market. This is because the patent literature represents a sizeable investment by a company or individual on what is believed will be desirable by customers and market competitors in the next-generation product. Furthermore, defense of a patent asset requires that the technology on which a patent is based must be protected by patent prior to its commercial introduction.

While a single patent may not be instructive as to the direction of future innovation, a clear trend in industrial patent activity can be unmistakable. Worldwide, over 4000 patents were sought or granted on the subject of polyurethane elastomers from 2000 to 2019 (Figure 9.27). Figure 9.28 shows an interesting geographical migration from Japan, which had in the previous decade been a very active accumulator of intellectual property, to China in polyurethane elastomers. In fact, no Japanese company for the years spanning 2014–2019 is in the top 10 of companies with applications or granted polyurethane elastomer patents. The bar graphs are in no way inclusive and Figure 9.29 shows a pie chart of the number of patenting companies

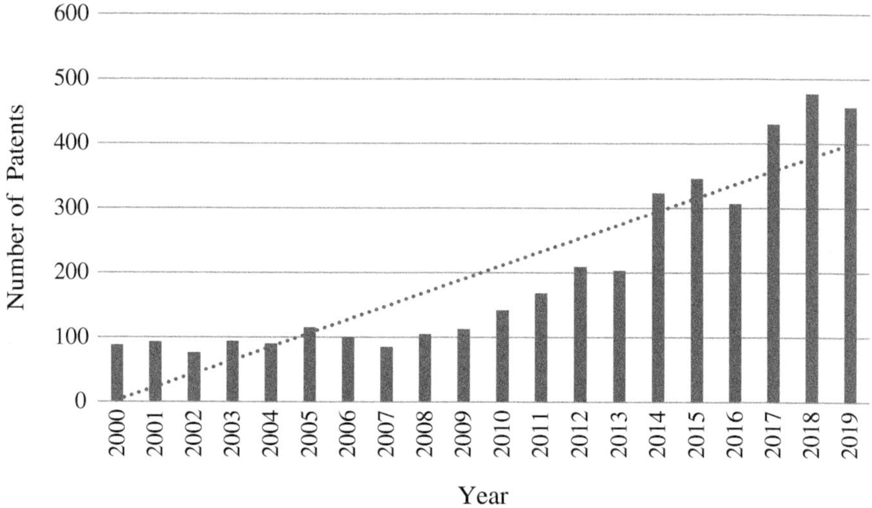


**FIGURE 9.27** Trend of increasing applied for and granted patents on polyurethane elastomers since 2000. There were over 4000 total patents published worldwide over this period.

(a)

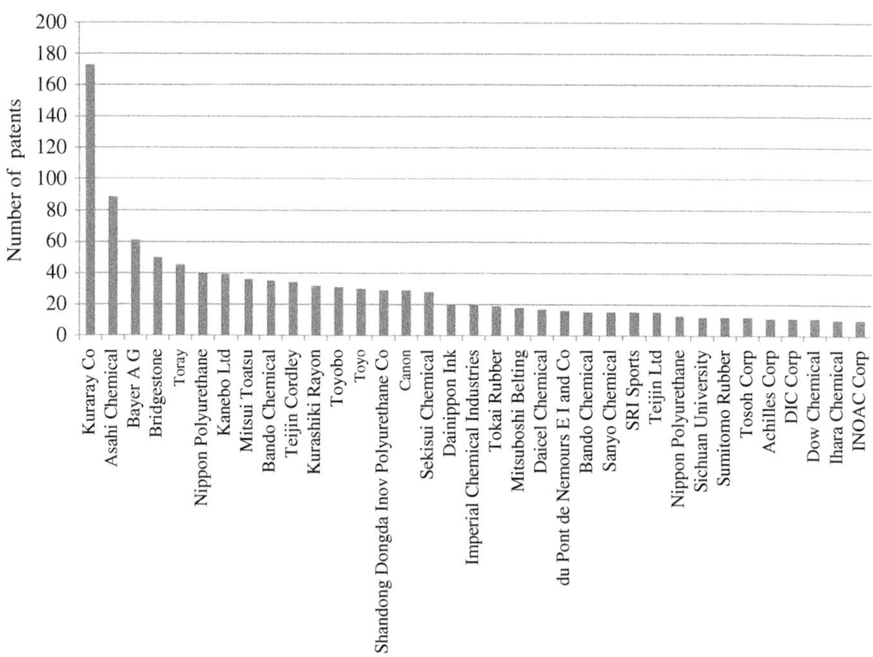


**FIGURE 9.28** (*Continued*)

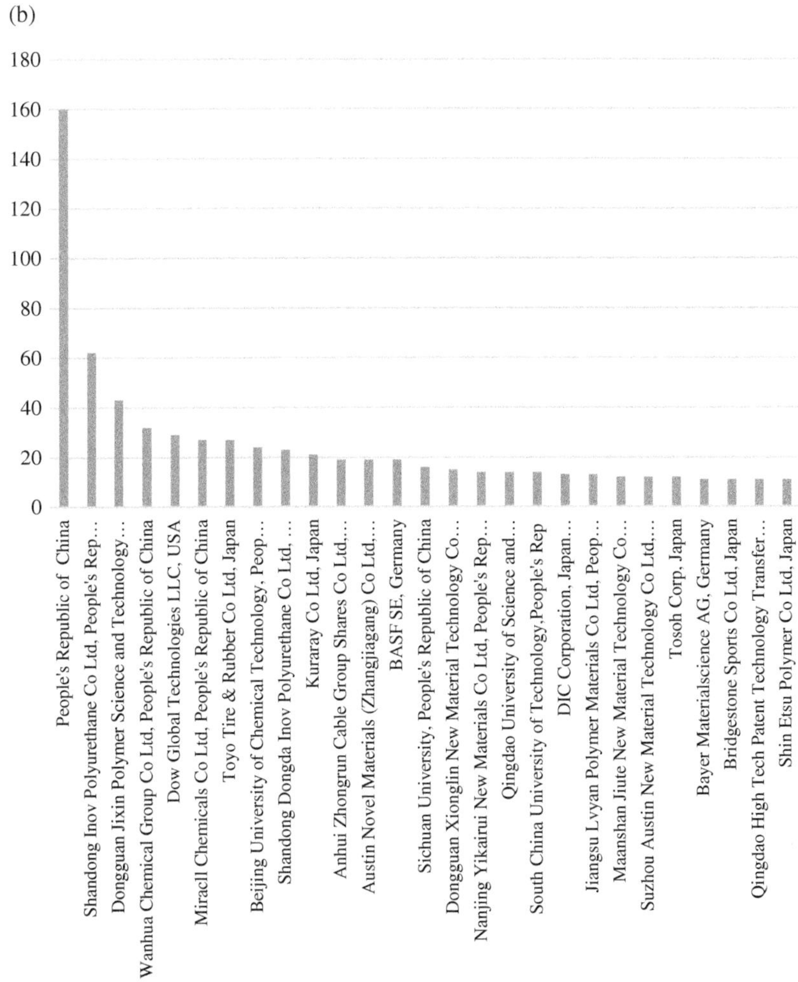


**FIGURE 9.28** (a) Patent activity of companies having at least 10 patents from 2000 to 2010. (b) Patent activity of companies from 2015 to 2019. About 700 patents were published between 2010 and 2013. Many Chinese companies and universities will stipulate their national origin in their names, which then gets recorded as the assignee by the search engine.

with greater than 2% of the patent share relative to the share patented by the top seven companies from the years 2015–2019.

Patent topics for the years 2015–2019 are shown in Figure 9.30. Patent activity has been broad and spans composition and application of polyurethane elastomers to specific market segments (rather than, for instance, production techniques or new compositions). Some of the categories shown in Figure 9.30 could rationally be combined, such as cross-linking agents, which would probably, at least in part, overlap with textiles and fibers because of the use of cross-linking agents to impart

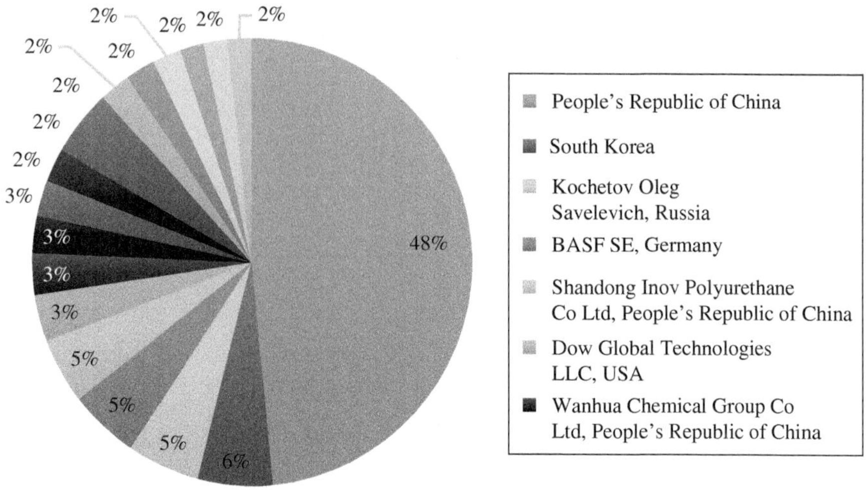


**FIGURE 9.29** Pie chart of patent activity filed from 2015 to 2019 of top 18 assignees relative to the activity of the top seven assignees. The top seven account for about 15% of all polyurethane elastomer patents filed. Many Chinese companies and universities will stipulate their national origin in their names, which then gets recorded as the assignee by the search engine. (*See insert for color representation of the figure.*)

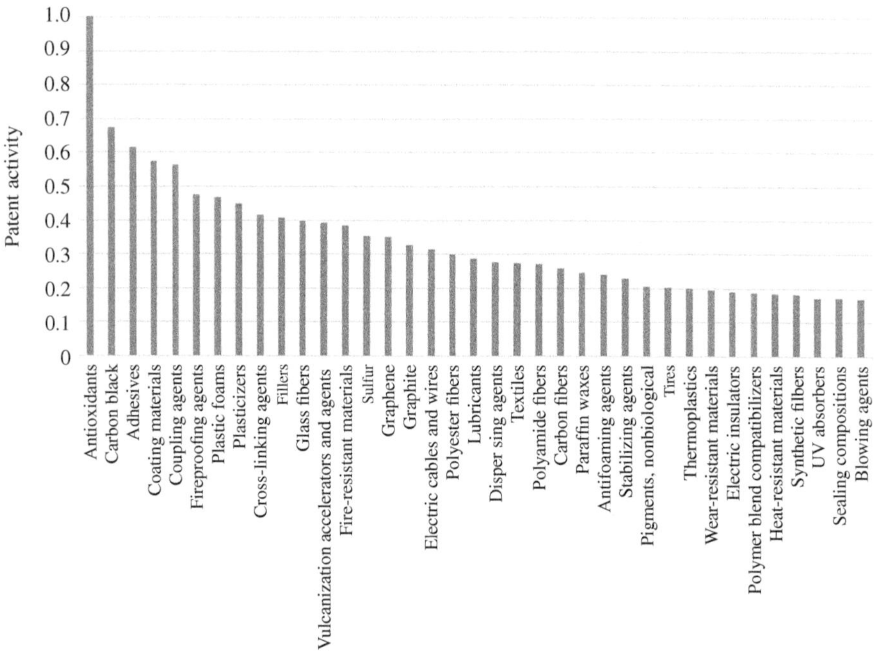


**FIGURE 9.30** Normalized patent activity during 2015–2019 based on application of polyurethane elastomers. All languages, all regions. The total number of patent documents is 8146.

wrinkle-free characteristics to cotton fabrics under challenging conditions for polyurethane bonds. The data suggest that, while there is some industrial application patent activity such as for various conveyor rollers and abrasion resistant compositions, the majority of activity protects final user applications such as for clothing (including shoes) and incremental performance enhancements for sporting goods, including golf balls and rollerblade wheels. This analysis suggests that current research can best be applied to satisfying the needs of the consumers/leisure economy and applications that do not bring a tangible value to the end user may have less long-term value.

## REFERENCES

[1] H Chinn, M Yoneyama, U Loechner, and X Xu, "Polyurethane elastomers," in *Chemical Economics Handbook*, IHS, Englewood, CO, 2017.
[2] M Reimann, O Schilke, and J Thomas, *International Journal of Research in Marketing*, **27**, (2010), 188–197.
[3] D Volz and K Elliott, *Environmental Science and Technology*, **47**, (2012), 6069–6070.
[4] *The Global Chemical Industry: Catalyzing Growth and Addressing our World's Sustainability Challenges*, Oxford Economics, London, UK, 2019.
[5] Y Shahid and K Nabeshima, *Changing the Industrial Geography in Asia*, The World Bank, Washington, DC, 2010, 29–122.
[6] D Manca and R Grana, *Computers and Chemical Engineering*, **34**, (2010), 656–667.
[7] J Charpentier, *Computers and Chemical Engineering*, **33**, (2009), 936–946.
[8] A Matta, Q Semeraro, and T Tolio, "A framework for long-term capacity expansions," in *Design of Advanced Manufacturing Systems*, A Matto and Q Semeraro, Eds., Springer, Dordrecht, the Netherlands, 2010.
[9] C Christensen and M Raymer, *The Innovator's Solution*, Harvard University Business Review Press, Boston, MA, 2013.
[10] H Chinn, M Yoneyama, U Loechner, and X Xu, "Polyurethane elastomers," in *Chemical Economics Handbook*, IHS, Englewood, CO, 2017, 31.
[11] H Chinn, M Yoneyama, U Loechner, and X Xu, "Polyurethane elastomers," in *Chemical Economics Handbook*, IHS, Englewood, CO, 2017, 45.
[12] Y Yi and R Ma, *Advanced Materials Research*, **630**, (2013), 67–70.
[13] E Corinti, L Zanacchi, and A Benvenuti, Oil resistant polyurethane shoe soles made from all-propylene oxide polyols. Assignee Dow Global Technologies. WO 2011157510 A1. 2011.
[14] E Geiger and D Hillshafer, Phthalic anhydride bases polyester–polyether polyols and urethane prepolymers and products. Assignee Stepan Co. WO 20022036656 A2. 2001.
[15] A Luximon, *Handbook of Footwear Design and Manufacture*, Woodhead Publishers, Philadelphia, PA, 2013.
[16] G Vairo, Polyurethane elastomers from renewable resources, their preparation and foamed polyester polyurethane thermoplastic elastomers for shoe soles. Assignee Dow Global Technologies. WO 2009129393 A1. 2009.
[17] R Goonetillake, Ed., *The Science of Footwear*, CRC Press, Boca Raton, FL, 2013.
[18] N Deans and J Bicerano, Footwear providing a near-barefoot experience. US patent 20130269211 A1. 2013.

[19] M Maiti, R Jasra, S Kusem, and T Chaki, *Industrial and Engineering Chemistry Research*, **51**, (2012), 10607–10612.
[20] C Huang and K Lee, Crosslinked thermoplastic polyurethane elastomers. Assignee Hwa Pao Resins Chemical Co. US patent 2012004351 A1. 2012.
[21] R Camargo, N Limerkens, and C Roberts. "Thermoplastic polyurethanes: new materials options for high-performance footwear," in *Defining the Future through Technology: Proceedings of the Polyurethanes Conference*, October 8–11, Boston, MA, 2000, 49–53.
[22] M Sonnenschein, B Wendt, A Schrock, J Sonney, and A Ryan, *Polymer*, **49**, (2008), 934–942.
[23] J de Verney, M Lima, and D Lenz, *Materials Research*, **11**, (2008), 447–451.
[24] A Batastini and G Perozzi, Thermoplastic compositions for footwear with no marking characteristics. Assignee Dow Global Technologies. WO 2009071549 A1. 2009.
[25] M Jarbas, Polymer composition, microporous rubber-like foamed vulcanizate, and microporous rubber-like foamed shoe sole. Assignee Dupe S/A Brazil. WO 2008083451 A1. 2008.
[26] M Navarro-Banon, M Pastor-Blas, and J Martin-Martinez, *Journal of Adhesion Science and Technology*, **19**, (2005), 1237–1247.
[27] J Kim and G Kim, *Elastomers and Composites*, **48**, (2013), 195–200.
[28] C Battilana, Sports footwear, such as a ski boot or suchlike, and corresponding method of production. Assignee Calzaturificio Dal Bello. EP 2591692. 2013.
[29] D Rosenvasser, J McCloud, R Camargo, and L Cao, "New polyurethane elastomers with increased bio-renewable content for footwear applications," in *Proceedings of the Polyurethane Technology Conference*, October 11–13, Houston, TX, 2011, 713–727.
[30] C Holmes, S Kerling, and H Le, Method for the manufacture of a shoe sole, shoe, with pre-manufactured TPU. Assignee Adidas AG. EP 3248493 A1. 2017.
[31] S Reinhardt, A Wardlaw, T Robinson, J Whiteman, and D Wood, Soles for sport shoes. Assignee Adidas AG. EP 2649896 A2. 2013.
[32] A Matheson, "A range of vinyl and rubber based TPEs offering cost effective solutions for component design," in *New Opportunities for Thermoplastic Elastomers* (unedited compendium of papers), Rapra Technology Publishers, Shawbury, UK, 1996.
[33] E Ormande, M Yoneyama, U Lochner, and X. Xu, "Styrenic block copolymers," in *Chemical Economics Handbook*, IHS, Englewood, CO, 2017.
[34] S Sum, Sport shoe outsole with anti-slip and anti-abrasion. Assignee Asia One Leatherwear Manufacturing. US patent 20130008059 A1. 2013.
[35] M Schuette, A Kappes, and M Ortalda, Low-density polyurethane foam materials and their use in shoe soles. Assignee BASF. WO 2005116101 A1. 2005.
[36] J Julia, "High performance TPU for conveyor belt applications," in *UTECH '96 Conference Book of Papers*, March 26–28, The Hague, the Netherlands, 1996, Paper 50.
[37] J Donnenwirth, Method and manufacture, refurbishment, and repair of non-moldable polyurethane/polyurea conveyor belt covers and related products. Assignee Garlock Rubber Technologies. US patent 7648021. 2010.
[38] A Veith, "Tires, roads, rainfall-vehicles: the traction connection," in *Friction Interaction of Tire and Pavement*, W Meyer and J Walter, Eds., American Society of Testing and Materials, Baltimore, MD, 1983, 20–21.
[39] I Clemmitson, *Castable Polyurethane Elastomers*, CRC Press, Boca Raton, FL, 2008, Chapters 4–8.

[40] L Majoros, B Dekeyser, R Hoogenboom, M Fijten, J Geeraert, N Haucourt, and U Schubert, *Journal of Polymer Science, Part A: Polymer Chemistry*, **48**, (2010), 570–580.
[41] A Heintz, D Duffy, S Hsu, W Suen, W Chu, and C Paul, *Macromolecules*, **36**, (2003), 2695–2704.
[42] M Sanchez-Adsuar, E Papon, and J-J Villenave, *Journal of Applied Polymer Science*, **76**, (2000), 1596–1601.
[43] D House, D Ilijevski, and R Scott, "A novel coating system utilizing stripped TDI-based prepolymers," in *Polyurethanes Expo'99, Proceedings of the Polyurethanes Expo'99*, September 12–15, Orlando, FL, CRC Press, Boca Raton, FL, 1999, 53–56.
[44] M Sonnenschein and W Koonce, "Polyurethanes," in *Encyclopedia of Polymer Science and Technology*, 4th Ed., H Mark, Ed., Wiley, Hoboken, NJ, 2012, 1–63.
[45] R Rosenberg, R Xie, and A Singh, High performance polyurethane elastomers from MDI prepolymers with reduced content of free MDI monomer. Assignee Dow Global Technologies. US patent 20030065124 A1. 2003.
[46] "Hazard information to consider when labeling of MDI, pMDI, and isocyanate terminated MDI prepolymers containing products for the consumer marketplace," Document No. AX 264, American Chemistry Council, Washington, DC, May 2012, 9.
[47] M Woolhiser, A Stephenson, and D Boverhof, "Sensitization potential of low monomer diisocyanate prepolymers," in *Proceeding of the UTECH Conference*, April 17–19, Maastricht, the Netherlands, 2012.
[48] A Maier, F Wolferstetter, H Winkelmann, J Weichmann, and A Kern, Solvent-free polyurethane dispersion. Assignee skw Bauchchemie. US patent 6635723. 2003.
[49] V Gajewski, "Chemical degradation of polyurethane," in *Failure of Polymer Products Due To Chemical Attack*, D Wright, Ed., Rapra Technology, Report 130, 2001. Originally in *Proceedings of the Polyurethanes '90*, September 30–October 3, Orlando, FL, 506–514.
[50] M Sonnenschein, V Ginzburg, K Schiller, and B Wendt, *Polymer*, **54**, (2013), 1350–1360.
[51] V Kapasi, "Thermoplastic polyurethanes," in *The Polyurethanes Book*, D Randall and S Lee, Eds., Wiley, Hoboken, NJ, 2002, 315–330.
[52] B Bantu, G Pawar, U Decker, K Wurst, A Schmidt, and M Buchmeiser, *Chemistry, a European Journal*, **15**, (2009), 3103–3109.
[53] D Huh and S Cooper, *Polymer Engineering and Science*, **11**, (1971), 369–376.
[54] M Horstmann, M Urbani, and W Veeman, *Macromolecular Symposia*, **205**, (2004), 129–142.
[55] S Levchik and E Weil, *Polymer International*, **53**, (2004), 1585–1610.
[56] H Kim, J Huh, E Kim, and C Park, *Journal of Applied Polymer Science*, **69**, (1998), 1349–1355.
[57] D Martin, G Meijis, P Gunitilake, S McCarthy, and G Renwicj, *Journal of Applied Polymer Science*, **64**, (1997), 803–817.
[58] O Rexin and R Mulhaupt, *Journal of Polymer Science Part A Polymer Chemistry*, **40**, (2002), 864–873.
[59] R Tuinman, T Lee, T Fishback, and C Reichel, Use of low unsaturated polyether polyols in slabstock foam applications. Assignee BASF Corporation. CA 2380301. 2008.
[60] C Wenning, A Schmidt, and M Leimenstoll, *Polymer International*, **67**, (2018), 481–489.

[61] D Park, W Wu, and Y Wang, *Biomaterials*, **32**, (2010), 777–786.
[62] P Petrini, M Tanzi, C Moran, and N Graham, *Journal of Materials Science: Materials in Medicine*, **10**, (1999), 635–639.
[63] S Clough, N Schneider, and A King, *Journal of Macromolecular Science – Physics*, **B2**, (1968), 641–648.
[64] M Sonnenschein, S Guillaudeu, B Landes, and B Wendt, *Polymer*, **51**, (2010), 3685–3692.
[65] M Sonnenschein, C Boyer, D Brune, B Wendt, G Meyers, and B Landes, *Macromolecules*, **44**, (2011), 865–871.
[66] C Macosko, *RIM: Fundamentals of Reaction Injection Molding*, Hanser Publishers, Munich, Germany, 1989.
[67] Z Tadmor and C Gogos, *Principles of Polymer Processing*, Wiley, Hoboken, NJ, 2006, 753–823.
[68] J Eckler and T Wilkenson, *Journal of Materials Shaping Technology*, **5**, (1987), 17–21.
[69] P Dufton, *Lightweight Thermoset Composites*, Rapra Technology, Shawbury, UK, 2000, Chapters 3 and 6.
[70] R Meyer *Handbook of Pultrusion Technology*, Chapman and Hall, New York, NY, 2013.
[71] H Chinn, M Yoneyama, U Loechner, and X Xu, "Polyurethane elastomers," in *Chemical Economics Handbook*, IHS, Englewood, CO, 2017, 25.
[72] R Gibson, *Principles of Composite Material Mechanics*, CRC Press, Boca Raton, FL, 2012.
[73] F Hupka, M Backer, and B Berstenhorst, Shaped parts made of reinforced polyurethane urea elastomers and uses thereof. Assignee Covestro AG. US patent 20170166719 A1, 2017.
[74] S Lindner, K Franken, D Passmann, and P Nordmann, Use of layer structures in wind energy plants. Assignee Bayer Materials Science AG. WO 2012113748. 2012.
[75] S Hu and C Chen, Glass fiber PU composite material for automobile outer decoration. Assignee Jiuda Vehicle Fittings-Making Co. CN 201192912. 2009.
[76] M Goldhawk, A James, G Klumb, and J Hetzner, Reaction injection molded polyurethanes made using high levels of natural oil-based polyols. Assignee Dow Global Technologies. WO 2009032789. 2009.
[77] M Malveda, B Sesto, S Passararat, and V Zhang, "Specialty organic fibers," *Chemical Economics Market Report*, IHS, Englewood, CO, 2019.
[78] Y Sahu and Y Doshi, Spandex Fiber Market by Type of Production Method and Application – Global Opportunity Analysis and Industry Forecast 2014-2022, Allied Market Research, Portland, OR, 2016.
[79] D Gorjanc, K Dimitrovski, and M Bizjak, *Textile Research Journal*, **82**, (2012), 1498–1506.
[80] I Shin, I Jung, and H Kim, High heat and chlorine resistant polyurethaneurea elastic fiber and preparation of thereof. Assignee Tae Kwang Ind. Co., Ltd. US patent 8623950. 2014.
[81] B Song, T Lee, J Kim, S Seo, and D Lee, High chlorine and heat resistant spandex fiber and manufacturing method thereof. Assignee Hyosung Corp. US 20030203199 A1. 2003.
[82] F Fourne, Ed., "Polymer-specific process," in *Synthetic Fibers – Machines and Equipment Manufacture, Properties*, Hanser Publishers, Munich, Germany, 1999, Chapter 2.

[83] M Sonnenschein, N Rondan, B Wendt, and J Cox, *Journal of Polymer Science Part A Polymer Chemistry*, **42**, (2004), 271–278.
[84] R Waldbauer and N D'Allura, Bioderived elastomeric polyurethane fiber. Assignee Invista North America. US patent 20170051436. 2015.
[85] H Jung and Y Kang, Polyurethane-urea elastomeric fiber with low-temperature workability. Assignee Hyosung Corp. WO 216085189. 2014.
[86] R Vedula and M-W Lee, Elastomeric fibers and methods of making and using thereof. Assignee Lubrizol Advanced Materials. US patent 20170127741. 2014.
[87] K Wilkenson, Process of preparing elastomeric thread. Assignee K Wilkenson. US patent 5616675. 1997.
[88] I Windschieg and T Dayarathne, Washable and reusable non-woven material. Assignee MAS Intimates. WO 2019112522. 2019.
[89] D Farmer, Garments including elastic composite fabric. Assignee Invista North America. US patent 20150351474. 2015.

# 10

# POLYURETHANE ADHESIVES AND COATINGS: MANUFACTURE, APPLICATIONS, MARKETS, AND TRENDS

## 10.1 ADHESIVES AND COATINGS INDUSTRIES: SIMILARITIES AND DIFFERENCES

Coatings and adhesives are regularly associated in polyurethanes technology because of the commonality of how the systems are manufactured and many of the technology trends. In both cases a film must be applied to a surface, wet the surface, form an integral film, and adhere to the substrate. In terms of raw materials, applications, and markets they are as different as people commonly understand their separate uses to be. More so than any other polyurethane-based technology, coatings and adhesives are formulated systems in which the polyurethane may represent a significant volume, but it often serves as a binder for all the other components put into the system. These additional components are things such as solvents, solid fillers, pigments and dyes, antioxidants, tougheners, surfactants, coupling agents, coalescing agents, catalysts, defoamers, moisture scavengers, water, plasticizers, and numerous hybridizing polymers.

The technologies for producing both coatings and adhesives can be simplified into variations of reactive and unreactive systems (Figure 10.1). The ratio of polyurethane sold into the coatings and adhesives markets is roughly 3:1 (Figure 10.2). The market position of polyurethane adhesives, sealants, and coatings is important to their

*Polyurethanes: Science, Technology, Markets, and Trends*, Second Edition. Mark F. Sonnenschein.

overall industries, but they are by no means volume leaders (Figure 10.3) [1–6]. Sealants for instance, is highly dominated by silicone backbone polymers.

Because of the highly formulated and often diluted nature of adhesive and coating products, it is challenging to obtain an accurate measure of total polyurethane consumption for these technologies. An analysis of the available market literature suggests that polyurethanes represent approximately 7–8% of all adhesive and sealant binders used, with about a total of 800–900 million pounds used globally. The global polyurethane coatings market grows with gross domestic product and has recently been around 4%. However, this number masks the broad variance of this growth from under 1% in Japan to over 6% in China. The coatings market accounts for about 3 billion pounds of polyurethane consumption. For adhesives and sealants the major markets are in order: construction ~ packaging > transportation > furniture > footwear.

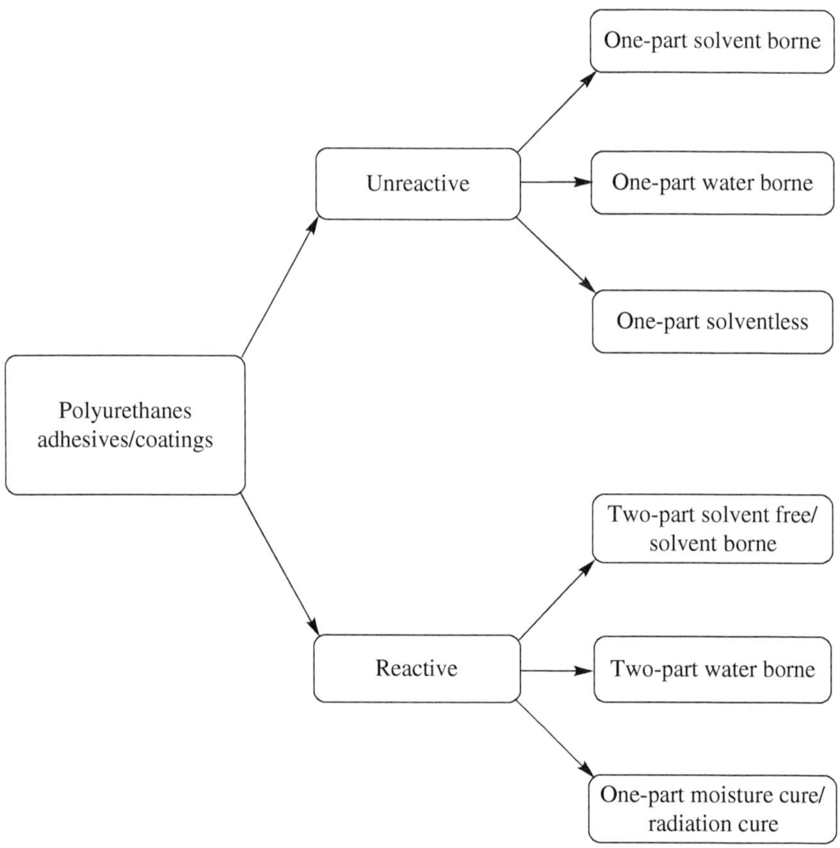


**FIGURE 10.1** Technology structure of polyurethane coatings and adhesives by format. This commonality of product offerings is the reason for grouping these technologies.

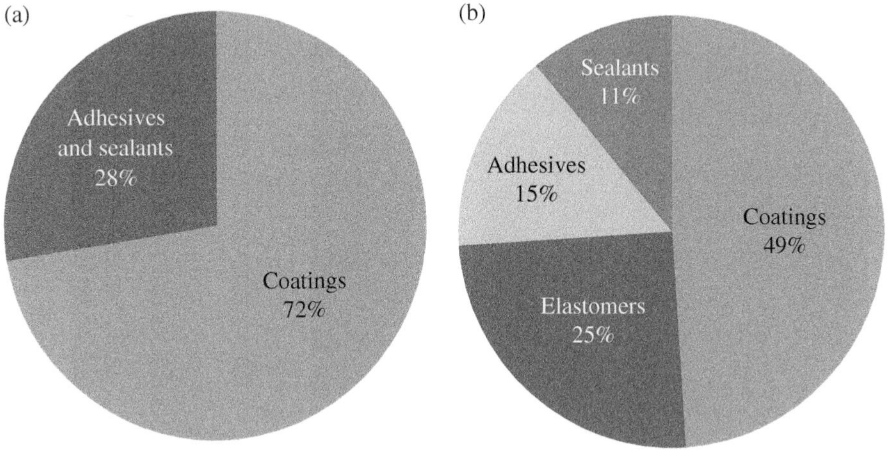


**FIGURE 10.2** Relative volume of polyurethane (a) volume and (b) value used in adhesive and sealants versus coatings and elastomers during 2018.

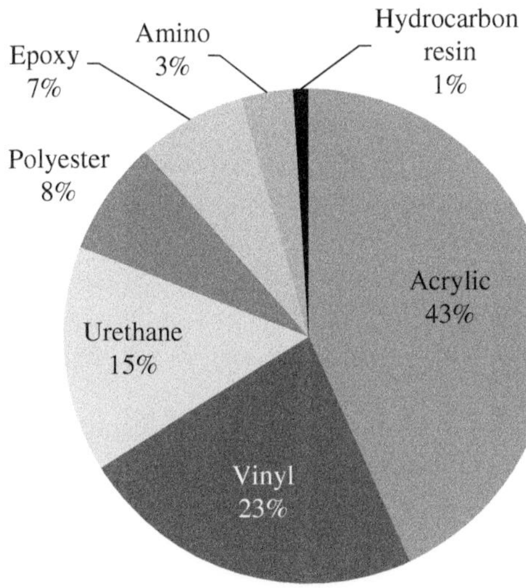


**FIGURE 10.3** Consumption of polyurethane resins in coatings relative to other resin options in the USA for 2019.

Consumption of polyurethane coatings is global and reflective of finished goods manufacturing that use polyurethane coatings (Figure 10.4). The global consumption reflects the convolution of industrial, architectural, and protective applications that use polyurethane coatings. The volume dominance of China is a relatively new phenomenon and reflects its position as a manufacturing center.

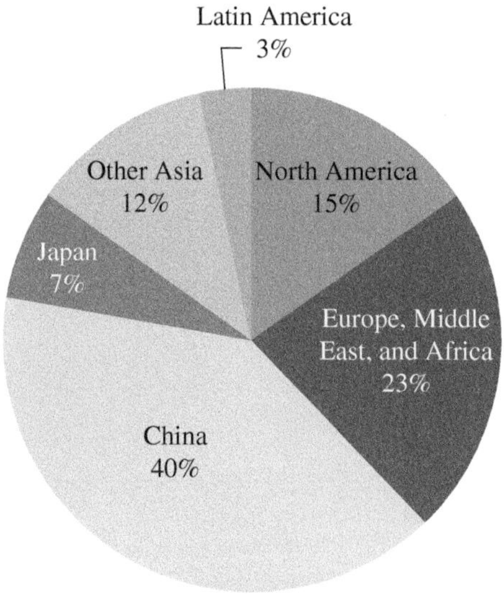


**FIGURE 10.4** Distribution of global consumption of polyurethane chemicals in coatings by region during 2018.

There are numerous applications consuming similar volumes, including automotive coatings, wood coatings, artificial leathers, and protective industrial coatings. This versatility is in part responsible for the global breadth of polyurethane adhesive and coatings markets, which does not favor consumption based on any single manufacturing center. Furthermore, because of highly formulated and customized solutions, each particular customer requires local technical service and production, again favoring a broad distribution of geographical consumption. For this reason, the general appearance of geographical consumption of polyurethane adhesives would have a qualitatively similar appearance to Figure 10.4.

The similarities in product format and global utilization belie the very different uses of adhesives and coatings and the resulting needs for specific polymer design and requirements [7–9]. In the next sections, adhesives and coatings will be treated separately to reflect this reality more fully.

## 10.2 ADHESIVES

As indicated in Figure 10.1, polyurethane adhesives are commercialized in numerous product forms. Each form exists because it caters to the needs of particular industries and customers. The technology used bespeaks the mode of adhesive action and the requirements of the application. The global use of formulated polyurethane adhesives according to product form is shown in Figure 10.5.

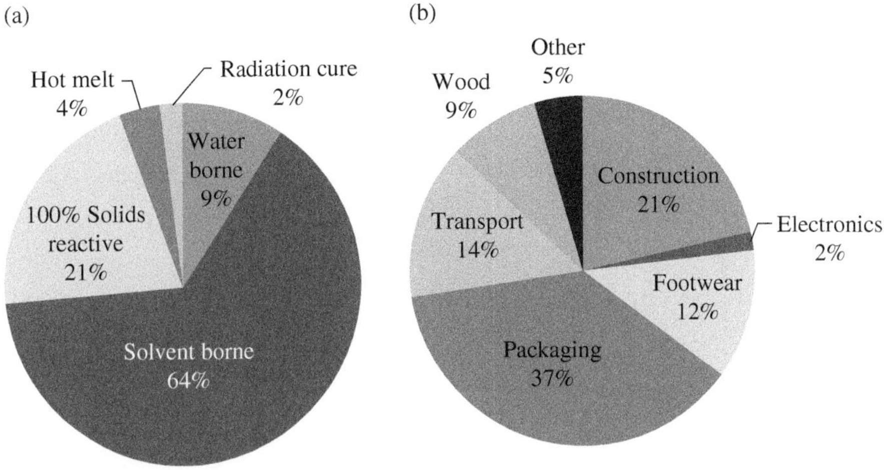


**FIGURE 10.5** Global consumption of polyurethane adhesives by (a) total formulated product and (b) application.

Of course, the fully formulated product is weighted by the accompanying formulation components, which in the case of solvent-borne adhesives, for instance, can actually outweigh the polyurethane polymer component. The distribution representing only the binder component looks quite different (Figure 10.6).

The large volume of solvent-borne polyurethane adhesive reflects the very large volumes applied for shoe sole construction. This is particularly evident in polyurethane adhesive consumption in China, where footwear manufacturing is dominant and an estimated 80% of solvent-borne polyurethane adhesive is consumed [2]. The geographical distribution of adhesive consumption by format is complex and changing based on downstream manufacturing. Because of the specialty- and customer-specific nature of the applications, the adhesive producers are located close to the assembly operations that utilize the particular adhesive. As manufacturers of finished goods move to more favorable geographies, the adhesive supply and portfolio of adhesive solutions tends to follow.

The large chemical producers supply the adhesive formulators with the typical building blocks used for other polyurethane applications. Because of the economic advantage brought by their back-integration, most large chemical producers have a significant adhesive product offering, either as fully formulated systems or via component sales to small and large adhesive producers. This situation can result in large producer companies being in direct competition with their customers in some markets.

While many small formulators have minimal technical capabilities beyond blending and packaging, other formulators are technically sophisticated, highly capable, and well known. Consumers of building blocks for adhesive formulation face common commercial challenges and must choose from similar building blocks to develop adhesive solutions. The effect of this constant evolution has resulted in 4,4′-methylene

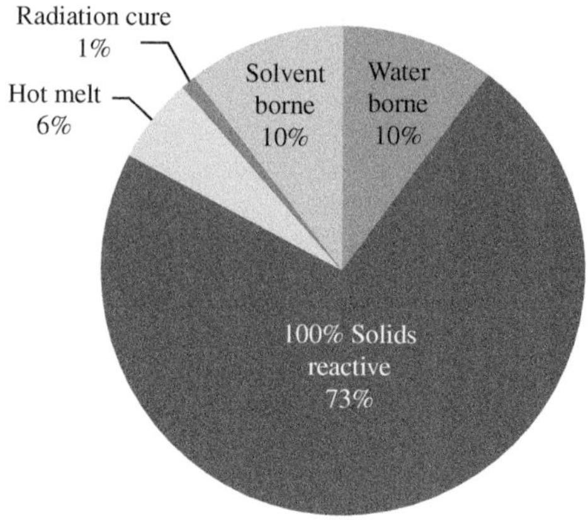


**FIGURE 10.6** Global consumption of polyurethane in polyurethane adhesives.

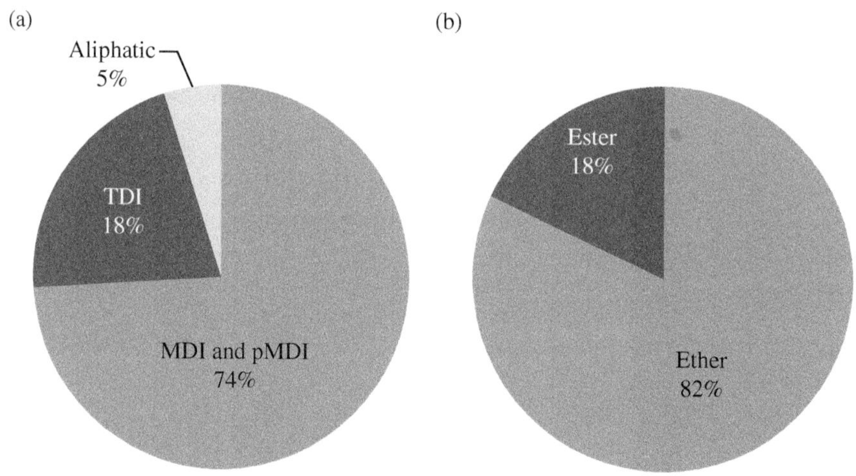


**FIGURE 10.7** Global consumption of (a) isocyanates and (b) polyols for polyurethane adhesives formulation. TDI, toluene diisocyanate.

diphenyl diisocyanate (MDI) and polymeric MDI (pMDI) enjoying approximately three-quarters of the overall polyurethane market for adhesives, and polyethers dominating polyester soft-segment volumes by 4:1 (Figure 10.7). The dominance of MDI/pMDI for adhesives reflects the uniform reactivity of the isocyanate groups; low volatility, resulting in easier workplace handling; and the fact that most adhesives do not require specific inhibition from light-stimulated degradation, such as that required for coatings. The dominance of polyether polyols reflects that they are liquids at room temperature, offer superior low-temperature flexibility, and are much less sensitive to hydrolysis than polyesters. Polyesters are more common in applications requiring faster set times or initial green strength, such as required for some hot-melt adhesives, and highly automated high unit volume applications.

Table 10.1 lists the various polyurethane adhesive formats and representative associated applications. Some of the different formats may be used for the same application and may find favor with specific manufacturers based on customer requirements, environmental health safety goals of the manufacturer, and preference of the manufacturer. There has been a general move towards water-borne formulations as manufacturers attempt to reduce volatile organic carbons (VOCs) in their workplace and customer venues [10–12]. Another characteristic of these categories is that two-part adhesives are higher performing than a corresponding one-part adhesive in terms of thermal and solvent resistance because of the ability to build hard-segment volume [13]. The growth of polyurethane *reactive* hot melts is recognition of their ability to generate lower viscosity at elevated temperatures and obtain excellent cohesive and adhesive strength as full cure is achieved, and the attractive properties of polyurethane backbones are fully realized [14, 15]. Thus, bookbinding and wood-bonding applications have particularly been able to generate customers for reactive hot melts owing to the long-term durability and flexibility of the product.

The large number of adhesive formats, substrates, applications, and application techniques has resulted in high reliance on standardized test methods to facilitate comparisons of adhesives and adhesion. A representative list of standardized test methods is provided in Table 10.2 [16]. Along with the standardized tests, it is also commonplace for industrial adhesive consumers to perform nonstandardized application testing, or to demand the same of their suppliers. Many of the tests put emphasis on sample preparation, test geometry, and conditions because of the difficulty in separating factors associated with overall adhesive performance. The large number of controls can help differentiate confusing variables that can mislead the designer from determining true performance.

**TABLE 10.1 Major applications for polyurethane adhesives as a function of format**

| Adhesive format | Applications |
|---|---|
| One-part solvent borne | Food packaging, footwear<br>May include blocked isocyanates requiring elevated temperature curing |
| One-part water borne | Footwear, bookbinding, fiber-glass sizing, textile laminates |
| One-part solvent-less | Hot melt – book bindings, furniture, apparel<br>pMDI – wood bonding<br>Food packaging |
| Two-part solvent free/solvent borne | Same as one-part but offering improved properties; flexible packaging, footwear, interlayer laminates for windows, flooring, metal-to-plastic bonding (often requires a primer) |
| Two-part water borne | Wood bonding, automotive dashboards and instrument panel linings, footwear, lamination of veneers and films for furniture, exterior sandwich panels for construction |
| One-part moisture cure/radiation cure | Reactive hot melt for packaging, furniture, bookbinding, wooden flooring<br>Moisture cure – automotive windshield<br>Radiation cure (acrylate tipped) – tougheners, structural adhesives |

pMDI, polymeric 4,4′-methylene diphenyl diisocyanate.

**TABLE 10.2 ASTM standard methods for testing adhesives**

| Test | Standard description |
|---|---|
| ASTM D4896 | Standard test for performance of adhesion in a single lap-shear geometry – one of the most common tests and also one of the most difficult to perform for unambiguous data |
| ASTM D1876 | Standard test for adhesion in a T-peel geometry. Primarily used for flexible substrates requiring flexible adhesive such as polyurethane; see also D2919 |
| ASTM D1144 | Standard test for determining strength development of adhesive bonds. This test is particularly relevant to polyurethane reactive adhesives, which require days to weeks to develop full adhesion and adhesive strength |
| ASTM D905 | Standard test method for strength properties of adhesive bonds in shear by compressive loading. Applies specifically to wood substrates, for which polyurethane adhesives are highly employed. Standard is very specific about all aspects of the test |
| ASTM D950 | Standard test method for impact strength of adhesive bonds. Uses a pendulum impactor on substrates in a shear geometry with a standard velocity of 3.4 m/s (7.6 mph) |
| ASTM D897 | Standard test method for tensile properties of adhesive bonds. Utilizes metal substrates and a strain rate of 0.05 in/min |
| ASTM D1151 | Standard practice for effect of moisture and temperature on adhesive bonds. Continuous exposure test for samples exposed to conditions from –57 °C to 316 °C and from 50% humidity to saturated by soaking. Samples at high-temperature conditions are aged in an oven with uncontrolled humidity |
| ASTM D7105 | Standard test method for determining the adhesive and cohesive strength between materials in roofing or waterproofing membranes and systems. These are applications for which polyurethane adhesives are commonly applied |
| ASTM D896 | Standard practice for resistance of adhesive bonds to chemical reagents. Quantification of adhesive performance following chemical exposure is a requirement for many applications |
| ASTM D4501 | Standard test method for shear strength of adhesive bonds between rigid substrates by the block-shear method. Specific to substrates with higher modulus and where machining of the substrate is impractical, such as with glass bonding – a common substrate for polyurethane adhesives for windshield adhesive/sealants |
| ASTM D3930 | Standard specification for adhesives for wood-based materials for construction of manufactured homes. Comprehensive standard for applications commonly employing polyurethane adhesives |
| ASTM D1337 | Standard practice for storage life of adhesives by viscosity and bond strength. A common qualifying attribute for all adhesives, including polyurethanes |
| ASTM C1523 | Standard test method for determining modulus, tear, and adhesion properties of procured elastomeric joint sealants |
| ASTM D3433 | Standard test method for fracture strength in cleavage of adhesives in bonded metal joints. Specifies metal but is applicable to all stiff adherends using structural adhesives not prone to elastic deformation. Employs a dual-cantilever beam geometry |
| ASTM D4497 | Standard test method for determining the open time of hot-melt adhesives. Very simple manual test |
| ASTM D4498 | Standard test method for heat-fail temperature in shear of hot-melt adhesives. A common test for hot-melt adhesive performance |

The measures of adhesion force using standards proscribed in Table 10.2 are the final result of interfacial and rheological forces that polyurethanes offer in building a joined system. Polyurethanes function well owing to their high internal cohesive strength within the adhesive layer, as a result of its block copolymer structure and internal hydrogen bonding. The same polymer material properties contribute to strong interfacial adhesion resulting from the same strong polar and hydrogen bonding interactions. Additionally, reactive polyurethane adhesives have the potential to chemically bond across the interface with substrates that offer active hydrogens on the surface. Detailed discussions of the specifics of adhesion and adhesives is available from numerous excellent source books devoted to the subject [17, 18].

### 10.2.1 Adhesive Formulations

***10.2.1.1 Solvent-Borne Adhesives*** There is a general industrial move away from application of solvent-borne polyurethane adhesives, but they retain high utility, particularly in the shoe sole market. The one-part adhesive is essentially a thermoplastic polyurethane (TPU) dissolved in a solvent. The solvated TPU is applied to the prepared surface as a contact cement, allowed to dry, and then adhesion generated by contacting the adherends with heat. The adhesive designer will design the system to meet the performance and process requirements of the application [19].

Although polyether soft segments dominate the adhesive application, many one-component solvent-borne polyurethane adhesives will employ a polyester soft segment such as polybutylene adipate for the simple reason that its ability to crystallize can increase the initial adhesion between adherends [20]. A typical formulation is suggested in Table 10.3. The TPU would normally be polymerized in a separate operation and then dissolved in the adhesive solvent. It is noticeable that the TPU has a relatively low hard-segment fraction, about 25%, and that the hard-segment length 1,4-butanediol/polybutylene adipate (PBA) is approximately 1. This assures that the resulting adhesive will have a low glass transition temperature ($T_g$) for good performance at low temperatures, and also that the adhesive will possess very good flexibility. If the designer determines that the process requires more rapid buildup of adhesion properties, hexanediol could be substituted for butanediol. It is known that, for systems that employ one-part solvent-borne polyurethane adhesives, pretreatment of the surface can be critical to performance [21]. This can involve extensive roughening of the surfaces to be joined, and also use of a primer coating that has preferential adhesion to the substrate and the TPU but does not possess sufficient properties as a sole adhesive layer to function in the application [22]. Typical primer polymers are substances such as chlorinated polyethylene at high dilution in an aromatic solvent like toluene. For these systems, the solvent can also be a major participant in the growth of adhesion by swelling both the substrate and the adhesive polymers to promote polymer chain entanglement or physical interlocking of larger polymer segments that form as the adhesive laminate swells and then contracts [23, 24].

**TABLE 10.3 Example formulation and properties of a one-part solvent-borne polyurethane adhesive**

| TPU component | pphp |
|---|---|
| 3000 molecular weight polybutylene adipate | 16 |
| Butanediol | 0.42 |
| 4,4′ MDI | 3.58 |
| Methyl ethyl ketone | 80 |
| *Properties* | |
| Viscosity (25 °C) | 2000 cP |
| $T_g$ | –46 °C |
| Engineering strain at break (of adhesive film) | 1200% |
| Tensile strength (of adhesive film) | 30 MPa |
| 90° T-peel (polyurethane to leather) | 30 lb/linear inch |

MDI, 4,4′-methylene diphenyl diisocyanate; pphp, parts per hundred parts; $T_g$, glass transition temperature.

A two-part solvent-borne version of polyurethane adhesives is typically an isocyanate-tipped prepolymer with relatively low isocyanate content (5–12%) on one side and chain extender (diol or diamine), catalyst, and often a cross-linker for the second component. A two-part solvent-borne system has the benefit of allowing the opportunity to tune adhesive modulus and improve solvent resistance via the cross-linking capability. However, beyond the obvious problem with VOCs, this format suffers from the complication of properties being dependent on accurate weighing of the adhesive parts, and also the unavoidable limitations on open time that come from a reacting system. In addition, polyurethane side reactions, particularly with ambient water, are capable of producing $CO_2$ gas. In the limit of high humidity this can have the very undesirable effect of foaming the adhesive layer. A merge of one-part and two-part adhesive technologies is achieved by combining all polyurethane components into a single solvent system but employing a blocked isocyanate. Blocked isocyanates are standard polyisocyanates prereacted with weakly bonded active hydrogen groups. When heated, the weakly reacted components decouple, releasing the original polyisocyanate and the blocking agent. This technology, of course, requires that the bonded substrates be heated under compression to effect the desired polymerization and bonding chemistry (see Section 10.3.1.5).

#### *10.2.1.2 Hot-Melt Adhesives*

*Nonreactive Hot-Melt Adhesives* Hot-melt adhesives are thermoplastic materials that are (i) heated and applied to a substrate in the molten state, (ii) the bonding substrates assembled, and (iii) the thermoplastic allowed to resolidify back to its thermoplastic state. If the substrate is wet by the adhesive, if the polar interactions across the adhesive–substrate interface are of sufficient number and strength, and if the adhesive has adequate cohesive energy, a durable bond can be established. TPUs have been employed for this application and have a market appeal for wood-bonding and shoe sole operations.

Polyester polyols are often used in these applications since crystallization of the soft segment can speed up TPU resolidification and so increase the rate of adhesive strength build. The performance of simple TPU hot-melt adhesives is usually worse than that of high-performance polyamide and polyester hot-melt adhesives. This is ascribed to the higher viscosity of TPUs at standard application gun temperatures, which limits substrate wetting. It is possible to reduce viscosity and so improve hot-melt adhesive properties by the addition of a monol (e.g. replacing some amount of butanediol with 1-butanol). Substitution of even 5% of the butanediol chain extender with 1-butanol can reduce the molecular weight of the chain-limited polymer by a factor of 2–3 and reduce the viscosity of the polymer melt by over an order of magnitude based on the $\eta \propto M_n^{3.5}$ (the number average molecular weight) rule [25]. This can also slightly reduce overall polymer tensile performance by reducing hard-segment formation and phase separation. An example of a polyurethane hot melt based on this design is provided in Table 10.4. The measures of quality for a hot-melt adhesive are unique to the technology, reflecting the viscosity of a polymer melt and its creep properties under a shear load. In addition, it is common to characterize the strength build of a hot-melt adhesive, referred to as its "green strength" and its $T_g$.

*Reactive Hot-Melt Adhesives* Reactive polyurethane hot-melt adhesives are a technology introduced in the 1990s [26, 27]. They are essentially a low percentage isocyanate prepolymer that is applied to a substrate and allowed to cure by reaction via adventitious moisture or by direct reaction with substrates possessing active hydrogen on their surface. Reactive hot melts possess a highly desirable grouping of properties, including low viscosity at application temperatures, rapid generation of green strength when using crystallizable soft segments, gradual development of very strong adhesive bonding, no VOCs, and the excellent material properties of polyurethane elastomers. These properties and performance have made polyurethane reactive hot melts very competitive with polyamide and polyester hot melts and allowed this polyurethane adhesive class to grow at twice the rate of polyurethanes as a whole [2]. Prepolymers and their preparation are covered in Chapters 2 and 9. Table 10.5 provides the composition and properties of a 4000 g/mol polycaprolactone-based MDI prepolymer with 2.3% isocyanate useable as a reactive polyurethane hot-melt adhesive. Polyether polyols can also be used to make very good reactive hot-melt adhesives, especially when the prepolymers are made of blended molecular weights to provide the mix of low viscosity and rapid green strength build.

***10.2.1.3 Water-Borne Polyurethane Adhesives*** Water-borne polyurethane dispersions (PUDs) have been an area of considerable interest and commercial growth since the first documented PUD in 1961 at E.I. du Pont de Nemours and Company [28]. The growth, as measured by patent activity, has been exponential, beginning in the mid-1970s (Figure 10.8). This growth has been driven largely by regulatory initiatives to reduce VOCs in coatings and adhesives. PUDs enable delivery of a low-viscosity urethane polymer to a substrate, while reducing or eliminating the use of fugitive organic solvents [29].

**TABLE 10.4 Formulation and properties of a nonreactive polyurethane hot-melt adhesive**

| Component | pphp |
|---|---|
| Polycaprolactone $M_n$ = 530 | 29.5 |
| 1,4-butanediol | 14.24 |
| 1-butanol | 0.62 |
| 4,4′ MDI | 55.64 |
| Viscosity at 190 °C | 10,000 cP |
| Melt index 44 psi at 190 °C | 160 |
| $T_g$ | 0 °C |
| Modulus of adhesive film | 4.3 × $10^7$ Pa/$cm^2$ |
| Elongation at break | 500% |
| 180° T-peel adhesion (canvas to Al) | 9 pli |
| Shear fail temperature | 160 °C at 2 lb/$in^2$ stress |
| | 106 °C at 14 lb/$in^2$ stress |

MDI, 4,4′-methylene diphenyl diisocyanate; $M_n$, number average molecular weight; pli, pounds per linear inch; pphp, parts per hundred parts; $T_g$, glass transition temperature.

**TABLE 10.5 Formulation and properties of a reactive polyurethane hot-melt adhesive**

| Component | Grams |
|---|---|
| 4000 molecular weight polycaprolactone | 100.0 |
| 50/50 *o,p′* MDI | 14.62 |
| *Properties* | |
| Viscosity at 121 °C/155 $s^{-1}$ | 72 cP (shear thinning) |
| $T_g$ (of the prepolymer) | 56 °C |
| Open time | >2 minutes |
| Green strength at 2/15/30 minutes ASTM D6746 | 0/5.4/6.8 lb/linear-inch |
| Green strength at 2/7 days | 32/>500 lb |

MDI, 4,4′-methylene diphenyl diisocyanate; $T_g$, glass transition temperature.

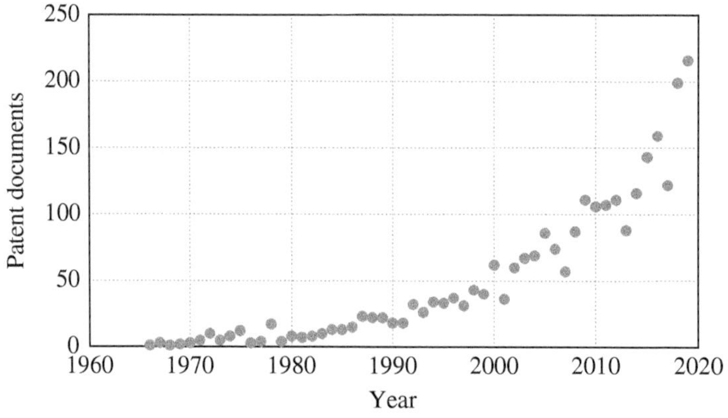


**FIGURE 10.8** Growth in PUD patent activity since development in the early 1960s.

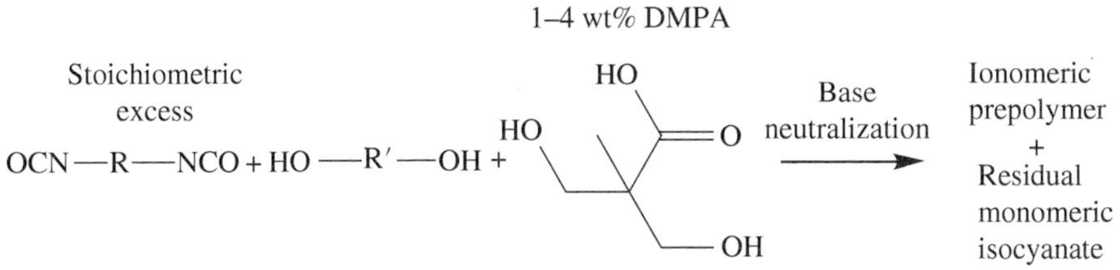


**FIGURE 10.9** Synthesis of ionomeric prepolymer for PUD based on DMPA.

As the name implies, PUDs consist of particles of urethane/urea polymer dispersed in a continuous phase of water. The most common technique for dispersing urethane in water involves the use of self-dispersing ionomeric polyurethane prepolymers. In this process, an ionizable moiety is reacted into a polyurethane prepolymer structure. The most common of these moieties is dimethylolpropionic acid (DMPA) (Figure 10.9) [30].

The polyol and isocyanate structures in PUD prepolymers can vary widely depending on final application and requirements. While PUD prepolymers are typically isocyanate terminated, they can also employ hydroxyl-terminated prepolymers for two-component water-borne urethanes.

Prior to dispersing in water, the carboxylic acid groups in the prepolymer are neutralized with a strong base such as triethylamine to convert the acid to the carboxylate. It is this hydrophilic neutralized carboxylate that serves to disperse and stabilize the particles in the aqueous phase. Immediately following this aqueous dispersion step, a chain extender is added to the dispersion to complete conversion of the remaining isocyanate groups to the final polymer within the particles. Because of the competing reactivity of water, the selection of chain extender is limited to chain extenders with higher reactivity toward isocyanates than water. Most commonly, these chain extenders are primary or secondary polyamines that chain extend the polymer via formation of urea bonds.

The PUD ionomer process is often done with low levels of solvent, such as acetone or *n*-methyl pyrrolidinone (NMP) to lower prepolymer viscosity in the process. In the case of acetone, the solvent is removed in a final step, while low levels of high-boiling solvents such as NMP may be left in the final product.

The use of aromatic isocyanates can also be challenging in this process owing to their high reactivity with water during the chain extension step. Low-residence-time dispersion technologies for PUDs employing mechanical dispersion are reported in the literature that address reactivity challenges associated with aromatic isocyanates [31].

In the case of *nonreactive* PUDs, the polymer is fully formed prior to application and properties develop solely on the basis of drying and coalescence of the polyurethane particles.

Two-component *reactive* PUDs employ two separate dispersed phases that react with each other during the drying phase. This type of technology is utilized in both coating and adhesive applications. Cross-linking of the polymer during the coating step can provide improved polymer properties versus a single-component PUD.

Apart from backbone alterations to make polyurethanes more water compatible, process innovation to obtain ever smaller dispersion particles and involving ever smaller amounts of organic solvents has been evident [32, 33]. As shown in Figure 10.9, modification of the backbone with a monomer building block such as DMPA can facilitate compatibilization of the polyurethane chain with water. However, this is usually not enough to result in a desirable adhesive product and the process of dispersion is highly optimized by manufacturers. Many of these production methods are batch operations where all the ingredients of the dispersion are added to a single reaction vessel (in the appropriate order and amounts), heated, mixed, degassed, filtered, etc. and stored for packaging. Continuous operations have also been developed that can minimize the need for storage of large amounts of inventory [34]. The ability to produce volumes at times and places near to their application is helpful in extending shelf life stability for the customer. Figure 10.10 is a simplified block diagram for the batch process.

The process in Figure 10.10 is not without complications. The polyols (including the DMPA) are added to the isocyanate, which is present in excess in a prepolymer process. Solvent, surfactant, and catalyst can be added according to the component and process requirements. The acid groups of the prepolymer are neutralized and mixed rapidly with added water. While the neutralizing agent (usually triethyl amine) imparts solubility to polyurethane in water, it also facilitates reactions of isocyanates with water and also potentially free isocyanate–isocyanate reactions (see Chapter 3). When water is added to the neutralized prepolymer (with energetic mixing) the uncontrolled isocyanate reactions can result in increased system viscosity and complicate the dispersion of the PUD particles into small particle sizes. Diamines such as ethylene diamine, 1,2-propanediamine, or hydrazine are added to the prepolymer to complete the polymerization process by chain extension, finishing formation of the polyurethane backbone. The role of added solvent (e.g. NMP, methyl ethyl ketone [MEK], toluene) is to compatibilize particle phase–continuous phase viscosities and facilitate shear-induced particle breakup into very small particles. Optimized processes can result in the formation of submicron particle sizes. The process solvent can then be removed in a postprocessing step, or be included in the final formulation to facilitate adhesive (or coating) application attributes. PUDs produced by this or a similar batch process using internal ionic stabilization can also suffer from instabilities caused by unintended excursions in pH, ionic strength, or temperature.

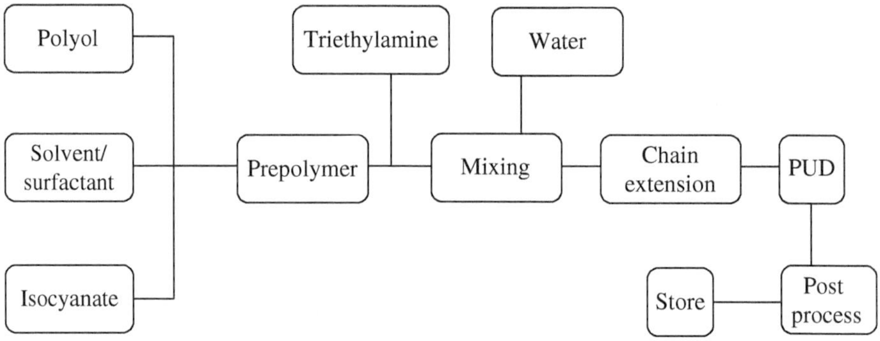


**FIGURE 10.10** Batch process for PUD production.

Some of the weaknesses of the batch process can be mitigated by a continuous process, as illustrated in Figure 10.11. The obvious difference between batch and continuous processes is the multistage mixing and the nonrequirement of solvent. In the continuous process, the prepolymer (ionic or nonionic) is formed and subsequently sheared with high energy concomitant with the addition of surfactant and a small amount of water. The process is controlled to invert an initial water-in-oil dispersion to an oil-in-water dispersion in spite of the small volume (~10% v/v) of water. Additional water and optional diamine chain extender can be added in a subsequent mixing step to dilute the dispersion to a desired solids content (Table 10.6).

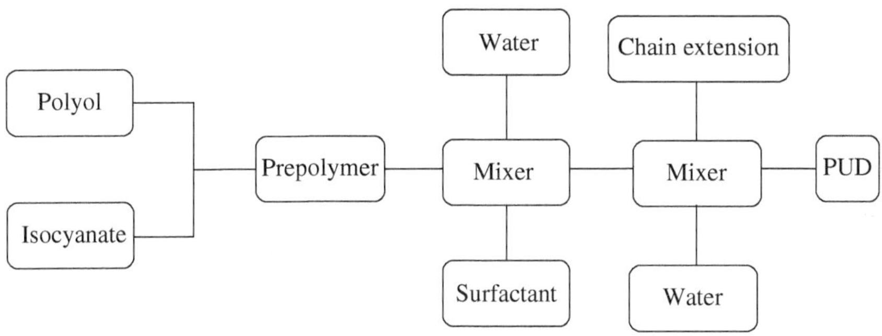


**FIGURE 10.11** Block diagram for a continuous process for making PUDs via a high-inverted-phase process.

**TABLE 10.6 Composition and properties of a PUD adhesive**

| Component | Amount | Comment |
|---|---|---|
| 2000 g/mol 12% EO-capped PO/EO triblock polyol | 41.3 | Prepolymer component |
| 2000 g/mol PTMEG | 36.1 | Prepolymer component |
| MDI | 18 | Prepolymer component |
| 1000 g/mol methoxy ethylene glycol | 2 | Surfactant |
| 1000 g/mol polyethylene glycol | 2 | Surfactant |
| Sodium dodecyl benzene sulfonate | 3 | Surfactant |
| % NCO (in prepolymer) | | 2.5% |
| Molecular weight prepolymer (g/mol) | 3350 | |
| 1,2-propane diamine | 2.45 | Chain extender |
| $M_w$ polyurethane polymer | 88K | Polydispersity $M_w/M_n = 4.8$ |
| % Solids | 39% | |
| Viscosity | 29 cP | |
| Particle size | 230 nm | |
| Adhesion nylon to PET | >4 lb/in | Cohesive substrate failure |
| Adhesion polyethylene to Al foil | 2.4 lb/in | Adhesive failure |

EO, ethylene oxide; MDI, 4,4′-methylene diphenyl diisocyanate; $M_n$, number average molecular weight; $M_w$, weight average molecular weight; PET, polyethylene terephthalate; PO, propylene oxide; PTMEG, polytetramethylene ether glycol.

## 10.3 TRENDS IN POLYURETHANE ADHESIVES

While the patent literature is notably poor at providing insight into fundamental science, it is the clearest way to understand the future of a technology and market. This is because the patent literature represents a sizeable investment by a company or individual on what is believed will be desirable by customers and market competitors in the next-generation product. Furthermore, defense of a patent asset requires that the technology on which a patent is based must be protected by patent prior to its commercial introduction. While a single patent may not be instructive as to the direction of future innovation, a clear trend in industrial patent activity can be unmistakable.

Adhesives in general, and polyurethane adhesives in particular, are highly formulated and can bring a high degree of value to specific customers with one formulation, while a small variation will meet the needs of another customer. This circumstance and the very keen level of competition among adhesive providers leads to an active patent landscape. Our previous category analyses by patent have focused on the activities of major feedstock providers. Major chemical producers are actively protecting their adhesive intellectual assets via patents, especially in defense of their systems house businesses. Figure 10.12 quantifies patent activity from 2000 to 2014, and separately from 2015 to 2019 of major international integrated manufacturers of polyurethane building blocks, as well some of the major manufacturers of formulated adhesive systems. While the number of patents among this group is high, it is useful to consider that, over the time frame characterized by Figure 10.12, more than 2200 patents dealt with polyurethane adhesives as either compositions of matter, methods of manufacture, or applications. Of the over 2200 patents, about half were solely concerned with adhesive composition of matter, about 700 were associated with the process of making or applying the adhesive, and the remainder were related to the use of the adhesive.

For additional perspective, a broader view of patent activity is provided in Figure 10.13, which illustrates the distribution of patent activity of all companies with more than five patent publications over the 2000–2014 and 2015–2019 time frames. Figure 10.13a shows that, for the former time period, the top six patenting companies were Japanese concerns. While representative of activity, this figure does not filter out kokai patent publications, which are unexamined applications. A kokai is often promulgated by a company for the purpose of destroying intellectual novelty rather than a sincere attempt to protect an economic asset. In any case, it is also notable that nonchemical companies appear to participate heavily in the landscape. This is misleading; many venerable and very sizeable Japanese concerns do not change their historical names to reflect the highly diversified portfolio of products they provide and, just as importantly in the adhesives field, utilize in their business. The latter time frame shows that, while Japanese companies are still active, pursuit of patents in the Chinese language is now very important and companies outside of Japan have been more active recently.

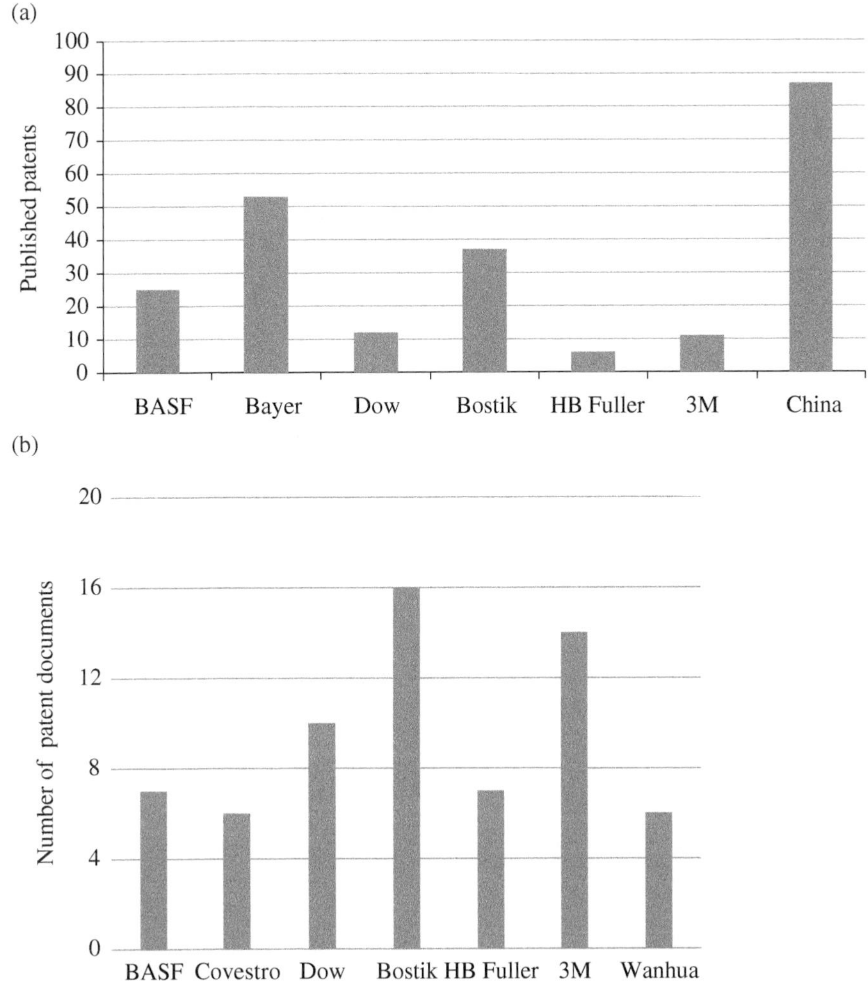


**FIGURE 10.12** Patent activities from (a) 2000–2014 and (b) 2015–2019 of major international polyurethane building block manufacturers and major manufacturers of formulated adhesives. Comparison is made with all patents on the same subject filed in the Chinese language in (a) substituted with Wanhua Chemical Group for latest data. For 2015–2019 there were 1843 patent documents filed in the Chinese language. Data on Bayer are substituted with data on Covestro, which was a spinoff company from Bayer.

As mentioned previously, about half of polyurethane adhesive patent activity is concentrated on new formulations and compositions for achieving new performance criteria. A partial review of composition of matter patents show that the majority of new formulations are in fact novel arrangements of conventional materials (Figure 10.14). No doubt the resulting performance is the basis for asset protection.

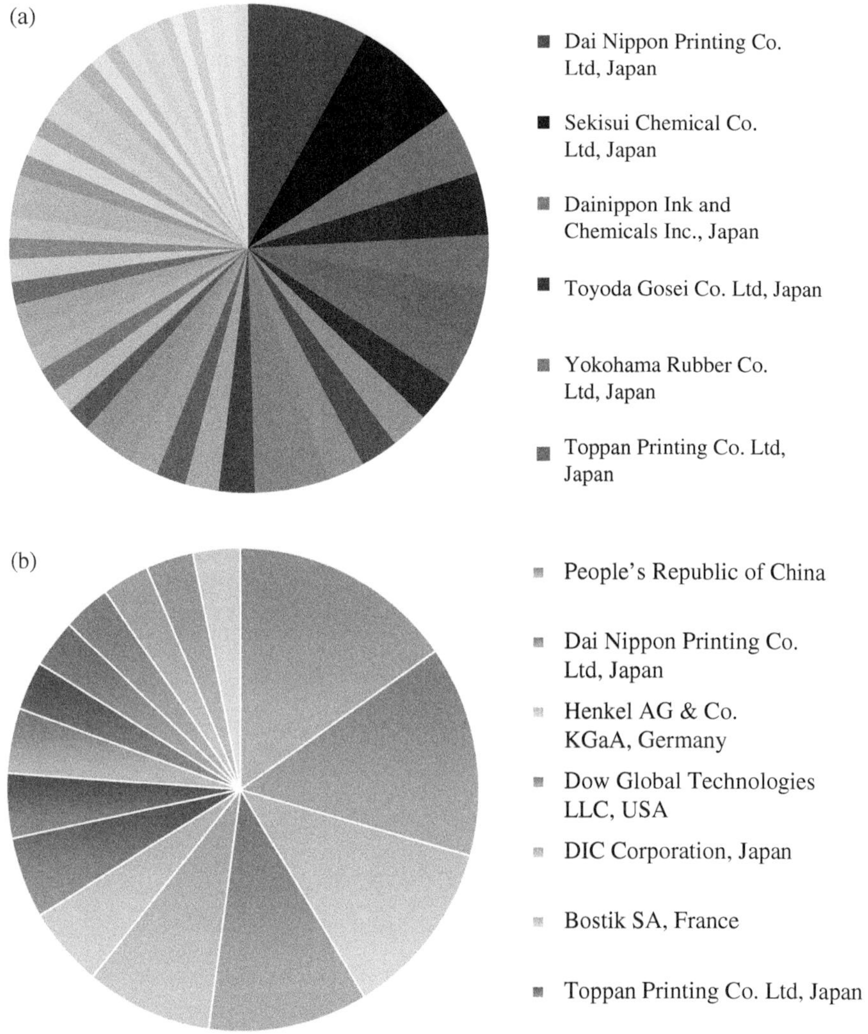


**FIGURE 10.13** Global patent activity for polyurethane adhesives for companies having more than five patents for (a) 2000–2013 and (b) 2014–2019. The top seven patenting companies (by number of publications are indicated). The graph shows the large number of formulators and manufacturers active in the competitive marketplace. For (b), many Chinese universities and companies identify "People's Republic of China" as part of their name and so get classified by search engines according to their origin rather than their name.

A similarly abridged analysis shows a varied list of well-known polyurethane adhesive applications (Figure 10.15). Any trend analysis would probably focus on the relatively large number of application patents focused on adhering articles of clothing. In fact, a general conclusion of the future trends might point to dissemination of conventional materials into conventional applications to achieve higher performance.

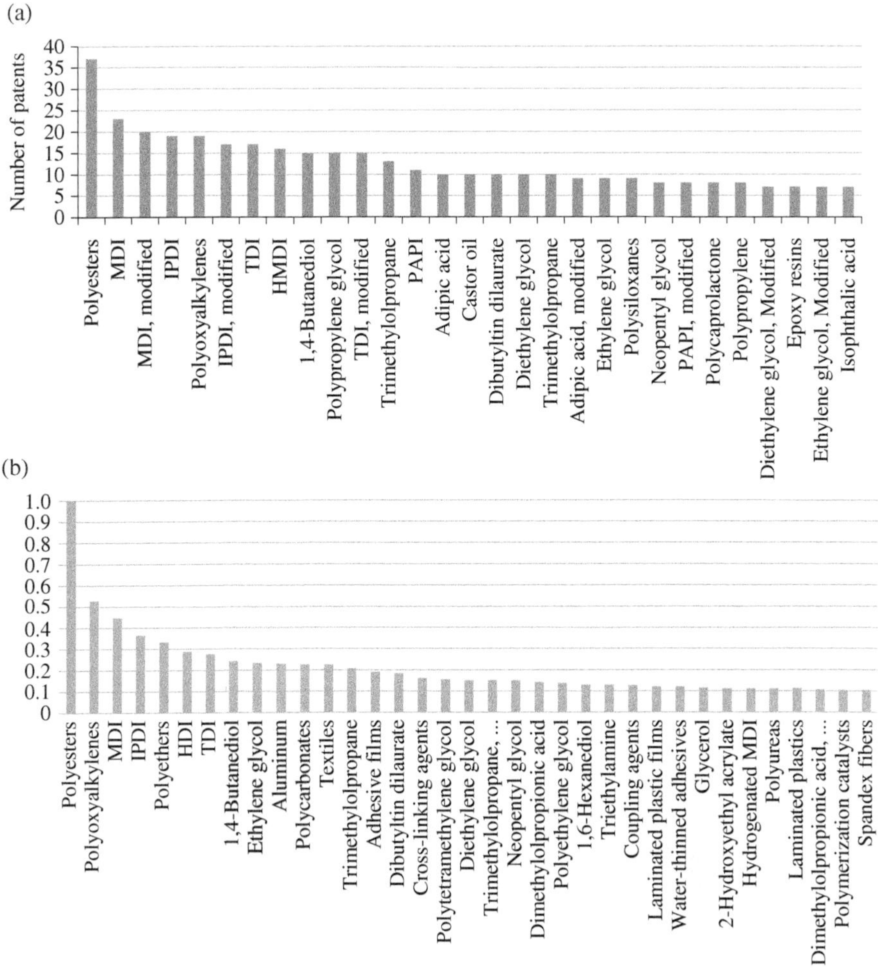


**FIGURE 10.14** Adhesive patent activity subject based on protecting specific polyurethane compositions of matter. Result indicates that innovation is based on new formulations of well-established materials. (a) 2000–2014; (b) 2015–2019. HMDI,; IPDI, isophorone diisocyanates; MDI, 4,4′-methylene diphenyl diisocyanate; PAPI,; TDI, toluene diisocyanate.

Patent activity in previous decades has been approximately evenly distributed between two-part and one-part formats; however, patent activity in two-component polyurethane adhesives has become more prominently pursued in China. This suggests that adhesive performance has recently become paramount to convenience. The range of patents pursued within the two-component space varies across packaging adhesives, flooring adhesives, solar cells, wind turbines, automotive adhesives with blocked isocyanate to provide thermally triggered cure, and apparel applications. One-component adhesive patent activity also spans a broad range defined across water-borne format,

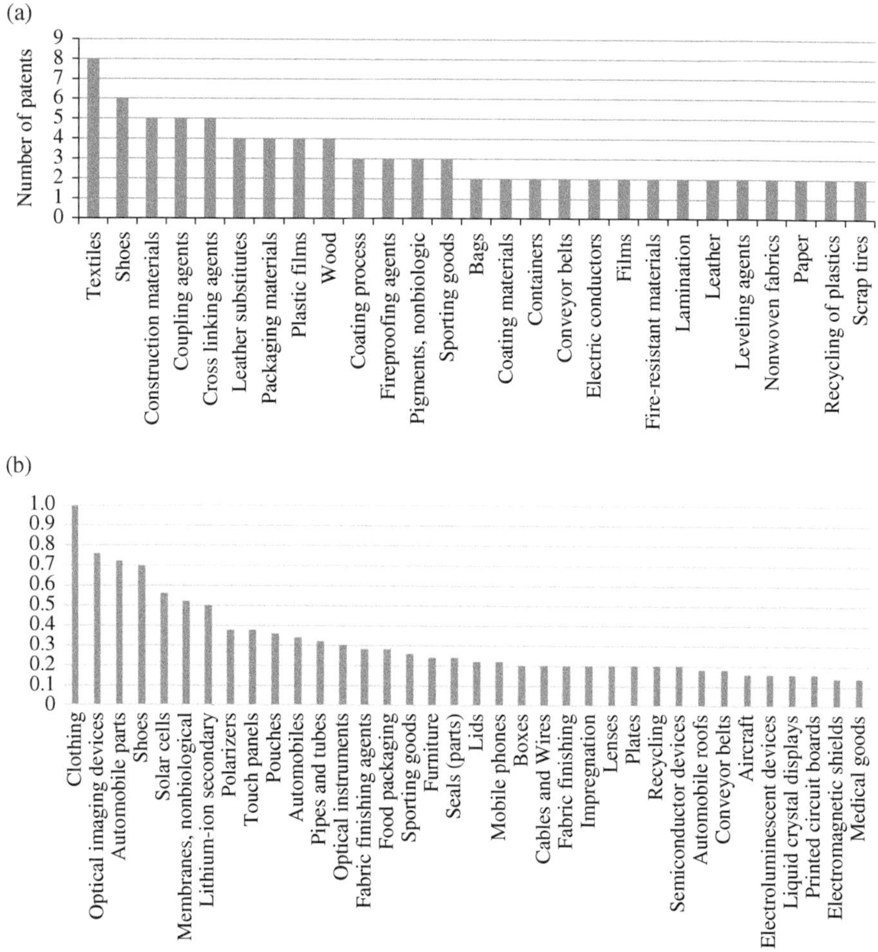


**FIGURE 10.15** Analysis of polyurethane adhesive applications patent activity for (a) 2000–2014 and (b) 2015–2019. Applications do not suggest any discontinuity with well-established applications.

moisture curable cure mechanism, adhesive with particularly good flame retardancy, and low VOC formulation. Furthermore, patent activity across adhesive technology platforms continues to suggest that, while water-borne and hot-melt platforms are the most active fields of innovation, there is substantial work across the spectrum and no particular platform is out of favor. The past 5 years of patent activity show about twice as much patent activity in water-borne than hot-melt technology.

Lastly, it is useful to consider that while polyurethane represents about 6% of all plastic manufactured it represents slightly less than 3% of all adhesives patents during the time frame 2000–2014 and similarly for the segment between 2014 and 2019,

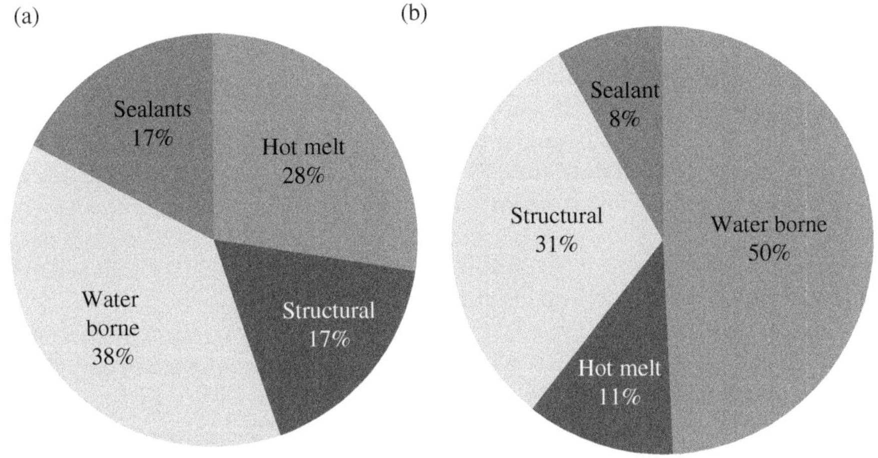


**FIGURE 10.16** Distribution of patents based on their delivery format and general application category for (a) 2014 and (b) 2018. The data reveal the global movement toward low VOC formulations.

suggesting that lower cost or specific applications (such as electronics and structural adhesives) may represent areas of more intense current activity. The highest growth in patent activity is in the fields of reactive polyurethane hot melts and in water-borne polyurethane adhesives (Figure 10.16).

## 10.4 COATINGS

Coatings is an important and rapidly growing value-added application of polyurethane chemistry [35, 36]. All the material aspects that make polyurethanes into successful elastomers, adhesives, and thermoplastics translate to making high-performance coatings as well. Thus, polyurethane coatings are valued particularly for durability, abrasion resistance, aesthetics, and formulation flexibility. In addition, like polyurethane adhesives, polyurethane coatings can be delivered in numerous formats to meet the process requirements of almost any coating operation.

Quantifying market consumption based on type and application is complicated by the varying classification of a coating. For example, should urethane modification of an acrylic or epoxy coating be counted as urethane? Should a sprayed truck bed liner be grouped with a sprayed clear furniture coating? Is the common means of application a sufficient qualification to group these very different products within the same category? Some analyses of usage do, and others do not. Variance in market data simply reflect an individual perception of what is a coating. Another quantitative variable is counting volumes of formulated (and diluted) coatings versus just the volume of polyurethane in that coating.

Another distinction of coating classifications are those that are defined as "architectural" and those defined as "industrial" [37]. Industrial coatings are all of those coatings applied at a manufacturing facility. Architectural coatings are those meant to be applied in or onto buildings. There is obviously overlap in the coatings themselves since wood coatings can be used in both places, as can paints and alkyd coatings. The distinction is merely the location, and in some cases formulation specification, consistent with use by unlicensed individuals in unventilated environments. Thus, it is reasonable to approach coatings market data with the perspective of viewing a snapshot that may reflect a set of definitional biases, along with the usual lack of precision accounting for volumes consumed by many thousands of manufacturers and billions of consumers.

While absolute volumes are potentially imprecise, the ratio of volumes has been relatively constant from 2012 until 2018 and can perhaps represent useful data for comparison. Figure 10.17 shows the proportional amounts of polyols and isocyanates used in coatings. For specialty performance products like coatings, the choice of building block is anything but arbitrary. The specific polyols and isocyanates are designed to meet stringent requirements for the application, and, while price is a consideration in every real function, performance will often be the driver of what particular polyurethane chemistry and product format will be used [38]. In this regard the applications that use polyurethane-based coatings and their approximate ratio is shown in Figure 10.18. Architectural coatings use little polyurethane resin and represent about 10% of total polyurethane resin used for coatings; these are not included in the figure. Most architectural polyurethane coatings are in the form of polyurethane-modified alkyd coatings for wood coverage. Additional architectural volume is polyurethane water-borne dispersions for wood coatings.

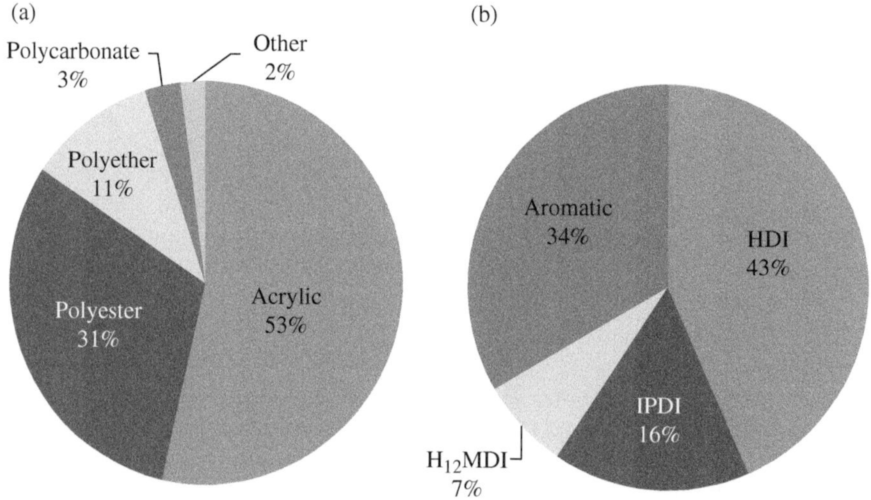


**FIGURE 10.17** (a) Distribution of polyol consumption in polyurethane coatings. (b) Distribution of isocyanate consumption in polyurethane coatings. HDI, hexane diisocyanates; $H_{12}$MDI, hydrogenated MDI; IPDI, isophorone diisocyanates.

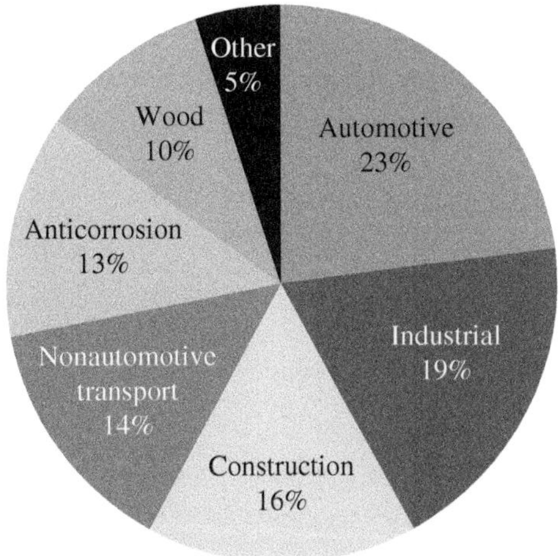


**FIGURE 10.18** Distribution of industrial applications employing polyurethane coatings in 2018. The importance of "other" is consistent with the high service requirements for participation in this market in providing custom solutions for smaller, customer-specific applications.

Table 10.7 lists polyurethane coating technologies as defined by ASTM conventions, including a list of applications for which those technologies are found useful. The technology choice (i.e. solvent borne versus water borne) reflects the demands and the conditions of application.

Almost all coatings that are applied to impart an aesthetic appeal utilize aliphatic isocyanates since their tendency to yellow is much reduced relative to aromatic polyisocyanates (see Chapter 2). However, aromatic polyisocyanates are used for coatings that are nondecorative, are part of an interior lining, have an underground function, or are part of a formulation that does not use a solvent.

Much as any technology, coatings are tested for their performance attributes. Given the range of environments and functions for coatings, the number of standard tests is quite large. Table 10.8 is an abridged list of tests commonly applied for coatings testing. The Standard Guide for Testing (D5324) should be consulted initially to provide a complete list of standards that apply to numerous applications for the specified technology. Since testing of coatings is largely a pursuit of industry, many individual tests are automated, or have commercially available equipment that is designed and sold for specifically meeting a given ASTM standard; this provides a modicum of standardization across different laboratories. Regardless, it is not uncommon for separate laboratories to achieve different results because of differences in surface preparation and differences in coating application, which every standard warns is critical to results [39, 40].

**TABLE 10.7 ASTM classification of polyurethane coatings based on format and method of cure and their predominant application areas**

| ASTM convention | Description of technology | Applications |
|---|---|---|
| Type I | One-part nonreactive, unsaturated aliphatic esters No free isocyanate groups in coating, cure by oxidative cross-linking of unsaturated groups and solvent evaporation. Uses soybean or linseed oil | Architectural floor and maintenance, topcoats. Better performance than unmodified alkyds |
| Type II | One-part moisture curing, contains free isocyanates, reacts with environmental water and isocyanate reactive groups on the substrate surface. Polyether or polyester soft segments. Can use various blocking techniques to preserve isocyanate reactivity for extended shelf life | Leather, concrete, maintenance |
| Type III | One-part heat cure. Uses blocked isocyanates, which are liberated upon heating to react with isocyanate reactive components in the formulation. Polyether or polyester, often phenol, caprolactam, or cresol-blocked aliphatic polyisocyanates or TDI | Coil and electric wire |
| Type IV | Two-part solvent borne. One part is the prepolymer polyisocyanate, while the other is the polyol(s), catalyst, and all other formulation components, such as solvents, pigments, and other additives. Acrylic and polyester polyols are common components. Also can use hydroxyl functional alkyds Ambient or heat cure. Pot life can be an issue and can be influenced by using aprotic complexing solvents | Plastics, wood furniture, marine exteriors |
| Type V | Two-part high-solids (>50%) coatings. One part is a prepolymer, the other is polyol, usually acrylic, polyesters, or polyethers. Considered high-performance coatings; can have limited pot life, complicating application | Leather, wood, automotive clear coats, refinishes, aircraft, bus, truck, industrial structure maintenance coatings |
| Type VI | One-part nonreactive low solids (<20%), solvent borne. Referred to as a lacquer. High-gloss film forms upon solvent evaporation | Textiles |
| Powder coatings | One-part reactive system using caprolactam or 1,2,4-triazole-blocked aliphatic isocyanates, especially IPDI and $H_{12}$MDI, polyester, and epoxy-modified soft segments. Viscosity must be controlled and kept low to ensure powder coalescence during application and bake | Automotive exterior panels and parts, wire, electrical transmission equipment (i.e. transformers), reflective surfaces, metal surfaces, outdoor lawn furniture |

**TABLE 10.7 (Continued)**

| ASTM convention | Description of technology | Applications |
|---|---|---|
| Radiation | High-solids coatings, rapid cure, high gloss, not practical for home use or complex shapes. Made by reacting isocyanate-capped prepolymer with hydroxyl-functionalized acrylate or methacrylate. Requires use of photoinitiators in formulation. Any isocyanate can function but usually aliphatic | Manufactured wood flooring, cabinets, metal surfaces, plastic |
| Water borne | Broadly applied to one- and two-part systems using aliphatic or aromatic isocyanates. Reduces VOC exposure, high growth of application in all geographies, can be used in hybrid technologies such as uralkyds and polyurethane acrylates. May be formulated with water-soluble organic solvents for better solubilization of components. May also use defoamer. Cure speed/drying is less of an issue as percentage of solids have increased | Wood coatings, decorative coatings, artificial leathers, textiles, plastics, inks, architectural, automotive |

Powder, radiation cure, and water-borne coatings do not yet fit into the ASTM conventions but are included here as well.

IPDI, isophorone diisocyanates; $H_{12}MDI$, hydrogenated MDI; TDI, toluene diisocyanate; VOC, volatile organic carbon.

The polyurethane coatings industry has much in common with the polyurethane adhesives industry, making it convenient to group them together. In both cases the application is a film that must adhere to a surface and cure within a predetermined time. The forms – solvent borne, water-borne dispersions, moisture cured, one and two part – are all in common with the adhesives industry. Also in common is the general similarity in industry structure depicted in Figure 10.19. This figure (simplified as it is) depicts the interrelationships along the supply chain in situations where participants can be simultaneously customers, suppliers, and competitors. This entangled system creates market opportunities and vulnerabilities that are very different from a landscape of limited and well-known competitors and customers [41]. For instance, the position of a coating formulator can be exposed to unusual pressures as they can be simultaneously squeezed by high raw material costs that can coexist with decreased consumer demand, depressing final product prices. While the coatings producer can purchase from alternative suppliers to create price competition, these suppliers are relatively few in number and their commodity feedstocks tend to have well-characterized supply, demand, and pricing. Like other products that may only have subtle differences (think of gasoline) their prices tend to move quickly and in unison up and down. Price differences such as exist are related to producer efficiency and spot supply conditions and tend to be relatively small. Even noncommodity raw materials such as aliphatic isocyanates exhibit less price elastic demand owing to limited volumes and limited producers resulting from exigencies of phosgenation (see Chapter 2) [42].

**TABLE 10.8 Abridged list of standardized tests and methods for characterizing coatings**

| ASTM test | Standard description |
|---|---|
| D5324 | Standard Guide for Testing Water-Borne Architectural Coatings includes a list of tests for flat interior and exterior latex paints, water-borne floor paints, and semi- and high-gloss latex paints. References 80 standards for characterization of architectural coatings |
| D6577 | Standard Guide for Testing Industrial Protective Coatings, specifically intended for paints applied to substrates on-site of structures and buildings, especially where subject to corrosive environments, such as industrial, urban, and marine environments. References 129 standards for characterization of industrial protective coatings |
| D5146 | Standard Guide to Testing Solvent-Borne Architectural Coatings includes interior low-gloss wall finishes, interior gloss and semi-gloss wall and trim enamels, exterior house and trim coatings, and exterior and interior floor enamels. References 75 standards for characterization of solvent-borne architectural coatings |
| D3451 | Standard Guide for Testing Coating Powders and Powder Coatings, particularly those applied by electrostatic, fluidized bed, or alternative methods. References 85 standards for characterization of powder coatings and coating characteristics |
| D3730 | Standard Guide for Testing High-Performance Interior Architectural Wall Coatings. "High performance" denoting coatings that are tougher, more abrasive abrasion resistant, and more stain resistant than usual coatings. References 49 standards for characterizing this class of architectural coatings |
| D2832 | Standard Guide for Determining Volatile and Nonvolatile Content of Paint and Related Coatings. Applicable to all ASTM coating types as well as powder coatings and water-based coatings. References 15 other standards for specific coating types that apply in conjunction with the present standard |
| D5178 | Standard Test Method for Mar Resistance of Organic Coatings. Coating is applied to a flat surface and progressively challenged with a weighted stylus that is incrementally loaded until coating failure |
| D3732 | Standard Practice for Reporting Cure Times of Ultraviolet-Cured Coatings. Specifies UV cure but is generally applicable. Involves application of the coating to a desired substrate at a desired thickness and curing for incrementally longer exposure. Cure is measured by property measurement, including impact resistance, hardness, solvent rub test, abrasion resistance, and stain resistance. The shortest time that yields a cured film is the cure time |

**TABLE 10.8 (Continued)**

| ASTM test | Standard description |
|---|---|
| D7394 | Standard Practice for Rheological Characterization of Architectural Coatings using Three Rotational Bench Viscometers. Applicable to water-borne architectural coatings. Instructions for using a Brookfield viscometer, a paddle viscometer, and a cone-and-plate-type viscometer for sampling behavior in low-, medium-, and high-shear rates |
| D4708 | Standard Practice for Preparation of Uniform Free Films of Organic Coatings. Useful procedure for direct determination of coating physical properties including permeability independent of substrate. Involves casting of film onto either fluorinated sheet, silicone-coated sheet, or halosilane-coated glass followed by cure and sample removal |
| B499 | Standard Test Method for Measurement of Coating Thicknesses by the Magnetic Method: Nonmagnetic Coatings on Magnetic Basis Metals. Nondestructive test method for measuring nonmagnetic coating thickness using commercially available device. Device monitors the magnetic flex density. Method cannot distinguish thickness of individual layers, only cumulative thickness |
| D2134/D4366 | Standard Test Method for Determining the Hardness of Organic Coatings with a Sward-Type Hardness Rocker. Common method in coatings industry for determining hardness and cure. Monitors hardness and cure by the amplitude of sinusoidal oscillation of a specific tester. Linear relationship between this technique and other techniques such as the Koenig pendulum hardness tester, also commonly used (D4366 withdrawn) |
| D6577 | Standard Guide for Testing Industrial Protective Coatings. Includes industrial and architectural coatings with reference to 126 related standards, including salt-spray testing, graffiti resistance, MEK double rub, etc. |
| D6943 | Standard Practice for Immersion Testing of Industrial Protective Coatings. Covers testing of samples at standard temperature and pressure, with a temperature gradient, and at constant temperature and increased pressure. Many applications including for can coatings |

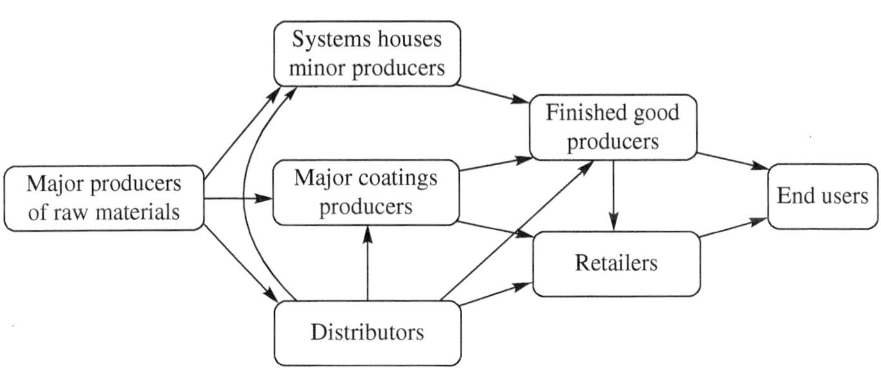


**FIGURE 10.19** Simplified supply chain and commercial relationships among polyurethane coatings market participants.

### 10.4.1 Polyurethane Coating Formulations

Coatings often have a decorative and a protective function. Since many coatings are exposed to the environment, especially to light, they are often formulated with aliphatic isocyanates to circumvent photolytic degradation mechanisms (see Section 2.2.3). Manufacturers of aliphatic isocyanates have modified aliphatic monomers to provide cross-linking potential for coatings by trimerizing isocyanates into isocyanurate rings. The resulting ring structures are very thermodynamically stable, have the potential to create three-dimensional networks through three-functional cross-links, improve solvent resistance, and promote formation of clear amorphous/glassy films by inhibiting phase separation of any chain extended phase. Figure 10.20 shows the structures of aliphatic isocyanate trimers commonly employed for polyurethane coatings. Hydrogenated MDI ($H_{12}$MDI), which is used in limited coating applications, is not commercially supplied as a trimer. The use of hexane diisocyanate (HDI) trimer or isophorone diisocyanate (IPDI) trimer is a matter of coating requirements, and it is not unusual for a coating formulator to mix them in a single coating to obtain optimized properties. The linear aliphatic HDI trimer is employed to impart flexibility and uniform reactivity. The cycloaliphatic IPDI trimer produces enhanced coating rigidity and higher film hardness. The secondary isocyanate of IPDI is slightly less reactive than the primary isocyanates of the HDI trimer and results in longer pot life and slower reactivity, as measured by tack-free times.

***10.4.1.1 Two-Part Solvent-Borne Coating*** The two-part solvent-borne coating formulation depicted in Table 10.9 is one that might be used for an automotive top coat, refinish, or general industrial coating. It utilizes a linear acrylic polyol with a binder $T_g$ of 48 °C, OH number of 59.4 (OH equivalent weight 944), an $M_n$ of 2941 g/mol dissolved in a xylene–*n*-butyl acetate mixture with a viscosity of 2200 cP at 40% solids. The total solids content of the formulated coating is 55% adjusted with added butyl acetate. Formulations were prepared with trimers of HDI and IPDI. Coatings were applied at room temperature and cured at 60 °C for 3 hours. Final coating thickness was 25 ± 2 μm for both coatings.

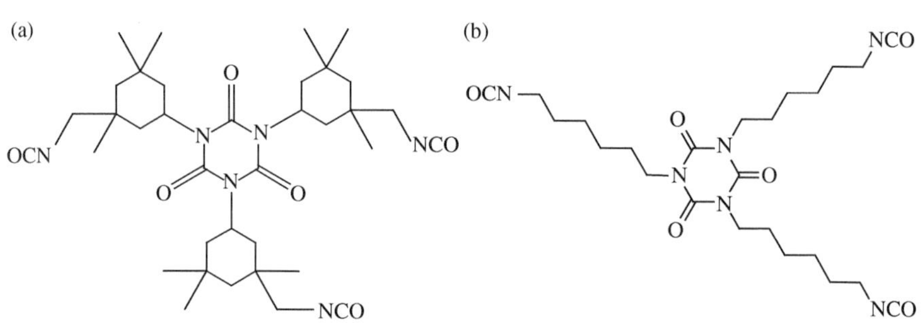


**FIGURE 10.20** Structure of aliphatic isocyanate trimers (a) isophorone diisocyanates (IPDI) and (b) hexane diisocyanate (HDI), which are commonly employed in decorative and protective polyurethane coatings.

**TABLE 10.9 Formulation and properties of a two-part solvent-borne polyurethane coating**

| Component | Formulation 1 (g) | Formulation 2 (g) |
|---|---|---|
| Acrylic polyol | 9.1 | 9.1 |
| Dibutyl tin dilaurate (catalyst based on polyol weight) | 100 ppm | 100 ppm |
| HDI trimer | NCO/OH = 1 | |
| IPDI trimer | | NCO/OH = 1 |
| Solvent *n*-butyl acetate | To 55% solids | To 55% solids |
| *Properties* | | |
| Gel time (min) | 35 | 140 |
| Gel content (%) | 91 | 70 |
| Pencil hardness (ASTM D3633) | H | 2H |
| Koenig hardness | 107 | 122 |
| Mandrel flexibility (ASTM 4338) (%) | 100 | 65 |
| MEK double rubs | >100 | 15 |

Coating properties are affected in predictable ways by choice of isocyanate cross-linking, and formulators often mix them to balance performance.
HDI, hexane diisocyanate; IPDI, isophorone diisocyanates; MEK, methyl ethyl ketone; ppm, parts per million.

The formulations of Table 10.9 are simplified to demonstrate coater binder properties separate from properties influenced by other components such as pigments or rheological and stabilization additives. The data illustrate the influence of aliphatic isocyanate on coating properties. The disparity of properties between trimers of HDI and IPDI naturally leads formulators to attempt the best of both worlds by blending [43, 44]. While blending of IPDI with HDI does moderate the coating performance between the extremes, it is often not a linear response, and the properties of different coating properties do not change with the same concentration function. Thus, while change in Koenig hardness may change linearly with isocyanate composition, the Mandrel test of flexibility may not change appreciably until the isocyanate composition is very close to a concentration limit. Additionally, change in the chemical resistance of the coating may follow a compositional "S" curve, suggesting some kind of required network formation to affect properties.

***10.4.1.2 Water-Borne Coatings*** As discussed in Section 10.2.1.3 there has been exponential growth in patent activity surrounding aqueous PUDs. A significant amount of this protection of intellectual assets has been the application of PUDs to coatings. Environmental and occupational safety and health regulations to reduce exposure to organic solvents has driven innovation and adoption of aqueous dispersion technology [45, 46]. Today, PUD coatings are accepted products for wood coatings, automobiles, textile fibers, leathers, fiber glass, floor polishes, and waxes, among others.

Batch and continuous methods of producing PUDs were introduced in Section 10.2.1.3. PUDs can also be made in small-scale batches for the purpose of trialing coating formulations. Such a procedure might be done on the subliter scale in a laboratory hood. An example of such a preparation is as follows.

A three-necked flask is placed in a heating mantle and the prepolymer components – polyol, DMPA, and diol chain extender in the solvent dipropylene glycol dimethyl ether – are weighed in. The solvent is added at one part to three parts polyol. The mixture is stirred with an overhead stirrer under a steady stream of $N_2$ and heated to a temperature of c. 70 °C. An isocyanate such as $H_{12}$MDI, IPDI, or HDI is added dropwise to the stirring mixture using an addition funnel, not allowing the reaction exotherm to increase the flask temperature more than c. 5 °C. The hard-segment components (isocyanate, diol, DMPA) are added in amounts to create a desired hard-segment volume – usually 40–55% of the final PUD. After allowing the urethane reactions to proceed for 3–5 hours at 70 °C, triethyl amine is added in a 1:1 stoichiometric ratio to DMPA and allowed to neutralize the acid for c. 15 minutes. The diluted polymer is poured into a plastic bottle and allowed to cool to c. 50 °C, after which the polymer solution is diluted with water (approximately one part water to one part polymer) and stirred at high-shear rate with a mixer such as a Cowles blade. The polymer is allowed to complete by addition of a diamine such as ethylene diamine or 1,2-propylene diamine and allowed to agitate for approximately 1 day. The dispersion is then filtered through an appropriate mesh filter. This dispersion is now ready for coating substrates. A rheology modifier may be added to improve coating aesthetics. The coating can be allowed to dry passively for 7 days. Properties of a PUD coating such as that described are given in Table 10.10.

***10.4.1.3 Water-Borne Hybrids*** While PUDs have demonstrated strong growth as a unique technology, it has been inevitable that innovation has explored PUDs blended with other dispersible polymers or hybridized with other polymer systems [47, 48]. One of the common blended or hybridized copolymer systems has been the polyurethane–acrylic water-borne dispersion. This category has been successful because (i) each system has properties that make up for shortcomings in the other, (ii) they have compatible chemistry, which makes them easily blended or hybridized,

**TABLE 10.10 Illustrative properties for a water-borne polyurethane coating obtained by a dispersion process such as specified in the text**

| Test | Value |
|---|---|
| MEK double rub | 140 |
| Koenig pendulum hardness | 130 |
| Taber abrasion (mg) | 11 |
| Pencil gouge resistance (ASTM 3363) | 10.5 |
| Impact resistance (ASTM 2794) (lb) | >160 |
| Mandrel flexibility | 100% |

MEK, methyl ethyl ketone.

(iii) the acrylic blend can be less expensive than the pure PUD, and (iv) they are already incumbent within the coating market, so barriers to acceptance by consumers of their combination is greatly reduced. Blends tend to exhibit better hardness and weatherability than pure PUDs and better flexibility and solvent resistance than pure acrylic dispersions.

Blending dispersions is the easiest way to make a new product, and this technique is certainly utilized industrially. The acrylate monomers making up the acrylic dispersion are usually formulated to optimize hydrogen-bonding opportunities with the PUD to enhance phase stability and coating performance. Hybridized or alloyed dispersions are produced by creating covalent bonding between the acrylate and urethane polymer chains. This has been done by reacting hydroxy functional acrylic monomers such as hydroxyl ethyl acrylate with diisocyanate and then performing the radical polymerization of the isocyanate–acrylate monomer with the other acrylate monomers for the dispersion [49]. The reactive acrylate dispersion is then reacted with the PUD prepolymer and chain extender to make the new urethane–acrylate polymer backbone. The hybrid backbone may obtain superior phase mixing and coating properties and commercial products do exist; however, the additional costs associated with the new product may, in some cases, create a barrier to implementation. An example of an acrylic–PUD system suitable for blending along with coating properties is given in Table 10.11.

**TABLE 10.11 Composition and coating properties of a blend of acrylic and polyurethane water-borne dispersions**

| | |
|---|---|
| 70% Acrylic component | 45% solids 100 nm average particle size, viscosity 80 cP 1% surfactant and 0.5% *n*-dodecyl mercaptan chain transfer agent employed to molecular weight and particle size |
| Acrylic dispersion part A 50% | 63% hydroxyethyl acrylate, 25% acrylonitrile, 8% styrene, 4% methacrylic acid |
| Acrylic dispersion part B 50% | 4% isobutylmethacrylate, 58% methylmethacrylate, 2% methacrylic acid, biocide (50 ppm) |
| 30% PUD 50% solid dispersion at 50% hard segment | 47% $H_{12}$MDI, 35% PTMEG (1000 g/mol), 7% 1,12-dodecane diol, 6% dimethylolpropionic acid, 5% 1,2-propylene diamine, antifoam |
| 2-butoxyethanol | Coalescent agent, solvent |
| *Properties* | |
| $T^g$ | 45 °C |
| Martens hardness (ASTM MNL-46) | 50 N/mm$^2$ |
| Taber abrasion | 90 mg |
| Mandrel bend test | 70 mm (out of 100) |
| Water resistance | Very good |
| 50% ethanol resistance | Very good |

$H_{12}$MDI, hydrogenated MDI; ppm, parts per million; PTMEG, polytetramethylene ether glycol; $T_g$, glass transition temperature.

***10.4.1.4 Ultraviolet-Cured Water-Borne Dispersions for Coatings*** While UV-cured polyurethane coatings are not a dominant coating technology, their use is growing for wood, metal, and plastic coatings in particular. UV cure technology will grow within its application space as the properties of the coatings improve. The concept of these coatings is that they allow for lower coating viscosity than two-part solvent-less formulations and emit less VOCs than solvent-borne formulations or those having unreacted monomers within the formulation. While radical-induced polymerizations resulting from UV irradiation can be fast, the need to evaporate water from the coating can slow down manufacturing line speeds [50, 51]. However, as formulations improve and solids levels increase, the water-borne coatings become an increasingly attractive option in some venues. In addition, the dual-cure nature of these coatings allows for tuned cross-link densities and the possibility of higher final coating cross-link densities at lower initial viscosity than obtained with other coating technologies [52].

The UV cure potential of polyurethanes is achieved by incorporating hydroxyl functional acrylate monomers into the prepolymer synthesis. There are many acrylates to choose from, but an available monomer that maximizes the available cross-link potential is the triacrylate of pentaerythritol. The reactive components are controlled to optimize structure and performance. An example of the polyurethane prepolymer components, including the polyacrylate monomer, is given in Figure 10.21. The composition is conventional with the possible exception of castor oil (the triglyceride of ricinoleic acid; see Chapter 2), which could potentially provide additional solvent resistance [53, 54]. The addition of the triacrylate needs to be controlled to prevent it from causing a maldistribution of molecular weights [55]. The isocyanate is added in amounts to control the final dispersion hard-segment volume.

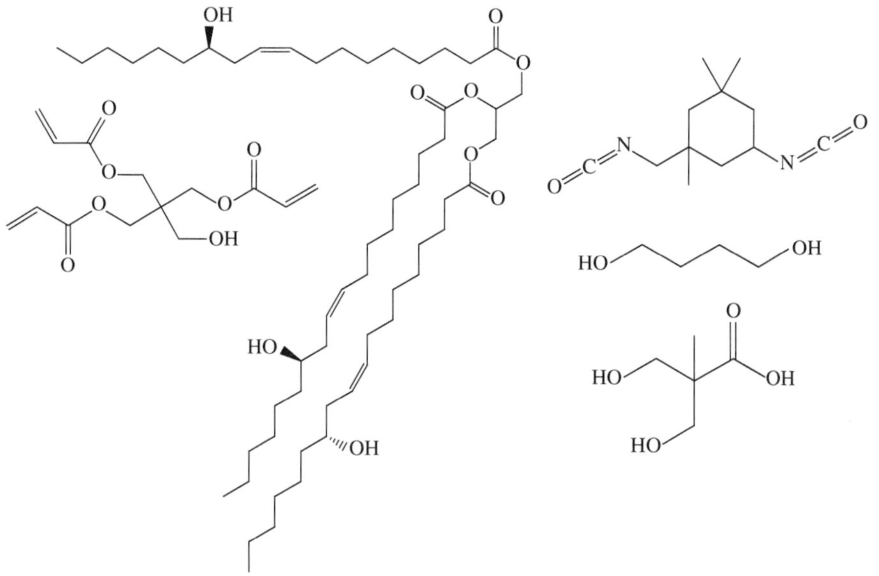


**FIGURE 10.21** Illustrative composition of building blocks for making a UV-curable polyurethane coating. The addition of fatty alcohol triglyceride might enhance solvent resistance.

**TABLE 10.12 Properties of UV-curable polyurethane coatings tested before and after curing**

| Test | No UV cure | UV cured |
|---|---|---|
| MEK double rub | 30 | 150 |
| Koenig hardness | 50 | 200 |
| Impact resistance | >160 ft lb | >160 ft lb |
| Taber abrasion (mg lost) | 22 | 7 |
| Pencil hardness | 5b | f |
| Pencil gouge resistance | 3b | 7h |
| Ethanol resistance (5 best; 0 worst) | 2 | 5 |

MEK, methyl ethyl ketone.

Upon completion of prepolymer formation, the prepolymer is added to cold water under high shear, as described in detail previously, and the diamine chain extender (i.e. 1,2-propylene diamine or ethylene diamine) added. An appropriate photoinitiator is added last and incorporated with agitation. Depending on the photoinitiator this may be more than 1% of the coating weight. A coated panel is then dried and subsequently passed under UV light exposure; time and energy are variables [56]. The properties of coated panels as a function of UV treatment versus no treatment are presented in Table 10.12. Coating performance data are a function of many parameters, including hard-segment volume and other characteristics of the polyurethane backbone, along with coating processing and substrate preparation.

***10.4.1.5 Polyurethane Powder Coatings*** While polyurethane coatings represent about 13% of the total industrial coatings market, polyurethane powder coatings represent about 5% of the total polyurethane volume applied to coatings. This value is consistent with powder coatings being about 5% of all coatings volumes independent of chemistry. While the size of the polyurethane powder coatings application is relatively small, it is a relevant technology because it does not require a solvent and has good coating properties. Like all polyurethanes, polyurethane powder coatings are valued for their durability, abrasion resistance, low-temperature flexibility, and good aesthetics. They find common use in decorative metal coatings, including automotive, lawn furniture, and home appliances.

The preparation of a powder coating is unlike any of the other technologies previously discussed. The common requirement for polyurethane and all other powder chemistries is that all of the components must be solids at room temperature [57]. They are subsequently blended into a homogeneous blend of materials in an extruder and then pulverized into a powder – typically of particle size tens of microns. The powder is then electrostatically sprayed at a substrate, or the preheated substrate is submerged into a fluidized bed of the powder. At this point the coating is often not yet finished. The coated part is then heated in an oven at a process-dependent temperature for a proscribed amount of time. During this oven heating, the particles will melt and coalesce into a fully integrated coating. It is usually during this heating step that polyurethane powder coatings will fully react to form a thermoset coating (Figure 10.22).

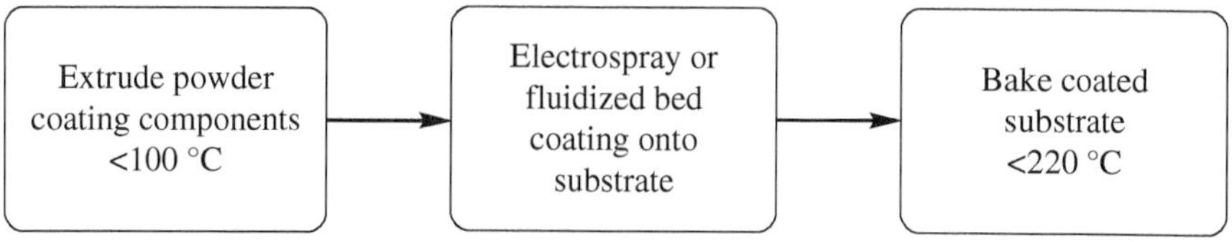


**FIGURE 10.22** Operations to a powder-coated item. The components of the powder coating must be blended (extruder operation), applied to the surface, and then fully integrated and cured in an oven.

To prevent the polyurethane reaction going to completion at any preliminary step such as extrusion, pulverization, or initial coating, the reactants must not be available to react. This reservation is achieved by protecting (or "blocking") isocyanate groups from reacting until they are made available at the oven temperature. In this regard, innovation of polyurethane powder coatings includes optimization of coating performance, minimization of VOCs, and minimization of the temperature at which the isocyanate deblocks, so reducing energy cost by reducing oven temperatures.

There are two primary methods of blocking isocyanates: (i) complexation of the isocyanate groups with a reversible leaving group (see Chapter 2) and (ii) internal blocking of the isocyanate linkage by reversible isocyanate dimer formation to the uretdione (see Chapter 3). Most formulators currently employ leaving group blocking technology since it is less complex and the deblocking temperature can in principle be controlled by the choice of deblocking agent. The blocking agent method suffers from the unavoidable loss of the blocking agent as a VOC in the oven cure step. To prevent unmitigated volatilization it is often required that the volatized blocking agent be trapped as it evolves. Several blocking agents have developed utility for powder coatings (Table 10.13). The temperature of deblocking can be influenced by the use of catalysts, which is an area of innovation and patent activity [58–60]. While caprolactam is one of the most cited blocking agents for making polyurethane powder coatings, it has been reported that color formation is a common unintended consequence of this adduct, and other agents have become more commonly encountered. Blocked isocyanate trimers are also available to increase crosslinking of the coating. Numerous metal catalysts have proven effective in lowering deblocking temperatures, and the lower end of the temperature ranges in Table 10.13 represent catalyzed values. Patented organometallic catalysts have included dibutyl tin dilaurate, bismuth tris(2-ethylhexanoate), cobalt bis(2-ethylhexanoate), zirconium bis(2-ethylhexanoate), zinc bis(2-ethylhexanoate), chromium tris(2-ethylhexanoate), and titanium (tetra(ethyl AcAc), for instance.

A second method of blocking isocyanate reactions has been by reacting polyisocyanates to form uretdiones (Figure 10.23). This chemistry is achieved efficiently using nitrogen- or phosphorous-based catalysts.

While reaction to uretdione occupies two of the isocyanates, there are obviously two isocyanates still available to react and that require blocking. It is possible to oligomerize the blocking agent, making a chain of uretdione structures, and these do form in

**TABLE 10.13 Structures of chemical blocking agents and their deblocking temperatures**

| Blocking agent | Structure | Deblocking temperature (°C) |
|---|---|---|
| ε-Caprolactam | 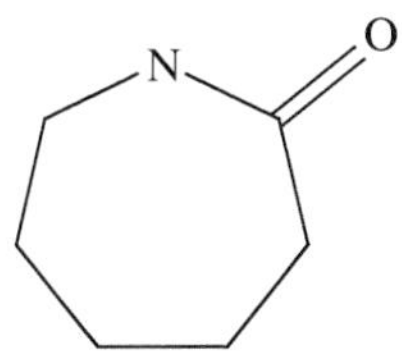 | 170–190 |
| Phenol | 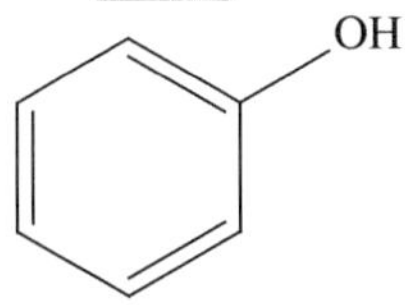 | 160–180 |
| Methyl ketone oxime | 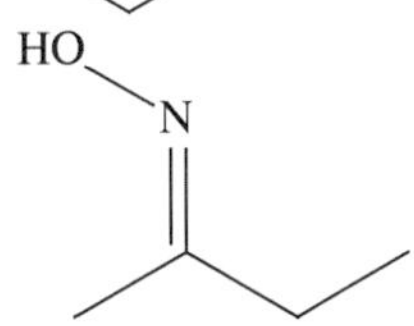 | 145–165 |
| 1,2,4-Triazole | 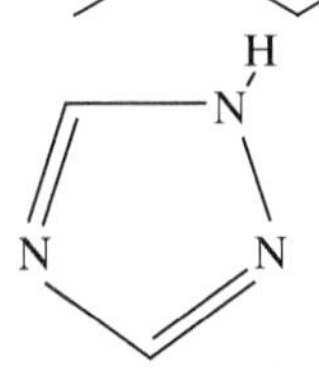 | 120–155 |
| Diethyl malonate | 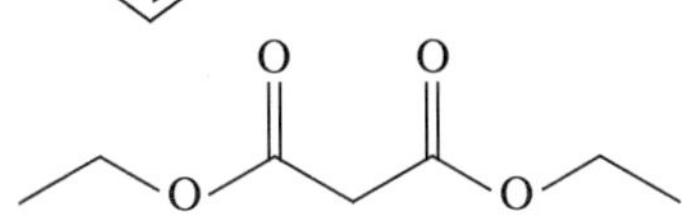 | 100–110 |

The use of a catalyst is required to achieve the lower end of the range.

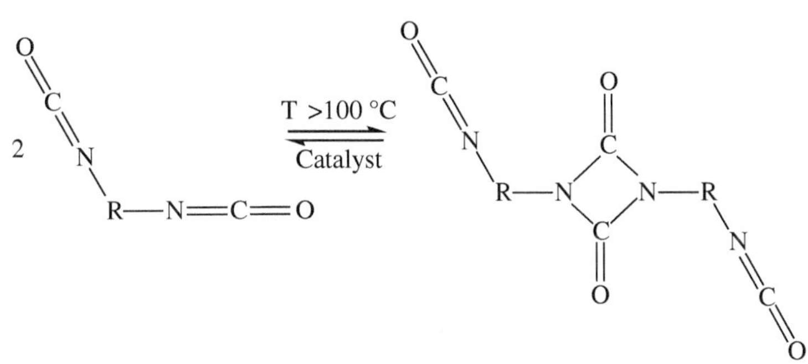


**FIGURE 10.23** Dimerization of a polyisocyanate to a blocked uretdione structure. Reversion back to monomers begins at about 100 °C but is not fast enough for industrial implementation until higher temperatures. Catalysts can lower the temperature of deblocking.

uretdione formation. More commonly, the unblocked isocyanate groups left from uretdione formation are reacted with the soft-segment component of the powder coating [61–63]. The soft segment is usually either a polyester or polyacrylic polyol – the choice of which depends on the requirements of the final product. A polyester-based polyurethane powder coating formulation, as might be encountered, is detailed in Table 10.14.

A typical preparation of this simplified powder coating (Figure 10.22) involves physical dispersion of the solid materials followed by extrusion on a low-shear extruder (~150 rpm) with a mixing zone temperature of c. 90 °C and an exit temperature of c. 110 °C. The extrudate is chipped and then finely ground (average particle size of tens of microns) with subsequent sieving to remove larger powder fractions. The resulting powder is then electrostatically sprayed onto a prepared steel substrate to a film thickness of 0.002 inch and baked at 180 °C for 10 minutes. Properties for this coating are given in Table 10.15.

### 10.4.2 Trends in Polyurethane Coatings

Quantifying trends in polyurethane coatings is complex for the same reasons as the same task for adhesives. The number of patenting companies is very large and the subject matter tends to be company specific. While many multinational polyurethane producers have coatings businesses, they do not all have developed patent strategies relative to coatings. This is certainly a reflection of the centrality of aliphatic isocyanates to coatings technologies, and the relatively few companies that produce aliphatic isocyanates. Figure 10.24 is an illustrative graph of large

**TABLE 10.14 Formulation of an illustrative polyurethane powder coating**

| Component | Amount (pphp) | Amount (wt%) |
|---|---|---|
| *Polyester component* | | 26.5 |
| Isophthalic acid | 54 | |
| Pentaerythritol | 23 | |
| Neopentyl glycol | 23 | |
| Catalyst | 100 ppm | |
| Acid number | 7 | |
| OH | 290 | |
| Viscosity at 175 °C (cP) | 150 | |
| $T_g$ (°C) | 63 | |
| *Isocyanate component* | | 40 |
| ε-Caprolactam-blocked IPDI | | |
| NCO content (%)/free | 15/<1% | |
| NCO (g/eq) | 275 | |
| $T_g$ (°C) | 50 broad | |
| $T_m$ (°C) | 70 broad | |
| $TiO_2$ | | 33 |
| Flow agent | | 0.5 |

IPDI, isophorone diisocyanate; ppm, parts per million; pphp, parts per hundred parts of polyol; $T_g$, glass transition temperature; $T_m$, melting point.

**TABLE 10.15 Properties of the polyurethane powder coating in Table 10.14**

| Property | Value |
|---|---|
| Gel time (at 180 °C) (s) | 134 |
| Direct impact (in-lb) (ASTM D2794) | 12 |
| Reverse impact (in-lb) | 8 |
| MEK double rubs | 100 |
| Cross-hatch adhesion (% adhering) ASTM D3359 | 99 |
| Pencil hardness | 2H |
| Gloss 60° (%) (ASTM 5767) | 85 |
| Gloss 20° (%) | 67 |
| Color b | –1.0 |

MEK, methyl ethyl ketone.

companies and their polyurethane coatings activity from 2000–2014, during which over 800 patents were filed worldwide specifically on protecting polyurethane coatings assets. For 2015–2019, the rate of patent-seeking on technologies that are specifically centered on polyurethane coatings slowed. This may reflect changes in the coatings business, and notable slowing of the polyurethane coatings growth rate in economies outside of China.

Covestro, and its original company Bayer, is by far the most active single filer of patents on polyurethane coatings extending beyond the years shown to 2018. This is a reflection of Covestro's wide portfolio of aliphatic isocyanates, including HDI, IPDI, and $H_{12}$MDI and blocked modifications and trimers of these materials. In fact, a look at the subject matter of coating compositions (Figure 10.25) shows that Covestro is highly focused on patenting applications and formulations associated with its product portfolio. Most recent focus has been on the use of polyurethane coatings for pultrusion operations, for improved scratch resistance for automotive applications, applications to artificial turf, and use of polyaspartic acid for polyurea coatings with improved workability. In contrast, French private company Roquette Freres had a concentrated patent strategy around the use of its isosorbide feedstock, and Evonik around its portfolio of silane additives. However, it is instructive that, for the nearly 160 patents between 2015 and 2019 specifically on polyurethane coatings, there are about 100 companies engaged in acquiring intellectual assets.

BASF is also a producer of aliphatic isocyanates but produces much smaller volumes than Covestro. However, BASF has widespread systems house operations. Thus, BASF protects primarily novel formulations as well as innovations pertaining to its building block manufacture. In contrast, Dow and Huntsman do not manufacture aliphatic isocyanates and their patent activity is concomitantly less (Figure 10.24).

An alternative means of exploring technology trends is to view patent activity on a regional basis. Figure 10.26 shows the intensity of polyurethane coatings patent activity based on filing language. For 2014–2019 polyurethane coatings patents have been most intensely pursued in China. This is a significant shift from previous decades in which

Japanese and English language patents were the most prevalent and follows the migration of manufacturing that includes coating operations. This premise is validated by quantifying the concentration of subject matter patent activity in China (Figure 10.27). The patents show a significant bias toward consumer goods, and especially textile applications.

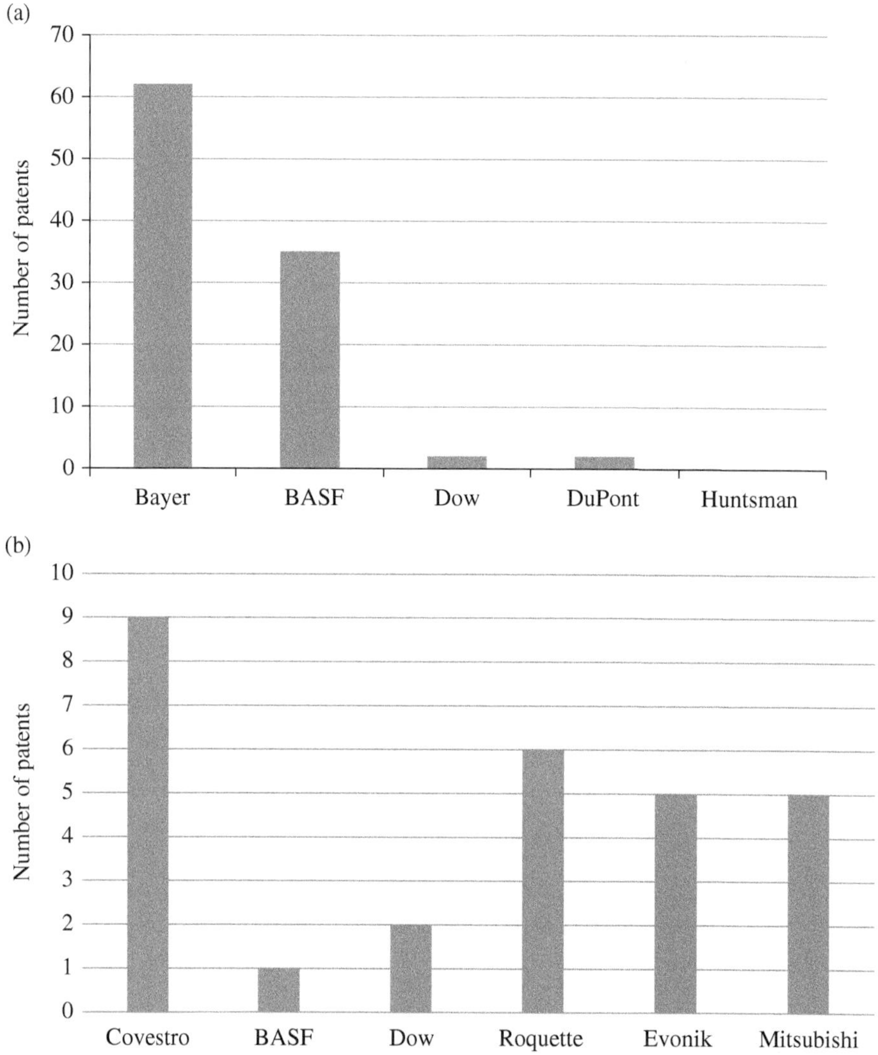


**FIGURE 10.24** Patent activity for (a) 2000–2014 and (b) 2015–2019 by large multination polyurethane manufacturers and DuPont de Nemours, a large multinational producer of coatings for the earlier time period. Patents are on compositions of matter, process, or applications of polyurethane coatings.

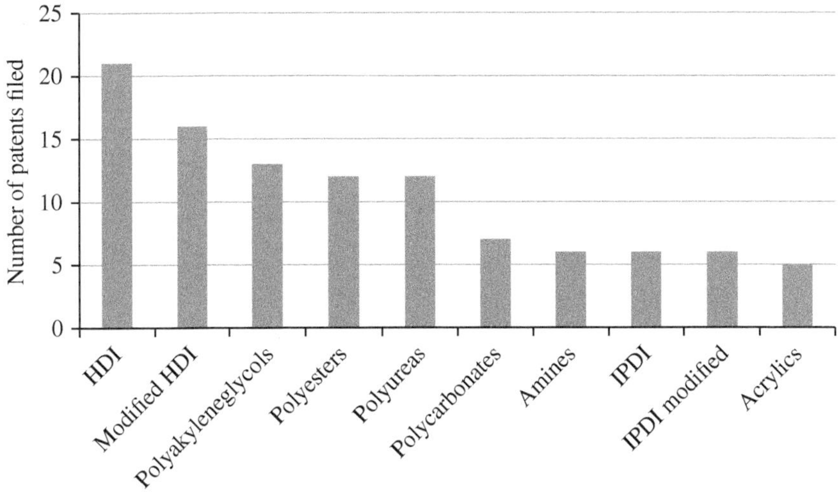


**FIGURE 10.25** Covestro's (formerly Bayer) activity during 2000–2014 on polyurethane coatings compositions of matter. IPDI, isophorone diisocyanates; HDI, hexane diisocyanate.

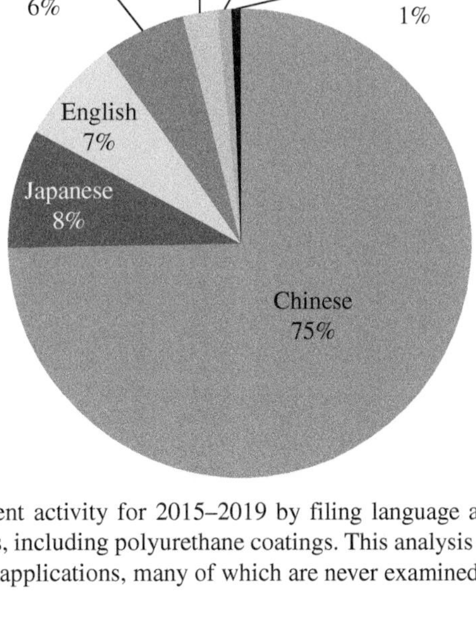


**FIGURE 10.26** Patent activity for 2015–2019 by filing language as a proxy for regional application of coatings, including polyurethane coatings. This analysis does not filter out filed Japanese kokai patent applications, many of which are never examined.

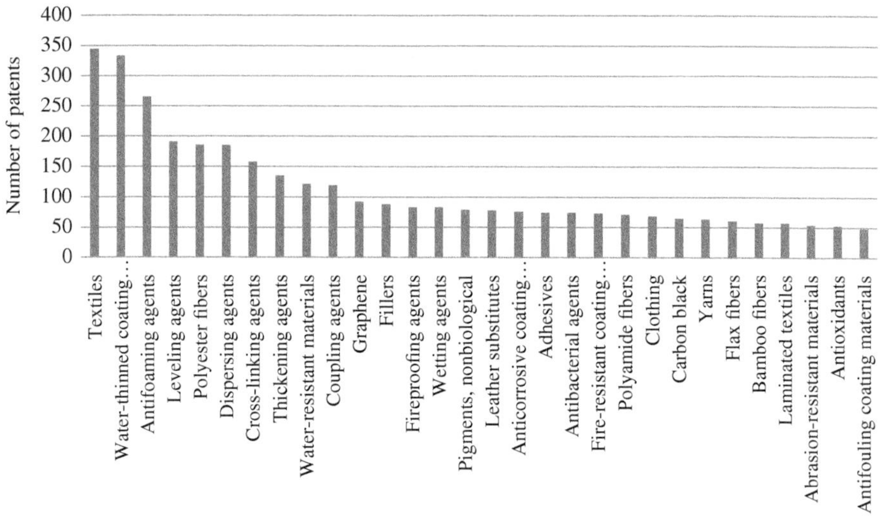


**FIGURE 10.27** Distribution of subject matter from filed patent documents on polyurethane coatings in China for 2015–2019. Many other subjects are also covered at lower levels of intensity.

## REFERENCES


[1] M Growney, M Kusumgar, and S Nerlfi, *The Global Adhesives Industry 2009–2014*, KNG Inc., Elmwood Park, NJ, 2010.

[2] E Linak, A Kishi, M Guan, and U.Buchholz, *Adhesives and Sealants*, IHS Chemical, Englewood, CO, 2016.

[3] E Linak, A Kishi, M Guan, and U Buchholz, *Urethane Surface Coatings*, Chemical Economic Handbook, IHS Chemical, Englewood, CO, 2019.

[4] E Linak, U Buchholz, M Guan, and A Kishi, *Paint and Coatings Industry Review*, Chemical Economic Handbook, IHS Chemical, Englewood, CO, 2019.

[5] *The Global Adhesive and Sealant Industry 2003–2006: A Strategic Market Trend Analysis*, 3rd Ed., Chem Research GmbH, Frankfurt, Germany, 2005.

[6] J Miron, A Brief, and M Savla, *Adhesives VIII*, Skeist Inc., Whippany, NJ, 2007.

[7] P Ghosh, *Adhesives and Coatings Technology*, Tata-McGraw Hill Publishers, New Delhi, India, 2008.

[8] A Tracton, Ed., *Coatings Technology: Fundamentals, Testing and Processing Techniques*, CRC Press, Boca Raton, FL, 2007.

[9] A Pocious, *Adhesion and Adhesives Technology*, 3rd Ed., Hanser Publishers, Munich, Germany, 2012.

[10] J Delgado-Saborit, N Aquilina, C Meddings, S Baker, and R Harrison, *Science of the Total Environment*, 409, (2011), 478–488.

[11] J Caro and M Gallego, *Chemosphere*, 77, (2009), 426–433.

[12] R Edwards, C Schweizer, M Jantunen, H Lai, L Bayer-Oglesby, K Katsouyanni, M Nieuwenhuijsen, K Saarela, R Sram, and N Kuenzli, *Atmospheric Environment*, 39, (2005), 2299–2307.

[13] E Kellar, "Key issues in selecting the right adhesive," in *Advances in Structural Adhesive Bonding*, D Dillard, Ed., Woodhead Publishing, Cambridge, UK, 2010.
[14] Q Tang, J He, R Yang, and Q Ai, *Journal of Applied Polymer Science*, 128, (2013), 2152–2161.
[15] D Chattopadhyay, B Sreedhar, and K Raju, *Polymer*, 47, (2006), 3814–3825.
[16] *Standard Test Methods for Testing Polyurethane Raw Materials: Determination of Hydroxyl Numbers of Polyols D4274*, ASTM, West Conshocken, PA, 2011.
[17] A Kinloch, *Adhesion and Adhesives: Science and Technology*, Springer Science, Berlin, Germany, 1987.
[18] S Ebnesajjad, *Handbook of Adhesives and Surface Preparation: Technology, Applications and Manufacture*, Elsevier, Amsterdam, the Netherlands, 2011.
[19] C Boër, S Dulio, and F Jovane, *International Journal of Computer Integrated Manufacturing*, 17, (2004), 577–582.
[20] L da Silva, A Ochsner, and R Adams, *Handbook of Adhesion Technology*, Springer Verlag, Berlin, Germany, 2011.
[21] M Grover, *Package Development and Systems*, 9, (1979), 20–22.
[22] R Clemens, G Batts, J Lawniczak, K Middleton, and C Sass, *Progress in Organic Coatings*, 24, (1994), 43–54.
[23] M Lee, J Lim, and B Lee, *Journal of Adhesion*, 89, (2013), 205–224.
[24] X Liu, D-K Sheng, B Li, and Y-M Yang, *Applied Surface Science*, 264, (2013), 61–69.
[25] J Ferry, *Viscoelastic Properties of Polymers*, 2nd Ed., Wiley, New York, NY, 1970, 268–270.
[26] H Stanley, I Davis, and W Chiao, Acrylic-modified reactive urethane hot-melt adhesive compositions. Assignee National Starch and Chemical Corp. US patent 5021507. 1991.
[27] H Oien, Reactive, hot-melt polyurethane-forming adhesive/sealing compositions. Assignee Minnesota Mining and Manufacturing Co. US patent 5525663. 1996.
[28] H Wolfe, Aqueous polyurethane dispersions. Assignee du Pont de Nemours, E. I., and Co. US patent 4183836. 1980.
[29] R Tennebroek, I Hoeven-van Casteren, R Swaans, S van der Slot, P Stals, B Tuijtelaars, and C Koning, *Polymer International*, 68, (2019), 832–842.
[30] Y Lee, T Kim, and B Kim, *Polymer International*, 28, (1992), 157–162.
[31] J Pate, R Lundgard, C Piechocki, P Keillor, N Lutenske, and I Thumma, Continuous method for preparing stable dispersion or emulsion by extruding and mechanically dispersing a polymer in aqueous or nonaqueous medium. Assignee Dow Chemical Company. US patent 6512024. 2000.
[32] D Lee, H Tsai, and R Tsai, *Journal of Applied Polymer Science*, 107, (2008), 1573–1577.
[33] A Mishra, P Das Purkayastha, J Roy, V Aswal, and P Maiti, *Journal of Physical Chemistry C*, 116, (2012), 2260–2270.
[34] B Erdem and D Bhattacharjee, An ultra-high solid content polyurethane dispersion and continuous production of ultra-high solid content polyurethane dispersions and article. Assignee Dow Global Technologies Inc. US patent 20100015341. 2007.
[35] M Bock, *Polyurethanes for Coatings*, Vincentz Verlag, Hannover, Germany, 2001.
[36] D Chattopadhyay and K Raju, *Progress in Polymer Science*, 32, (2007), 352–418.
[37] Z Wicks, F Jones, S Pappas, and D Wicks, *Organic Coatings: Science and Technology*, 3rd Ed., Wiley, Hoboken, NJ, 2007.
[38] W Chang, R Scriven, J Peffer, and S Porter, *Industrial Engineering Chemistry and Product Research Development*, 12, (1973), 278–288.

[39] CMH Hagen, A Hognestad, O Knudsen, and K Sørby, *Progress in Organic Coatings*, 130, (2019), 17–23.

[40] K Lee, Y Chung, C Chan, I Bello, S Lee, A Karimi, J Patschneider, M Delplancke-Ogletree, D Yang, B Boyce, et al., *Surface and Coatings Technology*, 168, (2003), 57–61.

[41] M Porter, *Competitive Strategy: Techniques for Analyzing Industries and Competitors*, The Free Press, New York, NY, 1980.

[42] R Daussin, P Cameron, L Chen, J Jimenez, and J Gehrke, Process for production of aliphatic isocyanates by rapid dispersion of primary amines or their salts in phosgene used as partial solvent. Assignee Dow Global Technologies Inc. US patent 20110275854. 2011.

[43] Z Cheng, Q Li, G Liao, B Zhang, Y Yu, C Yi, and Z Xu, *Progress in Organic Coatings*, 127, (2019), 194–201.

[44] N Heller, C Clayton, S Giles, J Wynne, M Walker, and M Wytiaz, *Colloid and Polymer Science*, 295, (2017), 2019–2030.

[45] C Wu, C Feng, Y Lo, T Lin, and J Lo, *Chemosphere*, 56, (2004), 71–80.

[46] Code of Federal Regulations 40 C.F.R. Part 60. *Standards of Performance for New Stationary Sources*, US Government Printing Office, Washington, DC, 2014.

[47] C Yuan, G Chen, X Zhang, Z Li, and Z Liu, Method for preparing aqueous polyacrylate modified polyurethane dispersions. Assignee The Hong Kong University of Science and Technology. US patent 20090111934. 2009.

[48] A Saeed and G Shabir, *Progress in Organic Coatings*, 76, (2013), 1135–1143.

[49] W Zom, H Goos, and J Stintjes, Aqueous dispersions of acrylic-urethane graft copolymers. Assignee Polyvinyl Chemie N.V. US patent 4,730,021. 1988.

[50] Y Qi, J Yang, Z Zeng, and Y Chen, *Advanced Materials Research*, 791–793, (2013), 175–178.

[51] E Zarshenas, S Bastani, and M Pishvaei, *Industrial & Engineering Chemistry Research*, 52, (2013), 16110–16117.

[52] S Mehravar, N Ballard, A Veloso, R Tomovska, and J Asua, *Langmuir*, 34, (2018), 11772–11783.

[53] M Kumar, M Satheesh, M Manjula, and K Siddaramaiah, *Journal of Applied Polymer Science*, 105, (2007), 3153–3161.

[54] P Nayak, D Mishra, D Parida, K Sahoo, M Nanda, S Lenka, and P Nayak, *Journal of Applied Polymer Science*, 63, (1997), 671–679.

[55] Y Lu and R Larock, *Biomacromolecules*, 8, (2007), 3108–3114.

[56] L Pichavant and X Coqueret, *Progress in Organic Coatings*, 63, (2008), 55–62.

[57] D Howell, *Powder Coatings: Vol 1. Technology, Formulation and Application of Powder Coatings*, Wiley/SITA Publishers, New York, NY, 2000.

[58] J Jimenez, D Bhattacharjee, and J Argyropoulos, Blocked isocyanates and their use in coating compositions. Assignee Dow Global Technologies Inc. US patent 20080033086. 2006.

[59] E Bossert, W Ranbom, and W Larkin, Catalysts for low temperature cure of blocked isocyanates. Assignee Elf Atochem North America, Inc. US patent 5718817. 1995.

[60] G Parekh, Powder coating compositions containing caprolactam-blocked isocyanates curable at low temperatures. Assignee American Cyanamid Co. EP 403779. 1990.

[61] A Malofsky and S Goldstein, Preparation of polyisocyanates containing uretidione and allophanate groups and their use in one- and two-component urethane coating compositions. Assignee Olin Corp. US patent 5461135. 1995.

[62] C Wamprecht, R Halpaap, H Meier-Westhues, and W Schultz, Powder coating compositions, a process for their preparation, and their use for the coating of heat-resistant substrates. Assignee Bayer A.-G. US patent 5508337. 1996.

[63] B Pilch-Pitera, L Byczynski, and B Mysliwiec, *Progress in Organic Coatings*, 113, (2017), 82–89.

# 11

# SPECIAL TOPIC: MEDICAL USES OF POLYURETHANE

The medical uses of polyurethanes are in many ways similar to more prosaic applications of foams, elastomers, coatings, and adhesives, and in certain ways very distinctive. Virtually every patent or academic publication on a medical application states that the material attributes of polyurethanes as a class makes them ideal for medical use. Among the required polyurethane properties valued for these applications are fatigue resistance, abrasion resistance, ability to design for biodegradability or chemical resistance, biocompatibility, wettability by bodily fluids, and the ability to be fabricated into useful shapes [1, 2]. However, medical uses of polyurethanes are also substantially more complex than other uses because of the numerous ways an organism has of rejecting a foreign body and the extreme wear potential in many uses. Polyurethanes can be susceptible to the numerous ways in which immune systems attack, such as by hydrolysis, oxidation, and various enzymatically driven chain scission processes [3]. In addition, an immune system may simply colonize and coat a foreign object, creating a source of thrombus (blood clots), sclerosis due to calcification, infection due to opportunistic growth, and occlusion of annular prostheses [4–6]. The criticality of performance creates a market for specialty monomers that would otherwise be too small volume, and, by extension, too expensive, to be commercially viable.

*Polyurethanes: Science, Technology, Markets, and Trends*, Second Edition. Mark F. Sonnenschein.

## 11.1 MARKETS AND PARTICIPANTS

The size of the medical polyurethane market is very small by the standards of the polyurethane industry. Accurate segmentation is difficult because of myriad small and specialty uses, but analysis of available estimates is that 2012 global volumes were approximately 35 million pounds of polyurethane building blocks sold for medical applications. However, not including items such as mattress covers, protective gloves, and condoms, the high-performance medical market is closer to 10 million pounds, with a compound annual growth rate of approximately 5% [7–9]. Volumes are split between North America, Europe, and Asia [10]. It is reasonable to expect that medical polyurethanes will find increasing use in the Asia Pacific region [11].

Because of stringent performance and regulatory requirements for medical applications, and the attendant risks for performance failures, the value of polyurethanes in the medical segment is approximately US$100 million [12]. The majority of applications for which urethanes find medical use are associated with the requirements of an aging population (i.e. pacemakers, catheters, valves, etc.) [13]. Furthermore, the amount of innovation required to meet medical requirements, and developed in anticipation of future requirements, is astonishing. Analysis of corporate activity over the past 20 years shows that, with the exception of Bayer (which has a very sizeable life sciences business), the large multinational polyurethane manufacturers of polyurethane components are not participating in the medical market. Bayer's polyurethane interests were spun off into a materials science business, Covestro AG, in 2015. Covestro participates in a very limited way in the biomedical space, much like other multinational producers of polyurethane building blocks. Instead, innovation is primarily an activity of medical device makers and, to a limited extent, universities and national laboratories (Figure 11.1). Almost all of this activity has occurred since 1990. Even more recent data suggest that, for 2014–2019, participation in innovation has narrowed even further and has primarily concentrated in academic environments (see Section 11.3).

The flexibility to make nearly any functional material with polyurethane chemistry is represented by the breadth of applications that have been patented. Segmentation of polyurethane applications according to function is given in Figure 11.2, which shows patent activity in virtually every medical device application. The intensity of coatings innovation for medical devices in many cases reflects innovation for coating a thermoplastic polyurethane (TPU) rather than a specific polyurethane coating [14, 15].

The activity of individual companies reflects the pointed need for corporate competitive advantage and defense against other market participants. The similarities and distinctions between companies actively innovating are interesting and instructive. For instance, Medtronic plc is a manufacturer of numerous implantable technologies such as heart pacemakers, stents, catheters, and other surgically important products. Figure 11.3 documents the proportional areas of Medtronic plc patent activity by technology segment for 1990–2014 and the numerical activity by year. Figure 11.4 is a similar analysis for activity by Bayer AG. As mentioned previously, much of the coatings innovation is related to nonpolyurethane coatings applied to a polyurethane

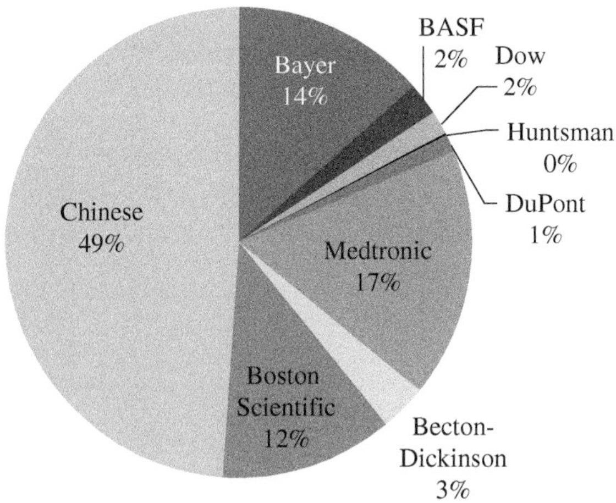


**FIGURE 11.1** Patent activity from 1990 to 2014 of various industrial polyurethane producers and producers of biomedical products. Except for Bayer, which has a large life sciences businesses, polyurethane manufacturers do not participate meaningfully in this market. There was a total of 2700 patents filed worldwide; 680 of these were filed by the participants shown here.

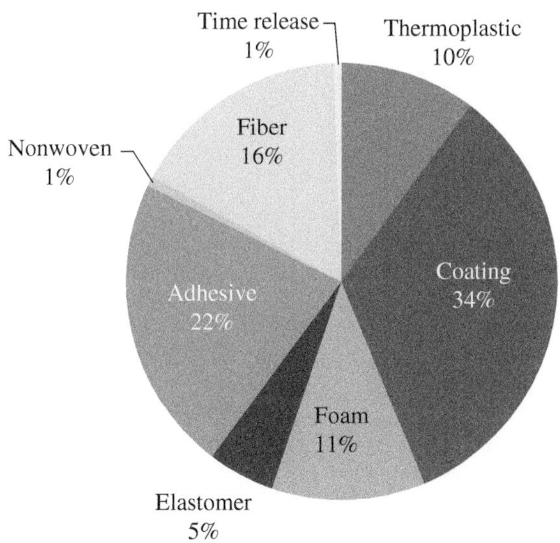


**FIGURE 11.2** Distribution of patents from 1990 to 2014 filed on biomedical uses of polyurethane by technology. A total of 2700 patents are represented in the distribution.

surgical product for a specific function. The same is true for adhesive activity. Many of the patents represent innovations improving adhesion of polyurethane to another specific substrate or to specific biological cells [16, 17]. An interesting phenomenon in the analysis is the increasing acquisition of medical polyurethane intellectual assets until 2009–2010, after which there was a pronounced decrease in activity. The causes for this may originate in economic or public policy concerns but are not mirrored in patent activity filed in the Chinese language on medical polyurethanes (Figure 11.5).

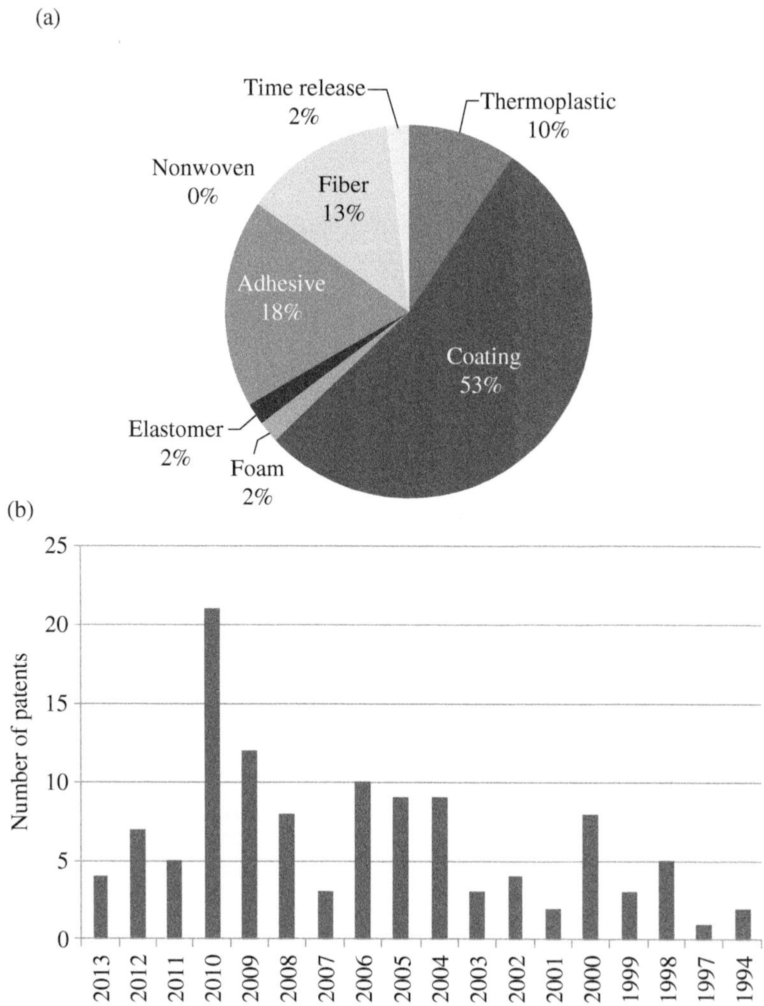


**FIGURE 11.3** Patent activity for Medtronic plc for 1994–2013 by (a) technology segment and (b) year. A total of 59 patents were filed during this time.

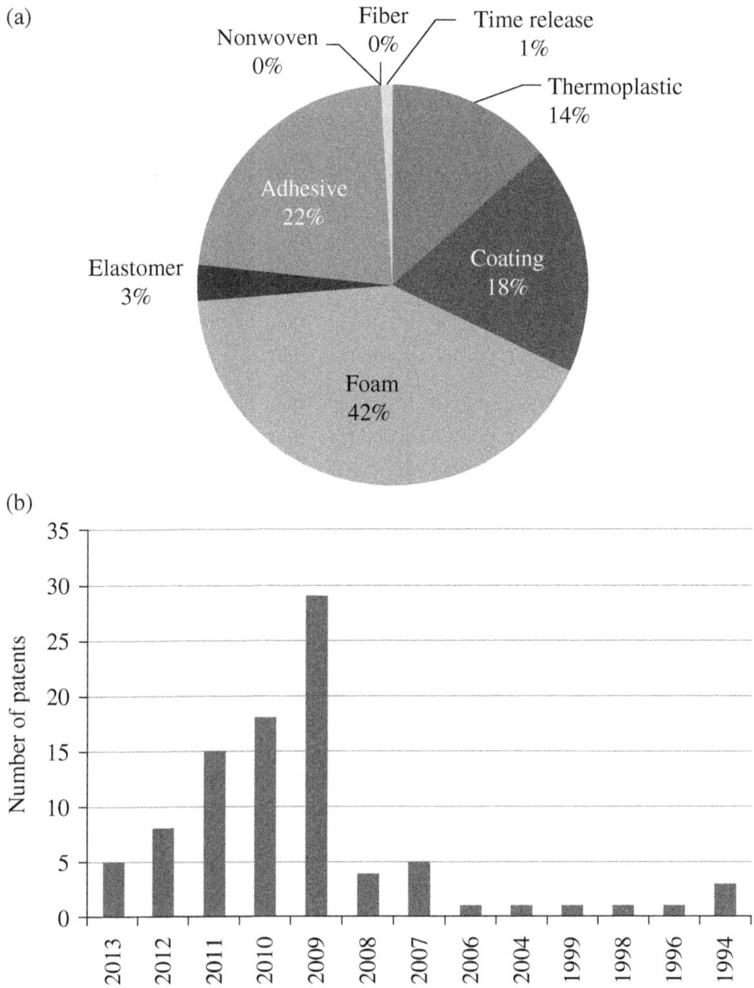


**FIGURE 11.4** Patent activity for Bayer AG for 1994–2013 by (a) technology segment and (b) year. A total of 92 patents were filed during this time.

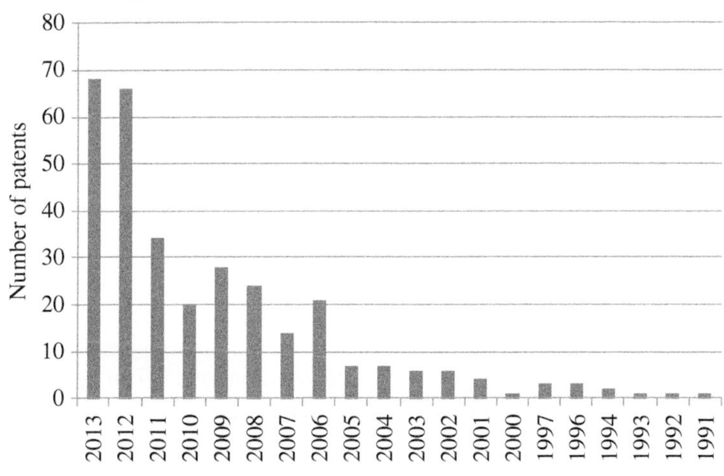


**FIGURE 11.5** Number of patents filed in the Chinese language from 1991 to 2013. A total of 317 patents were filed during this time.

## 11.2 TECHNOLOGY

### 11.2.1 Catheters

Approximately 3 million pounds of polyurethane are consumed per year in the making of catheters [18, 19]. A catheter is a hollow tube inserted into the body for any number of procedures, including removal of fluids such as urine or draining fluids resulting from infection, such as would be encountered from an internal abscess. Additionally, they are used for remote positioning of cardiac stents or balloon angioplasty. In addition, they can be critical components of other cardiac procedures, including ablative procedures and electrophysiology testing. Many materials have been employed for catheter construction, including polyvinyl chloride, polyethylene, silicones, perfluorinated polyethylene, and nylons. Polyurethanes are among preferred materials because of their flexibility, ability to engineer to the correct stiffness without kinking in use, ease of processing to appropriate form, low coefficient of friction when wet with water, and relatively low susceptibility to bacterial colonization. The reduced susceptibility to bacterial colonization is particularly important since infection is one of the most prevalent complications associated with cardiovascular procedures [20, 21]. It has been cited that between 30% and 40% of infusion therapy-related infections are complications associated with catheters; these begin as local infections and potentially progress to systemic infections. In this regard, it has also been reported that polyurethane has one of the lowest susceptibilities to *Staphylococcus* infection, possibly because of low surface adhesion energies [22, 23]. Coatings for catheter tips with biocides for the purpose of reducing the risk of infection is a common area of patent activity [24, 25].

With the significant demands placed on a material for catheter application it is perhaps surprising that thermoplastic polyurethanes utilized for catheters are relatively mundane. It is common for patents to specify the brand name and grade without mention of composition. In fact, it is common in medical literature and patents for the authors to simply cite "a urethane" or "polyurethane" catheter available from a specific manufacturer. Typical medical grades of thermoplastic polyurethane are simple and examples of their formulation are given in Table 11.1. The preference of polyether polyols over polyester polyols for catheter applications is associated with their superior low-temperature flexibility and their superior resistance to hydrolysis [26]. The simplicity of polymer design creates robustness and performance reliability. It also makes for a predictable and reproducible substrate for subsequent surface treatments to improve specific catheter functions and antibiotic performance [27].

### 11.2.2 Wound Dressings

Approximately 5 million pounds of polyurethanes per year are consumed making wound dressings [28, 29]. The role of a wound dressing is to provide protection to damaged skin following an operation, a burn, or a laceration, for instance. The dressing can be occlusive, such that the wound is completely isolated, or it can be nonocclusive, such that there is a semipermeable membrane or foam capable of absorbing fluids produced by the wound and transmitting gases. Wound dressings can also come with various surface treatments to prevent infection, prevent adhesion, or promote more rapid reendothelialization (skin regrowth) [30]. Polyurethanes find use in nearly every facet of wound dressing, but are by no means the only material used for this purpose.

For flexible solid substrate a thermoplastic material such as the Shore A 80 material defined in Table 11.1 is often used. Softer versions can also be employed by reducing the amount of hard segment in the formulation. In some cases, it is desirable to make a clear substrate through which the surface condition of the wound can be inspected without having to completely remove the dressing. For increasing film transparency there are many means of proceeding, including reducing hard segment, using an aliphatic hard segment, using a polyester soft segment, mixing chain extenders, or processing in such a way as to minimize hard-segment phase growth [31, 32].

For nonocclusive dressings the semipermeable component is sometimes a foam or nonwoven structure. There has been recent work on making nonwoven dressings by an electrosprayed process [33, 34]. Electrospray can make fibrous threads with a very small diameter, and allow addition of functional components during the spinning process. Conventional flexible polyurethane foams used for comfort applications are not used for wound dressings because they do not provide adequate hydrophilicity to sufficiently absorb bodily fluids. This is overcome by maximization of the polyethylene glycol content in the foam soft segment [35]. The increased hydrophilicity also provides an opportunity to act as a high surface area substrate for wound-healing functional additives.

**TABLE 11.1 Formulation and properties of thermoplastic polyurethane commonly employed for catheters with a Shore D hardness of 75 and 55 and Shore A 80**

| | Molar ratio | | |
|---|---|---|---|
| Component | Shore D 75 | Shore D 55 | Shore A 80 |
| Polytetramethylene glycol $M_n = 650$ | 1.0 | 1.0 | |
| Polytetramethylene glycol $M_n = 1000$ | | | 1.0 |
| 1,4-Butanediol | 3.5 | 2.9 | 1.3 |
| 4,4′ MDI | 4.5 | 3.9 | 2.3 |
| 50% stannous octoate in dioctylphthalate catalyst | 100 ppm | | 100 ppm |
| *Properties* | | | |
| $T_g$ (°C) | 35 | −14 | −48 |
| Shear modulus (MPa) | 500 | 70 | 10 |
| Diffusion coefficient $H_2O$ ($cm^2/s$) | $5 \times 10^9$ | $1 \times 10^9$ | $7 \times 10^7$ |
| Tensile strength (psi) | | | 7000 |
| Elongation at break (%) | | | 520 |
| $M_w$ (g/mol) | 115,000 | | 105,000 |
| % water content at equilibrium | 1.5 | 1.6 | 1.3 |

MDI, 4,4′-methylene diphenyl diisocyanate; $M_n$, number average molecular weight; $M_w$, weight average molecular weight; ppm, parts per million; $T_g$, glass transition temperature.

Polyurethane foam structures can be used for a wide variety of wounds, but, because of their absorptive capacity, they are primarily used in cases where a relatively large amount of fluid must be imbibed. Wounds where foam dressings are chosen include burn sites, sutured sites, skin grafts, and skin graft donor sites; these dressings are also used to cover tube sites that transit through to the body exterior (e.g. tracheostomy). The foam is usually covered by a solid or air-permeable film structure on the outside and a liquid-permeable membrane facing the wound to prevent abraded urethane from entering into the wound. A negative pressure is sometimes employed by evacuating the foam structure through a solid-surface/airtight bandage covering. The partial vacuum draws the exudates into the bandage to more efficiently prevent festering [36].

Examples of polyurethane foam hard segment are toluene diisocyanate–water, 4,4′-methylene diphenyl diisocyanate (MDI)–water, and various aliphatic structures. The soft segment is usually designed to provide the wetting interface, either for the fluids or for specialized functional coatings. Common soft-segment compositions are based on propylene oxide (PO) and ethylene oxide (EO), with EO reported to be as high as 75%. Froth foams have also been prepared for wound dressings using all polyethylene glycol soft segment, 1,4-butanediol (BDO), glycerine, and water as a chain extender/blowing agent and hexane diisocyanate for the polyisocyanate. The foaming is assisted by whipping large volumes of air into the reacting mass at elevated temperature and then allowing the blow reaction to expand the foam to its final potential. Such pads can be relatively soft, made at relatively higher densities, and are, of course, very hydrophilic [35].

Foam dressing size is also optimized for various functions. Foam pad thickness is usually in the range of 1.0–10.0 mm to optimize for capacity and facile attachment to the wound and surrounding skin. In contrast to a foam prepared for comfort where foam densities may be less than 2 $lb/ft^3$, a wound dressing density can be more than 6 $lb/ft^3$ to facilitate contact and wettability. The upper density is usually controlled such that the open cell content is very high and the foam is not perceived to be hard.

Recent work has made electrosprayed nonwoven absorbent layers with commercial equipment especially equipped to heat a molten polymer to a high enough temperature to reduce viscosity. Low viscosity is required so the thermoplastic polyurethane can exit a small orifice spinneret. Alternatively, the polyurethane elastomer can be dissolved in a solvent at a concentration such that the viscosity is optimal for extrusion and removal of the solvent. Nonwoven absorbent TPUs have been reported to be electrospun from polymer melts [33]. With specialized equipment this can be achieved using a standard TPU formulation. A TPU with Shore A 85 and composition by weight of 60% polytetramethylene ether glycol (PTMEG) 1000, 34% 4,4′ MDI, and 6% BDO has been demonstrated to extrude fibers less than 10 μm in diameter. In contrast, a Shore A 97 TPU based on a polyester soft segment was dissolved at 10% (weight) in a 50:50 mixture of dimethylformamide and methylethyl ketone to electrospray fibers less than 1 μm in diameter [34]. A similar result was obtained with a 50:50 solvent mixture of dimethylformamide and tetrahydrofuran.

### 11.2.3 Bioabsorbable Polyurethanes

Volumes of polyurethanes used for medical implant applications requiring polyurethane to be absorbed or degraded by the body are very small. However, about 10% of all patents filed in the biomedical polyurethane field relate to tissue scaffolds, sutures, bioresorbable polyurethanes, or biodegradation of the polyurethane. In addition, there has been steady patent activity in the field of polyurethane hydrogels for medical applications. The concentration of this work in the past 5 years may indicate that medical device manufacturers and doctors are perceiving that performance gaps in current materials can be filled with designed polyurethanes.

Degradation mechanisms in vivo are often consigned to hydrolytic, enzymatic, and oxidative processes [37]. In these applications, the standard thermoplastic polyurethane composed of 4,4′ MDI, BDO, and a polyether or polyester soft segment will be unacceptable. Part of the resistance is a result of biodegradation pathways that lead from the urethane bond to the formation of aromatic amines when using an aromatic polyisocyanate [38–40].

The highly specialized nature of applications requiring bioabsorbable polyurethanes removes many of the economic constraints associated with almost all other applications. In these applications it commonplace to copolymerize or transesterify two soft-segment polyols to provide the optimum balance of material, biocompatibility, and degradation performance to achieve all requirements. The simple use of polyethylene oxide polyols or PO/EO block or random copolymer polyols is in principle feasible from the perspective of biodegradability, but in fact is *not* usually employed [41]. PO/EO polyols can exceed 60% biodegradation in 10 days and up to 99% biodegradation in 28 days in the standard Organisation for Economic Cooperation and Development (OECD) 301B test [42]. The test measures the $CO_2$ evolution based on that predicted on the amount of carbon in the tested material. The microbial population can come from numerous sources, but a common one is from waste treatment plants. The level of degradation obtained by some polyalkylene glycols can normally be considered to result in nontoxic end products. Hydrolysis of the polyalkylene oxide chain is not commonly encountered (to produce two alcohols) but numerous studies have determined the ready oxidative and enzymatic degradation pathways for these materials [43, 44].

Aliphatic polyester polyols are also common substrates for designing biodegradable polyurethanes. Virtually all polyester polyols are prone to hydrolytic mechanisms of degradation, as well as oxidative and enzymatic mechanisms. This chemical susceptibility is usually disqualifying for applications where the urethane must be implanted and durable, and has been demonstrated to result in extensive material degradation when used for those applications. On the other hand, this susceptibility makes polyesters the usual choice for maximizing the rate of biodegradation of a polyurethane for which biosorption is desirable. Since the value-added aspects of optimum design outweigh most purely economic considerations, soft-segment design usually reflects a balance of tensile performance, hydrophilicity, biodegradability, and processability (in cases where porosity or surface texture is desired).

For optimum hydrolysis and hydrophilicity a semicrystalline polyester diol based on the naturally derived chemical lactic acid to form polylactic acid (PLA) is usually explored (Figure 11.6). With a large percentage of ester linkages, PLA is exceptionally susceptible to hydrolysis, but it also possesses forms having a glass transition temperature ($T_g$) of 65 °C and a crystalline melting temperature of 175 °C, making it quite brittle [45, 46]. PLA can be, and often is, copolymerized or transesterified with other polyester polyols, such as polycaprolactone, with more desirable material properties (apart from biodegradability) to make block copolymer soft segments with balanced properties (Figure 11.7). Blends of TPU having a polyester soft segment with PLA have also been demonstrated to have good physical properties and are capable of being extruded and injection molded into a scaffold for living tissue growth [47, 48].

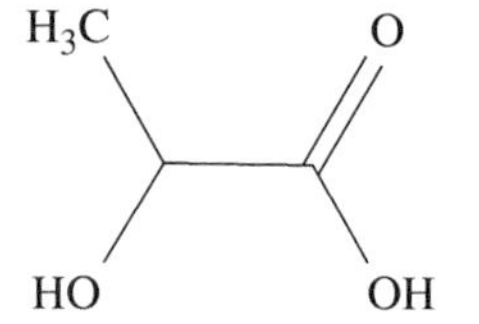


Lactic acid

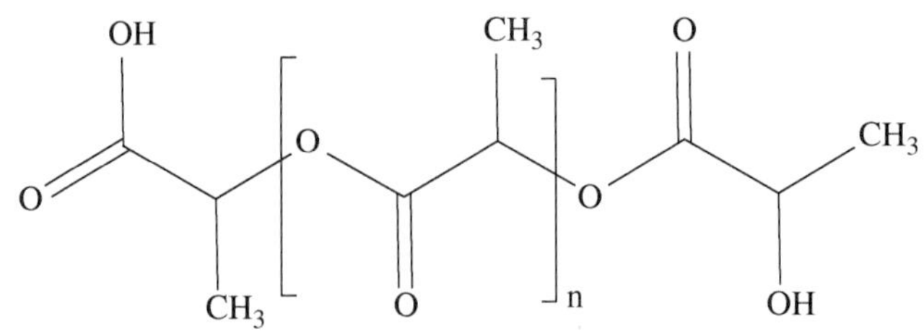


Polylactic acid

**FIGURE 11.6** Structure of lactic acid and the derived biorenewable and biodegradable polymer polylactic acid.

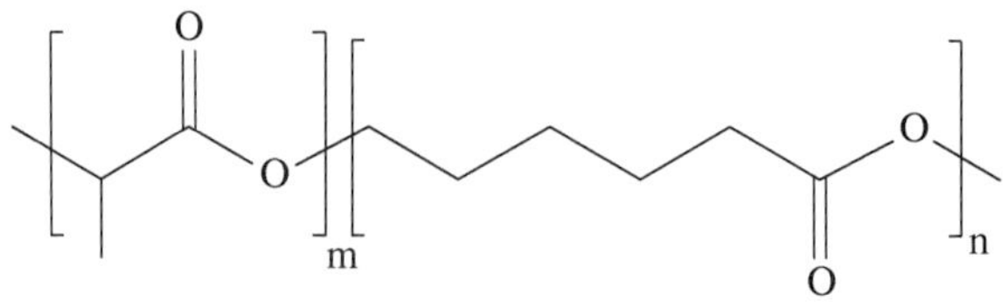


**FIGURE 11.7** Copolymer of polycaprolactone and polylactic acid, as might be prepared via transesterificaton.

Hard-segment design is also a part of developing biosorbable polyurethanes. Although it is certainly controversial, the use of aromatic isocyanates is almost never used in these applications for fear of biodegradation resulting in carcinogenic aromatic amines, such as toluene diamine or methylene dianiline [49]. Aliphatic diisocyanates have found some use in these applications owing to their enhanced biodegradation kinetics. To be sure, this is largely a result of the lower $T_g$ of aliphatic hard segments, resulting in significantly higher water diffusion rates within the matrix. The more rapid degradation rate obtained with aliphatic isocyanates makes them less desirable for durable applications, but the measured faster biodegradation rate of aliphatic hard segments is established. Thus, there is substantial literature associated with biosorbable materials prepared from relatively standard aliphatic hard-segment structures [50, 51].

The specialty nature of biosorbable materials has also spawned exploration of novel isocyanate structures for improved performance either from cell growth compatibility or from biosorption standpoints. One class of aromatic isocyanates is based on phosgenation of polyester–anilides. These materials can be prepared from commercially available polyols (ether or ester) and benzocaine (ethyl 4-aminobenzoate) followed by phosgenation (Figure 11.8). It is believed that these structures can biodegrade into nonharmful side products while providing polymer properties of related polyester–anilide structures used for high-performance elastomers [52].

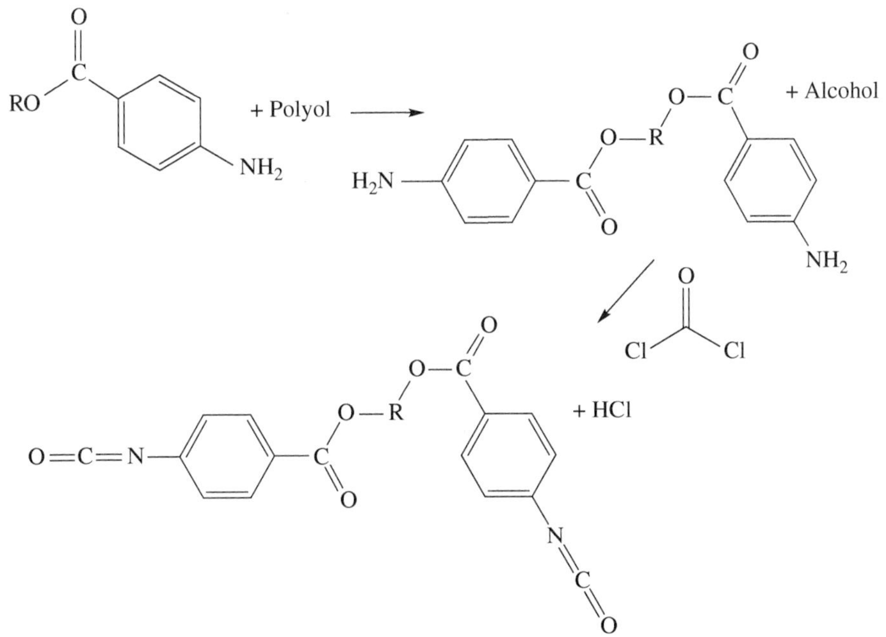


**FIGURE 11.8** Preparation of a polyester–anilide and subsequent conversion to the polyisocyanate via phosgenation. Nonphosgene routes could also be considered (see Chapter 2).

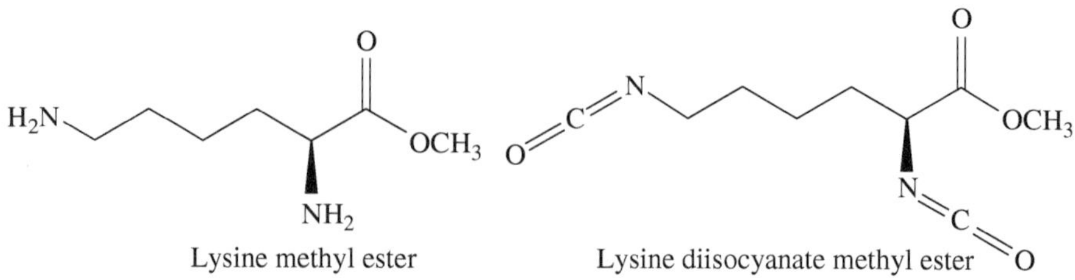


**FIGURE 11.9** Structure of the polyamine lysine and the derived biodegradable polyisocyanate.

Since the conversion chemistry of amines to isocyanates (Chapter 2) can be generally applied, there are almost as many possible polyisocyanate structures as there are polyamines. One such structure is based on lysine – an essential dietary amino acid [53]. While lysine is extractable from foods, it is commonly available via fermentation with total available volumes of approximately 1 billion pounds per year. Although this availability makes it potentially attractive as a specialty chemical feedstock, difficulties permitting new plants for producing and utilizing phosgene can make any new isocyanate a daunting barrier to commercial entry. Figure 11.9 shows the structure of the lysine diisocyanate and its parent amino acid.

### 11.2.4 Hydrogels

Typical durable thermoplastic polyurethanes will absorb between 1% and 3% by weight water under ASTM (formerly American Society for Testing and Materials) conditions depending on polyurethane composition. In contrast, some biomedical applications may require that a polymer matrix contain between 20% and 40% by weight water. Such applications may include soft contact lenses, wound dressing adhesive components, implant coatings, and media for drug delivery [54]. Polymer systems that can swell with water to this extent are often termed hydrogels. The design latitude that polyurethanes possess makes them effective hydrogel-forming polymers. A challenge for polyurethanes, and all polymers, is that the material must maintain substantial physical properties even when highly swollen. In addition, they should not present a fertile environment for microbial growth. Polyethylene glycol soft segments are commonly used in hydrogel design to provide hygroscopic base properties and a microbe-resistant interface [55, 56]. Polyurethane chemistry also presents numerous design options for developing cross-links to buttress swollen polymer physical properties. Cross-linking agents may be relatively small, like trimethylolpropane or diethanolamine. Alternatively, they can be higher molecular weight materials such as ethoxylated glycerine. The hard segment for hydrogels is invariably aliphatic – usually hexane diisocyanate or isophorone diisocyanate (Chapter 2). The equilibrium water content and the physical properties are mutually dependent on chemistry and the degree of cross-linking.

Cross-link density in the swollen matrix is designed from the building blocks, but the actual molecular weight between cross-links must be determined empirically. A convenient method for calculation of cross-link density is the Flory–Rehner equation (Equation 11.1), which is based on measurements of equilibrium matrix swelling by a solvent [57]:

$$M_c = \frac{v_s d_p \left( V_p^{\frac{1}{3}} - \frac{V_p}{2} \right)}{\ln\left(1 - V_p\right) + V_p + X_{12} V_p^2} \tag{11.1}$$

In Equation 11.1 $M_c$ is the molecular weight between cross-links, $v_s$ is the molar volume of the solvent, $d_p$ is the density of the polymer, $V_p$ is the volume fraction of the swollen polymer in the solvent, and $X_{12}$ is the polymer–solvent interaction parameter (see Chapter 4). As discussed in detail in Chapter 4, the interaction parameter can be calculated from Equation 11.2:

$$X_{12} = \frac{v_s \left(\delta_p - \delta_s\right)^2}{\mathrm{R}T} \tag{11.2}$$

In Equation 11.2 $\delta_p$ and $\delta_s$ are the solubility parameters of the polymer and solvent, respectively, R is the ideal gas constant, and $T$ is the absolute temperature. A discussion and table of solubility parameters useful for polyurethane polymers can be found in Section 4.1. A table of solubility parameters for standard solvents relevant to measuring cross-linking in polyurethanes is given in Table 11.2 [58].

The cross-link density ($d_c$) is calculated from Equation 11.3 following calculation of $M_c$ from Equation 11.1:

$$d_c = \frac{d_p}{M_c} \tag{11.3}$$

Since the volume of polyurethanes used for hydrogels is inconsequential, the significance of this technology is the fundamental science it offers, the investment in future therapies consequential to human wellbeing, and as a signpost for future high-technology research in polyurethane science. Activity on medical polyurethane hydrogels is approximately 52 patents from 1990–2014 and 44 from 2015–2019 (31 of Chinese origin, 13 by the Hunan University of Science and Technology). Only one company has filed more than two patents in this time, while over 500 publications in the open literature have been published.

**TABLE 11.2 Solubility parameters of representative solvents useful for determining cross-link density and molecular weight between cross-links for a polyurethane foam using the Flory-Rehner equation (Equation 11.1)**

| Solvent (at 25 °C) | $\delta_s \frac{cal^{1/2}}{cm^{3/2}}$ |
|---|---|
| Toluene | 8.9 |
| Acetone | 9.9 |
| 1,4-dioxane | 10 |
| *N*-methyl pyrrolidone | 11.3 |
| Methanol | 14.5 |

### 11.2.5 Gloves and Condoms

While not high-technology medical applications, surgical/examination gloves and condoms represent about 10 million pounds of combined polyurethane material volume. The two applications are about equal in size and they are quite similar in technology. Additionally, both applications face varied competition from alternative polymer substitutes [59]. Polyurethane gloves, in particular, have growth potential resulting from substitution of nonpolyurethane latex products owing to allergy concerns and powder coatings interfering with sterile operations [60].

The production processes for medical gloves and condoms are quite similar, differing primarily in film thickness (glove thickness about 120–150 μm and condoms about 60 μm) [61]. The greater thickness of gloves also necessitates a coagulation step using calcium nitrate as a precipitant. The precipitant acts by substituting $Ca^{2+}$ ions from the coagulant for $Na^{+}$ ions in the dispersion, which initiates precipitation [62]. A simplified version of the dip procedure process for these materials is shown in Figure 11.10. A heated ceramic form or mold of the item is dipped into a bath of the polyurethane dispersion. For thin films, a double dip is performed without coagulant. For a thicker film, a single dip is applied with the coagulant preapplied to the mold. The coated forms are then passed through a curing/annealing oven and washed thoroughly of coagulants and migratory chemicals from the dispersion bath. There are many critical process parameters in this process, including the cure step, which has significant influence on final tensile strength, and the cleanliness of the form/mold, which can affect the development of pinholes in the films.

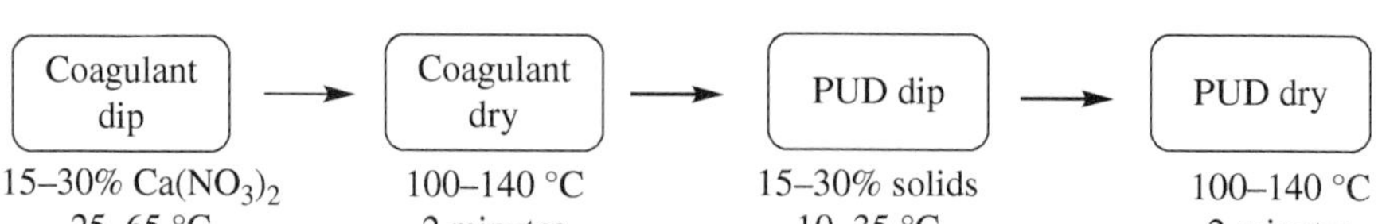


**FIGURE 11.10** Simplified process flow for making medical films such as examination/surgical gloves from aqueous polyurethane dispersion (PUD).

**TABLE 11.3 Composition of a polyurethane prepolymer/curative that could be made into a dispersion using procedures outlined in Chapter 10 to subsequently make medical gloves**

| Component | Amount | Comment |
|---|---|---|
| 2000 g/mol molecular weight polyether polyol, 88% PO, 12% EO triblock | 52 | Prepolymer component |
| 425 g/mol molecular weight polypropylene oxide diol | 14.7 | Prepolymer component |
| 4,4′ MDI 250 g/mol | 29.1 | Prepolymer component for higher modulus |
| 50/50 4,4′/4,2′ MDI 250 g/mol | 4.2 | Prepolymer component – lowers modulus slightly |
| 230 g/mol molecular weight α,ω-polyoxyproylenediamine | 16.8 | Chain extender |
| *Properties* | | |
| Tensile strength (psi) | 3000 | |
| Elongation at break (%) | 780 | High toughness |
| Stress at 100% elongation (psi) | 200 | Low stress at low elongations increases comfort |
| Tensile set | 1% | Low deformation after donning |

Full formulations will often include surfactant and biocide additives to aid processing and film formation.
EO, ethylene oxide; MDI, 4,4′-methylene diphenyl diisocyanate; PO, propylene oxide.

An example polymer composition of a glove-forming material and properties is presented in Table 11.3. The preparation of the dispersion made from the components in Table 11.3 would be similar to that shown in Figure 10.11.

## 11.3 FUTURE TRENDS

Recent trends have suggested a significant reduction of interest in pursuit of patent assets directly associated with biomedical applications of polyurethanes. The same cannot be said of academic interest, which remains fertile (Figure 11.11). Analysis of academic subject matter (Figure 11.12) published in English shows that the uses of polyurethanes are varied, making use of the full spectrum of properties polyurethanes have to offer. The overall subject matter appears to be primarily focused on implantable applications and on wound care. Similarly, patent applications (USA only) (Figure 11.13) have also seemed to focus on wound care with almost disappearance of interest in catheter and implantable electronics, which had in prior decades been fields of greater activity. Reduction in activity utilizing polyurethanes is correlated with the use of alternative materials such as polyethylene, polypropylene, polyvinylchloride, and silicones.

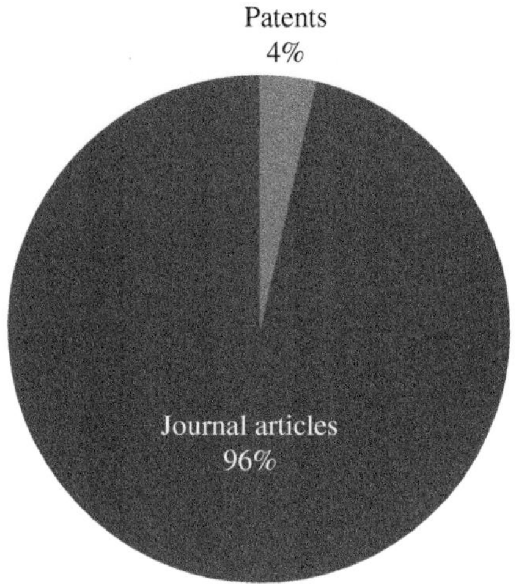


**FIGURE 11.11** Ratio of published patents to journal articles in 2014–2019 on the subject of biomedical applications of polyurethanes. English language only.

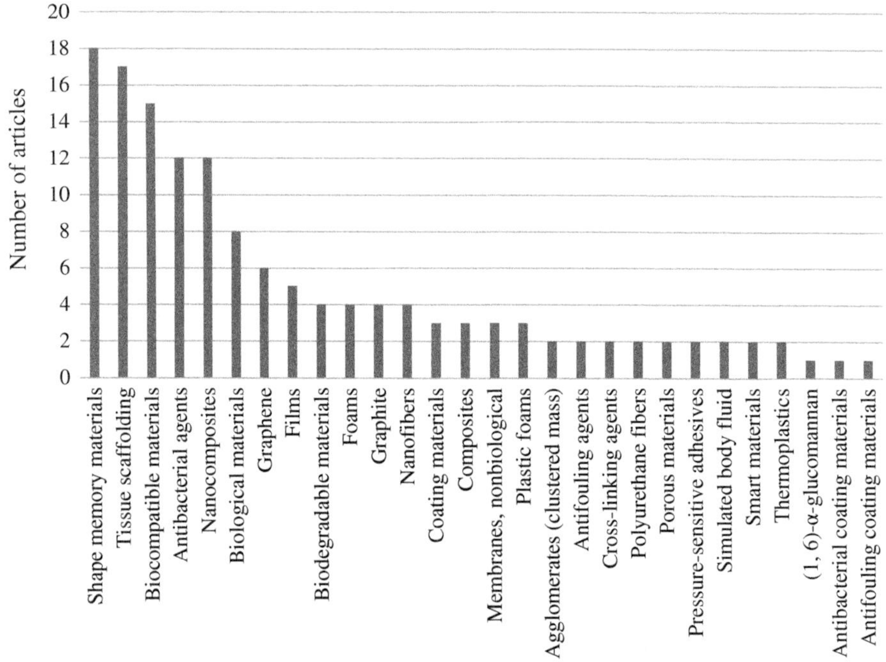


**FIGURE 11.12** Analysis of open literature (English language only) topics where the primary subject is biomedical applications of polyurethane. Journal articles in 2014–2019.

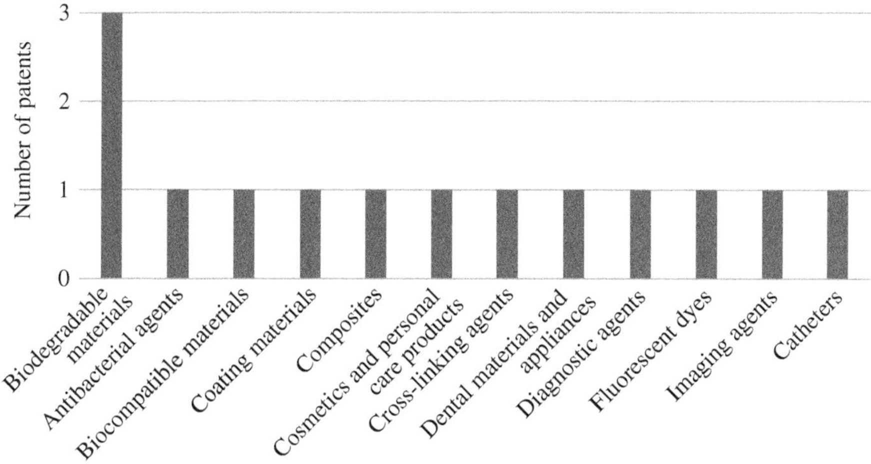


**FIGURE 11.13** Analysis of the patent literature (USA only) subject matter where the subject of the patent was biomedical applications of polyurethane. Patents in English in 2014–2019.

## REFERENCES


[1] S Kumbar, C Laurencin, and M Deng, Eds., *Natural and Synthetic Biomedical Polymers*, Elsevier, Burlington, MA, 2014.

[2] N Lamba, K Woodhouse, and S Cooper, *Polyurethanes in Biomedical Applications*, CRC Press, Boca Raton, FL, 1998.

[3] D Dempsey, H Benhardt, and E Cosgriff-Hernandez, "Resorbable polyurethanes," in *Szycher's Handbook of Polyurethanes*, 2nd Ed., M Szycher, Ed., CRC Press, Boca Raton, FL, 2013, 671–710.

[4] Y Marois, R Roy, M Marois, R Guidoin, W Maltzahn, R Kowligi, and R Eberhardt, *Clinical and Investigative Medicine*, **15**, (1992), 141–149.

[5] E Christenson, A Hiltner, and J Anderson, *Corrosion Engineering Science and Technology*, **42**, (2007), 312–323.

[6] J Anderson, *Annual Review of Materials Research*, **31**, (2001), 81–110.

[7] J Miron, A Brief, S Epstein, and C Robinson, *Polyurethane IV*, Skeist Inc., Whippany, NJ, 2006.

[8] M Schlechter, *Medical Plastics Market Research Report # PLS009F*, BCC Research, Wellesley, MA, 2010.

[9] M Schlechter, *Plastics for Heathcare Packaging #PLS007E*, BCC Research, Wellesley, MA, 2013.

[10] *Polymers in Catheters and Medical Tubing #N498-39*, Frost and Company, Palo Alto, CA, 2008.

[11] J Chen, *The Chinese Market for Medical Polymers #PLS074A*, BCC Research, Wellesley, MA, 2013.

[12] *Global Overview of the Thermoplastic Polyurethane Market*, IAL Consultants, London, UK, 2005.
[13] M Schlechter, *Plastics for Heathcare Packaging #PLS007E*, BCC Research, Wellesley, MA, 2013, 219.
[14] J Li and W Work, Hydrophilic and nonthrombogenic polymers for coating of medical devices. Assignee Biocoat Inc. US patent 20130323291 A1. 2013.
[15] J Peterson, J Mitchell, A Schlicting, N Glucklich, J Berglund, and Y Guo, Apparatus and methods for filling a drug eluting medical device via capillary action. Assignee Medtronic Vascular Inc. US patent 20130284311 A1. 2013.
[16] S Fung, D Amos, J Liu, J Junkang, T Dietz, J Ma, and C Opp, Adhesive articles for medical applications and fabricating adhesive constructions. Assignee 3M Innovative Properties Co. US patent 10456499. 2013.
[17] X Qu, R Gruening, K Merritt, P Chen, and V Falevich, Non-leaching surface-active polymer film compositions for microbial adhesion prevention. Assignee Hydromer Inc. Canadian patent CA 2636975. 2007.
[18] W Martineau and K Nolan, *Disposable Medical Supplies Industry Study #2196*, The Fredonia Group, Cleveland, OH, 2007.
[19] P Lehr, *Global Market for Catheters: Market Research Report #HLC019E*, BCC Research, Wellesley, MA, 2012.
[20] K Stokes, R McVenes, and J Anderson, *Journal of Biomaterials Applications*, **9**, (1995), 321–354
[21] D Maki and M Ringer, *Journal of the American Medical Association*, **258**, (1987), 2396–2403.
[22] T Franson, N Sheth, L Menon, and P Sohnle, *Journal of Clinical Microbiology*, **24**, (1986), 559–561.
[23] L Martinez, A Pascual, and E Perez, *Journal of Hospital Infection*, **16**, (1990), 311–318.
[24] P Hermans, M Wilhelmus, and J Van Eldere, Prevention and treating Staphylococcus biofilm formation by using antibodies to and vaccines. Assignee Katholieke University, Leuven, the Netherlands. US patent 20100291177. 2008.
[25] M Mansouri, D David, and O Rabih, Novel modification of medical prostheses by coating with therapeutic agents. Assignee Baylor College of Medicine. US patent 7238363. 2005.
[26] S Mondal and D Martin, *Polymer Degradation and Stability*, **97**, (2012), 1553–1561.
[27] (a) O Mrad, J Saunier, C Chodur, V Rosilio, F Agnely, P Aubert, J Vigneron, A Etcheberry, and N Yagubi, *Radiation Physics and Chemistry*, **79**, (2010), 93–103.(b) P Alves, P Ferreira, and M Gil, "Biomedical polyurethane based materials," in *Polyurethanes: Properties, Structure, and Applications*, L Cavaco and L Melo, Eds., Nova Publishers, Haupauge, NY, 2012.
[28] *US Moist Wound Dressings Markets #F874-54*, Frost and Company, Paolo Alto, CA, 2006.
[29] M Elder, and T Cotthoff, *Markets for Advanced Wound Care Technologies #PHM011G*, BCC Research, Wellesley, MA, 2011.
[30] T Pivec, M Sfilgoj, P Gasparic, and K Kleinshek, *Techstilec*, **60**, (2017), 182–197.
[31] M Sonnenschein, N Rondan, B Wendt, and J Cox, *Journal of Polymer Science Part A: Polymer Chemistry*, **41**, (2004), 271–278.

[32] J Cox, F Lerma, and M Sonnenschein, Low haze thermoplastic polyurethane using mixtures of chain extenders. Assignee Dow Global Technologies. US patent 8242228. 2008.
[33] C Hacker, Z Karahaliloglu, G Seide, E Denkbas, and T Greis, *Journal of Applied Polymer Science*, **131**, (2014), 40132.
[34] A Unnithan, G Gnanasekaran, Y Satishkumar, Y Lee, and C Kim, *Carbohydrate Polymers*, **102**, (2014), 884–892.
[35] S Oh, S Kim, H Jeong, J Lee, J Cho, W Jae, and J Park, *Fibers and Polymers*, **14**, (2013), 173–181.
[36] P Croizat, A Eckstein, and A Weiss, Wound dressing containing foam material and an ointment base and a swelling agent for negative pressure therapy. Assignee P Hartmann AG. US patent 20120316517. 2012.
[37] G Howard, *International Biodeterioration and Biodegradation*, **49**, (2002), 245–252.
[38] K Michail, N Aljuhani, and A Siraki, *Canadian Journal of Physiology and Pharmacology*, **91**, (2013), 228–235.
[39] T Darby, H Johnson, and S Northrup, *Toxicology and Applied Pharmacology*, **46**, (1978), 449–453.
[40] M Szycher, *Journal of Biomaterials Applications*, **3**, (1988), 297–402.
[41] R Smith, D Williams, and C Oliver, *Journal of Biomedical Materials Research*, **21**, (1987), 1149–1166.
[42] E Beran, S Hulland, and M Steininger, *Journal of Polymers and the Environment*, **21**, (2013), 172–180.
[43] B Ratner, K Gladhill, and T Horbett, *Journal of Biomedical Materials Research*, **22**, (1988), 509–527.
[44] S Phua, E Castillo, J Anderson, and A Hiltner, *Journal of Biomedical Materials Research*, **21**, (1987), 231–246.
[45] S Wang and T Su, *Advanced Materials Research*, **622**, (2013), 99–102.
[46] A Gleadall and J Pan, *Journal of Biotechnology and Biomaterials*, **3**, (2013), 1–5.
[47] H Mi, M Salick, X Jing, B Jacques, W Crone, X Peng, and L Turng, *Materials Science and Engineering C: Materials for Biological Applications*, **33**, (2013), 4767–4776.
[48] M Savaris, CSC Garcia, M Roesch-Ely, JAP Henrique2, V Dos Santos, RN Brandalise, *Materials Science and Engineering C*, **96**, (2019), 539–551.
[49] G Wang, J Santerre, and R Labow, *Journal of Biomedical Materials Research*, **35**, (1997), 371–381.
[50] B Barrioni, S de Carvalho, R Orefice, A Oleveira, and M Pereira, *Materials Science and Engineering C*, **52**, (2015), 22–30.
[51] K Seidler, K Ehrmann, P Steinbauer, A Rohatschek, OG Andriotis, C Dworak, T Koch, H Bergmeister, C Grasl, H Schima, et al., *Journal of Polymer Science Part A: Polymer Chemistry*, **56**, (2018), 2214–2224.
[52] R Bezwada, Bioabsorbable polymers from bioabsorbable polyisocyanates and uses thereof. Assignee Bezwada Biomedical LLC. US patent 2009/0292029. 2009.
[53] Y Li, B Noordover, R van Benthem, A Rolf, and C Koning, *ACS Sustainable Chemistry and Engineering*, **2**, (2014), 788–797.
[54] K Mequanint, A Patel, and D Bezuidenhout, *Biomacromolecules*, 7, (2006), 883–891.
[55] A Sirkecioglu, H Mutlu, C Citak, A Koc, and F Guner, *Polymer Engineering and Science*, **54**, (2014), 1182–1191.

[56] R Paris, A Fernandez, and I Garrido, *Polymers for Advanced Technologies*, **24**, (2013), 1062–1067.
[57] P Flory and J Rehner, *Journal of Chemical Physics*, **11**, (1943), 512–521.
[58] J Brandrup, E Immergut, and E Grulke, *Polymer Handbook*, 4th Ed., Wiley, Hoboken, NJ, 2003.
[59] R Frezieres and T Walsh, *Contraception*, **61**, (2000), 369–377.
[60] N Matsoane, M Bester, E Pretorius, and P Becker, *Human and Experimental Toxicology*, **22**, (2003), 659–664.
[61] F Parks, D Bhattacharjee, and T Sanchez, Polyurethane films useful for medical applications and aqueous dispersions thereof. Assignee Dow Chemical Co. US patent 20020028877. 2002.
[62] W Keberle, H Hornig, and G Sinn, Production of elastic polyurethane filaments. Assignee Bayer AG. US patent 3686387. 1972.

# 12

# SPECIAL TOPIC: NONISOCYANATE ROUTES TO POLYURETHANES

## 12.1 GOVERNMENTAL REGULATION OF ISOCYANATES

The growth of commercial polyurethane building blocks and products is a testament to their efficacy and their efficiency. From the standpoint of both cost and processing, polyurethanes are attractive alternatives to many thermoplastic and thermoset materials. The specificity of isocyanate reactivity with active hydrogen atoms, and the large number of available building blocks that have active hydrogen atoms, results in enormous potential for polymer innovation. However, this specific reactivity also makes isocyanate a potential hazard to living organisms ubiquitous with active hydrogen in living tissues and water [1–3]. Industrial consumers of isocyanates have adapted to the challenges associated with handling isocyanates by protecting workers from inhalation and skin contact. This is usually achieved by installing necessary and commonsense safeguards for worker protection [4]. However, products with unreacted isocyanates mean that consumers, unregulated contractors, and also unknowing bystanders have the potential for unintentional primary exposure or incidental secondary exposure with the possibility of harm. Occupational exposure has been documented to result in dermal and inhalation sensitization (rashes and asthma-like symptoms). Chronic exposure to isocyanates can additionally cause permanent harm to lungs, while of course acute exposure can in most severe cases be fatal [5].

In 2011 the US Environmental Protection Agency (EPA) produced an action plan (RIN 2070–ZA15, April 20, 2011) for possible future regulation of methylene

*Polyurethanes: Science, Technology, Markets, and Trends*, Second Edition. Mark F. Sonnenschein.

diphenyl diisocyanate (MDI) and related compounds (4,4′, 2,4′, 2,2′ isomers, polymeric MDI, and MDI dimers and trimers). The conclusion of the review resulted in the following action plan:

*EPA intends to:*

1. *Issue a data call-in for uncured MDI under TSCA section 8(c) to determine if there are allegations of significant adverse effects and initiate a TSCA section 8(d) rulemaking for one-time reporting of relevant unpublished health and safety studies for uncured MDI.*
2. *Consider initiating a TSCA section 4 test rule to require exposure monitoring studies on uncured MDI and its related polyisocyanates in consumer products and exposure monitoring studies in representative locations where commercial products with uncured MDI and its related polyisocyanates would be used.*
3. *Consider initiating rulemaking under TSCA section 6 for*
   a. *Consumer products containing uncured MDI, and*
   b. *Commercial uses of uncured MDI products in locations where the general population could be exposed.*
4. *Consider identifying additional diisocyanates and their related polyisocyanates that may be present in an uncured form in consumer products that should be evaluated for regulatory and/or voluntary action.*

As of the end of 2019 there does not appear to have been additional action taken by the US EPA to limit environmental exposure to isocyanates. In 2015, the US EPA promulgated a new rule requiring notification of new uses for isocyanates (80 FR 845 and 81 FR 21830). Also, additional guidance on personal protective equipment was offered in 2018 [6, 7].

The US Occupational Health and Safety Administration (OSHA) has issued Directive Number CPL 03-00-017 (June 30, 2013) to initiate a National Emphasis Program "to identify or eliminate the incidence of adverse health effects associated with occupational exposure to isocyanates." As mentioned above, occupational exposure to isocyanates in the developed economies, including the USA, is not previously a prominent issue, but the growing use of uncured isocyanate products may create a problem. OSHA "targets" for particular scrutiny are: general industry, construction, and maritime industries "where exposure to isocyanates are known or are likely to occur."

As of the end of 2019 it does not appear that the US OSHA has further regulated occupational exposure to isocyanates.

The State of California's Department of Toxic Substances Control (March 25, 2014) listed spray polyurethane systems containing unreacted polyurethane (along with fire-retardant tris(1,3-dichloro-2-propyl) phosphate and surface cleaners containing methylene chloride) as a priority product under the Green Chemistry Law of 2008. The covered materials for review include all isomers and mixtures of MDI (presumably including polymeric MDI, which contains MDI monomer), all isomers and mixtures of toluene diisocyanate (TDI), and hexamethylene diisocyanates

(HDIs). Other isocyanates or isocyanate trimers are specifically not included. While being listed as a priority product is not itself a ban, it is a step in the state process toward regulation. A subsequent revision in September 2014 narrowed the scope of regulation to exclude from regulation one-component canned spray foams and TDI or HDI two-component systems for roof coatings. It was emphasized in the revision that the initial priority profile document has no force in law.

In 2016 unreacted methylene bisphenol diisocyanates were given "final" designation by California as a Priority Product (Office of Administrative Law Notice File Number: Z–2017–0307–02) [8]. The designation requires manufacturers to determine if a safer alternative exists or that the product be reformulated to exclude the priority product – in this case isocyanates. The American Chemistry Council filed an informal protest against the priority product designation in 2018, resulting in a delay of implementation of the regulations [9]. However, the product categories excluded from regulation in September 2014 remained excluded.

This activity follows the original activity of the European Commission (No. 552/2009, June 22, 2009; updated May 25, 2011, to specify isomers) that

> *MDI shall not be placed on the market after 27, December 2010, as a constituent of mixtures in concentrations equal to or greater than 0.1% by weight of MDI for supply to the general market, unless suppliers shall ensure before the placing on the market that the packaging:*
>
> *a. contains protective gloves which comply with the requirements of Council Directive 89/686/EEC;*
>
> *b. is marked visibly, legibly and indelibly as follows and without prejudice to other Community legislation concerning the classification, packaging and labelling of substances and mixtures*
>
> *'– Persons already sensitized to diisocyanates may develop allergic reactions when using this product.*
>
> *Persons suffering from asthma, eczema or skin problems should avoid contact, including dermal contact, with this product...'*

In 2017 the European Committee for Risk Assessment (RAC) and the Committee for Socio-economic Analysis (SEAC) agreed on proposed restrictions in support of the proposal by Germany to restrict the use of diisocyanates in the workplace. The main goal of this restriction proposal is to prevent new cases of respiratory sensitization among all workers and professionals who may be exposed to diisocyanates in the workplace. Diisocyanates are used in a wide range of sectors and applications (e.g. foams, sealants, coatings) throughout the European Union (EU), with a total of about 2.5 million tonnes per year. SEAC concluded that the proposed restriction is the most appropriate EU-wide measure to address the identified risks in terms of the proportionality of its socioeconomic benefits [10].

It should be noted that European commissions and committees offer proposals and technical and sociological opinions on which subsequent laws can be based but are not themselves law-making institutions, suggesting that there has not been significant new European restrictions on isocyanates since 2011.

With the governmental regulatory environment and the general European goal of using industrial solutions employing the least toxic but still effective components available, there has been increasing industrial and academic emphasis on obtaining urethane properties from systems that do not employ isocyanates. This chapter covers nonisocyanate chemistries that are actively explored for their potential to functionally substitute in polyurethane applications, regardless of economic barriers. The increased activity can be illustrated by analysis of the number of publications since 1990 on an emblematic nonisocyanate chemistry: reacting polycyclic carbonates and polyamines to make polyurethanes (Figure 12.1). While the overall activity is relatively small, encompassing all academic and patent publications, the exponential increase during the period of increased scrutiny is quite dramatic.

Like many new areas where there is perceived to be an opening for acquiring intellectual assets, during the years denoted in Figure 12.1 there were numerous forays into patent space with concentrations particularly in the area of inks and toners. However, many of the patents represent an attempt to stake out a particular product based on specific material, rather than part of a concentrated strategy.

During the years encompassed by Figures 12.1 and 12.2 attention paid to cyclic carbonate transformations to polyurethanes was the area of most prolific publication and patenting. Other areas have also been developed using completely unrelated raw material and building blocks. These will be taken separately.

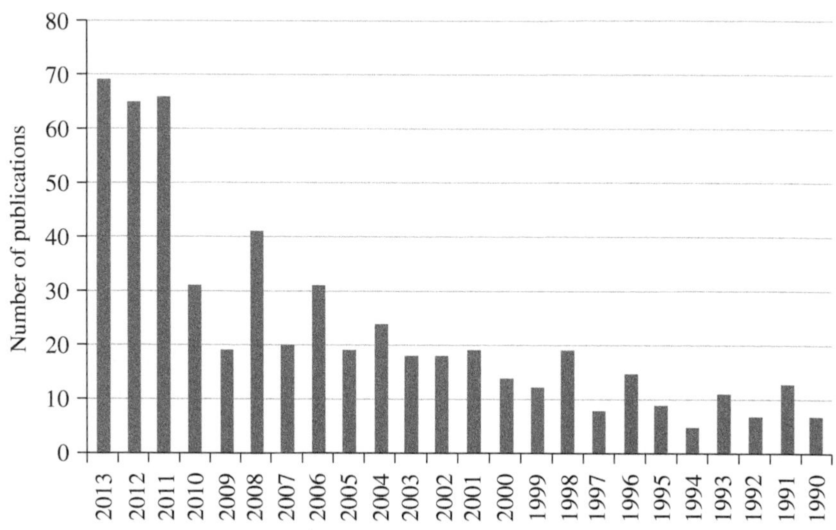


**FIGURE 12.1** Growth in public documents exploring or protecting the production of polycyclic carbonates, their polymerization to polyurethanes, or their uses. From 1990 to 2013, there was a total of 635 references.

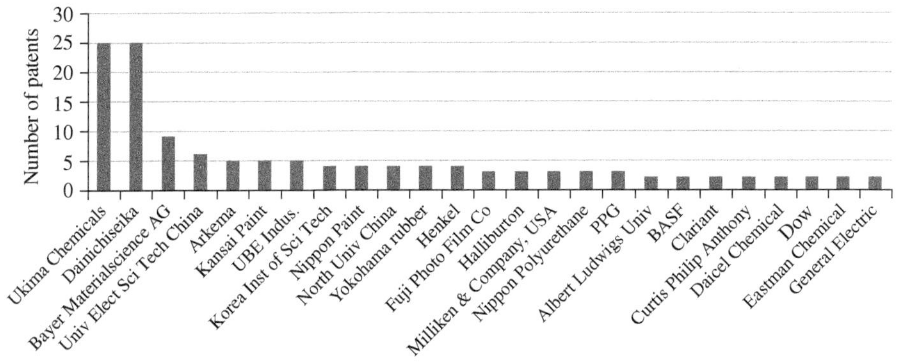


**FIGURE 12.2** Patent activity of top filers of polycyclic carbonate to polyurethane technology. From 2000 to 2014, there was a total of 282 patents.

While the time period surrounding proposed government regulation of isocyanates spawned significant activity, more recent years have seen reduced innovation. From the years 2015–2019 there were 60 publications of all types where the prime focus was on the development of urethane structure without the use of a polyisocyanate (Figure 12.3). During that time, there were 56 journal articles, four conference proceedings, and no found patents. Also, during this time period there was no new nonisocyanate route to urethane structure described. All the research described novel applications of feedstock using several of the techniques described within this chapter.

## 12.2 NONISOCYANATE ROUTES TO POLYURETHANES

There has been development of cross-linked and noncross-linked chemistries advertised as nonisocyanate where an isocyanate is actually used in the chemistry and simply reacted prior to application. In this case there is no isocyanate reaction subsequently involved in final part fabrication. An example might be the reaction of diisocyanate with polyol and the subsequent reaction of a hydroxy-silane or amino-silane to make a silane-tipped polyurethane backbone (Figure 12.4a) [11, 12]. The final product in its use is isocyanate free but there was at some point an isocyanate in its construction. A similar situation can exist in the formation of acrylate-tipped hybrid polymer building blocks (Figure 12.4b) [13, 14]. While these materials may be represented in some venues as "nonisocyanate" this is simply a matter of semantics and is not the intent of this chapter. This chapter presents synthetic routes to making urethane structures without ever employing isocyanate-based building blocks in their construction. The intent of such an exercise should, in principle, not replace one problematic building block (an isocyanate) with an equally (or more so) problematic substitute. In selective cases such a result is achieved, while in others the substitution

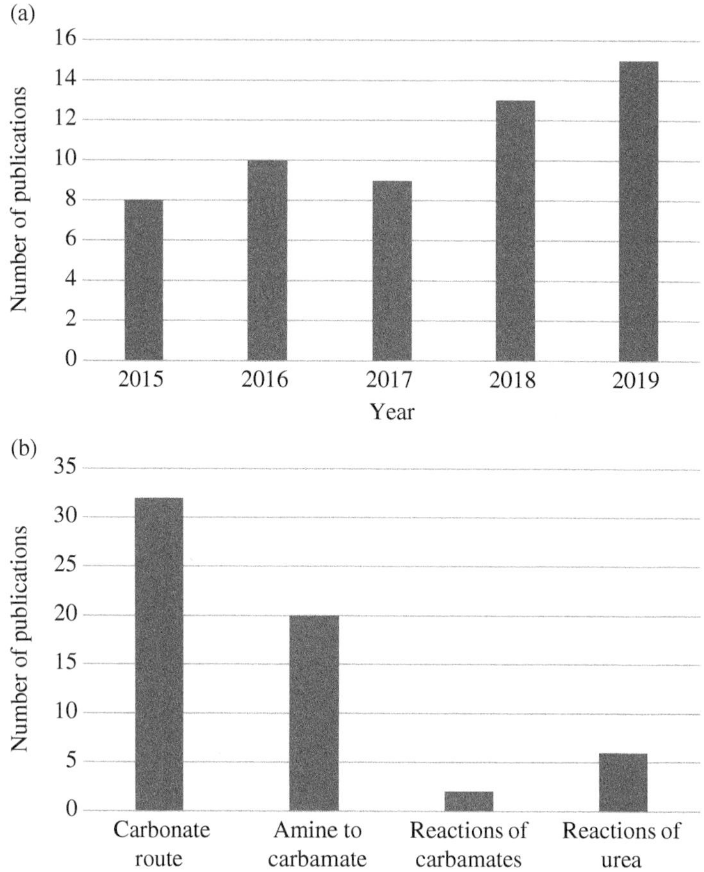


**FIGURE 12.3** (a) Activity intensity of research on the production of polyurethane structure without isocyanates for 2015–2019 and (b) routes to achieving urethane structure without polyisocyanate molecules. See text for detailed exposition of these routes.

may be just another target for future regulation. Additionally, there is nonisocyanate chemistry based on blocked isocyanates (see Chapters 2 and 10) [15]. These are also isocyanates that are utilized in their urethane form, reverting back to isocyanates upon heating. However, their "nonisocyanate" status is perhaps even more suspect.

### 12.2.1 Reactions of Polycyclic Carbonates with Polyamines

One route to polyurethanes without the benefit of polyisocyanate building blocks is via the reaction of polycyclic carbonates with polyamines [16–18]. As shown in Figure 12.1, research activity on this reaction grew at a high rate. To some extent this was due to the availability of reagents and the relative ease of their transformation to

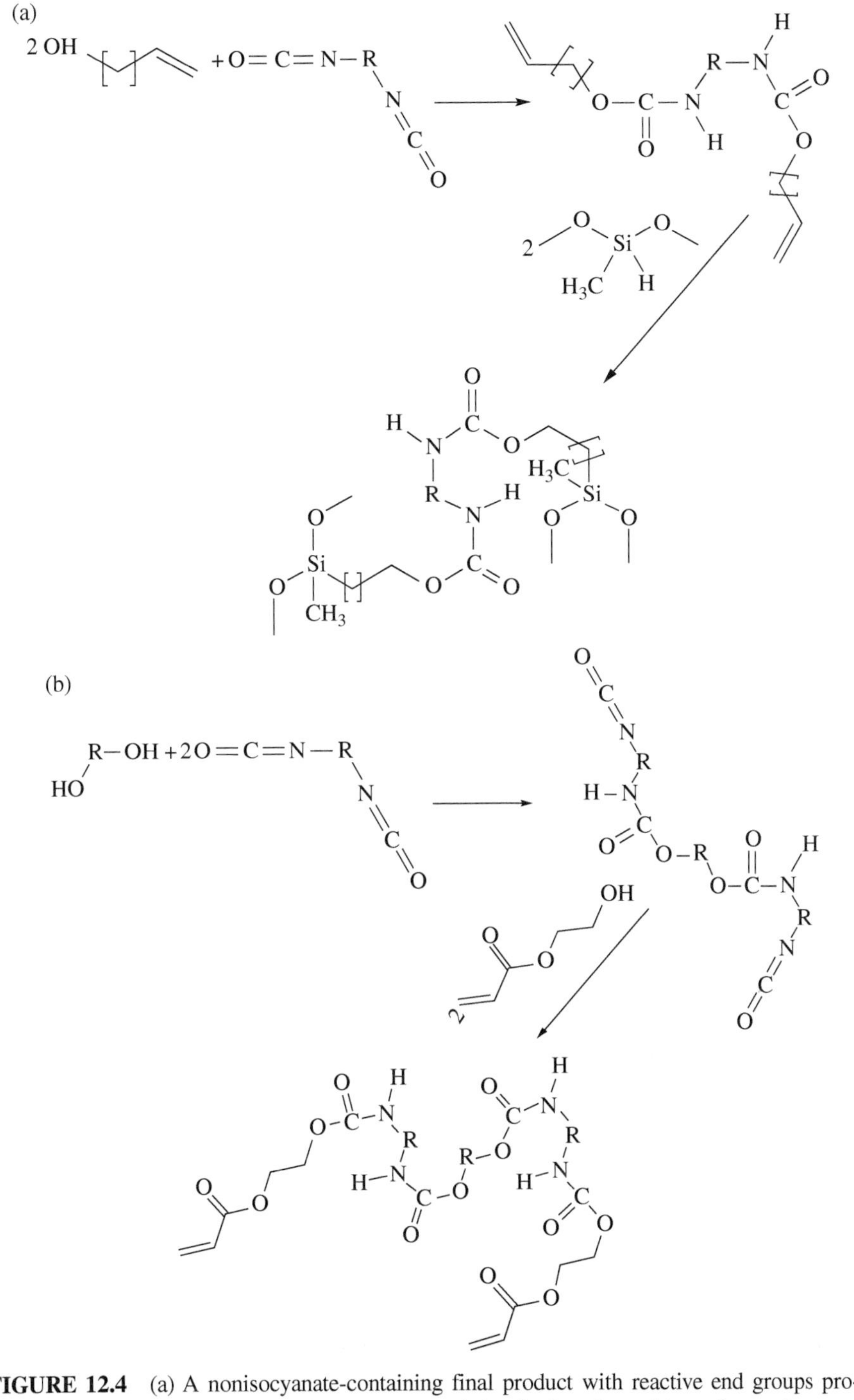


**FIGURE 12.4** (a) A nonisocyanate-containing final product with reactive end groups produced by hydrosilylation. (b) A nonisocyanate-containing final product with acrylate end groups.

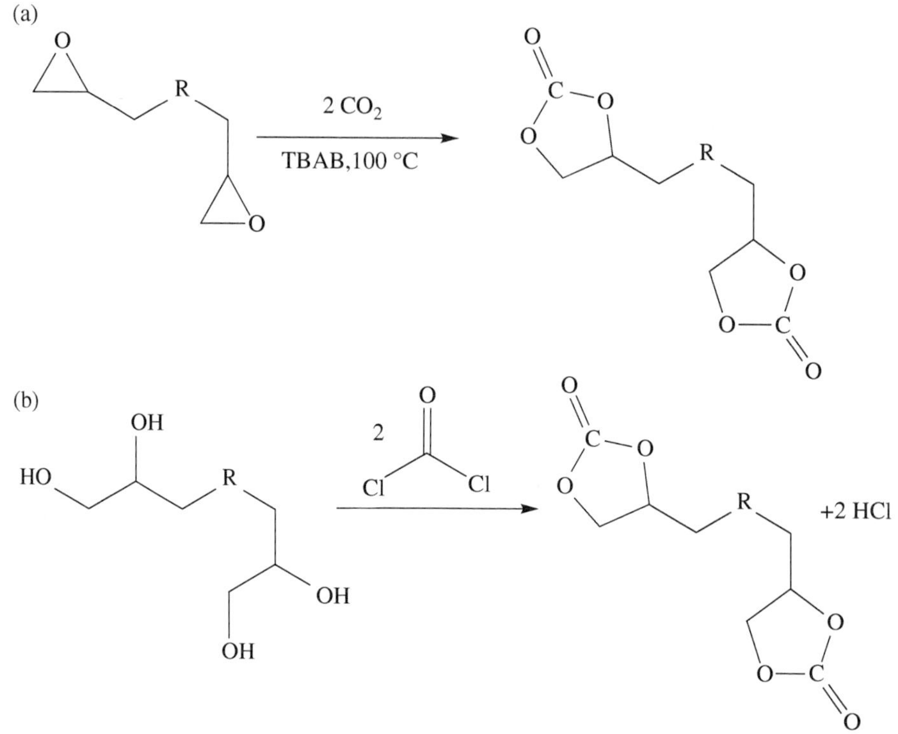


**FIGURE 12.5** Direct transformation of polyoxirane (epoxy) to a polycyclic carbonate building block by (a) $CO_2$ insertion or (b) reaction of the hydrolyzed oxirane with phosgene. TBAB, tetra *n*-butylammonium bromide.

polyurethane building blocks. Formation of cyclic carbonates is most easily achieved by the catalyzed reaction of an oxirane (epoxy) functionality with $CO_2$ (Figure 12.4a). The transformation can also be performed using phosgene or triphosgene (Figure 12.4b), as well as halohydrins or halogenated carbonates [19].

Figure 12.5a depicts the most prevalent route found in the literature and has been optimized to a great extent. Conversion temperatures are in the range of 25–150 °C, $CO_2$ pressures are 1–21 bar, and reaction times greater than 3 hours. Reaction times and extent of conversion are functions of reaction conditions, oxirane substrate, and various cross-influences. Conditions for nearly quantitative selectivity and yield have been presented using heterogeneous catalysis under supercritical $CO_2$ conditions and extended reaction times. Conversion of polyoxirane to polycyclic carbonate is accompanied by a significant increase in viscosity, which can be used to indirectly track conversion. Conversion of the oxirane to the cyclic carbonate is also easily followed directly by Fourier transform infrared spectroscopy (see Chapter 5) with the disappearance of the epoxy ether at 800 $cm^{-1}$ and the growth of the carbonyl peak at about 1780 $cm^{-1}$.

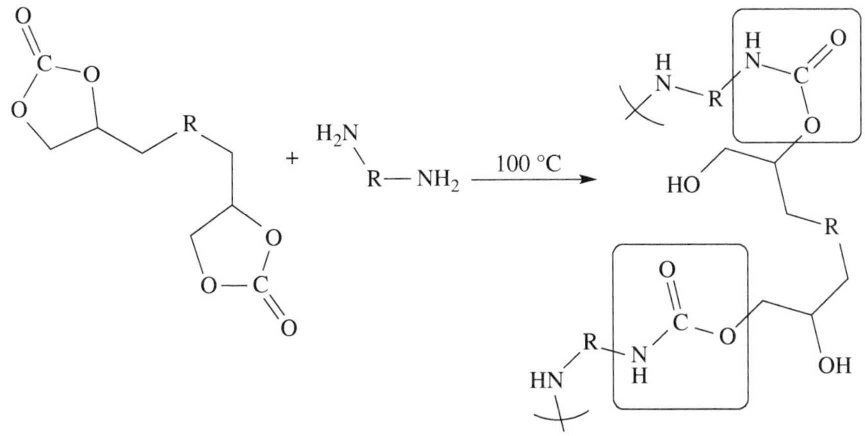


**FIGURE 12.6** Illustrative reaction of a polycyclic carbonate with a polyamine to make a polyurethane. Boxes indicate the urethane group.

Polyurethane formation is achieved by reaction of polycyclic carbonate with polyamines having primary amine functionality (Figure 12.6). The figure shows the potential for ring opening to create either primary hydroxyl or secondary hydroxyl products. The ratio of primary to secondary hydroxyl formation along the backbone is driven by substituent groups on the ring [17]. In the presence of highly electron-withdrawing groups secondary hydroxyls predominate. As shown in Figure 12.6 the ratio is approximately 1:1. The potential for intrachain hydrogen bonding between the backbone hydroxyls and the urethane carbonyls has been cited to bestow improved hydrolytic stability [18]. Evidence for this improvement is not yet presented in the literature.

It has been reported that cure of polyurethanes produced by reaction of cyclic carbonates and amines can occur at room temperature when allowed to proceed for extended times [18]. A similar experiment pursued with the parent polyepoxy and polyamine building blocks is not cured to the same degree, suggesting that electrophilicity of the carbonate linkage for the nucleophilic amine may predominate over ring strain promoting reactivity of epoxy moieties. As observed for conventional urethane synthesis, Lewis acids (such as tin octanoate) and Lewis bases (such as 1,8-diazabicycloundec-7-ene (DBU)) are very effective, but do not reportedly result in room temperature cure for normal processing times, instead requiring heating to 70 °C. However, while cyclic carbonates can be more reactive than epoxy functionality, they are by no means as reactive as isocyanates, which exhibit diffusion-limited reactivity with amines. An additional influence on reactivity is the volume functional group density. This means that reactivity of building blocks to form polymer will be influenced by the equivalent weight of the reactive components, with lower equivalent weight translating into higher reactivity. This is primarily an influence of the exothermic ring-opening reaction and the activation energy of reaction.

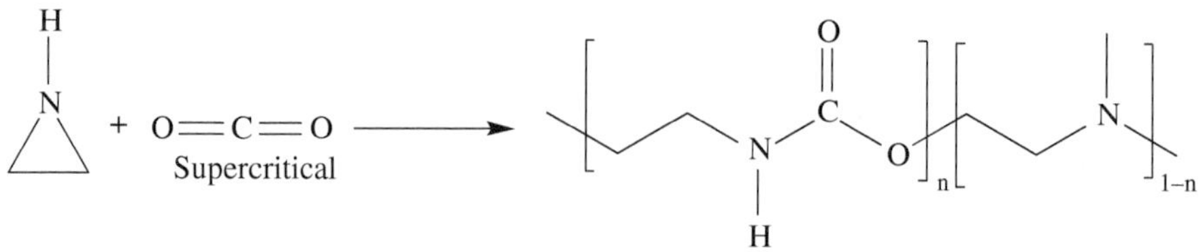


**FIGURE 12.7** Illustrative reaction of aziridine with super critical $CO_2$ to make a polyurethane–amine.

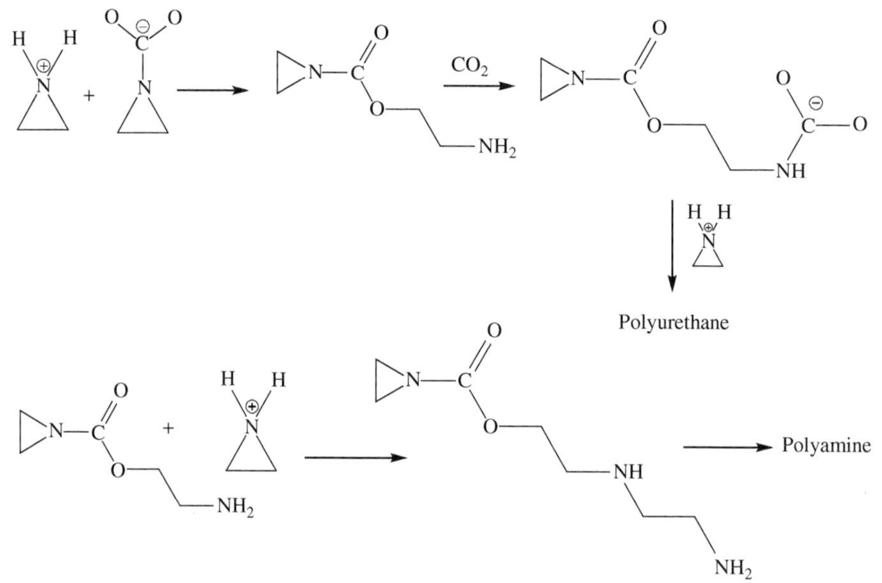


**FIGURE 12.8** Two potential mechanisms for formation of the polyurethane–amine backbone from aziridine polymerization with $CO_2$.

A related reaction of aziridines with $CO_2$ can lead directly to polyurethanes (Figure 12.7). This reaction can result in the formation of a urethane–amine with the relative ratios strongly affected by polymerization conditions [20–22].

The urethane formation reaction occurs via nucleophilic attack of the carbamate anion on the aziridinium cation. The competitive reaction forming the polyamine can occur via ring opening of the aziridinium cation by the primary amine (Figure 12.8).

Industrial and commercial adoption of cyclic carbonate chemistry for making polyurethane backbones will depend on several factors: (i) the ability to make useful materials, (ii) the ability to make those materials economically, and (iii) environmental, health, and safety advantages relative to isocyanate-based urethanes [23–26]. The following analysis for a polycyclic carbonate–amine reaction would apply equally to each other pathway discussed herein.

Taking each factor in consideration:

1. In principle, cyclic carbonate chemistry can be nearly as flexible as that of polyurethane. While it would appear that polycyclic carbonate chemistry is less able to form phase-separated materials, it may be possible to create hard block/soft block polymer structures, as made in polyurethane chemistry. This could be achieved by employing high-molecular-weight soft-segment diamines and preparing hard segments from cyclic carbonate reactions with very low equivalent weight chain extenders such as ethylene diamine. The ability to form intrahard-segment hydrogen bonding might create sufficient cohesive energy density to result in a durable physical cross-link. With sufficient phase volume it may also be possible to create a cocontinuous structure within the elastomer and observe traditional polyurethane tensile properties. However, it remains to be seen if the components can produce the required reaction-induced phase separation prior to gross phase separation of the reacting components because of reactivity, viscosity, and solubility differences between the components. A possible solution for this may come by preparation of polycyclic carbonate–polyamine prepolymers. Prepolymer preparation prior to making the hard segment may provide adequate compatibilization to maintain a homogeneous initial state from which microphase separation into hard and soft segments can occur. Additionally, owing to the relatively slow reactivity of polycyclic carbonates and incompatibility with blowing agent technology, this technology may not be suitable for flexible or rigid foam purposes.
2. Polyisocyanate prices are generally lower than those of polyoxirane (epoxy resins) because of fundamental costs associated with industrial manufacture. The price difference is as much as 30%. Subsequent addition of $CO_2$ to epoxy resin will further increase differential cost depending on the process chosen, the reaction time, the need to purify/separate products, and the volume produced. In addition, the cost of polyamines is highly variable but substantially more expensive than polyether or polyester polyols. The difference can be 100% or more. Thus, it can be anticipated that the cost of a polyurethane elastomer produced by reaction of a polycyclic carbonate with polyamines may be twice as high as the same elastomer produced by isocyanate-based chemistry. This may not be the dominant factor if regulatory agencies decide that the risks of polyisocyanates outweigh the benefits.
3. The cost–benefit ratio will require that polyamines employed in making nonisocyanate polyurethanes are substantially less problematic than polyisocyanates [27–29]. While the toxicity of any particular polyamine relative to a particular polyisocyanate will depend on numerous structure–property relationships, it is unavoidable that amines are strong bases and are therefore dangerous by skin contact and inhalation. It will remain to be seen if the potential benefits of this nonisocyanate route to polyurethanes can withstand a health and safety examination on the same grounds.

### 12.2.2 Direct Transformations of Amines to Urethanes

Section 2.2.1.2 presents nonphosgene routes to isocyanates by reactions of amines with dimethyl carbonates. In that process isocyanate is created without phosgene, but rather obtained by reversion of urethane to an isocyanate and an alcohol. In that method, the point is to obtain an isocyanate without employing the dangerous phosgene reagent. Work along these lines has also produced urethane functionality without isocyanate, which in principle can be extended to produce polyurethanes via simplified reaction and reagent choices.

One such route converts amines, alcohols, and $CO_2$ directly to urethanes [30–32]. The reaction is catalyzed and performed at elevated temperature and pressure (Figure 12.9). Given the instability of the carbamic acid and the relative reaction rates, it is possible that this reaction actually proceeds through an isocyanate intermediate before forming the urethane. The reaction has been optimized to produce high yields and selectivity; however, reaction times tend to be long. The use of mixed polyamines and polyols offers the potential to produce polymers with hard segments.

A related mechanism employs an -onium salt (in the example presented in Figure 12.10, an ammonium salt) with an alkyl halide to produce a urethane group [33].

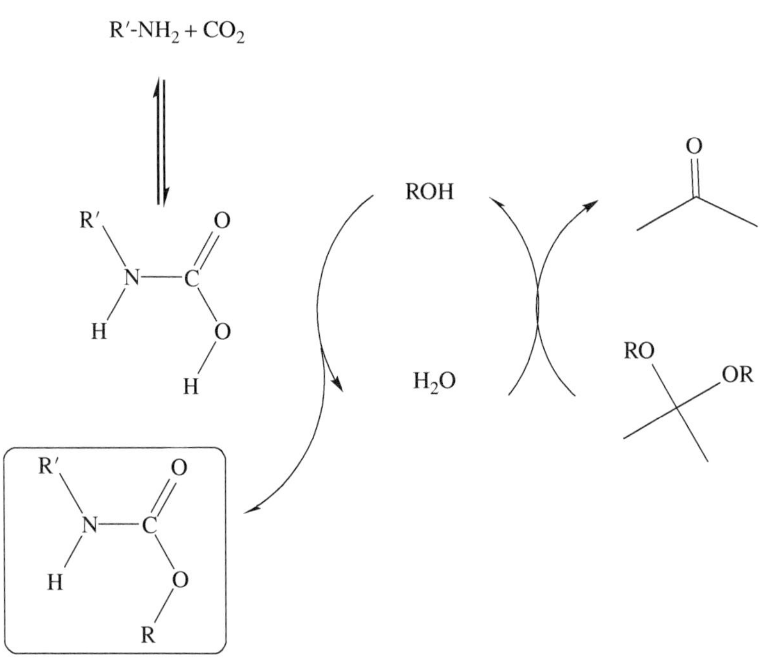


**FIGURE 12.9** Conversion of amine to carbamate by reaction with $CO_2$ and subsequent reaction with an alcohol. Polymers can be formed when the amine is a polyamine and the alcohol is a polyol. Box indicates the urethane product.

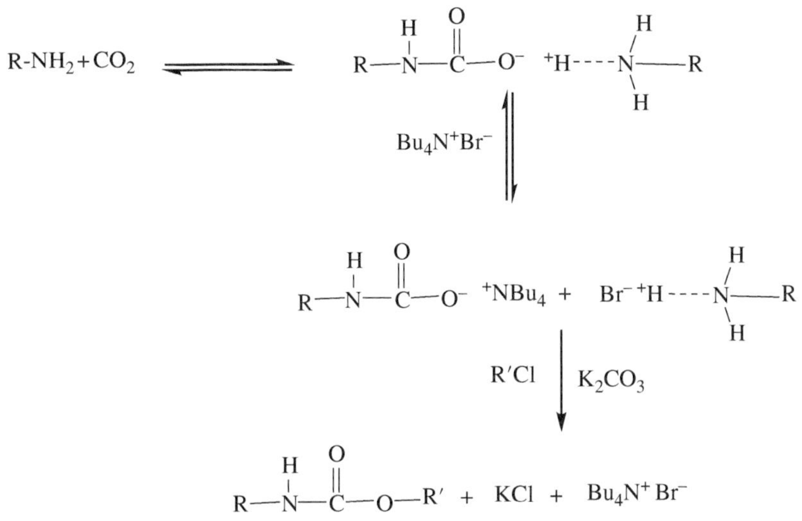

**FIGURE 12.10** Reaction of an amine with $CO_2$ to make urethane via an -onium salt complex. Polymer can be formed when a polyamine is employed and R′ is capable of transurethanification.

As before, the use of mixed polyamines and alternative alkyl halides offers the potential for polymerization and design flexibility.

Beginning with the same first step, urethane has also been produced with good yield and milder conditions using Mitsunobu reagent [34, 35]. The relatively easier reaction conditions are partially offset by consumption of the reagent, making the reaction potentially expensive (Figure 12.11). In fact, all reactions transforming amines to urethanes by reaction with $CO_2$, and a source of organic moiety to reduce the carbamic acid to the urethane, involve the loss of reagents and the need to separate the desired urethane from these compounds. Like the reactions of cyclic carbonates and amines to form urethanes discussed in the preceding section, conversion of amines and $CO_2$ to urethanes employs reagents with potentially troublesome safety profiles, but in this case the materials are consumed in the industrial manufacturing step and would not provide a possible route of amine exposure to consumers or unregulated contractors.

### 12.2.3 Reactions of Polycarbamates

Carbamates can be prepared by reaction of alcohols with other carbamates or by reactions of alcohols with urea [36]. The simplest carbamate is methyl carbamate and this is prepared by reaction of methanol with urea (Figure 12.12). The carbamate can be subsequently reacted with a single aldehyde to produce a hemiaminal, or with two aldehydes to produce an aminal (Figure 12.12). As noted in an early patent, the initial reaction with methanol can be equally performed on a polyol to produce a polycarbamate

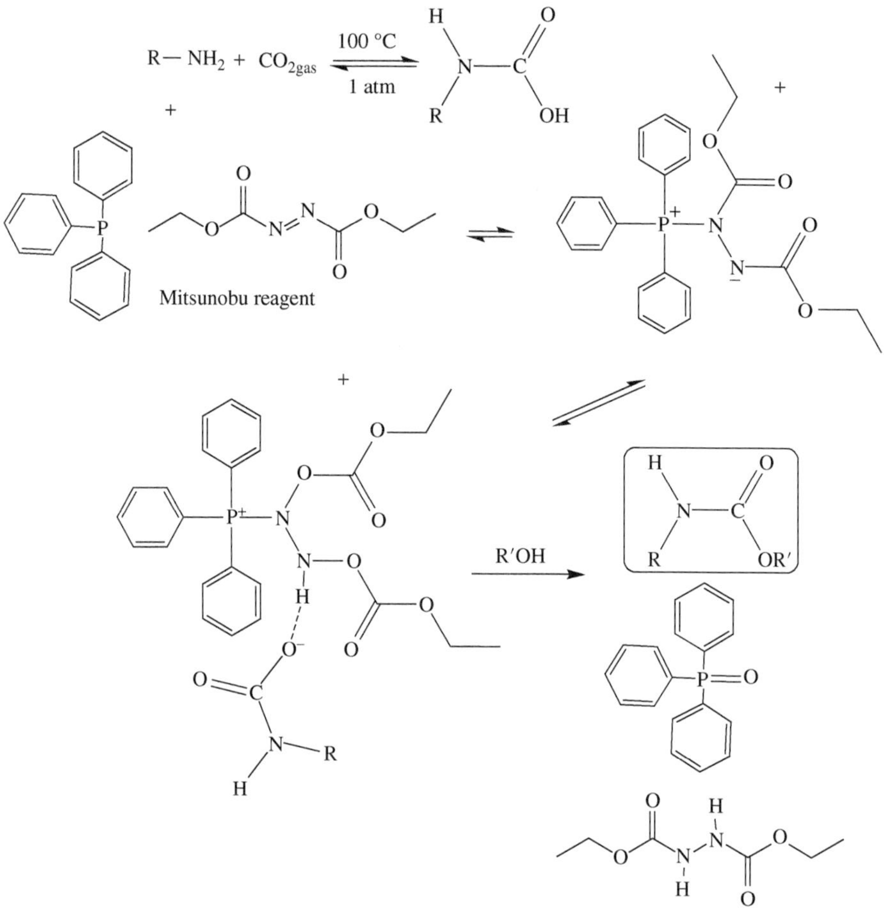


**FIGURE 12.11** Reaction of amine with $CO_2$ to form urethane using Mitsunobu reagent. A polyurethane can be produced when a polyamine is employed.

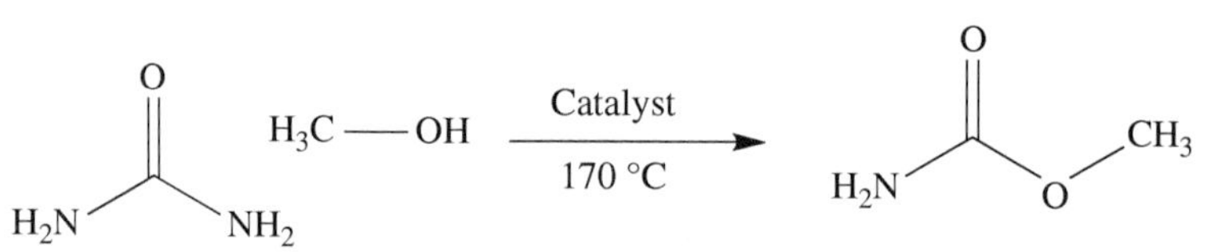


**FIGURE 12.12** Reaction of urea and methanol to form methyl carbamate.

(Figure 12.13), which can be subsequently reacted with a polyaldehyde to create a three-dimensional polyurethane cross-linking structure (Figure 12.14) [37]. The catalyzed reaction can proceed at room temperature, which makes the reaction particularly suitable as a potential replacement for some isocyanate-derived products.

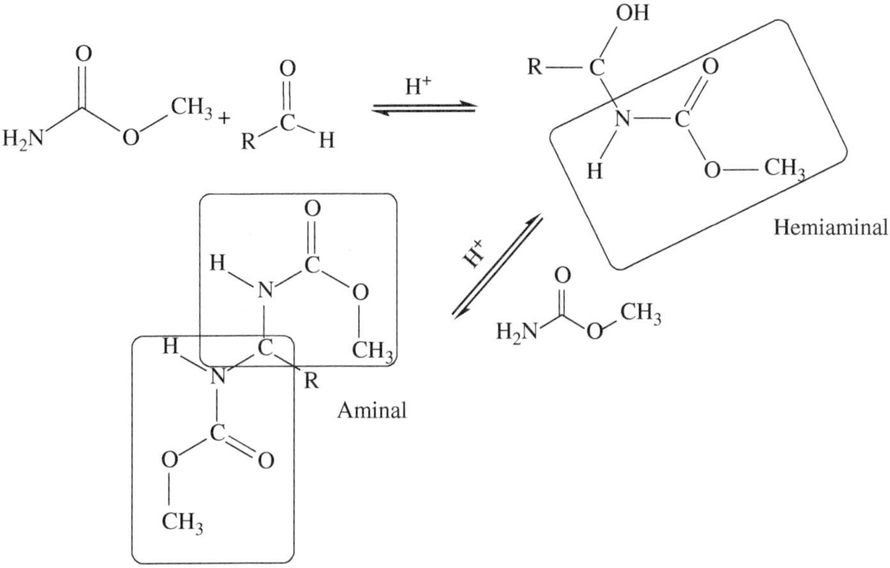


**FIGURE 12.13** Reaction of methyl carbamate and formaldehyde to make the hemiaminal and subsequently the aminal. A polyurethane can be formed when a polycarbamate and a polyaldehyde are reacted together. Boxes indicate the urethane groups.

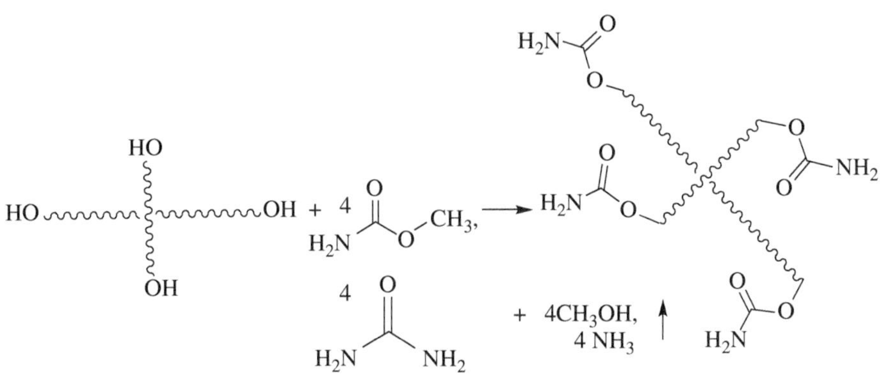


**FIGURE 12.14** Reaction of a polyol with methyl carbamate or urea to produce a multifunction polycarbamante that can form a cross-linked polyurethane when reacted with a polyaldehyde.

The range of potential polycarbamate building blocks via aminal chemistry is limited only by imagination since the number of possible polyol reagents is immense and the urea feedstock is relatively cost-effective. On the other hand, the number of polyaldehydes is relative small, and therefore potentially expensive. Furthermore, aldehyde propensity to oxidize in the presence of oxygen is a potential complication and must be accounted for in reactant stoichiometry in a similar manner that MDI dimerization must be accounted for in preparation of MDI-based foams.

### 12.2.4 Conversion of Hydroxamic Acids to Polyurethane

Conversion of diesters to dicarbamates via formation of the hydroxamic acid is a relatively seldom traveled route to polyurethanes. In part this is surely because the synthesis is somewhat laborious with relatively modest yield. In part it may also be because the route to urethane via hydroxamic acid is actually through an isocyanate intermediate. The reaction begins with displacement of the methyl ester with the hydroxamic acid by reaction with hydroxylamine hydrochloride (Figure 12.15). The hydroxamic acid is converted to isocyanate by Lossen rearrangement (Figure 12.16). Rapid reaction with methanol leads to the bis-carbamate [38, 39]. Subsequent reaction with a polyol results in the polyurethane polymer. High-functionality polyols and diesters of varying molecular weights can result in block copolymer formation.

### 12.2.5 Conversion of Hydroxylamines to Polyurethanes

The conversion of hydroxylamines to urethanes and polyurethanes has been described in the literature [32]. The carbonyl of the urethane linkage provides a ready coreactant for the hydroxyl moiety and high molecular weights have been reported using

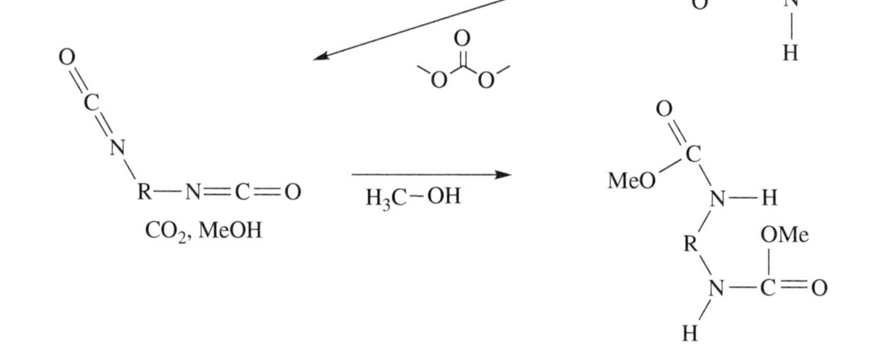


**FIGURE 12.15** Reaction of a dimethyl ester with hydroxylamine hydrochloride to form the dihydroxamic acid, which through subsequent reactions forms a dicarbamate. Reaction of the dicarbamate with a polyol can form a high-molecular-weight polyurethane.

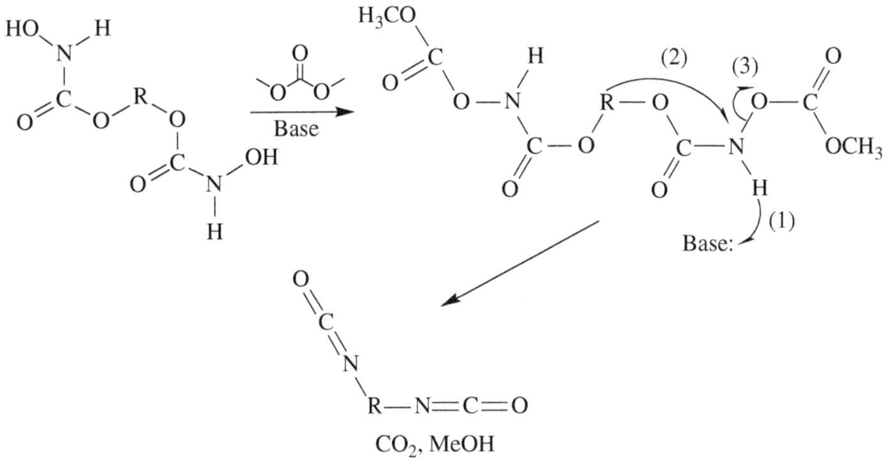


**FIGURE 12.16** Reaction of the polyhydroxamic acid with dimethyl carbonate, resulting in a latent polyisocyanate through the Lossen rearrangement.

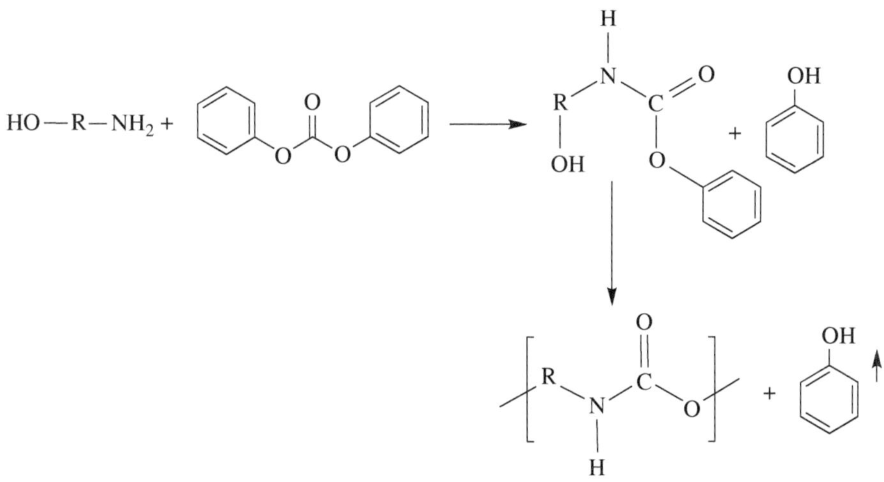


**FIGURE 12.17** Illustrative reaction of a hydroxylamine and diphenyl carbonate to form a polyurethane with phenol by-product. Removal of the phenol is required for attaining high molecular weight.

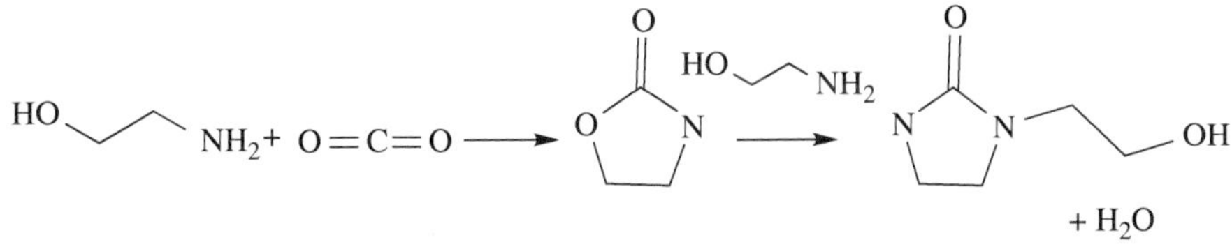


**FIGURE 12.18** Direct reaction of ethanolamine and carbon dioxide to form oxazolidinole and subsequent reaction with ethanolamine to form hydroxy ethyl imidazolidinone. This side reaction can be avoided by using higher molecular weight alcohol amines or having branched structures that inhibit intramolecular cyclization.

this procedure (Figure 12.17), including removal of the phenol coproduct and design of the R group to inhibit formation of the cyclic urethane. The cost of diphenyl carbonate, its high unit ratio, and the large volume of phenol coproduct could be an inhibiting factor in implementation. An alternative direct reaction of phosgene or $CO_2$ with hydroxylamines would yield mixed cyclic products such as 2-oxazolidionone or hydroxy ethyl imidazolidinone (Figure 12.18) or mixed urethane/carboxylate chain functionality owing to the reactivity of both amines and hydroxyls for the electrophilic carbonic acid derivatives, depending on initial reagents.

## REFERENCES

[1] J Lesage, I DeGraff, and R Danchik, *Isocyanates: Sampling, Analysis and Health Effects*, ASTM International Publishers, West Conshohocken, PA, 2002.

[2] H Ulrich, *Chemistry and Technology of Isocyanates*, Wiley, Hoboken, NJ, 1997.

[3] B Donovan, *A Summary of Health Hazard Evaluations: Issues Related to Occupational Exposure to Isocyanates, 1989–2002,* Centers for Disease Control and Prevention, National Institute for Occupational Safety and Health, Atlanta GA, 2004.

[4] *British Rubber Manufacturers Association Ltd, Toxicity and Safe Handling of Diisocyanates and Ancillary Chemicals*, 2nd Ed., Rapra Technology, Shawbury, UK, 2001.

[5] T Landry and C Steffens, "Health safety and environmental aspects," in *Reaction Polymers*, W Gum, W Riese, and H Ulrich, Eds., Hanser Publishers, Munich, Germany, 1992, 746–765.

[6] US Environmental Protection Agency, *Significant New Use Rule on Certain Chemical Substances*, Document number 2016-25933, 2016. Available at: https://www.federalregister.gov/documents/2016/10/27/2016-25933/significant-new-use-rule-on-certain-chemical-substances (accessed June 9, 2020).

[7] US Environmental Protection Agency, *Spray Polyurethane Foam (SPF) Insulation and How to Use it More Safely*, 2017. Available at: https://www.epa.gov/saferchoice/spray-polyurethane-foam-spf-insulation-and-how-use-it-more-safely (accessed June 9, 2020).

[8] Department of Toxic Substances Control, *Safer Consumer Products Regulations – Listing Spray Polyurethane Foam Systems with Unreacted Methylene Diphenyl Diisocyanates as a Priority Product*, 2017. Available at: https://media2.mofo.com/documents/170324-isor.pdf (accessed June 9, 2020).

[9] Chemical Watch, *ACC Appeals Against California Spray Foam Priority Product Designation: Methylene Chloride Paint Strippers added to SCP Programme*, 2019. Available at: https://chemicalwatch.com/73040/acc-appeals-against-california-spray-foam-priority-product-designation (accessed June 9, 2020).

[10] European Chemicals Agency (ECHA), *Restriction Proposal on Diisocyanates and Several Authorisation Applications Agreed by RAC and SEAC*, ECHA/NA/17/29, 2017. Available at: https://echa.europa.eu/-/restriction-proposal-on-diisocyanates-and-several-authorisation-applications-agreed-by-rac-and-seac (accessed June 9, 2020).

[11] M Kohl, M Proebster, R Ge, M Etzelstorfer, and C Manegold, Catalysis of silane-crosslinkable polymer compositions. Assignee Henkel AG and Dow Europe. WO 2014029837. 2014.

[12] C Cao, J Cheng, X Liu, R Wang, J Zhang, J Qu, and U Jaeger, *Journal of Adhesion Science and Technology*, **26**, (2012), 1395–1405.
[13] S Oprea, S Vlad, A Stanciu, and M Macoveanu, *European Polymer Journal*, **35**, (1999), 1269–1277.
[14] C Ha, S Jung, E Kim, W Kim, S Lee, and W Cho, *Journal of Applied Polymer Science*, **62**, (1996), 1011–1021.
[15] E Delebecq, J Pascault, B Boutevin, and F Ganachaud, *Chemical Reviews*, **113**, (2013), 80–118.
[16] J Guan, Y Song, Y Lin, X Yin, M Zuo, Y Zhao, Z Tao, and Q Zheng, *Industrial and Engineering Chemistry Research*, **50**, (2011), 6517–6527.
[17] H Tomita, F Sanda, and T Endo, *Journal of Polymer Science Part A: Polymer Chemistry*, **39**, (2001), 3678–3685.
[18] O Figovsky, Hybrid non-isocyanate polyurethane network polymers and composites formed therefrom. Assignee Euotech Ltd. US patent 6120905. 2000.
[19] R Burk and M Roof, *Tetrahedron Letters*, **34**, (1993), 395–398.
[20] O Ihata, Y Kayaki, and T Ikariya, *Macromolecules*, **38**, (2005), 6429–6434.
[21] O Ihata, Y Kayaki, and T Ikariya, *Angewandte Chemie International Edition*, **43**, (2004), 717–719.
[22] R Lundberg and D Montgomery, Carbon dioxide polymers. Assignee Union Carbide Corporation. US patent 3523924. 1970.
[23] B Thorpe and M Rossi, *New Solutions: A Journal of Environmental and Occupational Health Policy*, **17**, (2007), 177–192.
[24] A Williams and K Dunwoody, *Chemical Engineering*, **119**, (2012), 34–43.
[25] P Anastas and J Warner, *Green Chemistry: Theory and Practice*, Oxford University Press, New York, NY, 2000.
[26] G Guillen-Gosalbez, J Caballero, and L Jiminez, *Industrial Engineering Chemistry Research*, **47**, (2008), 777–789.
[27] A Pegg, *Chemical Research in Toxicology*, **26**, (2013), 1782–1800.
[28] F Cavender, "Aliphatic and alicyclic amines," in *Patty's Toxicology*, Volume **2**, E Bingham and B Cohrssen, Eds., Wiley, Hoboken, NJ, 2012, 433–516.
[29] V Pankaj, G Mayank, P Shikha, and K Prem, *Materials Science Forum*, **712**, (2012), 147–173.
[30] M Abla, J Choi, and T Sakakura, *Green Chemistry*, **6**, (2004), 524–525.
[31] M Abla, J Choi, and T Sakakura, *Chemical Communications*, **6**, (2001) 2238–2239.
[32] B Bhange, S Fujita, Y Ikushima, and M Arai, *Green Chemistry*, **5**, (2003), 340–342.
[33] M Yoshida, N Hara, and S Okuyama, *Chemical Communications*, **2**, (2000), 151–152.
[34] D Cane-Honeysett, M Dowle, and M Wood, *Tetrahedron*, **61**, (2005), 2141–2148.
[35] D Chaturvedi, A Kumar, and S Ray, *Tetrahedron Letters*, **44**, (2003), 7637–7639.
[36] M Unverferth, O Kreye, A Prohammer, and M Meier, *Macromolecular Rapid Communications*, **34**, (2013), 1569–1674.
[37] E Greyson, J Argyropoulos, P Foley, A Swartz, N Kamber, C Slone, and M Sekharan, Crosslinkable composition and method of producing the same. Assignee Dow Global Technologies. US patent 20130172472. 2013.
[38] L Bauer and O Exner, *Angewandte Chemie International Edition*, **13**, (1974), 376–384.
[39] S Neffgen, J Kusan, T Fey, H Keul, and H Hacker, *Macromolecular Chemical Physics*, **201**, (2000), 2108–2114.

# 13

# POLYURETHANE HYBRID POLYMERS

## 13.1 INTRODUCTION

Hybrid materials refer to compounds of two dissimilar materials that, in combination, exhibit unique properties. The primary motivation is to achieve positive attributes of the two or more components while mitigating their individual weaknesses by artful backbone or formulation design. Simple blends are not usually included among the materials deemed to be "hybrids." Instead, "hybrid" is usually applied to the groupings including polymer–inorganic mixtures and the subset of materials also termed interpenetrating networks (IPNs) [1–3]. Hybrid materials are composites consisting of two constituents mixed at the nanometer or molecular level. They typically differ from traditional composites, where the constituents are at the macroscopic (micrometer to millimeter) level. Mixing at the microscopic scale leads to a more homogeneous material that can demonstrate nonrule of mixture characteristics between those of the two original phases, or potentially even new and unanticipated properties.

Inorganic hybrid material properties can exhibit new properties that will depend on the length scale of interactions between the organic and inorganic phases. In the realm of molecular composites, the inorganic domain is in contact and modifying the organic components via short-range interactions [4–6]. In this way one can try to modify the electrical, optical, and transport properties of the organic part of the hybrid. In nanocomposite hybrids, the interaction between the inorganic and organic component is of the order of the molecular chain length. This level of interaction can influence the processes that are cooperative within the bulk organic polymer, and can

*Polyurethanes: Science, Technology, Markets, and Trends*, Second Edition. Mark F. Sonnenschein.

affect properties such as strength, toughness, and glass transition temperature [7–9]. Larger scale inorganic–organic hybrids have more well-known influences and properties and are well predicted by the numerous micromechanical models that have been developed for the purpose [10, 11].

An IPN is defined as a "polymer comprising two or more networks that are at least partially interlaced on a molecular scale but not covalently bonded to each other, and cannot be separated unless chemical bonds are broken," as per the International Union of Pure and Applied Chemistry (IUPAC) [12]. Therefore, mixtures of two or more polymers cannot be termed IPNs; rather, they would be called multicomponent polymer materials [13]. The main advantage of IPNs is their mechanical strength and stability. Also, IPNs provide a great opportunity to have two or more polymers with distinguishing properties; for example, one may be thermally responsive while the other provides mechanical strength. Thus, by modifying the interaction of the IPNs, a synergy can be achieved, which results in enhanced performance that surpasses that of either of the original polymers [14]. The related area of copolymers can be similar to IPNs but may have covalent grafting points connecting the two polymer chains.

In the past 10 years there have been nearly 4000 journal articles and patent documents on the subject of nanocomposites in which polyurethanes are either the polymer matrix or are one of several subject polymer mediums discussed. Of these nearly 4000 documents, over 660 are solely about nanofillers in a polyurethane matrix (Figure 13.1). However, of these 660 documents only 15 are patents, suggesting that

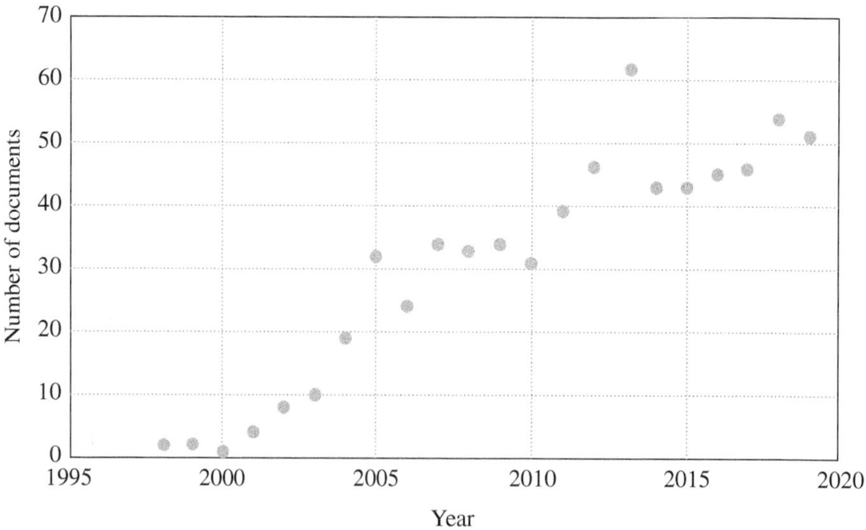


**FIGURE 13.1** Growth in total documents where the primary focus of the work is polyurethane nanocomposites, from 1995 to 2019.

interest at this point is largely fundamental and not yet commercial. Given the number of inorganic additives and applications, nanocomposites are probably worthy of their own book and constitute too large a subject for a chapter to do justice.

Polymeric copolymers and IPNs constitute an area of relatively low but consistent interest, both academically and industrially. Figure 13.2 shows that, with the exception of a burst of activity in the 1990s, the number of documents where the subject is copolymers and IPNs of polyurethanes is low but consistent. However, unlike nanocomposites, the distribution of patents to journal articles is about 450:70 (polyurethane-specific only), indicating a greater industrial perception of the commercial value of copolymers and IPNs of polyurethanes relative to nanocomposites. The distribution of copolymers and networked polymers combined with polyurethanes is given in Figure 13.3. The data in this instance do not cover copolymer polyols, which are covered extensively in Chapter 2 and include nearly 700 documents by themselves.

Figure 13.3 shows that almost 60% of polyurethane IPN and copolymer attention is directed toward acrylic and epoxy hybrids. This result is consistent with the relative importance that these networks are afforded in affecting the properties of coatings and adhesives (Figure 13.4).

For IPN materials it is important for the noncovalent networks to form nearly simultaneously [15–17]. For a copolymer it is important to establish a cross-linking chemistry that will provide a covalent cross-link between the two polymer backbones.

In the preparation of simultaneous networks to make IPNs, it is critical to control the polymerization rates and mixing of each forming network. Systems with mismatched polymerization rates are prone to uncontrolled polymer precipitation as the

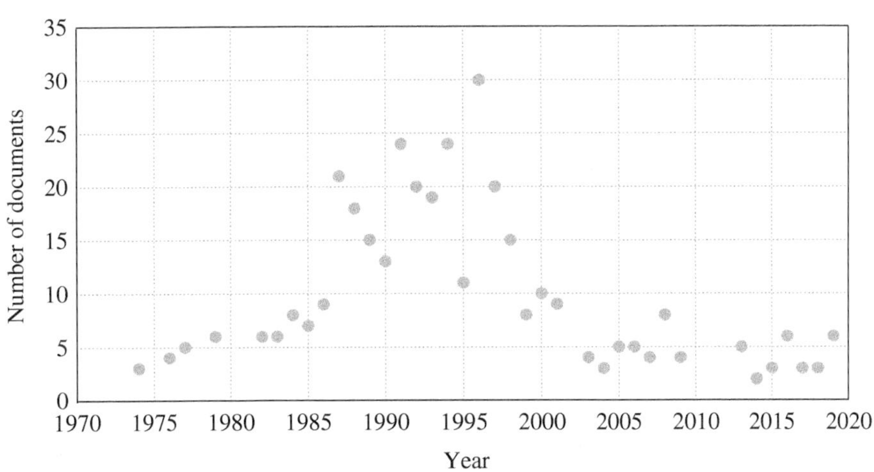


**FIGURE 13.2** Documents (both patent and open literature) where the subject is specifically on IPNs or copolymers of polyurethanes, from 1970 to 2019.

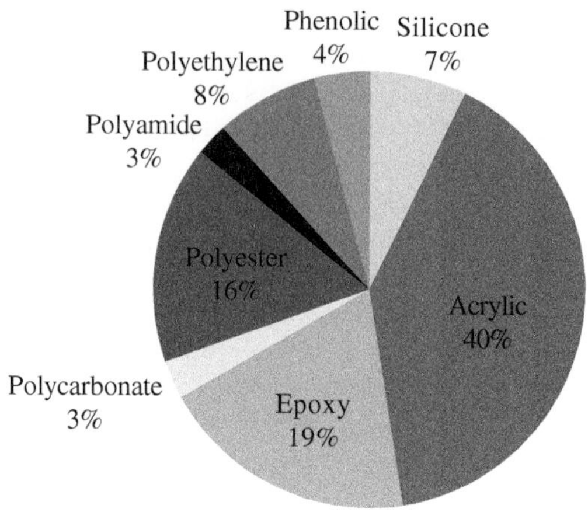


**FIGURE 13.3** Distribution of polymers networked or copolymerized with polyurethanes for the years shown in Figure 13.2.

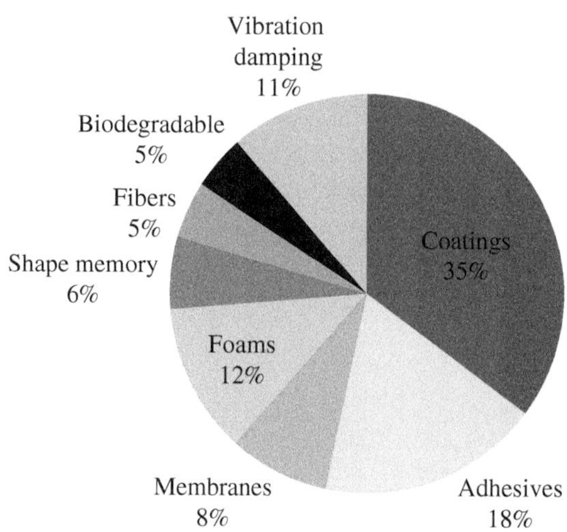


**FIGURE 13.4** Relative importance of polyurethane IPN and copolymer hybrid applications from 1990 to 2019.

two polymers are commonly not miscible [18, 19]. In practice, the kinetics of polymerization are difficult to control, so the success of the simultaneous polymerization approach rests on the careful selection of the monomers and reaction conditions [20].

In contrast to IPNs, preparing copolymer networks can be simplified. This is because there is not a requirement for the order of network preparation: it is only required that the two polymer domains be cross-linked or coreacted. This can, in principle, be achieved by reaction during backbone polymerization, but it may also be performed from polymer oligomers such as prepolymers or even with the fully formed polymers that will, upon cross-linking, become progressively more miscible as the molecular weight between cross-links becomes smaller. The rest of this chapter will cover methods and materials for hybridizing polyurethanes with other polymer backbones by covalent cross-linking or other reactions (such as transesterificiation/transurethanification). The work will in particular focus on network hybridizations with acrylics, epoxies, silicones, polyolefins, and networks built upon chain-segment scrambling.

## 13.2 POLYURETHANE–ACRYLATE HYBRIDS

As discussed in Chapters 2 and 10, acrylic hybrids with polyurethanes are well established in the coatings industry. Networks can be readily established by reaction of acrylate alcohols such as hydroxyethyl (meth)acrylate into the polyurethane reaction and subsequent reaction of the resulting terminal vinyl groups with acrylate or styrenic monomers (Table 13.1) [21]. As mentioned in Chapter 10, it is sometimes practiced to simply mix acrylate and polyurethane water-borne dispersions to achieve the benefits of both polymers in a single coating. However, it has been reported that this technique can lead to films of lower quality when the components are not appropriately matched for compatibility [22]. It has also been noted by numerous laboratories that the formation of the covalent hybrid system in which the different polymer chains are cross-linked to each other leads to superior mixing of component polymers and also, in many cases, superior final material properties [23, 24].

Most (but not all) work on polyurethane–acrylic hybrids and mixtures is for the purpose of coating applications. Much of this work is making water-borne latex systems. In the case of simple mixtures, most workers observe the presence of two distinct particle types characterized by their size distribution and appearance [25]. These two populations are ascribed to the respective polyurethane and acrylic particles. Their sizes, characteristics, and performance in application are strongly dependent on the individual polymer compositions and how interfacial properties between the polyurethane and acrylic particles have been designed in.

In copolymer hybrid systems the materials produced will again depend on the polymer components and the relative amount of the polyurethane and acrylic monomers. However, when copolymer hybrids are prepared, a "core–shell" morphology is often reported [26]. In this case a high concentration of one of the polymer backbones will be encapsulated in the center of the particle, while the other polymer backbone will predominate as a shell. By transmission electron microscopy this may appear as a two-phase particle with a central region ("core") presenting different contrast than the outer portion or shell [27].

In application, core–shell morphology means that, in coatings, the adhesive properties to the covered substrate, and the environmental stability provided by exposure to the exterior face, will be predominately determined by a single component of the

**TABLE 13.1 Cross-linking agents for copolymer hybridization of urethanes and acrylates**

| Agent | Structure |
|---|---|
| Hydroxyethyl acrylate | 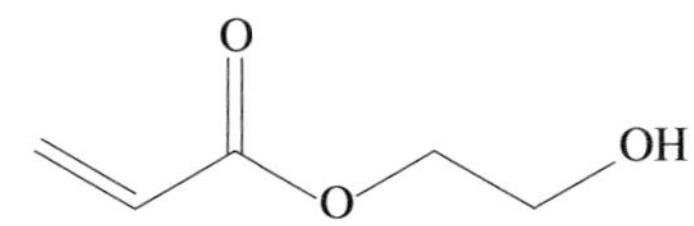 |
| Hydroxyethyl methacrylate | 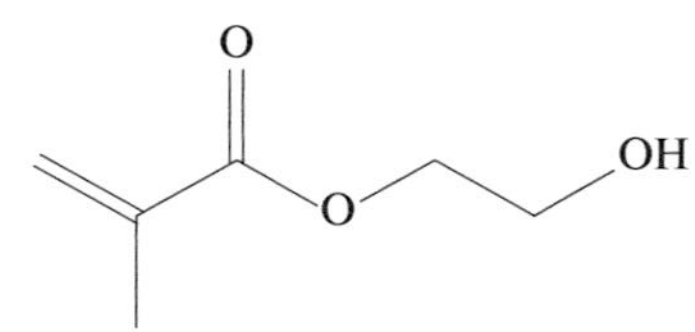 |
| 4-Hydroxybutyl acrylate | 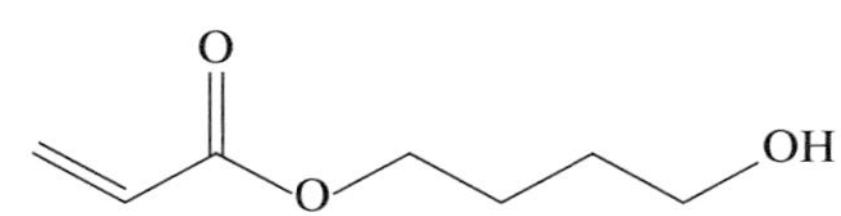 |
| Allyl glycerine | 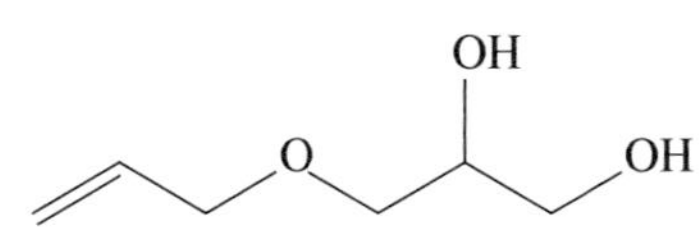 |

| | |
|---|---|
| Trimethylolpropane diacrylate | 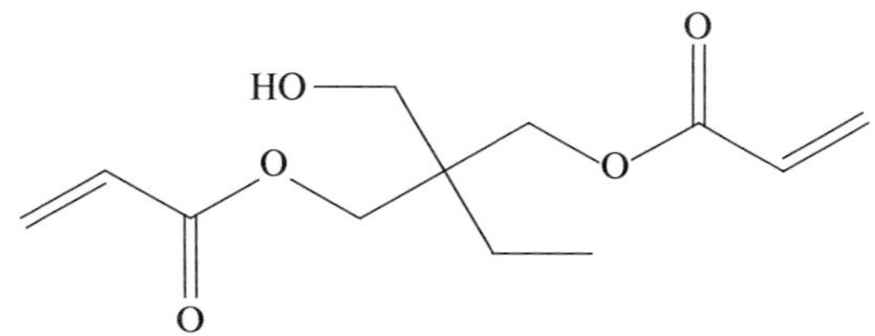 |
| Pentaerythritol triacrylate | 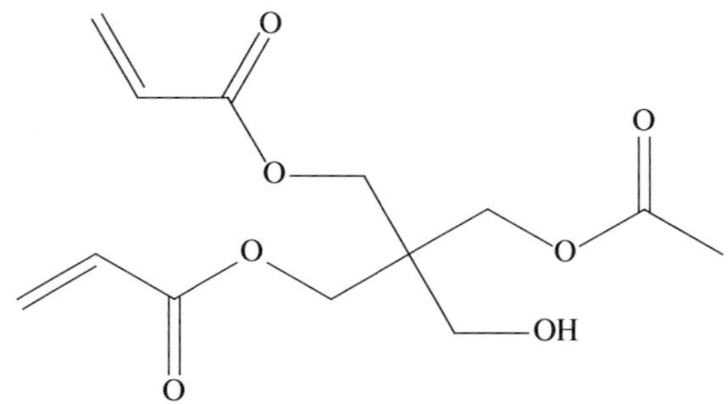 |
| 3-Phenoxy-2-hydroxy propyl methacrylate | 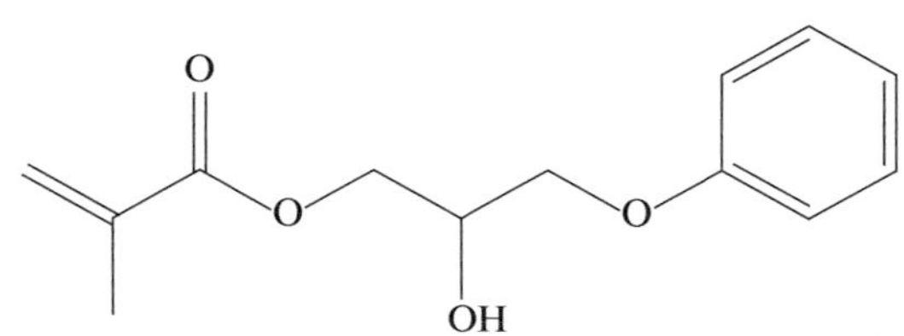 |

(*Continued*)

**TABLE 13.1 (Continued)**

| Agent | Structure |
| --- | --- |
| Bisphenol A-glycidylmethacrylate | 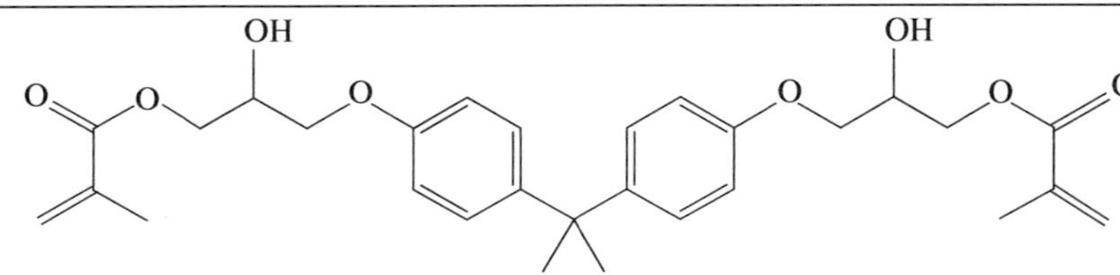 |
| 2-Ethylhexyl acrylate | 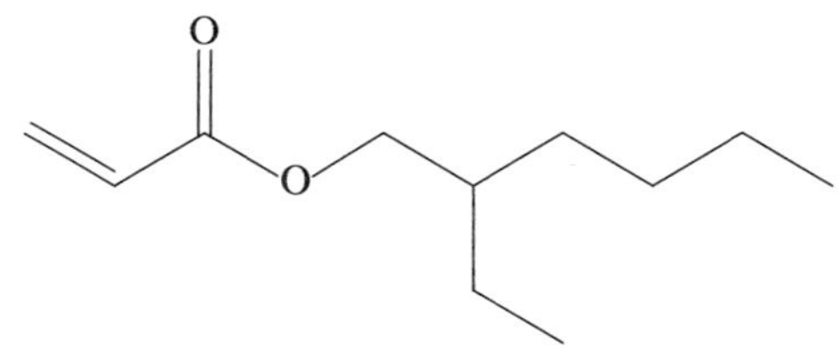 |
| 2-Hydroxypropylmethacrylamide | 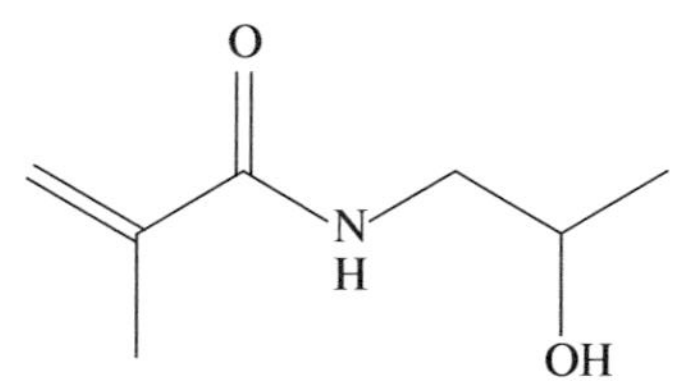 |

These are examples, but many other structures can be imagined and employed. Hydroxy ethyl and butyl methacrylates are very commonly used, but all have been employed depending on the desired performance.

hybrid – the shell [22]. At the same time, the core property can influence many mechanical properties of the coating. In aqueous dispersions, the polyurethane dispersion is made water stable by incorporation of dimethylol propionic acid (DMPA; see Chapter 10), followed by neutralization of the acid with an amine. The very high hydrophilicity of this unit usually assures that the shell of a core–shell urethane–acrylate copolymer will be urethane. However, several compositional details are known to affect this. For instance, a high ratio of acrylic to urethane in the composite is known to decrease the core–shell attributes in the latex, or can create two populations of particles consisting of core–shell hybrids and pure acrylic particles. Also, increasing the amount of cross-linking between the urethane and acrylic chains can inhibit the ability of the chains to sufficiently partition within the individual particles [23]. An interesting observation is that increasing the amount of DMPA in the urethane component of the hybrid reduces particle size since its presence favors increased water interfacial interactions at the ionic site. However, it has also been observed that beyond a certain amount of DMPA (that amount dependent on the overall polymer design, surfactancy, solvent, and other latex formulation aspects) the particle size can begin to increase. This has been postulated to be caused by enhanced particle coalescence owing to increased indifference to surfactants as the polyurethane chains become more hydrophilic [24].

## 13.3 URETHANE–EPOXY HYBRIDS

Polyurethane and epoxy polymer hybrids are likely materials to seek the benefits of hybridization since the positive features of polyurethanes are ideally suited to compensate for the negative attributes of epoxies (Table 13.2). The problem with formation of urethane–epoxy hybrids is similar to the formation of all hybrids – how to compatibilize the fundamentally immiscible polymer backbones? An approach described in Section 13.2 for polyurethane and acrylate hybrids is to include functionality of one backbone to react with functionality of the other. For epoxies and polyurethanes this can be challenging since many components that can react with epoxies also react into urethanes – for instance, amines. However, the reaction rate of amines with isocyanates is orders of magnitude faster than amines with epoxies. Thus, coreacting isocyanate and epoxy functional materials with amines can provide macrophase-separated materials with a low interfacial cross-link density. An alternative is to react polyurethane prepolymers with glycidol (Figure 13.5). This can create an epoxy reactive end group at the end of a polyurethane backbone and, in an approach similar to that used for polyurethane–polyacrylate hybrids, create a self-compatibilizing and designable template for hybrid synthesis. This approach has been reported [28]; however, it has also been reported by several laboratories that the glycidol end cap undergoes a competitive rearrangement to form a 3,4-oxazolidinone after initial formation of the carbamate [29, 30].

**TABLE 13.2 Comparison of material performance attributes of polyurethanes and epoxies**

| Attribute | Relative strength of epoxy | Relative strength of polyurethane |
|---|---|---|
| Crack growth inhibition | ↓ | ↑ |
| Adhesion | ↑ | ↓ |
| Peel strength | ↓ | ↑ |
| Lap-shear strength | ↑ | ↓ |
| Elongation | ↓ | ↑ |
| Chemical resistance | ↑ | ↓ |
| Thermal shock resistance | ↓ | ↑ |
| Thermal stability | ↑ | ↓ |

There is wide variation of performance across the two families, but, for modification of adhesives, this would be the basis for pursuing a hybrid strategy.

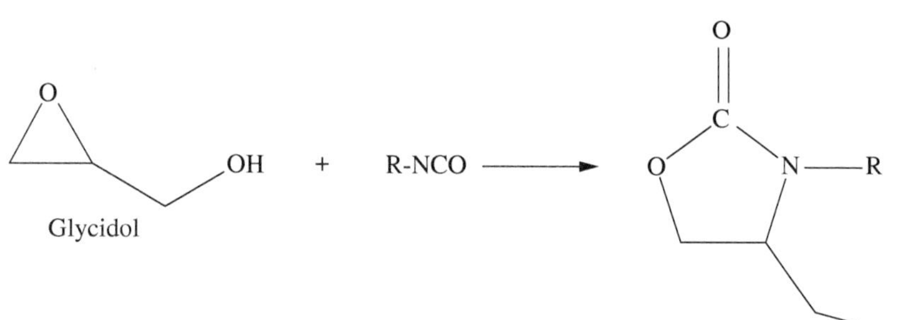


**FIGURE 13.5** Possible rearrangement during utilization of glycidol to compatibilize urethane prepolymer for reactions to form epoxy hybrids.

An alternative route to compatabilize epoxy and urethane networks is to advance a polyepoxy with a polyphenol. This results in production of secondary hydroxyls along a highly aromatic epoxy backbone (Figure 13.6). The end groups can be designed to be epoxy by careful control of stoichiometry, leaving an isocyanate reactive oxirane to form the stable oxazolidinone [31] and the secondary hydroxyls along the backbone to cross-link with isocyanates participating in the formation of the polyurethane phase. Alternatively, the phenol groups can be terminal on the chains

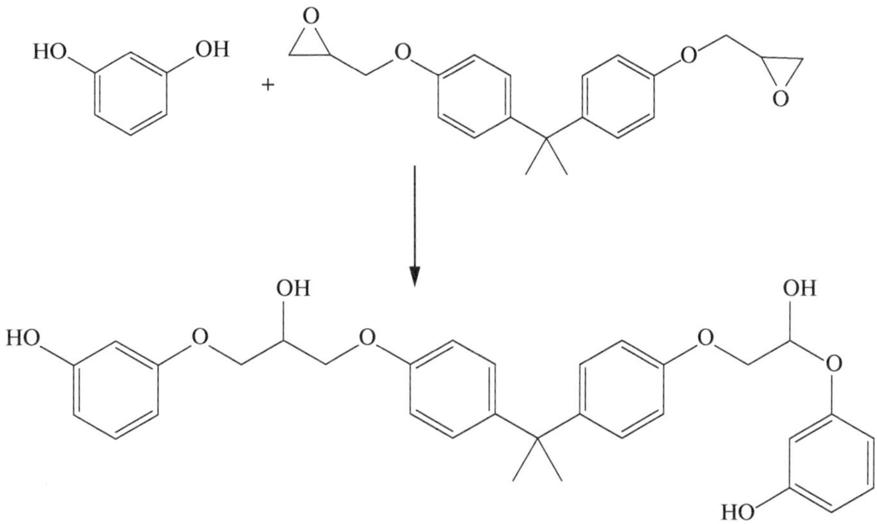


**FIGURE 13.6** Reaction of diglycidyl ether of bisphenol A with a diphenolic for reaction into a urethane network. Reaction would proceed through the backbone secondary hydroxyls with isocyanates during urethane polymerization.

and form the phenyl urethane end group, which is relatively less stable and prone to reversion back to phenol and isocyanate (see Chapter 10). This is why phenol is often termed a blocking group for protecting the isocyanate functionality rather than a reactant.

A different approach to cross-link epoxy and urethane networks has been to slow the kinetics of isocyanate with amine reaction to a rate more compatible with the reaction of amine with epoxy. Improved kinetics were achieved by using a latent source of amines (ketamine), which upon encountering water produces a polyamine (Figure 13.7). This reactant could then diffuse to an epoxy or isocyanate reaction site and force growing chains to form a cross-linked network. In principle, this gives either epoxy or isocyanate functionality an opportunity to react with the primary amine and there are reports of this working to produce finely dispersed epoxy in a urethane matrix [32]. In these reactions it may be that the by-product ketone has an effect on the product properties [33].

In addition to these methods of direct hybridization of urethane and epoxy backbones, there are numerous other approaches that first modify the urethane or epoxy backbone separately to make their copolymerization more facile. Such approaches include taking an acrylate-functionalized urethane and reacting the available epoxy and acrylate with polyamines to form an amino ester linkage, via Michael addition with the acrylate, and amino alcohol linkage with the epoxy [34]. Others have functionalized urethane and epoxy backbones separately with silicone functional amines and then linked the two separate backbones though silicone chemistry [28].

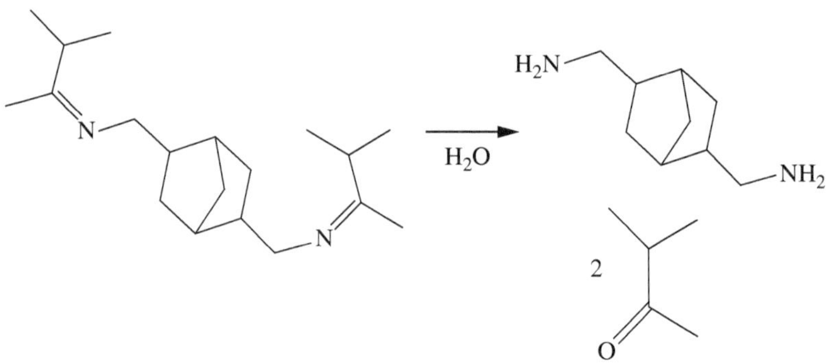


**FIGURE 13.7** Hydrolysis of a diketamine to produce the latent amine, which can then cross-link into polyurethane and epoxy networks.

These materials could also be looked at as derivative hybrids since the urethane backbones must first be prehybridized with acrylates or silicones to prepare them for the desired hybridization that hopefully melds the desired physical properties.

## 13.4 URETHANE–SILICONE HYBRIDS

### 13.4.1 Silicone-Modified Prepolymers

Polyurethane–silicone hybrids have received somewhat less attention through the years than other copolymer systems (Figure 13.3), but recently this system has been the most attractive industrially. The commercial success is a result of the silicone–urethane hybrids finding a value that customers can observe, especially in the industrial sealants market. Individually, polyurethane and silicone sealants are considered commercially acceptable materials in numerous applications. The trend toward their hybridization is to exploit the desirable properties of both these systems. Table 13.3 is a performance scale of the unhybridized material sealant properties and comparisons with hybrid materials. Of course, in this kind of comparison, the examples chosen for contrast can be selected to be consistent with the hybrid value proposition. Regardless, the market has decided that polyurethane–silicone hybrids exhibit superior adhesion relative to polyurethanes and superior paintability relative to silicones. These undeniable differential properties are sufficient to create a product category and stimulate significant patent activity.

One aspect that commercially available silicone–urethane prepolymer hybrids have in common is that they are oligomeric, and after application become fully cured polymers following exposure to moisture. An early process for making silicone–urethane hybrids was filed by Kaneka Corporation [35]. The process converts the hydroxyl end groups of polyalkylene glycols to allyl end groups following creation of the alkoxide (Figure 13.8). The allyl end groups are subsequently hydrosilylated

**TABLE 13.3 Property comparison (10 high to 0 low) of polyurethane, silicone, and urethane–silicone hybrid sealants**

| Property | Polyurethane | Silicone | Hybrid |
|---|---|---|---|
| Health and safety | 5 | 10 | 10 |
| Bubble producing | 3 | 10 | 10 |
| Ease of application | 10 | 10 | 10 |
| Rheology control | 10 | 10 | 10 |
| Cure speed | 10 | 10 | 10 |
| Shelf life | 10 | 10 | 10 |
| Weatherability | 7 | 10 | 8 |
| Adhesion | 8 | 10 | 10 |
| Thermal stability | 10 | 10 | 10 |
| Soil resistance | 10 | 5 | 10 |
| Paintability | 10 | 2 | 10 |
| Tear resistance | 10 | 4 | 8 |
| Price | 10 | 5 | 5 |

with trimethoxyhydrosilane using standard platinum catalysis. The trialkoxy silane end groups are readily cured by contact with atmospheric moisture in a reaction that can occur without catalysis or that occurs even faster with Lewis acid catalysts (Figure 13.9). The process to make this material is not without complications, including a significant amount of salt production and a broad product molecular weight distribution. Kaneka subsequently made many modifications on backbone structure and process [36, 37].

The Kaneka backbone is widely considered a polyurethane–silicone hybrid; however, it is notable that there are in fact no urethane linkages in this sealant. The simplified structure results in a material with lower intrinsic viscosity, which can be subsequently formulated with dry inorganic fillers, such as $CaCO_3$, to alter physical properties while maintaining easy gunnability. However, it is a relatively simple matter to begin with a polyol and then react with an isocyanato trimethoxy silane (Figure 13.10) or methyl dimethoxy isocyanato silane to quickly and quantitatively convert all the terminal isocyanates of the prepolymer to hydrolyzable methoxy silane end groups. The dimethoxy variation slows the water reaction down and can improve shelf life [38]. This approach has several advantages, and the true urethane hybrid structure afforded these structures improves material properties. However, this approach also has a relatively higher price to produce because of the isocyanate reagent; it also suffers from issues with high-molecular-weight dispersity and with short shelf life owing to very high reactivity in the presence of moisture. This can be ameliorated by the use of methyl dimethoxy silane end groups [39].

It is also possible to take the commercially available, or easily prepared, isocyanate-terminated polyurethane prepolymer and react with aminotrimethoxy silane or aminomethyldimethoxy silane [40].

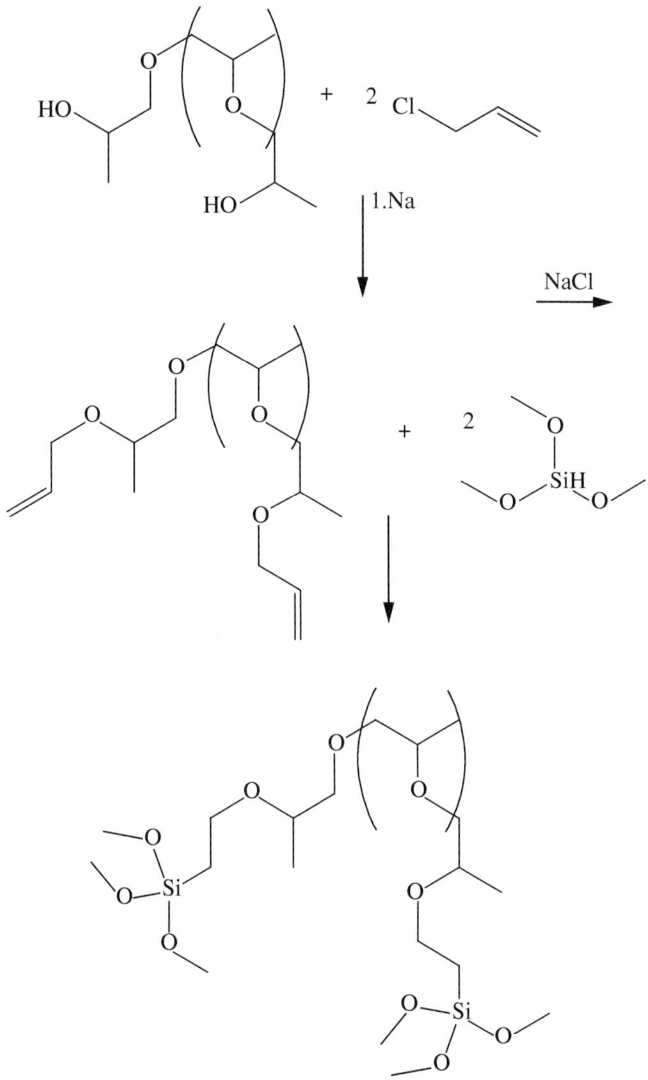


**FIGURE 13.8** Reaction pathway to produce methoxysilanes grafted onto a polyol to form a moisture-curable sealant. The final structure is analogous to a urethane prepolymer, which is also capable of moisture cure.

### 13.4.2 Urethane–Silicone Hybrids Produced Using Diblock Compatibilizers

A simplistic approach is to blend the polyurethane and siloxane polymers. This has been shown to form macrophase-separated domains independent of whether the polyurethane or polydimethylsiloxane (PDMS) is the majority phase. Such blends are qualitatively similar in appearance and scale to the PDMS acrylate mixtures shown

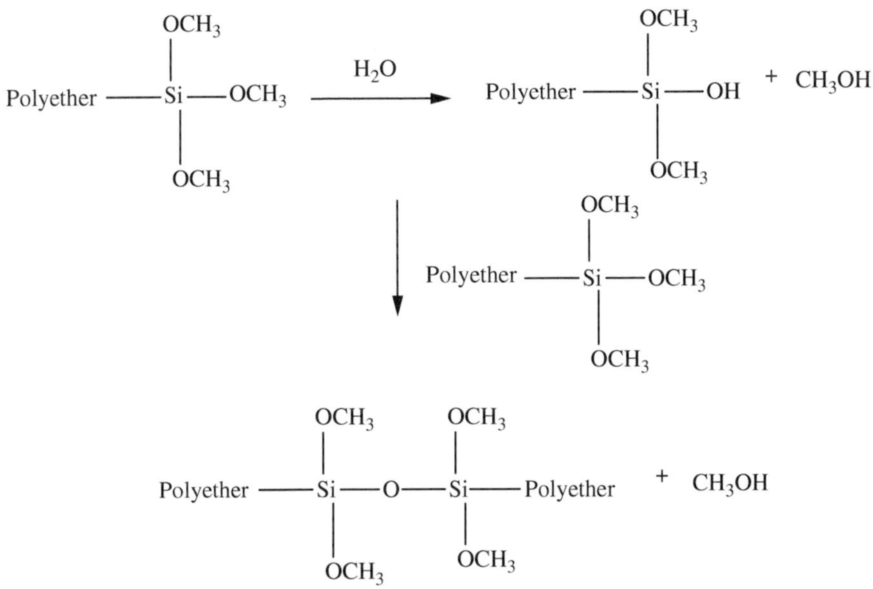


**FIGURE 13.9** Hydrolysis of alkoxysilanes for cross-linked network formation.

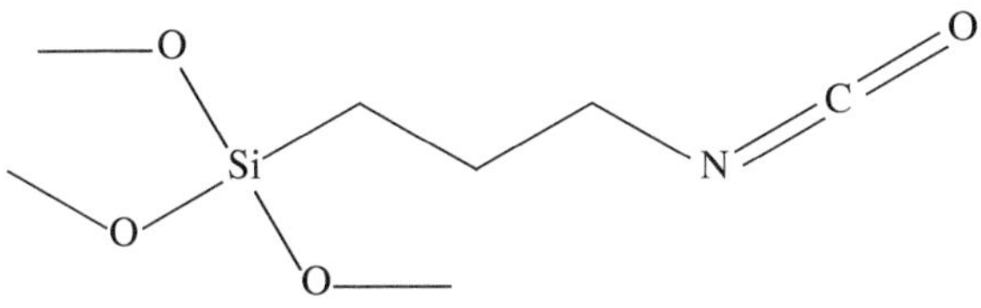


**FIGURE 13.10** An isocyanato-alkoxysilane for forming a polyurethane–silicone hybrid.

in Figure 4.3. Diblock copolymers of two immiscible phases have been shown to reduce surface tension between the individual polymers and improve mixing. Diblock copolymer for compatibilizing polyurethane and polysiloxanes can be produced beginning with reaction of polypropylene oxide end-capped on one end with a methyl group and on the other end with a primary amine. At low temperature (–40 °C) an aliphatic isocyanate is added in a standard prepolymer process (see Chapter 2), as shown in Figure 13.11. To the prepolymer produced in this prior step is added difunctional hydroxyl-terminated PDMS. The use of the block copolymer with a number average molecular weight ($M_n$) of ~7000 Da dispersed the phase domain size by an order of magnitude and exhibited good interphase adhesion (by scanning electron microscopy) at 2% incorporation and erased apparent phase boundaries at 5% incorporation. Maximum in tensile modulus for the polymer blend was observed at 2% incorporation [41].

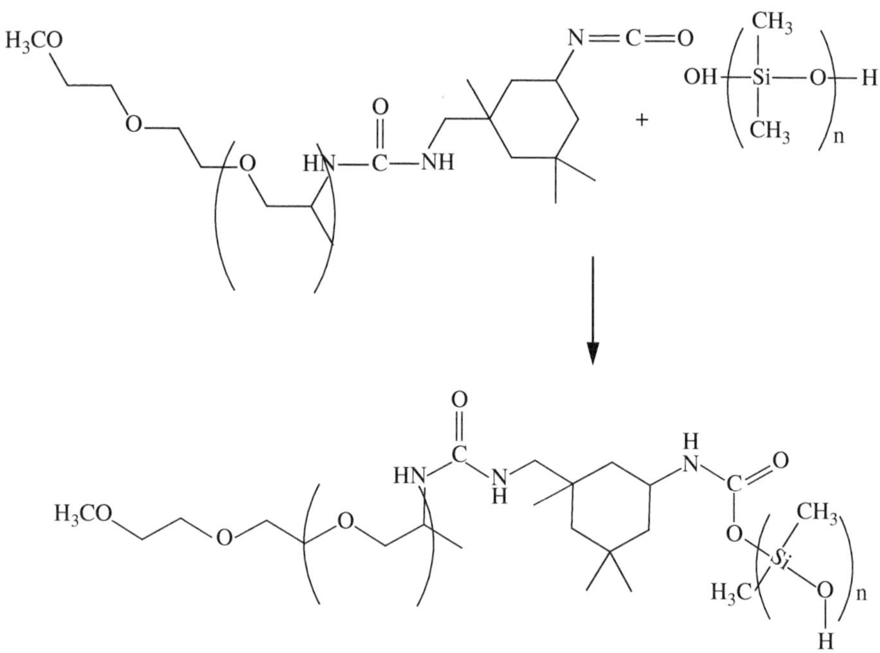


**FIGURE 13.11** Reaction scheme for preparation of a polyurethane–PDMS diblock copolymer used for compatibilizing a polyurethane–polysiloxane mixture.

### 13.4.3 Hybrids Employing Covalent and Hydrogen-Bonded Cross-Links

While preparation of a hybrid polymer using diblocks to compatibilize the otherwise immiscible phases is shown to form a new bulk phase, a more conventional copolymer hybrid has been realized through the use of polymer backbone design. An example of such a polymer design can be through the use of carboxyl-terminated PDMS followed by reaction of polyamines. An example [42] is provided in Figure 13.12, but the design attributes are clearly not limited to what is shown. A novel hybrid is formed by including durable chemical cross-links thorough diisocyanates bonding across the secondary amines and also a system of physical cross-links to hydrogen bond across chains using monoisocyanates such as *m*-toluylisocyanate. It was shown that this system could provide some self-healing character to an elastomer as hydrogen bond recovery can regain significant levels of properties existing before damage

### 13.4.4 Polyurethane Hybridization with Polyhedral Oligomeric Silsesquixanes

The last urethane–silicone copolymer hybrid to be covered here incorporates functionalized polyhedral oligomeric silsesquixanes (POSSs) into a polyurethane matrix. POSSs are cage-like structures where each Si sits at a tetrahedral vertex

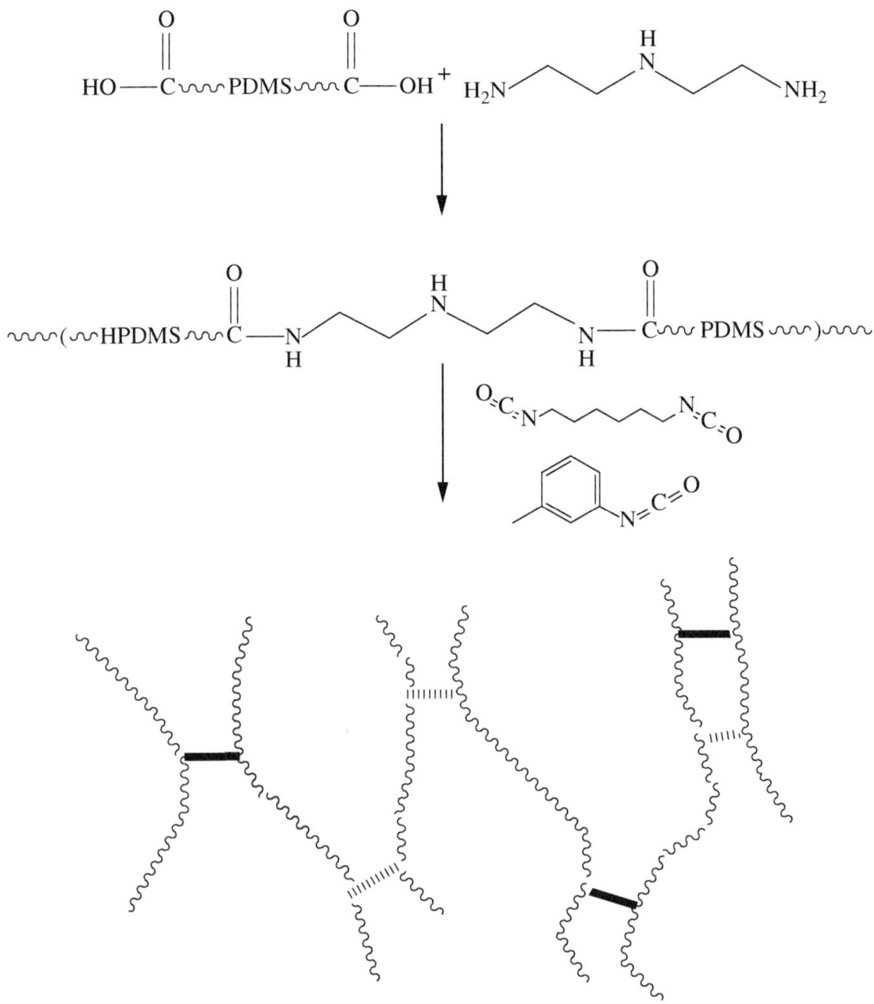


**FIGURE 13.12** A reaction scheme for making a polyurethane–PDMS hybrid using polyisocyanates to cross-link functionalized PDMS for making a cross-linked network with covalent and hydrogen bond cross-links.

connected to three oxygens and an organic moiety. The size of POSS molecules is of the order of a few nanometers, which provides an interesting intermediate between inorganic fillers and copolymer networks. The R moiety may be H or an organic moiety, including one that will provide a tethering site to a polyurethane matrix. In the absence of such tethering, POSS units can phase separate from a polymer matrix much like any PDMS might [43]. Figure 13.13 presents example of tethers to a polyurethane network.

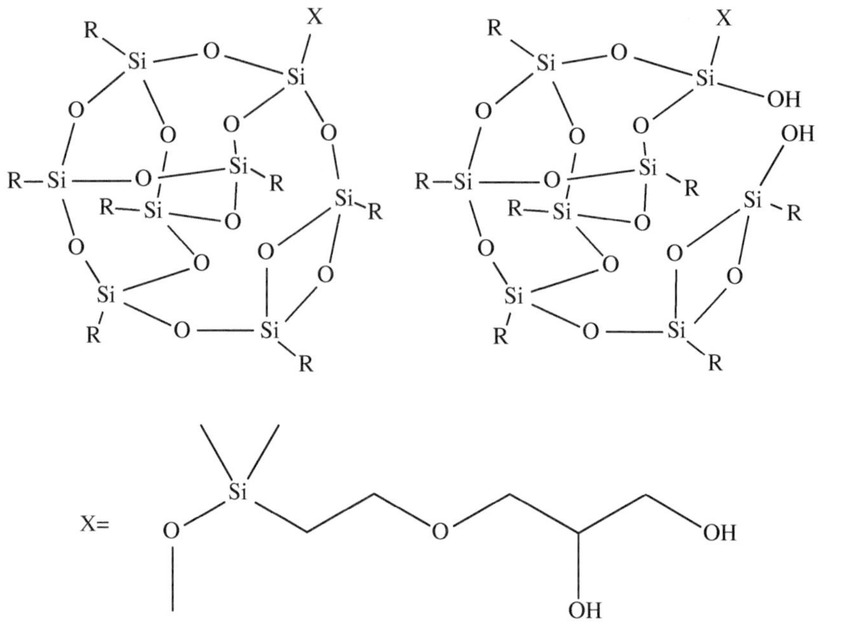


**FIGURE 13.13** An illustration of a POSS structure and examples of the kinds of functionalization that would allow tethering to a polyurethane backbone.

Prior work has shown that POSS materials can perform a similar function to styrene–acrylonitrile copolymer polyol particles (Chapter 2), enhancing phase separation of polyurethane hard and soft segments and acting as a physical reinforcement [44, 45]. In addition, it has been shown that the presence of POSSs in a polyurethane network can provide some flame retardance by both improving char formation and modifying the composition of gases during burning [46].

## 13.5 POLYURETHANE–POLYOLEFIN HYBRIDS

Polyurethanes and polyolefins might be considered among the most difficult blend materials to hybridize since the solubility parameters are relatively far apart (about 16 J/cm$^3$ for polyethylene and about 21 J/cm$^3$ for a polyurethane) and the drive to phase separate quite high. For hybridization there is very little on a polyolefin for a polyurethane to bond to or even interact with. Certainly the formation of a polyurethane–polyolefin IPN would be a significant challenge since the polymerization processes are incompatible with one another, and the building blocks for the polymers are immiscible. In contrast to IPNs, copolymer hybrids have been developed by employing functionalized polyethylene with the ability to covalently bond to elements of the polyurethane. The most notable among these functionalized polyolefins are those that

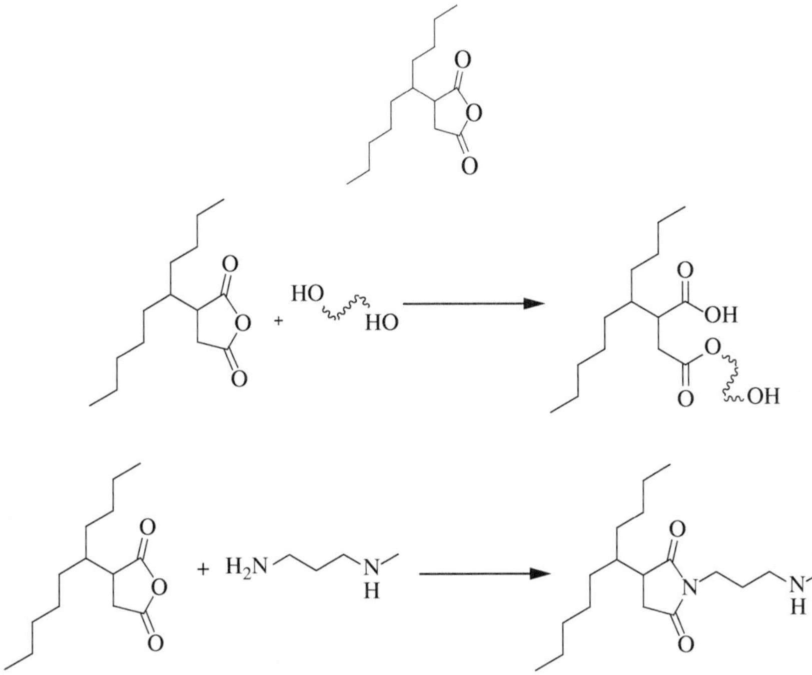


**FIGURE 13.14** Structure of maleic anhydride-modified polyethylene and its reactions with polyol to form a pendant ester/acid and a polyamine to form an imide with a pendant secondary amine.

are grafted with maleic anhydride (Figure 13.14). Anhydride-functionalized polyolefin is prepared commercially in a twin-screw extruder by mixing polyethylene with maleic anhydride and a peroxide to perform extraction of a backbone hydrogen and allow addition of the anhydride. Anhydrides react very well with water, alcohols, and amines. In the presence of water the anhydride will hydrolyze to form two carboxylic acids. With an alcohol the anhydride will open to form an ester on one branch and a carboxylic acid on the other. In the presence of a primary amine, an imide will form. If a polyamine with a primary and a secondary amine is used the primary amine will react exclusively with the anhydride [47]. In both reactions with alcohols and a polyamine as shown, a reactive end group remains to tether the olefin to the polyurethane backbone. The reaction with polyols has been shown to produce stable olefin dispersions up to 25% solids content and to produce significant effects on polyurethane elastomer properties [48]. When functionalized polyolefin–polyol copolymers are reacted into a polyurethane matrix a very finely dispersed polyolefin phase is observed by transmission electron microscopy (see Chapter 5) with excellent phase adhesion (Figure 13.15). Alternatively, the anhydride-modified polyolefin can be added directly to an already formed thermoplastic polyurethane in a dispersive mixer to react into the

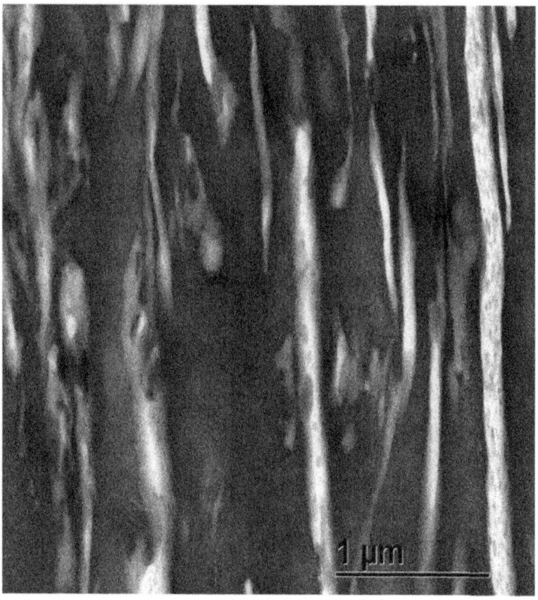


**FIGURE 13.15** Transmission electron micrograph of maleic anhydride polyethylene copolymerized with a polyol dispersed into a thermoplastic polyurethane. The white zones are polyethylene and the intimate nature of the dispersion is revealed by the intermingling of the phases. Apparent orientation of the polyolefin phase is an artifact of the molding process.

polyurethane melt as thermal reversion of the urethane bond produces alcohol groups capable of bonding directly to the anhydride-modified polyolefin. For this reaction to proceed, it is imperative for the maleic anhydride-grafted polyolefin to be in the anhydride form and not have hydrolyzed to the diacid form as a result of extended exposure to atmospheric moisture.

## 13.6 HYBRIDIZATION VIA TRANSURETHANIFICATION

The last hybridization technique covered will be copolymer hybridization of polyurethane with polyester or polycarbonates. This can be achieved by these backbones' ability to revert back to reactive groups at elevated temperatures. The reactive groups allow for scrambling of the polymer segments. In this case, the amount of phase dispersion will depend on the backbones, the temperature of hybridization, and the length of time the scrambling reactions are allowed to proceed. This scrambling reaction is depicted in Figure 13.16, which shows how the alcohol groups made in polyurethane synthesis and polyester synthesis can exchange between backbones and create essentially a new polymer. This hybridization method can create dispersions fine enough (Figure 13.17) to result in optically transparent materials exhibiting properties that are not predicted based on two-phase composite modeling [49].

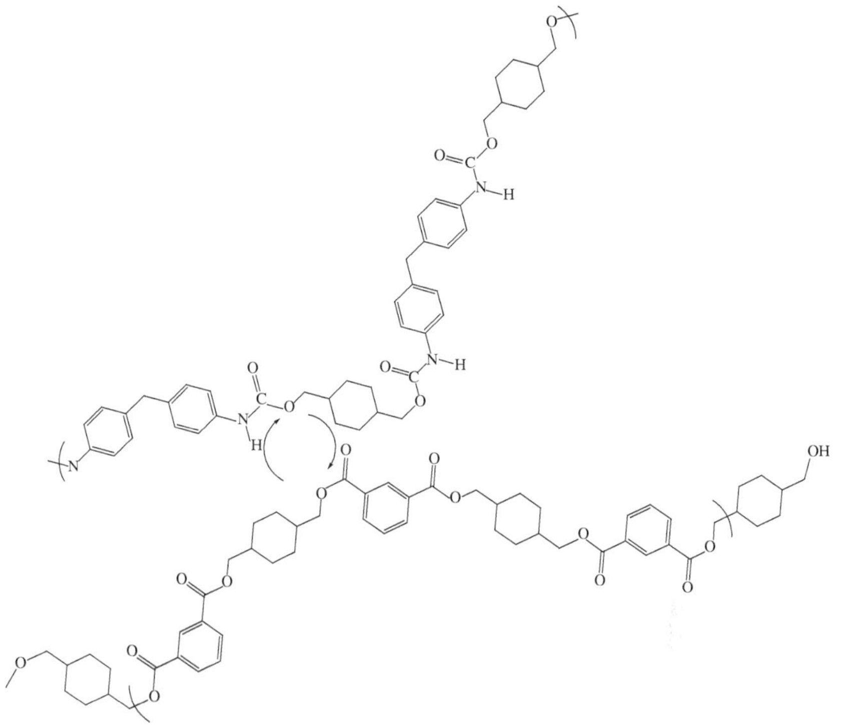


**FIGURE 13.16** A polyurethane and polyester backbone in position to facilitate a scrambling reaction at elevated temperatures.

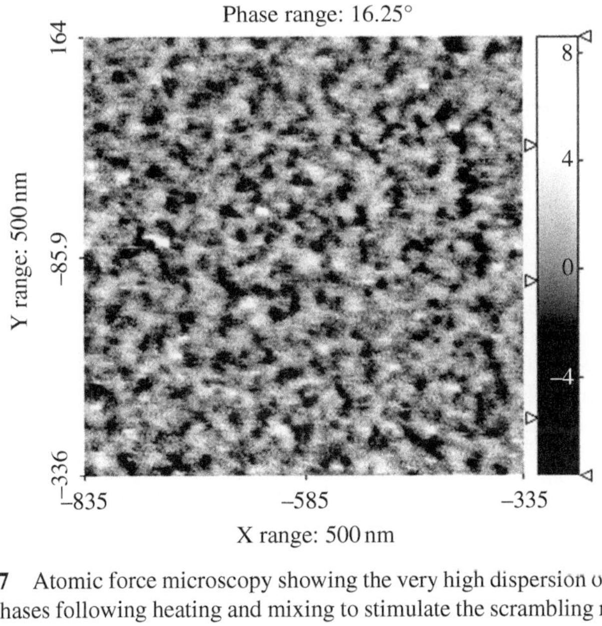


**FIGURE 13.17** Atomic force microscopy showing the very high dispersion of polyester and polyurethane phases following heating and mixing to stimulate the scrambling reaction shown in Figure 13.16.

## REFERENCES

[1] D Klempner, L Sperling, and L Utracki, *Interpenetrating Polymer Networks*, ACS Publications, Washington, DC, 1994.
[2] C Roland, "Interpenetrating polymer networks: structure and mechanical behavior," in *Encyclopedia of Polymeric Nanomaterials*, Springer Verlag, Heidelberg, Germany, 1993, 1–10.
[3] H Lu, M Wang, X Chen, B Lin, and H Yang, *Journal of the American Chemical Society*, **141**, (2019), 14364–14369.
[4] P Martins, M Silva, S Reis, N Pereira, H Amorin, and S Mendez, *Polymer Chemistry*, **9**, (2017), 62.
[5] X Liu, G Ding, X Zhou, S Li, W He, J Chai, C Pang, Z Liu, G Cui, *Journal of Materials Chemistry A* **5**, (2017), 11124–11130.
[6] B Chan, Y Qi, S Liow, and X Loh, *Polymer RCS Advances*, **6**, (2016), 34946–34954.
[7] M Canalp and W Binder, *RSC Advances*, **10**, (2020), 1287–1295.
[8] Y Wang and J Wang, *Journal of Applied Polymer Science*, **136**, (2019), 47481.
[9] B Hollo, I Ristic, J Budinski, S Calic, I Szilagyi, and K Szecsenyi, *Journal of Thermal Analysis and Calorimetry*, **132**, (2018), 215–224.
[10] E Thostenson and T Chao, *Journal of Physics D: Applied Physics*, **36**, (2003), 573.
[11] V Tac and E Gurses, *Journal of Composite Materials*, **53**, (2019), 4337–4348.
[12] International Union of Pure and Applied Chemistry, *Compendium of Chemical Terminology, 2nd ed. (the "Gold Book"), Compiled by AD McNaught and A Wilkinson*, Blackwell Scientific Publications, Oxford, UK 1997.
[13] G Zaikov, N Lekishvili, and Y Medvedevskikh, *Multicomponent Polymeric Materials: From Introduction to Application*, CRC Press, Boca Raton, Fl, 2013.
[14] J Kim, S Thomas, and P Saha, *Multicomponent Polymeric Materials*, Springer, Dordrecht, the Netherlands, 2016.
[15] M Sonnenschein, S Webb, and B Wendt, *International Journal of Adhesion and Adhesives*, **28**, (2008), 126–134.
[16] M Sonnenschein, S Webb, R Cieslinski, and B Wendt, *Journal of Polymer Science Part A: Polymer Chemistry*, **45**, (2007), 989–998.
[17] J Turner and Y Cheng, *Macromolecules*, **33**, (2000), 3714–3718.
[18] D Paul, "Polymer blends, phase behavior and property relationships," in *Multicomponent Polymer Materials*, D Paul and K Sperling, Eds., ACS Publications, Washington, DC, 1985, 3–19.
[19] R Koningsveld and W Macknight, *Polymer International*, **44**, (1997), 356–364.
[20] A Pojman, W Elcan, A Khan, and L Mathias, *Journal of Polymer Science Part A: Polymer Chemistry*, **35**, (1997), 227–230.
[21] T Hu, T Tian, J Pojman, and. Chen, *Journal of Polymer Science Part A: Polymer Chemistry*, **44**, (2006), 3018–3024.
[22] B Li, H Liu, B Xu, T Wu, P Wang, X Yu, and Y Yu, *Progress in Organic Coatings*, **112**, (2017), 263–269.
[23] P Peruzzo, P Anbinder, O Pardini, J Vega, C Costa, F Galembeck, and J Amalvy, *Progress in Organic Coatings*, **73**, (2011), 429–437.
[24] L Wu, B You, and D Li, *Journal of Applied Polymer Science*, **84**, (2002), 1620–1628.
[25] R Brown, D Fortier, R Coogan, and M Reeve, *Progress in Organic Coatings*, **52**, (2005), 73–84.

[26] J Zhang, X Li, X Shi, M Hua, X Zhou, and X Wang, *Progress in Natural Science: Materials International*, **22**, (2012), 71–78.
[27] M Hirose, J Zhou, and F Kadowaki, *Progress in Organic Coatings*, **31**, (1997), 157–169.
[28] L Byczynski, M Dutkiewicz, and H Maciejewski, *Progress in Organic Coatings*, **112**, (2017), 118–126.
[29] T Baronsky, C Beattie, R Harrington, R Irfan, M North, J Osende, and C Young, *ACS Catalysis*, **3**, (2013), 790–797.
[30] J Routier, M Calancea, M David, Y Vallée, J-N Denis, *European Journal of Organic Chemistry*, 2008, (**2008**), 5687–5691.
[31] B Krishnan, L Grier, W Heath, A Colson, M Gelfer, and G Jialanella, *Polyurethane-acrylate-epoxy adhesive*. Assignee Dow Global Technologies. WO 2015/164031 A1. 2015.
[32] H Okihira, N Iwamoto, M Ochi, and H Takeyama *Journal of Polymer Science Part B: Polymer Physics*, **42**, (2004), 1137–1144.
[33] H Okuhira, T Kii, M Ochi, and H Takeyama, *Journal of Adhesion Science and Technology*, **18**, (2004), 205–211.
[34] M Sonnenschein, K Patankar, J Virgili, T Collins, and B Wendt, *Polymer*, **179**, (2019), 121663.
[35] K Isayama and I Hitano, *Vulcanizable silylether terminated polyether*. Assignee Kanegafuchi Kagaku Kogyo KK. US patent 3971751. 1975.
[36] S Inoue, T Aida, M Honma, and K Isayama, *Process for preparing polyalkylene oxide having unsaturated end groups*. Assignee Kanegafuchi Kagaku Kogyo KK. US patent 4774356. 1985.
[37] H Iwakiri, M Fujita, and T Hasegawa, *Curable composition*. Assignee Kanegafuchi Kagaku Kogyo KK. US patent 5286780. 1992.
[38] B Arkles, J Steinmetz, J Zazyczny, and P Mehta, *Journal of Adhesion Science and Technology*, **6**, (1992), 193–206.
[39] R Griswold, *Silylated polyurethane compositions and adhesives therefrom*. Assignee Momentive Performance Materials. US patent 8247514. 2012.
[40] S Mohamed, S Jaisankar, M Rajavelu, and A Mandal, *Journal of Physical Chemistry B*, **117**, (2013), 2682–2694.
[41] P Knaub, Y Camberlin, and J Gerard, *Polymer*, **28**, (1988), 1365–1377.
[42] Y You, W Huang, A Zhang, and Y Lin, *Journal of Polymer Science Part A: Polymer Chemistry*, **54**, (2016), 3760–3768.
[43] K Raftopoulos, C Pandis, L Aprkis, P Pissis, B Janowski, K Pielchowski, and J Jaczewska, *Polymer*, **51**, (2010), 709–718.
[44] BX Fu, BS Hsiao, S Pagola, P Stephens, H White, M Rafailovich, J Sokolov, PT Mather, HG Jeon, S Phillips, et al., *Polymer*, **42**, (2001), 599–611.
[45] V Bliznyak, T Tereschenko, M Gumenna, Y Gomza, A Shevchuk, N Klimenko, and V Shevshenko, *Polymer*, **49**, (2008), 2298–2305.
[46] J Pagacz, E Hebda, B Janowski, D Sternik, M Janicia, and K Pielchowski, *Polymer Degradation and Stability*, **149**, (2018), 129–142.
[47] H Silvis, S Hahn, D Pawlowski, P Ansems, L Mergenhagen, H Lakrout, A Batra, L Weaver, L Effler, and K Oswald, *Functionalized olefin polymer, compositions and articles prepared therefrom, and methods for making the same*. Assignee Dow Global Technologies LLC. US patent 8981013. 2015.
[48] M Sonnenschein, J Virgili, D Arian, and B Wendt, *Polymer*, **76**, (2015), 80–88.
[49] M Sonnenschein, C Boyer, D Brune, B Wendt, G Meyers, and B Landes, *Macromolecules*, **44**, (2011), 865–871.

# 14

# RECYCLING OF POLYURETHANES

## 14.1 INTRODUCTION

Modern drilling and recovery methods have assured plentiful and reliable petrochemical feedstocks. The result is that commodity chemicals, including those used for making polyurethanes, are attractively priced. However, global growth and increasing prosperity have made plastics, including polyurethanes, more abundant, resulting in more abundant waste products. Since most plastic waste is not easily degraded in the environment, it fills up landfill sites or, worse, is disposed of irresponsibly, creating an eyesore and possible public health issue. This issue has increased visibility from both regulatory activity and public perception, particularly in the case of plastics that are single use, such as disposable containers and packaging. This has led to initiatives and innovations on the reuse of these plastics that are prevalent to the single-use category. The developed world has witnessed sophisticated collections systems to aggregate and separate these single-use plastic waste streams for effective reuse. In addition, societies have been willing, and even enthusiastic, participants in creating a "circular economy," supporting the collection and reuse of postconsumer waste.

In contrast to single-use containers and packages separated by consumers and municipal waste handlers with relative ease, polyurethanes have a much longer in-service lifetime. For instance, mattresses and furniture foams can last 10 years or more. Home and appliance insulation will presumably last the lifetime of the item or structure it inhabits. In addition, while consumers and municipalities have formed a partnership to collect recyclable household waste that is produced on a regular basis, no such mature system exists to collect or separate the polyurethane components

*Polyurethanes: Science, Technology, Markets, and Trends*, Second Edition. Mark F. Sonnenschein.

from their composite structures. Regardless, there is growing demand on the part of governments, especially in Europe, that sourcing industries take end-of-life responsibilities for these long-lived plastics [1, 2]. Regulation will, through a system of incentives and penalties, create the structure enabling the reuse of polyurethanes in ways that chemistry and engineering innovation can accommodate. The economics of the system implementation and the value of the integrated recycle polyurethane stream are not yet accurately understood. If there is a significant mismatch between cost and performance at maturity, then it may result that only the regulatory edifice will support commercial success of recycle feedstocks.

Reuse of consumer plastics requires some understanding of where to find them as well as understanding of how they are disposed. Figure 14.1 shows data for the environmental fate of consumer waste.

The data in Figure 14.1 show that, in general, developed societies primarily try to bury waste out of sight. However, in both Europe and the USA there is increasing effort

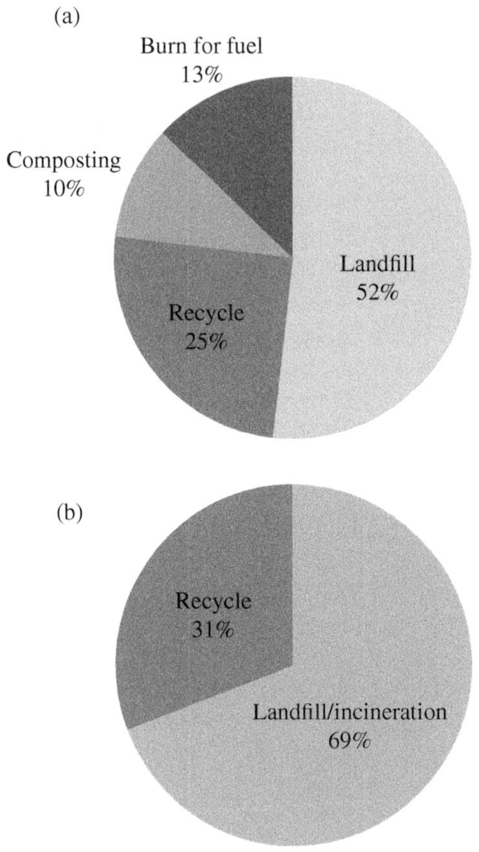


**FIGURE 14.1** Environmental fate of consumer waste in (a) the USA for 2017 (*Source:* Modified from [3]) and (b) Europe for 2016 (*Source:* Modified from [4]).

to find new ways to utilize the waste in all of the ways available. This includes burning waste as a solid fuel, much as one would burn gas or coal. Polyurethane waste is not like polyethylene or polyethylene terephthalate (PET) waste, which has many options for reuse, including chemical degradation back to building blocks that can, in principle, be reused to make virgin materials. As shown throughout this book, polyurethanes are highly inhomogeneous materials. Even if a stream of foams from just mattresses could be accessed, there is so much diversity within the mattress category that the final use would have to be insensitive to the very wide variety of component materials. Modern science has not turned away from this significant challenge: as Figure 14.2 shows, the amount of research and innovation on the subject of recycling polyurethane has always been high and has been increasing dramatically since 2010.

The subject of documentary materials produced on the topic of polyurethane recycle is shown in Figure 14.3. Over the past 10 years the vast majority of work (~90%) has involved a "lysis" chemistry developed to break the polyurethane down to building blocks exclusively to recover the soft-segment component. However, the fact that an insignificant amount of new work is being paid to adhesive rebonding should not be construed to mean that this is a fruitless path to polyurethane recycle. In fact, much available commercial carpet underlayment that is available on the market is produced by adhesive rebonding of chopped up recycled polyurethane integrated into the multicolored roll stock that is quite familiar to consumers [5]. However, the amount of effort expended in the last 10 years to reclaim polyurethane building blocks rather than just hide used polyurethane in another form points to the very low value of these recycle techniques and the limited number of applications for which ground or chopped up polyurethane foam can find use. Recent similar ideas include incorporating waste polyurethane into road asphalts, for instance.

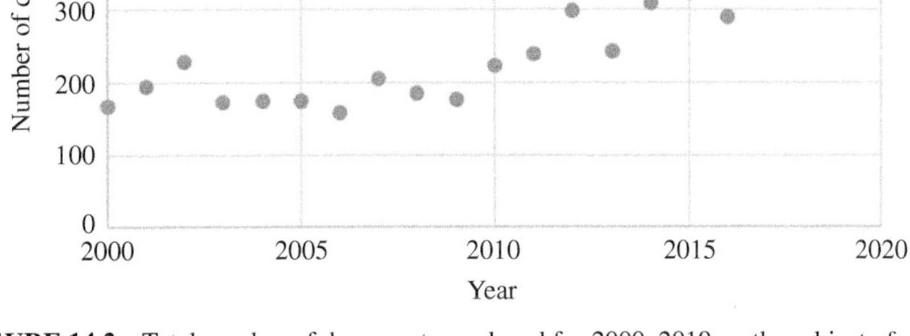


**FIGURE 14.2** Total number of documents produced for 2000–2019 on the subject of polyurethane recycle, including all document types, all languages, and all recycle modes.

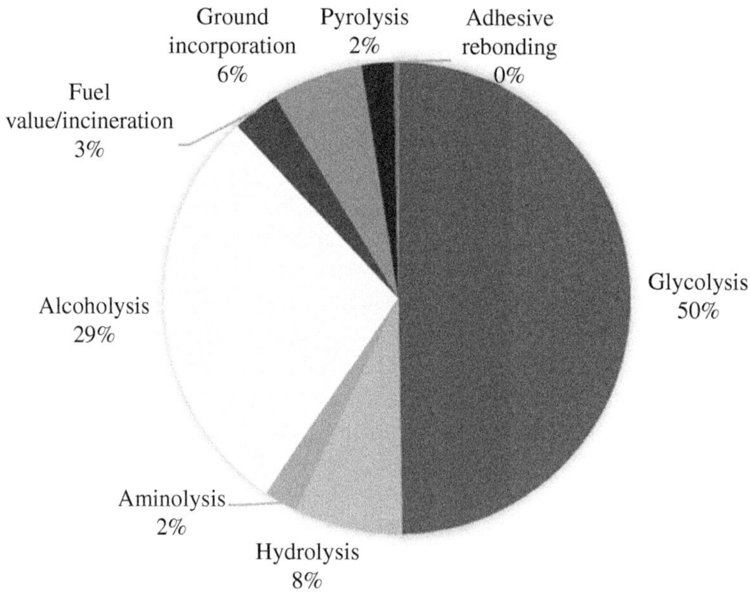


**FIGURE 14.3** Analysis of documents detailing the mode of recycle in all languages on the subject of polyurethane recycle for 2010–2020.

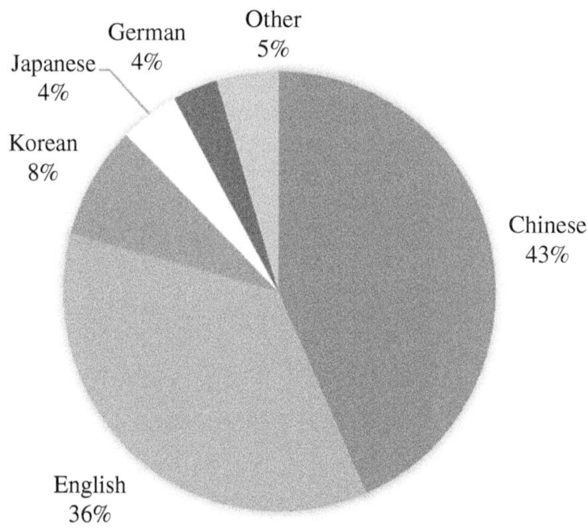


**FIGURE 14.4** Language of document publication for those in Figure 14.3.

Figure 14.4 shows by language proxy the geographical location of the work being done to innovate on the reuse of polyurethanes. China, which produces a significant amount of polyurethane and also produces a significant amount of waste, is also very

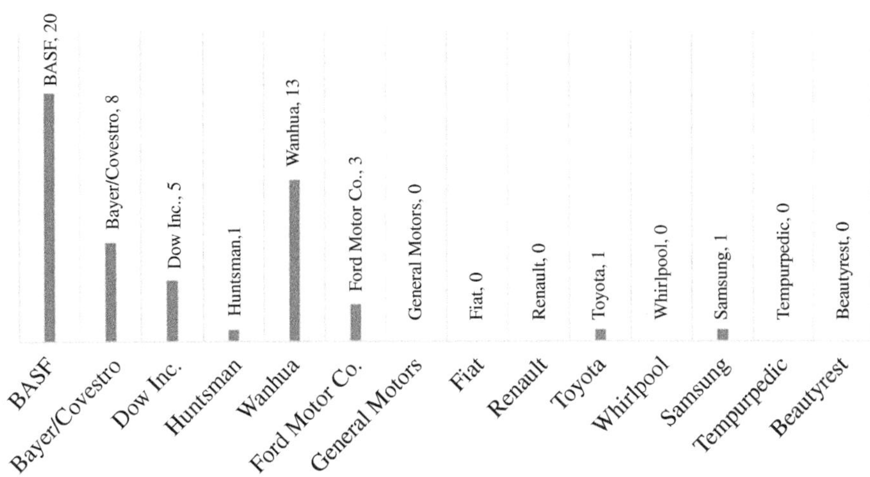


**FIGURE 14.5** Companies patenting, and their number of patents, for 2010–2019 on the subject of polyurethane recycle using two patent search engines.

active in pursuit of solutions to the problem. The large amount of work in English does not necessarily imply that only English-speaking countries are also very active, since much research in non-English-speaking countries will publish and patent in the English language as a language of commerce and science.

Lastly, Figure 14.5 shows the activity of industries producing documents on the subject of polyurethane recycle. It is clear that the great majority of work that is being done is by the chemical companies that produce polyurethane building blocks in the first place, and that industries which then use these building blocks in their devices and structures are relying on the chemical companies to develop the procedures and infrastructure to recycle. The rest of this chapter will cover in detail the methods of polyurethane recycle depicted in Figure 14.3.

## 14.2 GLYCOLYSIS, HYDROLYSIS, AMINOLYSIS, AND ACIDOLYSIS

Recycle of polyurethanes can be broken into a least three categories covering the mode of waste handling (Figure 14.6). The direct use category takes the waste polyurethane and uses the material just as it is, except for perhaps some kind of mechanical operation such as cutting or grinding. Figure 14.3 shows that, for the past 10 years, about 6% of research and innovation has explored this method, including use as an oil spill sorbent [6] and as an additive to road asphalts [7, 8]. While about 13% of waste polyurethane is currently disposed of either for fuel value or simply incinerated to reduce volume, about 6% of new work is exploring optimization or understanding of this approach to recycle. The vast majority of work and innovation is studying chemical methods of converting polyurethanes back to constituent building blocks, which could, in principle, be reused as an unadulterated feedstock.

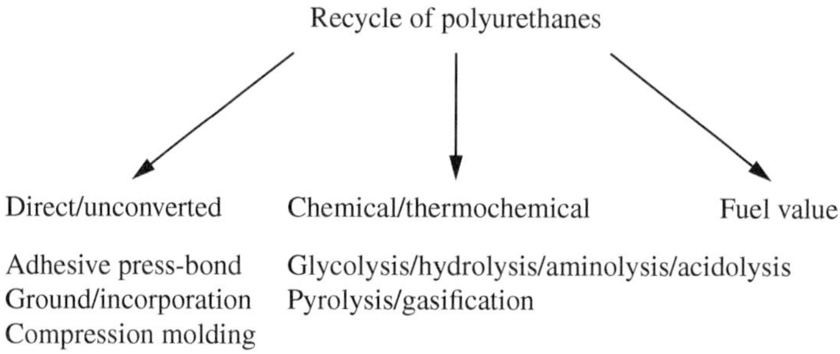


**FIGURE 14.6** Simplified classification of manner of polyurethane recycle.

As shown in Figure 14.3, the primary means of attack by researchers are glycolysis and the closely related alcoholysis. Both glycolysis and alcoholysis take advantage of the ability of the urethane bond to revert to original reagents (see Chapter 3) at elevated temperature and reform back to the urethane upon cooling. When the urethane bond has reverted back to the isocyanate and polyol in the presence of other alcohols, the isocyanate may reform the urethane with any alcohol it encounters. In the presence of small-molecule glycols such as ethylene glycol or diethylene glycol the isocyanate may substitute the polyol with the small glycol and release the polyol. This substitution of the polyols for the small-molecule glycol is termed "glycolysis." When a monoalcohol is used, this process is termed alcoholysis, but the process is in many regards identical. The process is shown explicitly in Figure 14.7. In the practice of glycolysis chemistry, the polyol is released from the urethane and, owing to density differences among the polyol, the small glycol, and the new glycol–isocyanate urethane, the system phase separates into either two or three phases. The relatively pure polyol phase rises to the top and the excess small-molecule glycol forms a second phase directly below that contains the glycol–urethane at its solubility limit. As the densest component, the urethane can form a third layer at the bottom of the reactor.

It has been reported that in adequate excess (150–200% molar excess) of the small-molecule glycol or alcohol, the freed polyol can be collected at or near the theoretical quantity available. In practice, it has been shown that the collected polyol is not in fact a drop-in replacement for an identical polyol that has never been previously used. This has been shown to be due to a small amount of the glycol present in the polyol at its solubility limit. This amount of contamination has been shown to cause foams made with recycled polyol to collapse during the foaming process, while a virgin polyol of the same nominal composition foams according to expectation [9].

In the glycolysis process there are several process design variables, such as reaction temperature, reaction time, which small-molecule glycol is employed, and which catalyst might be used to speed the process and optimize the economics of the recycle product. Many researchers have by experiment found that temperatures

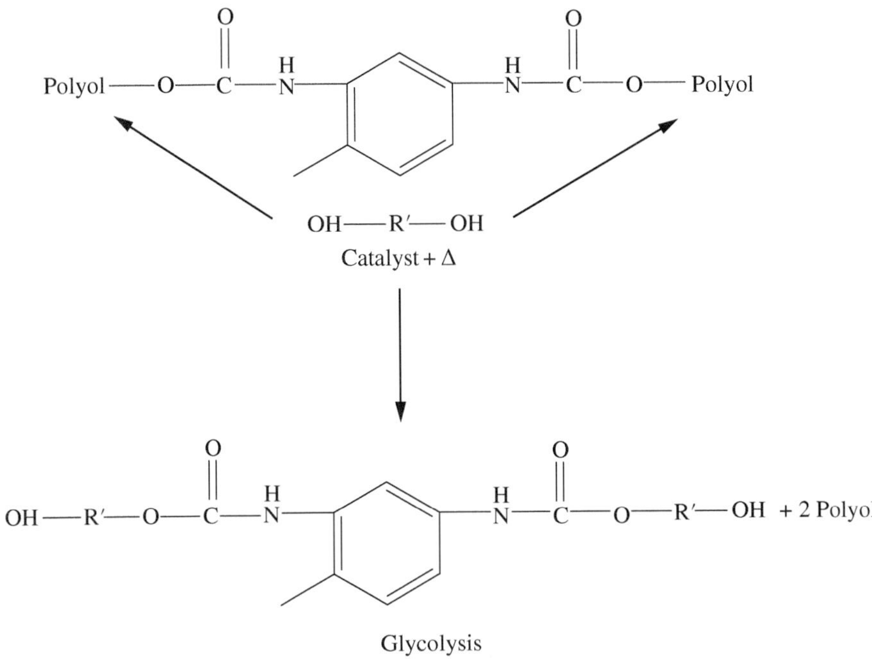


**FIGURE 14.7** Chemistry of glycolysis. The large desirable polyol moiety is substituted with a relatively low-molecular-weight glycol molecule.

around 200 °C can obtain the desired polyol rapidly (less than 90 minutes) in the presence of catalyst. Temperatures greater than 220 °C have been associated with greater levels of thermal decomposition products and undesirable side reactions of polyols/glycols and the carbamate waste materials [10]. The choice of glycol has consequences for the utility of the reaction products. It has been found that if the glycol has too high a molecular weight it may inhibit formation of a clean polyol product phase and not result in a split phase; the latter allows for easy decanting of the desired polyol phase. More polar choices can result in multiphasic reaction products and relatively pure polyol product, but may increase the solubility of the carbamate waste in the excess small-molecule glycol and complicate its recovery for subsequent use [11]. Diethylene glycol is now considered the small-molecule glycol that produces the purest polyol product most quickly and with the least interphasic contamination.

The economics of the recycle process will have a significant effect on the speed of adoption of products in current and new applications. Just as catalysts help speed the reactions in making polyurethane products by lowering the energetic barriers for reaction, the same is true for reactions designed to break the polyurethane bond and produce free reagents. Numerous catalysts have been explored for their performance in the glycolysis process. Base catalysts such as amines, hydroxides, and alkoxides have been shown to be effective [12]. It has been reported that some of these catalysts also promote undesirable hydrolysis reactions (see below) that produce primary aromatic amines [13], many of which are determined to be carcinogenic and are controlled as a class [14]. Many acetate salts have been described for their effectiveness as transesterification catalysts, but only potassium and sodium acetate are routinely utilized [15–17] in the glycolysis reaction. Octoate salts have also received a high level of recognition for their effectiveness in the glycolysis reaction [18]. The lithium and potassium octoate salts were especially noted for their efficiency and superior economics. Among transition metal catalysts, titanium butoxide has been noted as effective. Like many transition metal catalysts, the mechanism is believed to follow an insertion/substitution reaction around the metal coordination sphere [18].

The hydrolysis reaction is related to the glycolysis reaction in that the desired result is the release and phase separation of polyol from polyurethane. Water superheated to 288 °C has been shown to rapidly produce a two-phase product with the desired polyol on top and an amine phase below (Figure 14.8). The desired polyol from this process has been shown to be effective in making new foams when diluted with virgin polyol. The disposal of the amine bottom phase remains a difficulty of

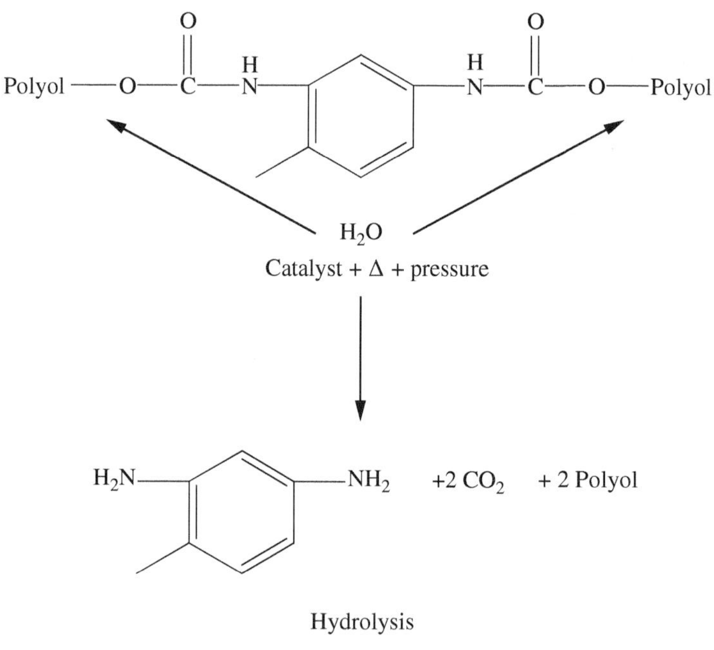


**FIGURE 14.8** Chemistry of hydrolysis. The large desirable polyol moiety is substituted with water, which converts isocyanate rapidly to an amine and $CO_2$ (see Chapter 3).

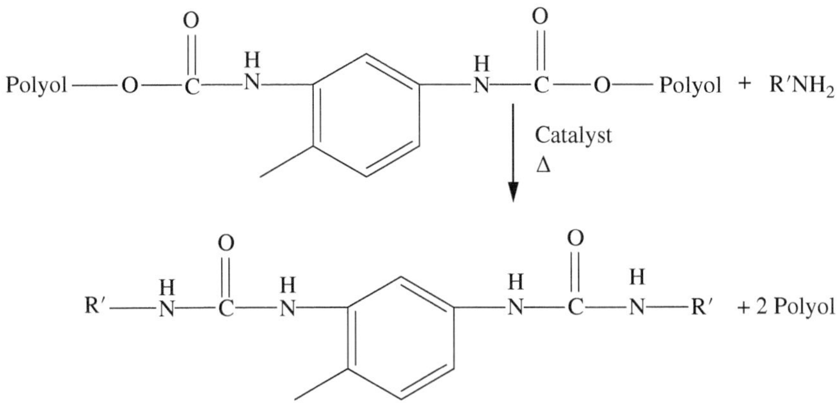

**FIGURE 14.9** Chemistry of aminolysis. The large desirable polyol moiety is substituted with a relatively low-molecular-weight glycol molecule.

this process and complicates its otherwise potentially attractive economics. Amended versions of the basic hydrolysis process have been proposed to alkoxylate the aromatic amines to produce valuable rigid foam polyols, but have so far not been commercialized [19].

As discussed in Chapter 3, upon heating, polyurethanes will revert back to starting reagents. The equilibrium relationship at moderate temperatures (~150 °C) is highly weighted towards products, but a small amount of isocyanate and polyol will exist. As temperature increases this equilibrium will become more highly weighted toward reactants. As previously discussed, this reversibility of the urethane bond creates an opportunity to substitute for the desired polyols with a less desired reagent. The fastest reactions for isocyanates at most temperatures is with amines reacting to form urea bonds, which are substantially more thermally stable than urethane bonds. In this process, urea will act as an irreversible kinetic sink, which assures quantitative recovery of the desirable polyol (Figure 14.9). A patent [20] describes the treatment of the soluble urea waste product with additional water at elevated temperature (up to 320 °C), whereupon the water hydrolysis reaction proceeds rapidly, releasing the original amine used for the aminolysis (in their examples, toluene diamine is employed). This provides a useful way of recovering the original amine for subsequent aminolysis processing of polyurethane waste. Aminolysis is not currently practiced owing to complications of the waste products and presence of soluble amine in the recovered polyol. The very high reactivity of alkyl amines, or the very high undesirability of aromatic amines, has led most industrial work away from aminolysis and towards glycolysis.

A last chemical transformation that holds potential for recovery of useful feedstocks for polyurethane is acidolysis. In this approach a carboxylic acid, or preferably a dicarboxylic acid, is utilized. As shown in Figure 14.10, the dicarboxylic acid can

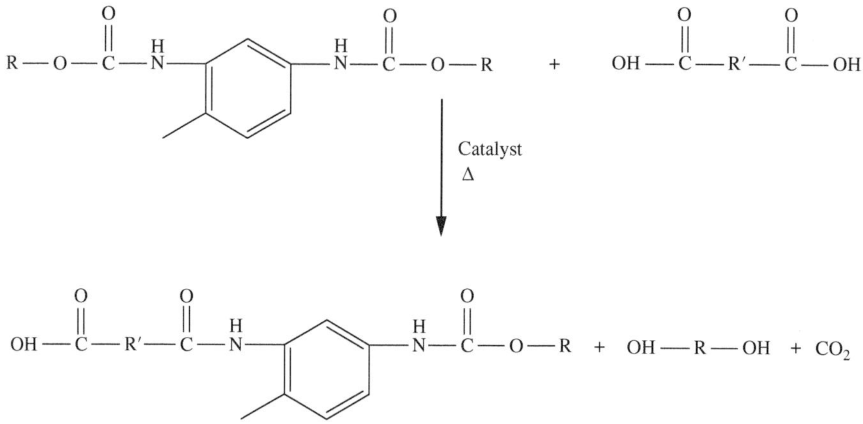


**FIGURE 14.10** Chemistry of acidolysis. The dicarboxylic acid progressively degrades the polyurethane backbone to produce glycols of broad molecular weight distribution.

substitute into the polyurethane backbone and stepwise degrade the backbone into smaller fragments as the dicarboxylic acid progressively inserts into the structure. The use of relatively small-molecule dicarboxylic acids such as oxalic, succinic, or adipic acid and their mixtures, for instance, also permits the tuning of recovered polyol molecular weight [21]. As presented, this route has the attractive feature of obviating the production of aromatic amines. It has been reported that reaction time and temperature are critical parameters to achieve a desirable result. It has been noted that, for acidolysis, low temperature and short reaction time yield results with higher acid number (more carboxylic acid end groups), while higher temperature and longer reaction time result in products with lower acid number (more alcohol end groups), lower molecular weight, higher hydroxyl number, lower viscosity, and darker color [22]. This technology is practiced commercially on a limited scale and has been tailored to produce polyols suitable for both rigid and flexible foams.

## 14.3 PYROLYSIS

Pyrolysis is the process of controlled heating of a material in a nonoxidizing atmosphere to convert plastics back to a useful feedstock. Research on pyrolysis of polyurethanes began in the early 1950s [23], whereby scrap polyurethane could be burned to produce a liquid that could in turn be used effectively in a new polyurethane synthesis. Subsequent work developed a continuous process for doing the same [24], and also engineering and chemical processes to optimize the result.

Thermogravimetric studies of polyurethanes in oxidizing environments such as air showed several decomposition processes. In air, decomposition commenced at approximately 220 °C, approached a maximum rate around 275 °C, and was

**TABLE 14.1 Reversion temperatures of commonly occurring polyurethane bonds**

| Linkage | Decomposition temperature (°C) |
|---|---|
| Allophanate aliphatic | 85–105 |
| Allophanate aromatic | 100–120 |
| Biuret aliphatic | 100–110 |
| Biuret aromatic | 115–125 |
| Urea aliphatic | 140–180 |
| Urea aromatic | 160–200 |
| Urethane aliphatic | 160–180 |
| Urethane aromatic | 180–200 |
| Disubstituted urea | 235–250 |

*Source:* Modified from [26].

complete by 350 °C. In an $N_2$ environment decomposition began about the same time as in air but proceeded more slowly and achieved maximum rate and complete combustion at higher temperatures. Many variables can influence these observations, including the temperature ramp rate and the physical form of the polyurethane (i.e. foam or elastomer).

The decomposition processes that occur upon pyrolyzing a polyurethane build progressively with increasing temperature. For the purpose of recycle it is desirable that thermolysis does not proceed all the way to formation of $CO_2$, $N_2O$, HCNO, etc., which will occur upon heating to 700 °C [25]. Instead, it is desirable to selectively break polymer backbone bonds, releasing desirable constituents. As detailed in Chapter 3, the bonds routinely found in a polyurethane and their reversible decomposition temperature are provided in Table 14.1.

However, the actual decomposition process is often not easily ordered and can be influenced by the residence time of the initial decomposition products in the reactor. It was demonstrated [25] that between 200 ° and 300 °C the cleavage reactions shown in Figure 14.11 can occur. The final products suggested the transition states provided in the figure. In addition, the simple reversion back to building blocks can produce new, less reactive products in side reactions. For instance, upon formation of diisocyanates rapid reaction with any opportunistic water will form aromatic amine, which will react with available diisocyanate to form a urea or urea trimer. These can appear as a yellow smoke product and not be easily accounted for in a mass balance [27].

## 14.4 RECYCLE FOR FUEL VALUE

As shown in Figure 14.1, the majority of polyurethane is still landfilled. As currently structured there is very little in the way of a collection mechanism for integrating polyurethane waste in order to deliver it to concerns capable of performing some

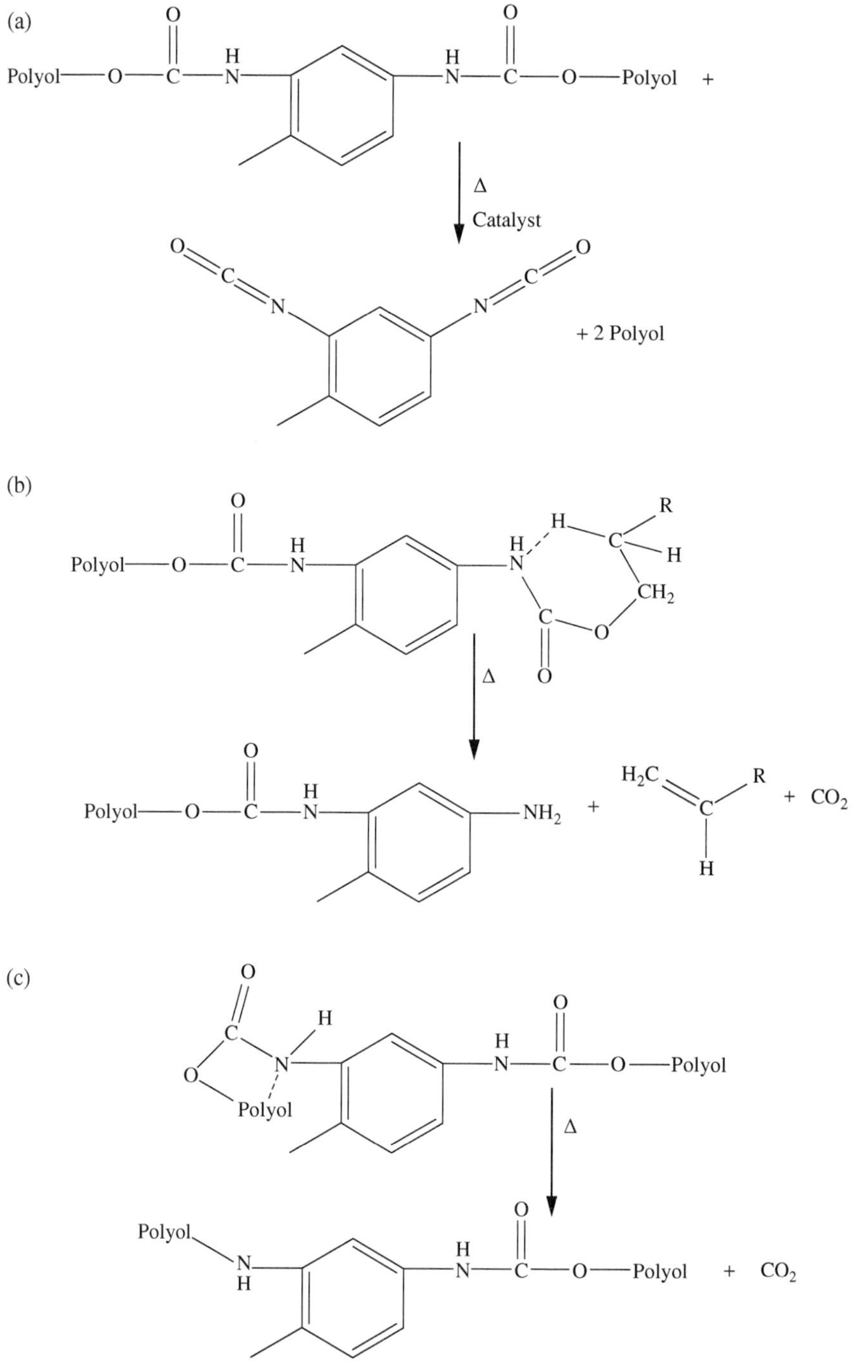


**FIGURE 14.11** Polyurethane decomposition reactions occurring between 200 °C and 300 °C. Transition states leading to detected products are suggested in the structures. (a) Reversible reversion back to starting materials; (b) irreversible partial decomposition to an aromatic amine, an olefin, and carbon dioxide; and (c) irreversible decomposition to an aromatic amine and carbon dioxide.

kind of value upgrade. In addition, there is little in the way of industrial processing to separate polyurethane foam from other components such as furniture frames, automotive seating, or mattress composites. However, as shown in Figure 14.1, methods have been found to contribute a significant quantity of waste polyurethane foam to the growing societal fuel-to-energy requirements. In 2012 it was estimated that about 1.3 billion pounds of waste polyurethane foam was produced from end-of-life furniture and mattresses alone [28]. Of this total about 40% was incinerated, including for energy recovery, and the rest was landfilled. From Figure 14.1 it is reported that about 13% of waste polyurethane is burned for its fuel value. While the market for polyurethane recycled products, both production and use, is still nascent, the ability to burn polyurethane foam and recover its fuel value is fully within current industrial capacity.

An underappreciated aspect of burning polyurethane foam, and all manner of recycle for that matter, is that the long service life of many polyurethane products insures that there are many constituents along with the polyurethane that are not acceptable for remanufacture. These include bromine- and phosphorus-containing flame retardants, mercury- and tin-containing catalysts, some chlorofluorocarbon blowing agents, and some phthalate ester plasticizers. In Europe it is not permitted in some cases to remanufacture with recycled materials containing banned substances [29]. This provides additional burden on fuel recovery but may also require that producers of recycled products from chemical methods will have to prove compliance with this directive. In addition, the presence of flame retardants in construction polyurethane foams can complicate their burning for energy recovery [30]. Various controlled trials have been run to determine the suitability and value capture of burning polyurethane foam waste in municipal burner facilities. It was determined that the heating value of the foam was 24–30 MJ/kg compared with pulverized coal at 17.6 MJ/kg and that the cost of the polyurethane foam waste for this application was 50 euros/tonne. Additional work was described to minimize the formation of CO and NO emissions. The presence of the foams furthermore increased the concentration of trace elements, primarily from resident tin catalysts from foam manufacture [31]. However, for a goal to reduce landfilling of polyurethane waste, conversion of plastic, including polyurethane, reduces volume by up to 99%.

## 14.5 REGRINDING AND INCORPORATION

An intuitively obvious approach to recycle is to take waste polyurethane and grind it to small dimensions, and then incorporate it as an inert filler into nominally new articles. Just as styrene–acrylonitrile (SAN) particles are incorporated through copolymer polyol technology into flexible foam applications (see Chapters 2 and 6) to

**TABLE 14.2 Comparison of properties of cold-mold methylene diphenyl diisocyanate (MDI) foam (see Chapter 6) with and without incorporation of reground foam from the same formulation**

| Property | Benchmark foam | Benchmark foam + 15% polyurethane regrind |
|---|---|---|
| Part weight (g) | 1320 | 1320 |
| Polyurethane powder (%) | | 15 |
| Indentation hardness (N) | 360 | 360 |
| Compression set (50%) | 5.6 | 7 |
| Tensile strength (kPa) | 152 | 144 |
| Elongation (%) | 132 | 103 |

produce higher load-bearing foams with little or no density increase, the incorporation of ground foam can serve as a low-cost filler. One difference with copolymer polyol, however, is that the powdered foam is not chemically grafted into the polymer backbone through covalent bonding, and offers a less efficient avenue to bear load. In addition, while waste polyurethane is a low-cost material, the process to powderize a flexible foam can be costly, requiring copious amounts of liquid nitrogen. In fact, as shown in Table 14.2, the properties of a molded flex foam with 15% ground polyurethane foam are modestly inferior to those of a foam without the recycled waste.

The implementation of this technology normally employs postindustrial waste (i.e. waste from the manufacturing process that has never seen a consumer) instead of postconsumer waste. This assures a constancy of product composition and negates the necessity to separate a polyurethane foam from other nonpolyurethane components that may accompany postconsumer products. In fact, it is possible for the foam producer to employ waste from its own plants into new products and so negate the entirety of the waste collection process. The waste polyurethane foam can be converted to a powder form by several processes, including using a counter-rotating two-roll mill with rolls run at different speeds [32, 33]. Optimization of this technique has produced useable powder at sizes between 50 and 200 μm, more than a factor of 10 larger than SAN copolymer polyol dimensions. Along with cost, the use of regrind flexible foam as a filler is also a challenge for its effect on formulation viscosity.

Reaction injection molding (RIM) products (see Chapter 9) have also been successfully formulated with RIM scrap up to 15% incorporation. Since RIM materials are much more rigid than flexible foams, the powderizing process is less stringent, though still a cost addition. Table 14.3 details comparative performance of a RIM formulation with and without 10% filler incorporation. While some of the properties are slightly lower for incorporation of reground materials, in general the properties are comparable.

**TABLE 14.3 Comparison of properties of a reaction injection molded (RIM) polyurethane composite (see Chapter 9) with and without incorporation of reground RIM composite from the same formulation**

| Property | Benchmark RIM | Benchmark RIM + 10% RIM regrind |
|---|---|---|
| Density (g/cm$^3$) | 1.18 | 1.2 |
| Shore D hardness | 59 | 63 |
| Flex hardness (psi) | $1.3 \times 10^5$ | $1.2 \times 10^5$ |
| Compression set (50%) | 5.6 | 7 |
| Tensile strength (psi) | 4100 | 3600 |
| Elongation (%) | 132 | 103 |

*Source:* Modified from [33].

## REFERENCES


[1] Directive 2000/53/EC of the European Parliament and of the Council of 18 September 2000 on End-of Life Vehicles, 2000. Available at: https://eur-lex.europa.eu/legal-content/EN/ALL/?uri=CELEX%3A32000L0053 (accessed June 10, 2020).

[2] Report from the Commission to the European Parliament, The Council, the European Economic and Social Committee and the Committee of the Regions on the Implementation of the Circular Economy Action Plan Brussels, COM(2019) 190 final, 2019. Available at: https://ec.europa.eu/environment/circular-economy/pdf/report_implementation_circular_economy_action_plan.pdf (accessed June 10, 2020).

[3] US Environmental Protection Agency, *Advancing Sustainable Materials Management-2017*, Washington, DC, 2019.

[4] European Commission, *European Strategy for Plastics in a Circular Economy,* Document SWD(2019)92 final, Brussels, Belgium, 2019.

[5] American Chemistry Council, *Polyurethanes Sustainability – Polyurethane Recycling*, Washington, DC, 2020.

[6] H Li, L Liu, and F Yang, *Procedia: Environmental Sciences*, **18**, (2013), 528–533.

[7] M Salas and H Perez-Acebo, *Gradevinar*, **70**, (2018), 403–412.

[8] S Adnan Raheel Shah, H Arshad, A Waqar, M Hasnain, S Hafeez, J Mansoor, A Sadiq, and M Malik, *Energies*, **11**, (2018), 3025.

[9] C Molero, A de Lucas, F Romero, and J Rodriguez, *Journal of Applied Polymer Science*, **109**, (2008), 617–626.

[10] D Simon, A Borreguero, A de Lucas, and J Rodriguez, *Polymer Degradation and Stability*, **116**, (2015), 23–35.

[11] C Molero, A de Lucas, and J Rodriguez, *Polymer Degradation and Stability*, **91**, (2006), 221–228.

[12] G Bauer, "Recycling of polyurethanes," in *Recycling and Recovery of Plastics*, E Weigand, Ed., Hanser Publishers, Munich, Germany, 1996.

[13] H Kriecheldorf and Z Denchev, "Interchange reactions in condensation polymers and their analysis by NMR spectroscopy," in *Transreactions in Condensation Polymers*, S Faikirov, Ed., Wiley-VCh, Weinheim, Germany, 1999, 1–78.

[14] S Crettas, P Kampfer, B Bruschweiler, S Nussbaumer, and O Deflorin, *Journal of Consumer Protection and Food Safety*, **15**, (2020), 49–61.
[15] C Wu, C Chang, C Cheng, and H Huang, *Polymer Degradation and Stability*, **80**, (2003), 103–111.
[16] Y Joo and K Dukjoon, *Journal of Applied Polymer Science*, **77**, (2000), 2646–2656.
[17] J Borda, G Pasztor, and M Zsuga, *Polymer Degradation and Stability*, **68**, (2000), 419–422.
[18] C Molero, A de Lucas, and J Rodriguez, *Polymer Degradation and Stability*, **94**, (2009), 533–539.
[19] J Schiers, *Polymer Recycling: Science Technology and Applications*, Wiley, Hoboken, NJ, 1998.
[20] K Murayama, K Kodama, and T Kumaki, Method of decomposing a polyurethane. Assignee Mitsui Chem. US patent 6515036. 2004.
[21] M Soltysinski, K Piszczek, D Romecki, S Narozniak, J Tomaszewska, and K Skorczewska, *Polimery*, **63**, (2018), 234–238.
[22] N Gama, A Ferreira, and A Timmons, *Materials*, **11**, (2018), 1841–1876.
[23] A Heinz and W Tacke, Method for reclaiming scrap polyurethane resin. Assignee Reeves Bros. Inc. US patent 2998395. 1961.
[24] R Hurley, Method for reclaiming scrap polyurethane resins. Assignee Reeves Bros. US patent 3143515. 1964.
[25] C Cullis and M Hirschler, *The Combustion of Organic Polymers*, Clarendon Press, Oxford, UK, 1981.
[26] M Szycher, *Szycher's Handbook of Polyurethanes*, 2nd Ed., CRC Press, Boca Raton, FL, 2013.
[27] M Ravey and E Pearce, *Journal of Applied Polymer Science*, **63**, (1997), 47–74.
[28] *Flexible Polyurethane Foam in Mattresses and Furniture: An Overview of Possible End-of-Life Solutions*. Europur, Brussels, Belgium, 2016. Available at: https://www.isopa.org/media/2763/flexible-foam-end-of-life-solutions.pdf (accessed June 15, 2020).
[29] Waste Framework Directive (2008/98/EC). Circular Economy Action Plan Update, 2019. Available at: https://ec.europa.eu/environment/waste/framework (accessed June 10, 2020).
[30] A Tokha and R Zevenhoven, *Processing Waste-Derived Fuels Containing Brominated Flame Retardants*, Helsinki University of Technology, Energy Engineering and Environmental Protection Publication TKK-ENY-7, Helsinki, Finland, 2002. Available at: http://users.abo.fi/rzevenho/TKK-ENY-7_print.pdf (accessed June 10, 2020).
[31] R Zevenhoven, *Treatment and Disposal of Polyurethane Waste: Options for Recovery and Recycling*, Helsinki University of Technology, Department of Mechanical Engineering, Energy Engineering and Environmental Protection Publication TKK-ENY-19, 2004. Available at: http://users.abo.fi/rzevenho/tkk-eny-19.pdf (accessed June 10, 2020).
[32] B Martell and R Villwock, Process for grinding mixtures of polyurethane foams. Assignee Mobius Technologies Inc. WO 2003066298, 2003.
[33] European Isocyanate Producers Association, *Fact Sheet: Recycling and Recovering Polyurethanes – Regrinding/Powdering*, Brussels, Belgium, 2001.

# INDEX



*Polyurethanes: Science, Technology, Markets, and Trends*, Second Edition. Mark F. Sonnenschein.